Biotechnological Innovations for Sustainable Biodiversity and Development

This book reviews the important role of biotechnological innovations in achieving sustainable development goals and conserving global biodiversity. It presents the latest biotechnological techniques used to identify and characterize various groups of plants and animals, such as genomic tools for animal identification, and DNA barcoding for precise plant characterization. It also reviews the utility of proteomics and metabolomics in enhancing our understanding of diverse species. The book also discusses responsible development and sustainable utilization of bioresources, including strategies for conserving and managing bioresources, bioprospecting for novel biodiscoveries, and sustainable agricultural practices to preserve agrobiodiversity. Further, the book addresses the pressing challenges faced by biodiversity, including the far-reaching effects of climate change, the threat posed by invasive species, the consequences of pollution on biodiversity degradation, and the interplay between diseases and biodiversity decline. Toward the end, the book analyzes the impact of environmental degradation on biodiversity and explores emerging technologies in biodiversity conservation, focusing on genetically modified organisms (GMOs). The book will interest scientists, researchers, policymakers, environmentalists, academics, and students involved in biotechnology, ecology, genetics, and conservation biology.

Key Features:

- Discusses cutting-edge biotechnological innovations for biodiversity conservation
- Presents cutting-edge biotechnological techniques for identifying and characterizing selected groups of plants and animals
- Addresses critical environmental challenges such as climate change, invasive species, and pollution

Biotechnological Innovations for Sustainable Biodiversity and Development

Edited by
Ranbir Chander Sobti

CRC Press
Taylor & Francis Group
Boca Raton London New York

CRC Press is an imprint of the
Taylor & Francis Group, an **informa** business

Designed cover image: Shutterstock

First edition published 2025
by CRC Press
2385 NW Executive Center Drive, Suite 320, Boca Raton FL 33431

and by CRC Press
4 Park Square, Milton Park, Abingdon, Oxon, OX14 4RN

CRC Press is an imprint of Taylor & Francis Group, LLC

ISBN: 9781032697529 (hbk)
ISBN: 9781032697536 (pbk)
ISBN: 9781032697550 (ebk)

DOI: 10.1201/9781032697550

Typeset in Times
by Deanta Global Publishing Services, Chennai, India

Contents

Editor

Ranbir Chander Sobti is an emeritus professor and Indian National Science Academy (INSA) senior scientist at Punjab University. He is a former vice chancellor of Panjab University, Chandigarh & Babasaheb Bhimrao Ambedkar University (Central University), Lucknow. He started his career as a cytogeneticist and then moved on to molecular biology, including genomics, to understand the susceptibility and disease processes of cancer, COPD, AIDS, metabolic syndrome, and kidney diseases. He has also used stem cells and nanoparticles to understand the process of tissue and organ development through a designed decellularization protocol. He has published more than 300 research articles in journals of international repute and has also published more than 40 books.

He is a Fellow of the Third World Academy of Sciences, National Academy of Sciences India, Indian National Science Academy, National Academy of Medical Sciences, National Academy of Agricultural Sciences, Canadian Academy of Cardiovascular Diseases, and a few others. He was the general president of the Indian Science Congress for the 102nd session held at the University of Jammu in 2013. Sobti is the recipient of many prestigious awards, including the INSA Young Scientist Medal, UGC Career Award, Punjab Rattan Award, JC Bose Oration and Sriram Oration Awards, and Lifetime Achievement Awards of the Punjab Academy of Sciences, Zoological Society of India, and the Environment Academy of India.

Preface

BIOTECHNOLOGICAL NOVELTIES FOR SUSTAINABLE DEVELOPMENT GOALS

Experiments for Biodiversity Conservation

Biotechnological innovations drive significant progress in achieving sustainability, ensuring food security, and eradicating hunger. These advancements enable the development of healthier, more productive crops, extend the shelf life of produce to reduce food waste, and improve child nutrition by enriching staple crops with essential vitamins and minerals. Genetically Modified (GM) crops with enhanced nutritional content are vital in addressing global malnutrition thereby providing sustainable solutions for undernourished populations.

Biotechnology also enhances crop resilience by equipping plants to withstand environmental stresses such as drought, salinity, and extreme temperatures. This ensures more stable yields, supporting food security even in unpredictable climates. In addition, biotechnological solutions contribute to animal health by reducing reliance on antibiotics, addressing antimicrobial resistance, and minimizing greenhouse gas emissions through sustainable livestock management practices. Fortified food ingredients derived from biotechnology provide essential micronutrients, supporting the global fight against malnutrition and hunger.

Higher agricultural productivity from GM crops leads to increased farm incomes, particularly for smallholder and subsistence farmers, thus reducing poverty and hunger. In many developing countries, GM crops have empowered resource-poor farmers by lowering production costs, improving yields, and offering economic stability. Additionally, crops engineered to be herbicide-tolerant encourage the adoption of no-till farming, which reduces soil erosion, preserves soil health, and mitigates water pollution by minimizing chemical runoff.

Insect-resistant crops such as *Bt* maize are crucial in combating pests that devastate crops. For example, the Fall Armyworm, identified in West Africa in 2016, threatens maize production across 39 African countries, impacting the livelihoods of over 300 million farm families. *Bt* maize has shown promising results in countries like Nigeria, Ghana, and Burkina Faso, while efforts are underway to develop *Bt* cowpea, resistant to the destructive pod borer insect, which can damage up to 90% of a farmer's crop.

Furthermore, biotechnological innovations address challenges posed by climate change. Drought-tolerant crops like Water Efficient Maize for Africa (WEMA) are designed to thrive under water-scarce conditions, helping farmers maintain productivity despite erratic weather patterns. Nutritionally enhanced crops such as Golden Rice, biofortified with beta-carotene, combat Vitamin A Deficiency (VAD), which affects millions of children globally. Further research aims to improve the bioavailability of iron and zinc, enhance pro-vitamin A stability, and boost protein digestibility in staple crops like sorghum and cassava. The Banana21 initiative exemplifies the potential of biotechnology to address nutrient deficiencies, with genetically engineered cooking bananas offering increased iron and pro-vitamin A content. Additionally, crops like GM soybeans have been developed to contain healthier oils. *Bt* brinjal (eggplant) has reduced pesticide usage in countries such as Bangladesh, benefiting both farmers and the environment. Virus-resistant squash and virus-resistant papaya (Rainbow Papaya) are transgenic crops that have restored crop yields by protecting plants from devastating viral infections.

Moreover, biotechnology promotes conservation by enabling sustainable farming practices. Herbicide-tolerant crops encourage conservation tillage, reducing soil erosion, preserving water quality, and limiting nutrient runoff. Innovations in microbial biotechnology also enhance the nitrogen-fixing capacity of crops, reducing the need for synthetic fertilizers, which aligns with

Sustainable Development Goal 15 (Life on Land). These practices not only protect terrestrial ecosystems but also contribute to the conservation of biodiversity.

Biotechnology minimizes deforestation and preserves biodiversity by reducing the amount of land required for crop production. For instance, GM crops have enabled a 92% reduction in pesticide use on eggplants, contributing to safer agricultural practices. Studies confirm that *Bt* crops do not negatively impact non-target organisms, making them environmentally sustainable. Biotechnology also saves endangered species, as demonstrated by the GM American chestnut tree developed by the State University of New York (SUNY), which aims to restore the species devastated by fungal infections. Similarly, Cornell University's development of Rainbow Papaya saved the crop from the ringspot virus, preventing its near-extinction in the 1990s.

In conclusion, biotechnological innovations are indispensable for sustainable development and zero hunger. They provide transformative solutions to agricultural challenges, improve nutrition, strengthen food security, and promote environmental conservation. Biotechnology offers a promising path forward as we face global challenges such as climate change, biodiversity loss, and food insecurity. By embracing these innovations, we can envision and build a future where hunger is eradicated, biodiversity is preserved, and sustainable practices ensure the planet's and its people's well-being. This book delves into the profound impact of biotechnology on achieving these goals, highlighting scientific breakthroughs that pave the way for a healthier and more sustainable world.

Contributors

K. Sankaraiyah
Faculty of Zoology, Kanchi Mamunivar Government Institute for Postgraduate Studies and Research, India

Neha Salaria
Department of Life Sciences, Arni School of Basic Sciences, Arni University, India

Neeraj
Department of Life Sciences, Arni School of Basic Sciences, Arni University, India

Indu Kumari
Department of Life Sciences, Arni School of Basic Sciences, Arni University, India

Bhawna
Drug Testing Laboratory, QMI CSIR-IIIM Jammu, India

Rajesh Kumar
Department of Biosciences, Himachal Pradesh University, India

Asha Poonia
Faculty of Life Sciences, Chaudhary Bansi Lal University, Bhiwani, India

Radhika Sharma
Department of Biology and Environmental Sciences, College of Basic Sciences, CSK Himachal Pradesh Krishi Vishvidayala, India

Seema Rani
Department of Zoology, Faculty of Life Sciences, Chaudhary Bansi Lal University, Bhiwani, India

Neetika Kimta
School of Biological and Environmental Sciences, Shoolini University of Biotechnology and Management Sciences, Solan, India

Nidhi Bhardwaj
School of Biological and Environmental Sciences, Shoolini University of Biotechnology and Management Sciences, Solan, India

Sunil Puri
School of Biological and Environmental Sciences, Shoolini University of Biotechnology and Management Sciences, Solan, India

Kawalpreet Kaur
SGGS College Sector 26, Chandigarh, India

Varruchi Sharma
SGGS College Sector 26, Chandigarh, India

Bhavna Kabila
K.M.V College, Jalandhar, Punjab, India

Malkiat Chand Sidhu
Department of Botany, Panjab University Chandigarh, India

Rakesh Kumar
Department of Life Sciences, Sant Baba Bhag Singh University Jalandhar, Punjab, India

Vikas Sharma
Department of Life Sciences, Sant Baba Bhag Singh University Jalandhar, Punjab, India

Sudesh Kumar
Department of Zoology, Govt. Degree College Kilhotran, Doda, J&K, India

Neha Thakur
Department of Botany, Panjab University, Chandigarh, India

Arshpreet Kaur
Department of Botany, Panjab University, Chandigarh, India

Jaspreet K. Sembi
Department of Botany, Panjab University, Chandigarh, India

Radha
School of Biological and Environmental Sciences, Shoolini University of Biotechnology and Management Sciences, Solan, India

Niharika Sharma
School of Biological and Environmental Sciences, Shoolini University of Biotechnology and Management Sciences, Solan, India

Neeraj Kumari
School of Biological and Environmental Sciences, Shoolini University of Biotechnology and Management Sciences, Solan, India

Suraj Prakash
School of Biological and Environmental Sciences, Shoolini University of Biotechnology and Management Sciences, Solan, India

Ashok Pundir
Department of Civil Engineering, Shoolini University, Solan, India

Shivani Guleria
Shoolini University of Biotechnology and Management Sciences, Solan, India

Kriti Singh
Shoolini University of Biotechnology and Management Sciences, Solan, India

Ranbir Chander Sobti
Department of Biotechnology, Panjab University, Chandigarh, India

Swati Yadav
Department of Zoology, School of Life Sciences, Babasaheb Bhimrao Ambedkar University, Lucknow, Uttar Pradesh, India

Suman Mishra
Department of Zoology, School of Life Sciences, Babasaheb Bhimrao Ambedkar University, Lucknow, Uttar Pradesh, India

Kamal Jaiswal
Department of Zoology, School of Life Sciences, Babasaheb Bhimrao Ambedkar University, Lucknow, Uttar Pradesh, India

Naresh Thakur
Shoolini University of Biotechnology and Management Sciences, Solan, India

Priyanka Kumari
Faculty of Basic Sciences, Shoolini University of Biotechnology and Management Sciences, Solan, India

Pardeep Kumar
Soil Ecosystem and Restoration Ecology Lab, Department of Botany, Panjab University, Chandigarh, India

Isha
Soil Ecosystem and Restoration Ecology Lab, Department of Botany, Panjab University, Chandigarh, India

Abhishek Kumar
Soil Ecosystem and Restoration Ecology Lab, Department of Botany, Panjab University, Chandigarh, India

Meenu Patil
Soil Ecosystem and Restoration Ecology Lab, Department of Botany, Panjab University, Chandigarh, India

Sabir Hussain
Soil Ecosystem and Restoration Ecology Lab, Department of Botany, Panjab University, Chandigarh, India

Sheenu Sharma
Soil Ecosystem and Restoration Ecology Lab, Department of Botany, Panjab University, Chandigarh, India

Anand Narain Singh
Soil Ecosystem and Restoration Ecology Lab, Department of Botany, Panjab University, Chandigarh, India

Nandini Rai
Shoolini University of Biotechnology and Management Sciences, Solan, India

Archi Sehgal
Shoolini University of Biotechnology and Management Sciences, Solan, India

Deepali Sambyal
Shoolini University of Biotechnology and Management Sciences, Solan, India

J.M Julka
Faculty of Basic Sciences, Shoolini University of Biotechnology and Management Sciences, Solan, India

Priyanka
School of Biological and Environmental Sciences, Shoolini University, Solan, India

Priyanka Chauhan
School of Biological and Environmental Sciences, Shoolini University, Solan, India

Meenu Chaurasia
Department of Zoology, University of Delhi, India

Sonia
School of Biological and Environmental Sciences, Shoolini University, Solan, India

Ritesh Banerjee
School of Biological and Environmental Sciences, Shoolini University, Solan, India

Aditi Fishta
School of Biological and Environmental Sciences, Shoolini University of Biotechnology and Management Sciences, Solan, India

R.C. Sobti
Department of Biotechnology, Panjab University, Chandigarh, India

Ruhi Thakur
School of Biological and Environmental Sciences, Shoolini University of Biotechnology and Management Sciences, Solan, India

Shivani Devi
Department of Biosciences, University Institute of Biotechnology, Chandigarh University, Gharuan, Mohali, Punjab, India

Gagandeep Singh
Department of Biosciences, University Institute of Biotechnology, Chandigarh University, Gharuan, Mohali, Punjab, India

Anita Rana
Department of Biosciences, University Institute of Biotechnology, Chandigarh University, Gharuan, Mohali, Punjab, India

Sakshi Sharma
School of Biological and Environmental Science, Shoolini University, Solan, India

Komal Raghuvanshi
School of Biological and Environmental Science, Shoolini University, Solan, India

Ritika Dadhwal
School of Biological and Environmental Sciences, Shoolini University, Solan, India

Pankaj Yadav
Fish Molecular Biology Laboratory, Department of Zoology, University of Delhi, India

Diksha Sharma
School of Biological and Environmental Sciences, Shoolini University, Solan, India

Sneha Siwach
Fish Molecular Biology Laboratory, Department of Zoology, University of Delhi, India

Sheenam Kamboj
Soil Ecosystem and Restoration Ecology Lab, Department of Botany, Panjab University, Chandigarh, India

Ambika Choudhary
School of Biological and Environmental Sciences, Shoolini University of Biotechnology and Management Sciences, Solan, India

Aarzoo Yadav
Fish Molecular Biology Laboratory, Department of Zoology, University of Delhi, India

Manish Thakur
Department of Microbiology, School of Bio Engineering and Biosciences, Lovely Professional University, Phagwara, Punjab, India

Sharali Sharma
Department of Biosciences, University Institute of Biotechnology, Chandigarh University, Gharuan, Mohali, Punjab, India

Rashi Miglani
Department of Zoology, DSB Campus, Kumaun University, Nainital, Uttarakhand, India

Gaurav Rawat
Department of Zoology, DSB Campus, Kumaun University, Nainital, Uttarakhand, India

Swati Joshi
Department of Zoology, DSB Campus, Kumaun University, Nainital, Uttarakhand, India

Vishal Singh
Department of Zoology, DSB Campus, Kumaun University, Nainital, Uttarakhand, India

Nagma Parveen
Department of Zoology, DSB Campus, Kumaun University, Nainital, Uttarakhand, India

Uzma Siddiqui
Department of Zoology, DSB Campus, Kumaun University, Nainital, Uttarakhand, India

Netrapal Sharma
Department of Zoology, DSB Campus, Kumaun University, Nainital, Uttarakhand, India

Ankit Kumar
Department of Pharmaceutical Sciences, Sir J.C. Bose Technical Campus, Bhimtal, Kumaun University, Nainital, Uttarakhand, India

Seeta Dewali
Department of Zoology, DSB Campus, Kumaun University, Nainital, Uttarakhand, India

Satpal Singh Bisht
Department of Zoology, DSB Campus, Kumaun University, Nainital, Uttarakhand, India

Tripti Singh
Department of Biotechnology, Jaypee Institute of Information Technology, Noida, Uttar Pradesh, India

Dr. Ashwani Mathur
Department of Biotechnology, Jaypee Institute of Information Technology, Noida, Uttar Pradesh, India

1 Conservation of Biodiversity
The Role of Biotechnology

K. Sankaraiyah[1]

[1]Faculty of Zoology, Kanchi Mamunivar Government Institute for Postgraduate Studies and Research, Pondicherry, India

1.1 INTRODUCTION

Before we move into the details of the role of biotechnology in conserving biodiversity, it will be appropriate to give a brief overview of biodiversity. "Biodiversity", a contracted term for biological diversity, was first used by Walter G. Rosen in 1985 and later it was popularised by the renowned American ecologist Edward O. Wilson. Even though there are many definitions for this popular term, the Food and Agricultural Organization (FAO) defined it, in 2019, as "the variability that exists among living organisms (both within and between species) and the ecosystems of which they are part". Biodiversity is a product of 3.5 billion years of evolution of life forms, and eukaryotes originated around 2 billion years ago. Since their origin, they have undergone a variety of changes in their structure and functions. For decades, scientists have continuously worked to estimate the total number of organisms on this planet. A recent estimation of the number of species by May (1988) was about 15 million species, and according to E.O. Wilson (1992), only 1.4 million species were described (Figure 1.1) until 1992, with almost 8 million species yet to be described. Mora *et al.* (2011) have estimated 1.2 million described species and a total of 8.74 million eukaryotes, with 10,100 species of prokaryotes living on planet Earth.

India alone has described 1,00,693 (ZSI2023) species of animals and 44,500 (BSI,2023) species of plants, which include bacteria. Twenty-eight per cent of Indian plants are endemic to the country. The famous ecologists Ceballos *et al.* (2017) have suggested that the diversity of global fauna and flora is underestimated, there are greater threats to ecosystem services and human well-being, and there is an immediate need to conserve the living resources of the Earth. Also, Paul R. Ehrlich emphasised the niche of each species in its ecosystem through the rivet popper hypothesis. I strongly believe that these are underestimated numbers and that species are added to the list continuously every year. E.O. Wilson, concerned for future generations, wrote that we share our planet with 10 million other species and that they are being wiped out at a rate of 100 to 1,000 times faster. It may lead to a loss of biodiversity before the end of the century. The senseless and inappropriate approaches to the conservation of biodiversity are unlikely to be forgiven by our descendants.

It becomes a matter when it comes to the status of endemic species, and we must remember that once these species are lost, they cannot be replaced. Unfortunately, in all parts of the world many endemic species are continuously under threat. The ecological role of these species is also poorly understood, and some of them are keystone species. The disappearance of species is not new; this mother planet has suffered major damage to species composition over the past 440 million years. There were five major extinctions in the evolutionary history, all of which were natural and are detailed in Table 1.1. The Earth has revived from mass extinctions on its own because all of them were natural phenomena.

It is strongly believed that the sixth mass extinction is on its way and it is anthropogenic. The golden toad (*Incilius periglenes*), dodo (*Raphullus cacullatus*), Tasmanian wolf (*Thylacinus cynocephalus*),

DOI: 10.1201/9781032697550-1

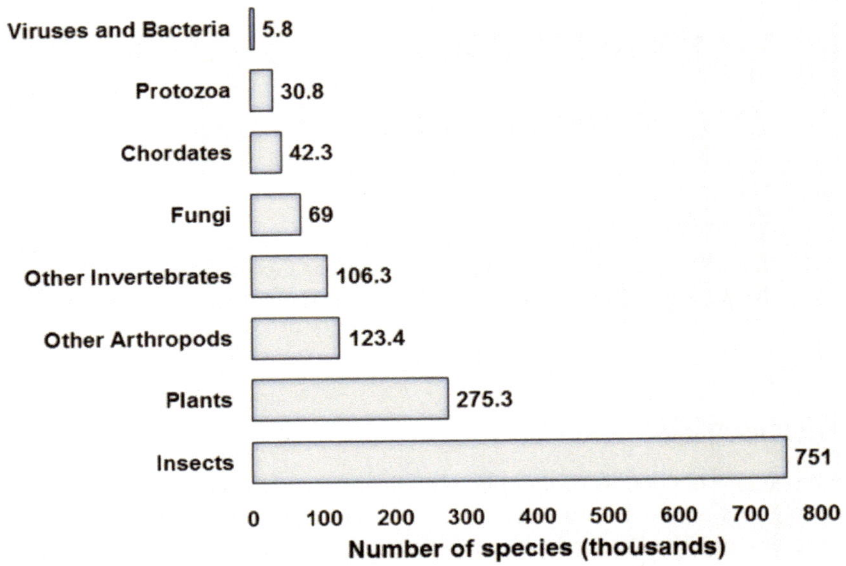

FIGURE 1.1 Distribution of described species within major taxonomic groups (Wilson, 1992).

TABLE 1.1
The Five Mass Extinctions

Name of the Extinction	Million Years Ago	Per Cent of Extinction	Reason for Extinction
Ordovician–Silurian	440	86	Decrease in CO_2 and glacier formation
Late Devonian	365	75	Excess nutrients in ocean and decrease in oxygen level
Permian–Triassic	252	96	Warming due to greenhouse gases (CO_2 and CH_4)
Triassic–Jurassic	201	80	Volcanic eruptions
Cretaceous–Paleogene	66	60–76	Asteroid impact

mamo (*Drepanis pacifica*), passenger pigeon (*Ectopistes migratorius*), Jamaican flightless ibis (*Xenicibis xympithecus*), quagga (*Equus quagga quagga*), Steller's sea cow (*Hydrodamalis gigas*), Eurasian aurochs (*Bos primigenius primigenius*), great auk (*Pinguinus impennis*), and woolly mammoth (*Mammuthus primigenius*) were made extinct by human activities and many of them were keystone species.

1.2 MAJOR THREATS TO BIODIVERSITY

There are three major levels of biodiversity, i.e., genetic diversity among individuals, species diversity among the species of a community, and ecosystem diversity among ecosystems. The number of species in any given ecosystem is referred to as alpha diversity, the difference in species composition is beta diversity, and the total diversity over a vast area is gamma diversity or the overall species pool. The percentage of endangered species is increasing continuously, as shown in Figure 1.2. More than 44,000 species are listed as threatened with extinction, accounting for 28 per cent

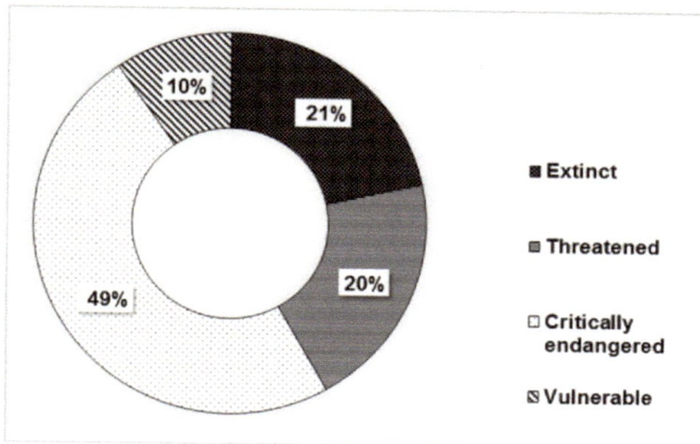

FIGURE 1.2 The percentage of categories of endangered species.

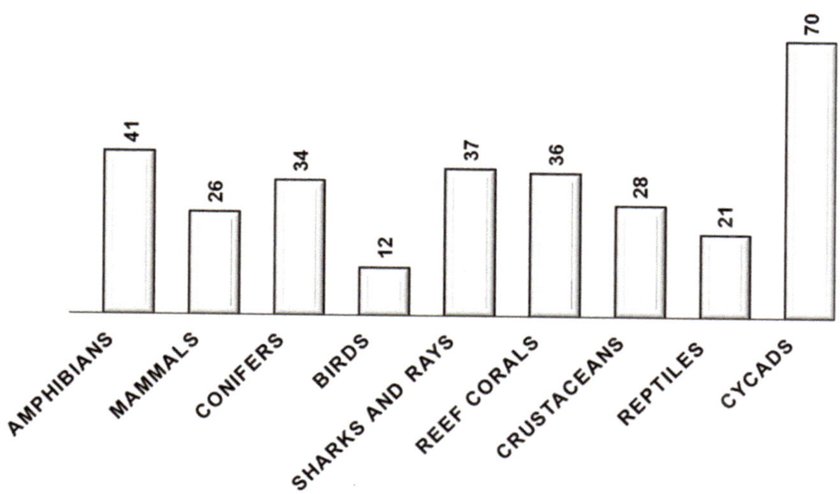

FIGURE 1.3 Percentage of threatened species.

of all assessed species, as shown in Figure 1.3. Hunting for common products, pollution, pet trade, pests and predator control, introduction of alien species, hunting for food, hunting for sport, and superstitious beliefs are the major causes of biodiversity threats. Among these causes, habitat loss is the leading contributor to major loss.

1.3 BIOTECHNOLOGICAL APPROACHES FOR CONSERVATION OF BIODIVERSITY

Conservation of biodiversity is inevitable for the stability and interaction in the environment along with its economic, cultural, scientific, aesthetic, and social values. The productivity of an environment increases with its increased biodiversity. After having understood the significance of biodiversity conservation, most countries of the world have come up with a variety of plans to conserve (Figure 1.4), and made stringent rules to protect endangered species, prevent habitat loss, and increase the population of endangered organisms. Gowthami *et al.* (2021) have estimated 2,143

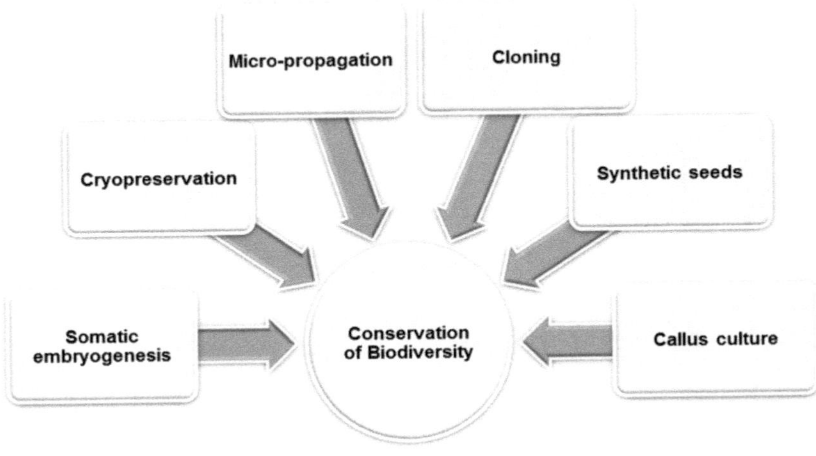

FIGURE 1.4 Biotechnological tools involved in biodiversity conservation.

species of endangered plants and 6,381 species of endangered animals. Some organisms are found to be extinct in the wild but are protected from extinction by modern technologies. These include angel's tears (*Brugmansia suaveolens*), Panamanian golden frog (*Atelopus zeteki*), Alula or cabbage on a stick (*Brighamia insignis*), butterfly goodeid (*Ameca splendens*), and *Ginkgo biloba.*

1.3.1 Micropropagation and Conservation

Micropropagation is a biological technique where the explant is taken from a plant that requires conservation. The tissue is then sterilised and inoculated into the appropriate medium. The tissue grows into a callus and it is used for the production of genetically identical plantlets. This technique is rapid and has been used for the production of many commercially important crops such as the banana, apple, pear, strawberry, cardamom, and many ornamentals (e.g., orchids) and other plants (Mandal and Datta, 2002). *Rauwolfia serpentina* is a medicinal plant considered to be a critically endangered species in North India (Kumar, 2019). Barik *et al.* (2018) have reported that 76 species of critically endangered plants have been conserved using micropropagation. *Bacopa monnieri*, *Celastrus paniculatus*, *Ginkgo biloba*, *Glycyrrhiza glabra*, *Gymnema sylvestre*, *Holostemma*, *Oroxylum indicum*, *Picorrhiza kurroa*, *Salacia oblonga*, *Saussurea lappa*, *Swertia chirata*, *Tinospora cordifolia*, and *Tylophora indica* have also been conserved through micropropagation (Sharma *et al.*, 2010a). *Crinum malabaricum* is an aquatic weed that is endemic to the Malabar region of the Indian subcontinent and its distribution is less than 500 sq m. It has been successfully grown in the laboratory and greenhouse using micropropagation and somatic embryogenesis (Priyadharshini *et al.*, 2020).

1.3.2 Somatic Embryogenesis

A somatic embryo is an embryo-like structure grown from somatic tissue of plants (Figure 1.5). It may be produced either directly or indirectly, and it is a non-sexual developmental process. *In vitro* somatic embryogenesis was conducted for the first time on *Daucus carota*, the carrot, by Reinert (1958). *Viola canescens,* a threatened species of medicinal herb found in the Himalayas at an altitude of 1500 to 2400 m, was regenerated through somatic embryogenesis by Khajuria (2021). Many other endangered Himalayan herbs such as *Podophyllum hexandrum* (Nadeem *et al.*,2000), *Picrorhiza kurroa* (Sharma *et al.*, 2010b), *Swertia chirata* (Jha *et al.*, 2011), *Angelica glauca* (Bisht *et al.*, 2015), *Gentiana kurroo* (Kaushal and Sidana, 2018), and *Viola odorata* (Narayani *et al.*, 2018) have

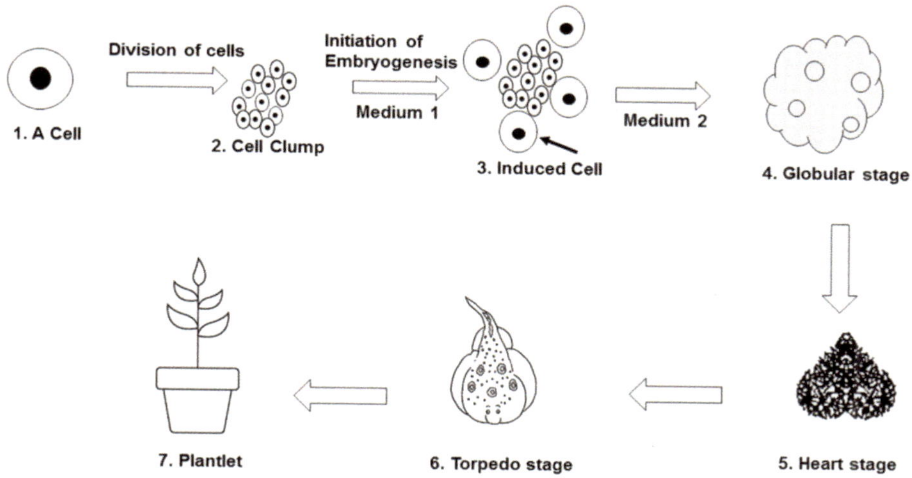

FIGURE 1.5 Steps involved in somatic embryogenesis.

been grown *in vitro* through somatic embryogenesis. Norway spruce (*Picea abies*) and Scots pine (*Pinus sylvestris*) are economically and ecologically important trees and have been produced using somatic embryogenesis (Hazubska-Przybył *et al.*, 2022). Trees such as the spruce, Douglas fir, and some pine species are commonly produced from somatic embryos. Genetically manipulated trees are produced most efficiently through somatic embryogenesis. This technique is applied to produce cell lines for genetic engineering and *ex situ* conservation of rare species or populations (Park and Klimaszewska, 2003). *Pulsatilla vulgaris,* a herbaceous perennial plant whose rhizome has some medicinal value, has become extinct in Poland and endangered in the central part of Europe. Using genotype-dependent mass somatic embryogenesis, an attempt has been made to revive the extinct population (Żabicka *et al.*, 2021).

1.3.3 Somaclonal Variation and Conservation

The plant cells cultured *in vitro* can be genetically modified, and such variants are known as somaclones. The somaclonal variations are generally studied using inter simple sequence repeat (ISSR) molecular markers and flow cytometry (Parzymies *et al.*, 2020). When plants are regenerated through callus culture over a long period, they often develop chromosomal variations (Figure 1.6). Such variations could also be induced by altering the composition of the medium or by irradiation. These variants develop favourable characters for ever-changing environmental conditions. Zabicki *et al.* (2019) regenerated *Viola stagnina,* an endangered European plant, with a number of favourable characters without altering the genome size. *Viola uliginosa*, a European violet endemic to the Baltic Sea region, is now endangered and threatened. Slazak *et al.* (2015) have successfully enhanced the biosynthesis of cyclotide through somaclonal variation. *Salix lapponum*, a downy willow, is a boreal relict species threatened with extinction due to environmental changes in Poland. This plant has been successfully micropropagated with improved quality of survival without affecting the genetic stability compared to that of the parent population by Parzymies *et al.* (2020).

1.3.4 Synthetic Seeds and Conservation

The encapsulated plant somatic embryos in sodium alginate along with mycorrhizae, insecticides, fungicides, and herbicides are known as synthetic or artificial seeds. These seeds are very easy to take to the field without any acclimatisation, can also be mass propagated, and hybrid varieties are

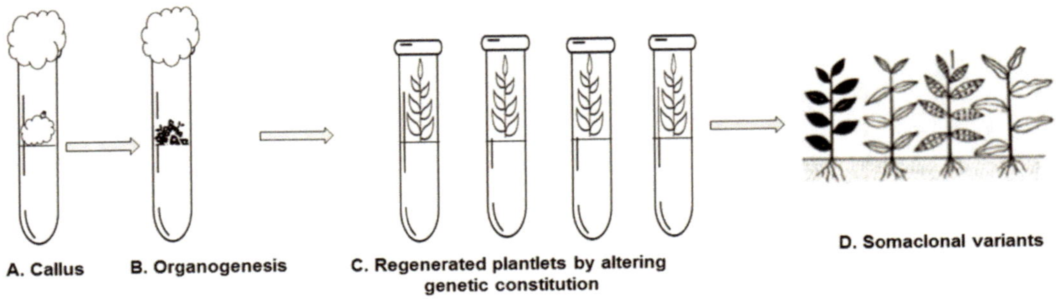

FIGURE 1.6 Steps involved in somaclonal variation.

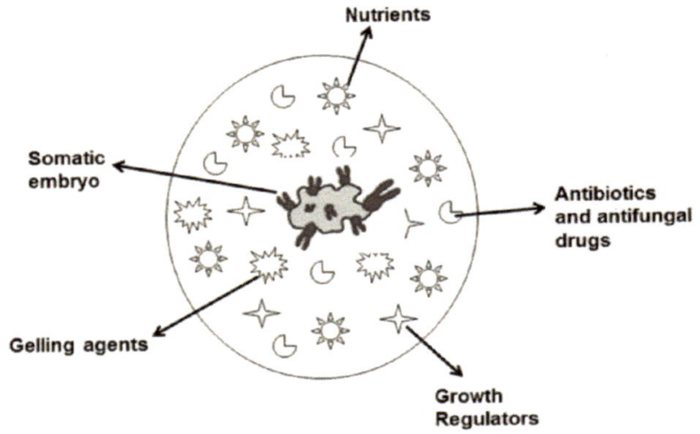

FIGURE 1.7 Composition of synthetic seeds.

also produced. Garg and Maheswari (2023) have emphasised the importance of synthetic seeds in the vegetative multiplication of plants and the conservation of endangered plants (Saxena *et al.*, 2019) and endangered medicinal plants (Ali and Ray, 2022).

1.3.5 Cryopreservation

Cryopreservation is the storage of biological samples at 196°C in containers with liquid nitrogen. The temperature could vary depending on the cryopreservative agent and is trusted as the most effective long-term preservation procedure for plant samples. Apart from liquid nitrogen, glycerol and dimethyl sulfoxide (DMSO) could also be used (Murray and Gibson, 2022). This is a sophisticated procedure for the preservation of mature seeds, apical buds, embryogenic tissues, pollen grains, embryos, root and shoot primordia, protocorms, protocorm-like bodies (PLBs), callus, protoplasts, spores, bulblets, microtubers induced from nodal segments, rhizome buds, dormant buds, plumules, and *in vitro*-derived explants (Kaviani and Kulus, 2022). The general procedure involved in the cryopreservation of plant tissues is explained in Figure 1.8.

1.3.6 Molecular Marker Technology

Biochemical markers like isozyme markers are powerful tools for studying genetic variability within and between populations (Khan *et al.*, 2012). Polymorphism is very common among populations

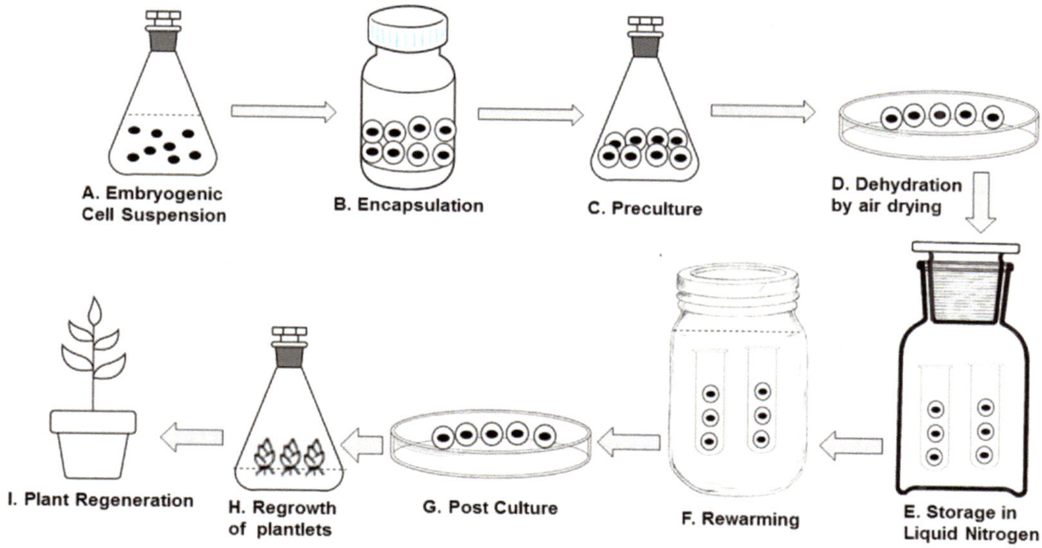

FIGURE 1.8 Cryopresrvation of plant cells and regeneration of plants.

and can be studied using specific DNA markers. The most frequently used markers are random amplified polymorphic DNA (RAPD), sequence characterised amplified region (SCAR), simple sequence repeat (SSR), inter simple sequence repeat (ISSR), restriction fragment length polymorphism (RFLP), amplified fragment length polymorphism (AFLP), selective amplification of microsatellite polymorphic loci (SAMPL), single nucleotide polymorphism (SNP), expressed sequence tagged (EST), single-strand conformation polymorphism (SSCP), cleaved amplified polymorphic sequence (CAPS), and spacer sequences. The genetic diversity of endangered plants in China, such as *Monimopetalum chinense*, *Gynostemma pentaphyllum, and Heptacodium miconioides,* has been studied using ISSR markers (Xie *et al.*, 2005, Wang *et al.*, 2008, Jin and Li, 2007).

1.3.7 Somatic Hybrids and Cybrids

This technique is otherwise known as somatic cell hybridisation, parasexual hybridisation, or protoplast fusion, which offers an alternative tool for making distant hybrids with desirable traits, particularly between species or genera, in cases where plant crossing by the conventional method of sexual hybridisation is not possible. This technology has not yet been utilised for the conservation of endangered plants.

1.3.8 Biotechnology and Conservation of Animals

The International Union for Conservation of Nature (IUCN) Green Status of Species is a complement to the IUCN Red List, which assesses the recovery population of a species and measures their conservation success. There are eight Green Status categories: Extinct in the Wild, Critically Depleted, Largely Depleted, Moderately Depleted, Slightly Depleted, Fully Recovered, Non-Depleted, and Indeterminate. A Green Status assessment examines how conservation strategies have affected the current Red List status. The Green Status of an organism assesses the following three facets of recovery. First, a species is fully recovered if it is present in all parts of its range, even those that are no longer occupied but were occupied prior to major human impacts/disruption; second, it is viable (i.e., not threatened with extinction) in all parts of the range; third, it is performing its ecological functions in all parts of the range. These factors contribute towards a "Green Score" ranging from 0 to 100 per cent, which shows how far a species is from its "fully recovered" state (Akçakaya *et al.*, 2018).

The following are the techniques used in the conservation of endangered animals.

1.3.9 Genetic Polymorphism Studies for Conservation Breeding

Rhinoceros unicornis is a critically endangered species found in the Northeastern states of India. In order to conserve this species, heterozygosity and genetic polymorphism were assessed using micosatellite DNA (Umapathy, 2017). Similar studies were conducted on the clouded leopard (*Neofelis nebulosa*), red panda (*Ailurus fulgens*), pheasants, snow leopard (*Panthera uncia*), and Tibetian wolf (*Canis lupus*) from Padmaja Naidu Zoological Park, Darjeeling; the Indian mouse deer (*Moschiola indica*) from Nehru Zoological Park, Hyderabad; the Asiatic lion (*Panthera leo*); and the lion-tailed macaque (*Macaca silenus*). These studies help to understand the genetic distance of endangered animals, and if the distance is greater, they are involved in breeding. If the distance is less, breeding among the individuals is avoided due to inbreeding depression. Microsatellite genotyping and mitochondrial DNA sequencing have been successfully employed as tools of molecular genetics for the conservation of tigers (Luo *et al.*, 2010).

1.3.10 Detection of Fertility Status in Endangered Animals Using ELISA and RIA

The pregnancy and reproductive potential of animals can be monitored by predicting the presence of circulating progesterone and relaxin in the blood through radioimmunoassays (RIAs) or enzyme-linked immunoassays (ELISAs) based on the measurement of steroids circulating in the blood. This is because the hormone concentrations in circulation are assumed to be accurate indicators of the physiological status of animals. Progesterone is metabolised into 5α-pregnan-3α–ol-20-one and excreted in the faeces, and this metabolite is also used to confirm pregnancy through these techniques. The pregnancy of big cats, musk deer (*Moschus* sp.), Indian mouse deer (*M. indica*), and red panda (*Ailurus fulgens*) is tested and monitored in captivity in zoos for conservation (Umapathy, 2017).

1.3.11 Diagnosis of Wildlife Diseases

Eighty per cent of endangered artiodactyls and carnivores are threatened by infectious diseases. Sometimes these diseases transfer to humans and become zoonotic. Such diseases are detected using polymerase chain reaction (PCR) from these animals to take appropriate steps to protect them. PCR is used to detect canine parvovirus (CPV), feline parvovirus (FPV), and infectious canine hepatitis (ICH). It is a routine procedure in zoos and captive breeding programmes of endangered animals.

1.4 CRYOBANKING

1.4.1 Rescue of Gametes from Dead Endangered Species

Oocytes can be recovered during post-mortem of endangered animals, and these oocytes could be activated through parthenogenesis (Umapthy, 2017). This technique has been successful in the conservation of barasingha, thamin deer, swamp deer, sambar deer, and spotted deer.

1.4.2 Maintenance of Cell Bank for Endangered Animals

Fibroblast cell lines have been developed and cryopreserved from chousingha, sambar deer, leopard, chimpanzee, and marmoset. The cells were collected during post-mortem. The cell lines developed with this procedure were cryopreserved. There are many assisted reproductive technologies (ARTs) for conservation strategies. Somatic cell nuclear transfer technology is one such technology to clone an endangered species. The somatic cells retrieved from dead animals can be cryopreserved and used for cloning if needed.

1.5 CONCLUSION

Biodiversity has been a topic of discussion in the last few decades due to the damage caused to it. Apart from natural phenomena, anthropogenic factors cause irreversible damage to biodiversity. It has become mandatory for every one of us to understand the concept of biodiversity and contribute to our maximum capacity to preserve and protect it. There is a serious demand for ways to conserve the biodiversity and especially endangered species. Biotechnology has played a significant role in the strategies of conservation strategies, and also many more techniques could be developed. The government, educational institutions, and other stakeholders should come forward with better solutions.

REFERENCES

Akçakaya, H.R., Bennett, E.L., Brooks, T.M., Grace, M.K., Heath, A., Hedges, S., Taylor, C.H., Hoffmann, M., Keith,D.A., Long,B., Mallon,D.P., Meijaard,E., Milner-Gulland,E.J., Rodrigues, A.S.L., Rodriguez, J.P., Stephenson,P.J., Stuart,S.N. Richard P. and Young, R.P. 2019. Quantifying species recovery and conservation success to develop an IUCN green list of species. *Conservation Biology. Open Access Journal.* (5): 1128–1138. https://doi.org/10.1111/cobi.13112

Ali and Ray, 2022. Synthetic seeds vis-a-vis cryopreservation: An efficient technique for long-term preservation of endangered medicinal plants. In *Medicinal Plants* (1st edition). Apple Academic Press.

Barik, S.K., Tiwari, O.N., Adhikari, D., Singh, P.P., Tiwary, R., Barua, S. 2018. Geographic distribution pattern of threatened plants of India and steps taken for their conservation. *Current Science.* 114(3): 470–503.

Bisht A.K., Bhatt A., Dhar U., Bhatt A., Dhar U. 2015. Note on somatic embryogenesis and synthetic seed production in *Angelica glauca*: A valuable medicinal plant of Himalaya. *Journal of Medicinal Plants Research.* 9: 419–425.

Ceballos G., Ehrlich P.R., Dirzo R. 2017. Biological annihilation via the ongoing sixth mass extinction signaled by vertebrate population losses and declines. *Proceedings of the National Academy of Sciences.* 2017 Jul 25; 114(30): E6089–E6096.

Garg, R., and Maheshwari, S. 2023. Synthetic seed technology, application and future trends. *International Journal of Agriculture and Environmental Research.* 9(1): 1–10

Gowthami, R., Sharma, N., Pandey, R., Agrawal, A. 2021 Status and consolidated list of threatened medicinal plants of India. *Genetic Resources Crop Evolution.* 68(6): 2235–2263.

Hazubska-Przybył, T., Wawrzyniak, M.K., Kijowska-Oberc, J., Staszak, A.M., Ratajczak, E. 2022. Somatic embryogenesis of Norway Spruce and Scots Pine: Possibility of application in modern forestry. *Forests,* 13: 155. https://doi.org/10.3390/f13020155

Jha, T.B., Dafadar, A. and Chaudhuri, R.K. 2011. Somatic Embryogenesis in *Swertia chirata* Buch. Ham. ex Wall.—A multipotent medicinal herb. *Asian Journal Biotechnology.* 3: 186–198. doi: 10.3923/ajbkr.2011.186.193.

Jin, Z., Li, J. 2007. Genetic differentiation in endangered *Heptacodium miconioides* Rehd. based on ISSR polymorphism and implications for its conservation. *Forest Ecology Management.* 245: 130–136

Kaushal, S., Sidana, A. 2018. Somatic embryogenesis and plant regeneration from cell suspension cultures of *Gentiana kurroo* Royle. *Annals of Plant Science.* 7: 2239–2246. doi: 10.21746/aps.2018.7.5.6.

Kaviani, B., Kulus, D. 2022. Cryopreservation of endangered ornamental plants and fruit crops from tropical and subtropical regions. *Biology (Basel).* 11(6): 847. doi: 10.3390/biology11060847.

Khajuria, A.K., Hano, C., Bisht, N.S. 2021. Somatic embryogenesis and plant regeneration in *Viola canescens* Wall. Ex. Roxb.: An endangered Himalayan Herb. *Plants (Basel).* 10(4): 761. https://doi.org/10.3390/plants10040761.

Khan,S., Qurainy,F,A., Nadeem, M. 2012. Biotechnological approaches for conservation and improvement of rare and endangered plants of Saudi Arabia. *Saudi Journal of Biological Sciences.* 19(1): 1–10

Kumar, A. 2019. *Rauwolfia serpentine* L. Benth. ex Kurz.: A rare and endangered species in India. *Agriculture & Food: E-Newsletter.* 1(10): 534–535.

Luo, S.J., Johnson W.E., Brien, S.J. 2010. Applying molecular genetic tools to tiger conservation. *Integrative Zoology.* 5(4): 351–362. https://doi:10.1111/j.1749-4877.2010.00222.x

Mandal, A.K., Datta, S.K. 2002. Introduction of Gerbera cultivation in Lucknow agro-climate through tissue culture of young flower head. *Indian Journal of Biotechnology.* 1(2): 212–216.

May, R.M. 1988. How many species are there on earth? *Science, New series.* 241(4872): 1441–1449.

Mora, C., Tittensor, D.P., Adl, S., Simpson, A.G.B., Worm, B. 2011. How many species are there on earth and in the ocean? *PLoS Biology.* 9(8): e1001127. https://doi.org/10.1371/journal.pbio.1001127

Murray, K.A., Gibson, M.I. 2022. Chemical approaches to cryopreservation. *Nature Reviews Chemistry.* 6: 579–593. https://doi.org/10.1038/s41570-022-00407-4

Nadeem, M., Palni, L.M.S., Purohit, A.N., Pandey, H., Nandi, S.K. 2000 Propagation and conservation of *Podophyllum hexandrum* Royle: An important medicinal herb. *Biology Conservation.* 92: 121–129. https://doi: 10.1016/S0006-3207(99)00059-2.

Narayani, M., Varsha, M.S., Potunuru, U.R., Beaula, W.S., Rayala, S.K., Dixit, M., Chadha, A., Srivastava, S. 2018. Production of bioactive cyclotides in somatic embryos of *Viola odorata. Phytochemistry.* 156: 135–141.

Park, Y.S., Klimaszewska, K. 2003. Achievements and Challenges in Conifer Somatic Embryogenesis for Clonal Forestry. In *Proceedings of XII World Forestry Congress*, 21–28 Sep. 2003. Quebec City, Canada.

Parzymies, M., Pogorzelec, M., Głębocka, K., Śliwińska, E. 2020. Genetic stability of the endangered species *Salix lapponum* L. Regenerated in vitro during the reintroduction process. *Biology (Basel).* 9(11): 378. https://doi: 10.3390/biology9110378.

Priyadharshini, S, M. Manokari, M., Shekhawat, M.S. 2020. In vitro conservation strategies for the critically endangered Malabar river lily (*Crinum malabaricum* Lekhak & Yadav) using somatic embryogenesis and synthetic seed production. *South African Journal of Botany.* 135: 172–180

Reinert, J. 1958. Uber die Kontrolle der Morphogenese and die Induktion Von Adventivembryonen and Gewebek Ulturen aus Korotten. *Planta.* 53: 318–333.

Saxena, A., Shukla, M., Saxena, P., 2019. Synthetic seeds: Relevance to endangered germplasm conservation in vitro. In *Synthetic Seeds,* Faisal, M., Alatar, A.A. (eds.), pp. 21–60. Springer Nature Switzerland AG 2019.

Sharma, S., Katoch, V., Rathour, R., Sharma, T.R. 2010a. In vitro propagation of endangered temperate Himalayan medicinal herb *Picrorhiza kurroa* Royle ex benth using leaf explants and nodal segments. *Journal of Plant Biochemistry and Biotechnology.* 19: 111–114. doi: 10.1007/BF03323447.

Sharma, S., Rathi, N., Kamal, B., Pundir, D., Kaur, B., Arya, S. 2010b. Conservation of biodiversity of highly important medicinal plants of India through tissue culture technology- A review. *Agriculture and Biology Journal of North America.* 1(5): 827–833.

Slazak, B., Sliwinska, E., Saługa, M., Ronikier, M., Bujak, J., Słomka, A., Göransson, U., Kuta, E. 2015. Micropropagation of *Viola uliginosa* (Violaceae) for endangered species conservation and for somaclonal variation-enhanced cyclotide biosynthesis. *Plant Cell, Tissue and Organ Culture* 120: 179–190. https://doi.org/10.1007/s11240-014-0592-3

Umapathy, G. 2017. Conservation of endangered animals based on genetic polymorphism studies and assisted reproduction. Report 2012–17. CSIR-CCMB-Laboratory for Conservation of Endangered Species, Hyderabad.

Wang, C., Zhang,H., Qian, Z.Q., Zhao, G.F. 2008. Genetic differentiation in endangered *Gynostemma pentaphyllum* (Thunb.) Makino based on ISSR polymorphism and its implications for conservation. *Biochemical Systematics and Ecology.* 36: 699–705

Wilson, E., 1992. *The Diversity of Life.* Harvard University Press, Cambridge, MA.

Xie, G.W., Wang, D.L., Yuan, Y.M., Ge. X.J. 2005. Population genetic structure of *Monimopetalum chinense* (Celastraceae), an endangered endemic species of Eastern China. *Annals of Botany.* 95(5): 773–777. https://doi: 10.1093/aob/mci087

Żabicka, J., Żabicki, P., Słomka, A. .2021 Genotype-dependent mass somatic embryogenesis: a chance to recover extinct populations of *Pulsatilla vulgaris* Mill.. *Plant Cell Tissue Organ Culture.* 146: 345–355. https://doi.org/10.1007/s11240-021-02074-7

Żabicki, P., Sliwinska, E., Mitka, J. *et al.* 2019. Does somaclonal variation play advantageous role in conservation practice of endangered species?: Comprehensive genetic studies of in vitro propagated plantlets of *Viola stagnina* Kit. (Violaceae). *Plant Cell Tissue Organ Culture.* 136, 339–352. https://doi.org/10.1007/s11240-018-1519-1

2 Utilization and Future Prospective of Next Generation Sequencing (NGS) in Biodiversity Conservation of Endangered Medicinal Plants

Neha Salaria[1] *Neeraj*[1], *Indu Kumari*[1], *Bhawna*[2]
Rajesh Kumar[3]

[1]Department of Life Sciences, Arni School of Basic Sciences, Arni University, Kathgarh Himachal Pradesh, India
[2]Drug Testing Laboratory, QMI CSIR-IIIM Jammu, India
[3]Department of Biosciences, Himachal Pradesh University, Shimla (H.P.)

2.1 INTRODUCTION

A significant portion of the flora of the planet comprises therapeutic plants. The use of medicinal herbs has been beneficial since humans were nomads transitioning from hunting to scavenging and production. Since ancient times, various medical, therapeutic, and dietary uses of plants have been made (Petrovska, 2012). Key historical writings from nations such as India, Japan, Egypt, and China have referenced the usage of plants or different compounds derived from plants for treating various ailments and disorders. Medical herbs still represent the lowest wave of domestication, despite the benefits and range of uses that we are aware of (Nogueira et al., 2018). Scientists' understanding of the value and potential applications of therapeutic herbs increased as time and technology progressed. There is a worrying decline in the number of therapeutic plants as a result of unethical and unsustainable wild harvesting practices and habitat destruction (Niazian, 2019). Considering the growing population, so has the use of these plants. With no human assistance, these plants can't reproduce at a rate that will meet our needs. It is essential to carefully domesticate and protect medicinal plants as a result. The expansion of the medicinal plant industry in underdeveloped countries will lead to a reduction in reliance on very expensive synthetic pharmaceuticals. In-depth study of medicinal plants is urgently required in order to create superior plant-based bioactive compounds. Comprehensive genome and transcriptome analyses of medicinal plants will improve our comprehension of the metabolic processes. Next-generation sequencing (NGS) is giving us the technology to quickly and inexpensively access the genomic, transcriptomic, and epigenomic properties of these non-model therapeutic plants (Gandhi et al., 2015). NGS, a massively parallel sequencing method, offers extraordinarily high speed, scalability, and throughput. By means of this technique, the nucleotide sequence of whole genomes or individual DNA as well as RNA segments can be determined. The biological sciences have changed significantly as a result of NGS, which allows labs to perform a number of tasks and investigate biota at previously unimaginable depths. Using NGS, genes in a number of metabolic pathways of medicinal plants were discovered. Due to the

DOI: 10.1201/9781032697550-2

development of high-capacity sequencing platforms, NGS was introduced (Edwards and Batley, 2010). The phrase "next-generation sequencing" has been used to describe techniques that generate information at or above the genomic level for more than a decade (Metzker, 2010). It is currently possible to analyze the entire genome through NGS to determine the genotype held accountable for phenotypic variation. This enables us to recognize and combine the qualities that are more significant. NGS is not just useful for this; furthermore, it offers species-specific information that can be utilized to create species-specific polymerase chain reaction (PCR) primers and assists in the entire-genome sequencing of closely related wild species. It is possible to investigate the genetic variation in several genes by sequencing the PCR amplification products of the target genes. Newer allelic differences in closely related species can be described by utilizing NGS. The advancement of genomics research through NGS has been crucial. The Human Genome Project utilized Sanger sequencing techniques and required 13 years to sequence and decipher the whole human genome. With the aid of NGS technologies, this entire accomplishment may now be completed in a single day. Different bioinformatics techniques are used to analyze the results. Following that, the individual reads are mapped and organized in accordance with the accessible reference genome. In addition to sequencing the entire genome, NGS methods can be implemented to sequence specific regions of interest in a few single genes or in nearly 22,000 coding genes (Behjati and Tarpey, 2013). Currently, NGS technologies are widely utilized to piece together the fragments of a medicinal plant's normal operation. Numerous plant science researchers and non-researchers are anticipated to benefit greatly from NGS.

2.2 THE DEMAND FOR RAPID, LOW-COST, AND RELIABLE DNA SEQUENCING TECHNOLOGIES

The human genomes, as well as a few other well-chosen plant and animal genomes, have been fully sequenced thanks to the automated Sanger method. The limits of the technology, however, highlighted the need for new and enhanced methods for the mass sequencing of human and other genomes. The automated Sanger technique, a "first-generation" technological innovation, has been the target of efforts to replace it in the early 21st century and the late 20th century. The adoption of more recent techniques, known as next-generation sequencing, has altered scientific methodologies both in fundamental and applied research across a wide range of scientific fields, particularly in many aspects of biology, such as the study of plant viruses and pathogens. The primary advantages of NGS include its capacity to generate massive amounts of information, sometimes reaching one billion short reads every instrument cycle, and its capacity to offer precise, timely, and reasonably priced genomic data (Barba et al., 2014).

2.3 DEVELOPMENT OF NGS PLATFORMS (2000–PRESENT)

Lynx Therapeutics (USA) launched massively parallel signature sequencing (MPSS), the very first NGS technique, in 2000. Later, Illumina purchased the brand (Barba et al., 2014).

A massively parallel edition of pyrosequencing was introduced in 2004 by 454 Life Sciences (Branford, Connecticut, USA). Their machine's initial iteration, which followed MPSS as novel sequencing technology's second generation, cut sequencing costs by a factor of six when compared to automated Sanger sequencing. Roche (Switzerland) acquired 454 Life Sciences. In contrast to Sanger sequencing, Illumina and SOLiD pyrosequencing offers intermediate read lengths, as well as base prices.

In 2005–2006, the introduction of Roche's 454 GS 20 sequencing technology, which produced 20 million bases (20 Mbp), revolutionized DNA sequencing. The GS FLX model, which could yield more than 100 Mbp of sequence in four hours, took its place in 2007; its yield increased to 400 Mbp in 2008. The 454 GS FLX+ Titanium sequencing technology, with Sanger-like read lengths of approximately 1,000 base pairs, can produce nearly 600 Mbp of sequence data in just one run.

The Roche GS Junior is a compact sequencing platform system that is used on a lab bench top. It is another platform system made by Roche. With a quick sequencing run, it offers lengthy 400 base pairs sequencing reads (Barba et al., 2014).

Solexa released the Genome Analyzer (GA) in 2005. Illumina acquired Solexa in 2007. The most recent model can generate 85 Bbp in a single run, but the GAIIx technology can produce a maximum of 50 billion bases of relevant data in just one run.

Illumina has spent the last three to four years creating the HiSeq platform series, which includes the HiSeq® 2500, HiSeq 2000, HiSeq 1500, and HiSeq 1000 sequencing devices. The HiSeq 2500 has a read length of 200 bp, which is the largest read length, whereas the HiSeq 1000 has the smallest read length. They have different maximum read lengths, paired-end reads, cluster generations, run times, and outputs. A human genome can be sequenced using the HiSeq 2500 technology in about 24 hours. Additionally, 30 RNA sequencing samples can be processed in just 5 hours, or 20 exomes in a day. Each run of the common HiSeq 2000 generates 600 Bbp of bases. HiSeq 2500 is capable of producing 600 Gbp. In 2011, Illumina also unveiled the benchtop platform MiSeq, which utilized many of the same technologies as the HiSeq series. In roughly 10 hours, it produces 1.5 Gbp per run (Barba et al., 2014).

SOLiD technology uses oligo ligation detection for sequencing. Sequences that compare in quantity and length to Illumina sequencing are the end result. By purchasing SOLiD, Applied Biosystems (developed in Foster City, California, in 2006, USA, and based in Carlsbad, California, USA) became a part of Life Technologies in 2008. SOLiD efficiency can possibly approach 99.99% using the two-base sequencing technique.

The SOLiD 5500 w as well as 5500 xlw sequencing technology were introduced in 2010, whereas Applied Biosystems manufactured the foremost SOLiD sequencing system during 2007. The SOLiD 5500 xlw offers 85 bp read lengths, 99.99% efficiency, plus 30 Gbp throughputs each cycle. Within a week, an entire cycle might be done. Analysis of entire genome clusters is one of SOLiD's applications. Other recently created NGS technologies include the Life Technologies Ion Torrent sequencer, which was published in 2011, and the Helicos sequencer, which was introduced in 2009. Single-molecule real-time (SMRT) sequencers with super long single-molecule reads became accessible between 2012 and 2013 from Oxford Technologies Nanopore (Oxford, UK), as well as Pacific Biosciences (Menlo Park, California, USA) in 2011 (Barba et al., 2014). Helicos Biosciences (Cambridge, Massachusetts, USA) declared bankruptcy in November 2012 and is actively restructuring its business. Its website is no longer accessible as a result.

Additionally, new NGS techniques were created. These techniques include sequencing of DNA Nanoball, VisiGen Biotechnologies sequencing, Polonator sequencing, and Polony sequencing. Table 2.1 displays the primary NGS technologies currently in use.

2.4 TECHNOLOGY ADVANCES IN MEDICINAL PLANT RESEARCH

The need for medicinal plants and the rising population make it imperative to better examine their genomes, transcriptomes, epigenomes, and interactomes (Hamilton and Robin Buell, 2012). To analyze the complexity of plant genomes, NGS is used. Medicinal plants were found to be considerably more complicated than model plants. Therefore, deeper research into this complexity would lead to a greater comprehension pertaining to the plant. NGS is mostly utilized for integrative genomics research (Hirsch and Robin Buell, 2013). By comparing a plant's genome with the genotype of genetically similar species, comparative genomics can help find unique chemical compounds in a given plant. The root of *Rehmannia glutinosa* is reported to contain a variety of bioactive compounds, primarily iridoids. *R. glutinosa*'s reference genome was discovered utilizing Hi-C sequencing, Nanopore, and Illumina. With a scaffold N50 size of 70 Mb and significant heterozygosity (2%), the assembled genome is 2.49 GB long (Ma et al., 2021).

NGS is utilized as well to identify novel and undiscovered bioactive components (Sharma and Shrivastava, 2016). Using Illumina paired-end sequencing data, the 152,469 bp long chloroplast

TABLE 2.1
Currently Used NGS Technologies

Sequencing System	Approach of Amplification	Length of Reading (Base Pair)	Speed of Sequencing per Hour	References
PacBio (Pacific Biosciences)	No amplification, single-molecule real-time (SMRT)	4,000–5,000	50–115 megabase pairs	(Carneiro et al., 2012)
Helicos (Helicos Biosciences)	No amplification, single molecule	25–55	83 megabase pairs	(Kircher and Kelso,2010)
SOLiD (Life Technologies)	Emulsion PCR	75–85	21–28 megabase pairs	(Kircher and Kelso, 2010)
Ion Torrent (Life Technologies)	Emulsion PCR	100–400	25 megabase pairs–16 gigabase pairs	(Bragg et al., 2013)
Illumina (Illumina)	Bridge PCR	100–300	25 megabase pairs	(Archer et al., 2012)
454 (Roche)	Emulsion PCR	400–700	13 megabase pairs	(Archer et al., 2012)

genome of *Ocimum gratissimum* was sequenced. The genome was determined to have 138 distinct genes, including 45 tRNA, 8 rRNA, and 85 protein-coding genes (Balaji et al., 2021). An essential component of the entire transcriptome analysis is the discovery of potential genes expressed along a certain bioactive compound route (Luo et al., 2012). The roots of *Panax notoginseng* are valued in particular for their triterpenoid saponin content. Also, 2,383 full-length non-chimeric (FLNC) reads were mapped using SMRT and Illumina sequencing (SGS) methods, 1,925 incorrectly identified genes were fixed, and 927 new genes were created (Zhang et al., 2020). Small and lengthy non-coding RNA species are found using NGS. It is well recognized that non-coding RNAs are crucial in the control of protein production.

It would be beneficial to learn more regarding the transcriptional control of biosynthesis cascade proteins if these RNAs were found in medicinal plants (Huang et al., 2013). The PacBio sequencing platform was used to examine the berberine production pathway in *Berberis koreana*. Also, 23,246 full-length unigenes were found after the cDNA pool was sequenced and expressed gene isoforms were found (Roy et al., 2021). The use of NGS technology offered the framework for the examination of the transcriptomics of medicinal plants in order to pinpoint the essential genes involved in the manufacture of bioactive substances (Chen et al., 2013). Recent research on medicinal plants using NGS technology is shown in Table 2.2.

2.5 SOME SELECTED NGS TOOLS AND THEIR UTILITIES IN CONSERVING ENDANGERED MEDICINAL PLANTS

Advances in next-generation sequencing technology have transformed biological science since 2005 (Su et al., 2011; Van Dijk et al., 2014). The utilization of unique gene markers, for instance, species-specific DNA barcodes, to analyze ambient DNA has been a fundamental use of next-generation sequencing technology in environmental and ecological investigation. Access to huge volumes of sequencing data in parallel, as well as subsequent advances in read length and throughput of

TABLE 2.2
Recent Findings Utilizing NGS Technologies on Therapeutics Medicinal Plants

Medicinal Plant	NGS Platform Used	Aims and Implementation	References
Coptisdeltoidea	PacBio, Illumina	Genes responsible for the biosynthesis of benzylquinoline alkaloids have been recognized	(Zhong et al., 2020)
Camellia sinensis	Nanopore	Analyzing of the transcriptome	(Wang et al., 2020)
Clitoriaternatea	Illumina	Identification of possible genes associated in cyclotide biosynthetic pathway	(Kalmankar et al., 2020)
Alismaorientale	Illumina	Comparative transcriptome as well as metabolic profiling	(Lin et al., 2019)
Swertiamussotii	Illumina	Recognition of genes responsible for the biosynthesis of secoiridoids	(Liu et al., 2017)
Aconitum carmichaellii	Illumina	Examine key genes associated in diterpene biosynthesis	(Rai et al., 2017)
Plantagoovata	Illumina	De novo transcriptome analysis	(Kotwal et al., 2016)
Panax ginseng	Illumina, Roche	Cultivar identification and analysis of genetic diversity	(Li et al., 2015)
Piper nigrum	Illumina	Biosynthesis gene characterization	(Hu et al., 2015)
Salvia miltiorrhiza	Illumina Roche	Recognition of genes associated to the biosynthesis of tanshinone	(Yang et al., 2013)

multiple sequencing technologies, is resulting in a more accurate depiction of sample diversity at a fair cost (Schadt et al., 2010). New technologies are fast emerging, with the prospect of significantly expediting environmental and ecological studies. The rapid development and advancement of next-generation sequencing technology may result in larger and more robust applications in environmental DNA research (Lodge et al., 2012; Hansen et al., 2018). In the present chapter, we describe the operation and application of various contemporary next-generation sequencing technologies for the preservation of threatened medicinal plants.

2.5.1 Roche 454 System

Roche 454 sequencing may produce significantly longer reads than other sequencers. It does this, like others, by sequencing several reads at once by scanning optical signals as bases are added (Zhou et al., 2010). It employs pyrosequencing technology for real-time sequencing-by-synthesis. Each nucleotide integrated by DNA polymerase results in the release of a pyrophosphate molecule in 454 pyrosequencing (Fakruddin and Chowdhury, 2012). This release starts a chain reaction that produces light through the action of the enzyme luciferase (Ahmadian et al., 2006). The count of nucleotides utilized directly relates to the amount of light generated. The RNA and DNA are split into smaller readings, up to 1 kb in this instance, similar to Illumina. Generic adaptors are added to the ends of the DNA fragments, which are then annealed to beads, one DNA fragment per bead (Kulski, 2016). After that, PCR is used to amplify the fragments employing adaptor-specific primers. After that, every bead is inserted on a slide in a single well. As a result, each well will have a single bead coated with multiple PCR copies of a single sequence (Gabriel et al., 2014). Additionally, the wells contain DNA polymerase along with sequencing buffers.

The Roche 454 sequencing system was one of the first commercially available systems for next-generation sequencing technology (Desai and Jere, 2012). The synthesis of DNA libraries in the 454 sequencing system differs from that of the Illumina system. It breaks down DNA material into small pieces of 300–800 bp and inserts various adaptors at both ends using the spray approach (Erlich, 2012). After DNA denaturation, utilize primers are utilized for amplification, cloning onto appropriate vectors, and finally forming a single-stranded DNA library. The development of a vast number of distinct reaction spaces for DNA amplification is the most notable aspect of emulsion PCR (Nocker et al., 2007).

The key technology is the separation of distinct beads utilizing emulsion properties. The fundamental procedure is as follows. When an aqueous solution with all PCR reaction components is infused into the surface of mineral oil with high-speed spinning, it creates numerous little water droplets enveloped in mineral oil (Caruccio, 2011). A small, microscopic droplet establishes its own PCR reaction area. One DNA template and one bead should preferably be present in each tiny drop of water. Each reaction in the 454 sequencing technique is carried out in an isolated nanopore of the PTP board, which considerably decreases mutual interference and sequencing bias (Shendure et al., 2011). The most significant advantage of 454 technologies is the capacity to get more extensive sequencing lengths of reads. The typical read length of 454 technologies is now 400 bp (Wommack et al., 2008).

2.5.2 AB SOLiD System

SOLiD technology was introduced by Applied Biosystems (Life Technologies) as its NGS platform in 2007 (Chi, 2008). SOLiD, unlike the other systems, uses a technique called sequencing-by-oligo ligation. This method binds oligo adaptor-linked DNA fragments to complementary oligos fixed on the surface of 1 mm magnetic beads (Ninomiya et al., 2014). It cost millions of dollars and years to develop the first draft of the human genome. Researchers have devised new approaches, such as clonal cluster sequencing, to achieve the same result in a fraction of the time and expense (Weisenfeld et al., 2017). The SOLiD System accomplishes highly parallel sequential ligation processes by using this and other innovative technologies, significantly boosting throughput above conventional techniques and competing technologies (ten Bosch and Grody, 2008).

The SOLiD System is a throughput revolution that distinguishes itself from other next-generation systems with unrivaled precision, system robustness, and scalability (Tripathi et al., 2016). High throughput systems are required for whole genome investigations. The SOLiD System can create more than six gigabytes of mappable data and over 240 million tags every run, which is more useful data than any other next-generation system in the industry today (Egan et al., 2012). The SOLiD System's huge throughput may be due to three features: (1) emulsion PCR for single-molecule amplification of DNA, which provides effective amplification of all molecules (Zhu et al., 2012); (2) a thorough bead enrichment process that eliminates unhybridized beads and thus maximizes slide space usage for productive beads (Ong et al., 2014); and (3) high-density bead deposition onto a glass slide rather than constraining flow channels or microtiter plates (Voelkerding et al., 2009).

2.5.3 HiSeq System

The HiSeq 2500 has emerged as Illumina's indisputable powerhouse for high-throughput sequencing applications. Using Illumina sequencing technology, millions of samples have been sequenced and thousands of articles have been published (Van Dijk et al., 2014). The HiSeq 2500 is the production platform of choice for a number of genome centers and universities across the world, with unequaled data quality, customizable run configurations, and a 10 gigabase (Gb) to 1 terabase (Tb) per run capacity (Mak et al., 2017; Klumpp et al., 2012). The HiSeq 2500 System has the greatest daily throughput and overall yield of any sequencer marketed separately on the market today (Eisenstein et al., 2012). With an extensive daily throughput of 160 gigabytes per day or 1 terabyte each run, extra samples may be sequenced at higher depths at the same time compared to the past, resulting in greater and more beneficial data. Large research studies may be done with less hands-on time and reagent expenditure (Zaura et al., 2021).

One hundred and fifty human exomes or eight human genomes with 30 times coverage can be sequenced using the HiSeq 2500 Platform in high-output mode, presuming 275 base pairs per run at 4 gigabytes per exome. In high-output mode, the HiSeq 2500 System is ideal for large-scale projects with huge genome sizes or research with hundreds of samples (Zook et al., 2016).

2.5.4 Compact PGM Sequencer

The Torrent Personal Genome Machine (PGM) and accompanying sample preparation equipment are intended for quick DNA or RNA sequencing (Quail et al., 2012). It can generate up to 2 gigabytes of information within a few hours, making it ideal for small genomics studies as well as the analysis of gene expression. The One Touch 2 template preparation system and its related equipment provide automated, scalable, and reproducible outcomes (Lohse et al., 2013).

The sequencing chips include up to 11 million micro-machined wells, enabling high-throughput parallel sequencing with the possibility of sample barcoding for cost-effectiveness (Wang, 2021). The PGM Sequencer uses semiconductor sequencing to detect nucleotide incorporation into the developing strand of DNA in each micro-well without using chemiluminescence or fluorescence, resulting in smaller, more readily handled datasets (Grada and Weinbrecht, 2013). The direct detection system generates data quickly, with run times ranging from 2 to 8 hours depending on the kind of run and application (Egan et al., 2012).

2.6 UTILIZATIONS OF NGS TOOLS FOR CONSERVATION OF ENDANGERED MEDICINAL PLANTS

Medicinal plants have been used as traditional herbal medicines in the treatment of various types of diseases (Muthu et al., 2006; Abera, 2014). The increased demand for these plants, however, emphasizes how crucial conservation is, especially for endangered species. The development of NGS technologies has markedly boosted exploration on medicinal plants while lowering time and cost requirements. NGS systems enable high-throughput whole genome sequencing as well as direct RNA sequencing and transcriptome analysis (Gullapalli et al., 2012; Chaudhary and Sharma, 2016). The sequence datasets created can be used in a variety of areas of study, including biodiversity conservation, comparative genomics, transcriptomic analysis, single-cell mining, metagenomics, epigenetics, molecular marker discovery, and multigenome sequencing (Hua et al., 2015).

The technique used to solve bioinformatics issues in the assessment of NGS data is constantly being enhanced by commercial sequencing service providers. Several genome sequencing projects on medicinal plants have been completed recently, and a few more are in the works (Ghazal et al., 2021). Massive NGS-based data have developed in some medicinal plants. We have made an effort to briefly discuss developments in NGS technology on medicinally important plants in this chapter. The application of NGS technologies to the investigation of the genomes of numerous threatened medicinal plants, whose genome sequences are not typically available, is another idea that has been proposed. This might provide important insights for the sustainability of these critically endangered species.

NGS can be utilized to create new microsatellite markers in endemic and threatened medicinal plants all over the world (Yun et al., 2015). Natural populations are rapidly dwindling due to their extremely limited range, overharvesting for traditional medicinal uses, and habitat destruction (Manish, 2013; Ahmad et al., 2013). All new microsatellite markers may be effectively genotyped by employing samples from populations in their natural habitat. The unique markers will be useful for analyzing the genetic diversity of diverse endangered medicinal plants as well as germplasm conservation of endangered species (Borah et al., 2021).

Medicinal plants from different bioresources are noted for their exceptional variety and significant secondary metabolites that are unique to dynamic geo-climatic areas (Kumar et al., 2017). These medicinal plants have been utilized for centuries for their healing properties. However, increased pharmaceutical demands, as well as illicit and uncontrolled commerce in these therapeutic plants, have exacerbated the rate of overexploitation in a non-scientific way during the previous two decades (Bodeker et al., 2014; Ganie et al., 2019). Furthermore, climate change and industrial activity have impacted their natural environment, causing the majority of these indigenous plant species to become severely endangered, putting ecoregions at risk of mass extinction (Erwin, 2009).

As a result, there is an immediate requirement to create sustainable techniques and strategies to exploit these natural bioresources while also ensuring concurrent conservation. With the lack of genomic information, the advent of sequencing-based transcriptome research considerably contributes to a better understanding of the backdrop of essential metabolic pathways and linked genes/enzymes of high-value medicinal herbs (Xiang et al., 2010). Several researchers in various areas have used comparative transcriptomics in combination with biochemical approaches over the last decade, with one of the major areas being the North-Western Himalaya, resulting in significant breakthroughs in identifying the molecular actors involved in the formation of secondary metabolic pathways (Kapoor et al., 2021). This knowledge might be utilized to additionally modify breed-specific breeding plans as well as metabolic pathways, eventually directing the establishment of methods used in vitro for the commercial manufacturing of secondary metabolites with potential for use in medicines. Consequently, the successful implementation of these measures may undoubtedly ensure the long-term exploitation of diverse bioresources by relieving pressure on the critically endangered medicinal plant populations in the wild.

2.7 LIMITATIONS OF NEXT-GENERATION SEQUENCING TECHNOLOGIES

NGS innovations are a collection of techniques that include template preparation, sequencing, imaging, and sequence assembly. Longer read lengths, better throughput, and enhanced sequence accuracy are the primary objectives of NGS platforms. Next-generation sequencing systems have increased our understanding of crop genome sequencing, as well as mapping, detecting causative mutations, gene regulation, and other topics (Schmitt et al., 2012). NGS-based technology has made outstanding breakthroughs in genomic studies, agricultural genomics, and plant breeding studies, and in the future has promising applications in other research areas (Voelkerding et al., 2010).

High-throughput sequencing techniques have the capacity to change not only comparative biology but also genetics by providing a large number of genomes quickly. Although de novo genome assembly can be built in a matter of months, less consideration has been given to the information missed as a result of using short sequence reads. Although the prices of NGS data generation have decreased and the size of data creation has expanded considerably over the last decade, the following technological restrictions remain in NGS (Mardis, 2017).

Most plant genomes are known to be complex, with a substantially larger amount of repetitive sequences and transposons. As a result, short sequence reads (35–700 bp) created by second-generation sequencing systems such as SOLiD, Illumina, Roche (454), and others are not always efficient, especially when the genome size is large (Le Nguyen et al., 2019; Alkan et al., 2011). Third-generation sequencing techniques, such as Oxford Nanopore and PacBio, seem to be more feasible because they can generate long-read sequencing (several Kb). This allows for the detection and expression of epigenetic traces, including DNA methylation, in extremely variable genomic areas.

It is also useful in building high-quality reference genomes as well as speeding up gene research in plants (Yuan et al., 2017; Jo et al., 2019). NGS techniques have become very favorable in genome-wide SNP identification, allele mining, and generating molecular markers, as well as genotyping in other species or non-model animals, allowing for rapid use in research operations (Henry, 2012) However, identifying uncommon point mutations in plant genomes using NGS remains difficult. To increase the precision of the traditional NGS approach (Ståhlberg et al., 2016; Ståhlberg et al., 2017), simple, multiplexed, PCR-based bar-coding of DNA for selective mutation detection utilizing sequencing was created as an upgraded form of sequencing (SiMSen-Seq). This can help find variations at 0.1% or lower frequency with limited DNA input, and similarly, in order to overcome the limits of mutation detection in compressed representation sequencing (Monson-Miller et al., 2012). To use a comparison of restriction enzyme sequences (RESCAN), single-nucleotide polymorphism (SNP) was found in rice as well as *Arabidopsis*. Though genotyping/sequencing has advanced significantly in the previous decade, phenotyping has not kept up. Automated high-throughput phenotyping platforms for greenhouses and the field can help to speed up gene discovery even more (Jo et al., 2019).

From relatively small insert libraries, NGS technologies often produce shorter sequences with higher error rates (Alkan et al., 2011). For instance, Illumina's synthesis-based sequencing from libraries generally yields read lengths of 75–100 base pairs and insert sizes of 200–500 base pairs, making it one of the most widely utilized technologies. Consequently, it is projected that the gathering of larger repetitions as well as duplications will suffer due to the short read length.

DNA contamination from other organisms is the prime concern for any sequence alignment project, even those that employ Sanger sequencing. Before evaluating the genomes, a study compared the repeat-masked YH genome to the US National Center for Biotechnology Information (NCBI) nucleotide (nt) database to look for any contamination (Zhang et al., 2000). It found 1,033 contigs as well as 166 scaffolds that were highly similar to other species. De novo sequence configurations are complementary to clone-based methods and could serve as a significant source for the exploration of insertion polymorphisms, but because they tend to augment contamination artifacts, they need additional screening and verification. Distinguishing of patterns prior to sequence arrangement becomes particularly challenging when the underlying sequence read data is small.

A de novo sequence assembly approach based on whole genome sequencing will cause similar repeats to collapse, decreasing or eliminating genomic diversity. In one investigation, 420.2 missing frequent repeated sequences from the YH assembly were discovered to match 173.6 megabase pairs of mislaid LINE1 (L1) as well as 159.2 megabase pairs lacking Alu repeats. Although extremely similar sequences will be difficult, the researcher evaluated this impact by analyzing this reduction as a result of sequence variation. Repetitive sequence reduction was more frequent in L1 classes by means of reduced sequence variation (R2 = 0.86). It was discovered that depletion rose dramatically (>50%) for L1 repeat belonging to families with sequence identities more than 85%. There were fewer Alu subfamilies in general; however, Alu repeats that were more similar to conserved regions had significant depletion rates. This tendency was modest (R2 = 0.02), probably as a result of the Alu repeat class's smaller sequence length. The YH genome had a low representation of the most prevalent repetition classes (Green, 2002).

The third-generation sequencing platform has a low GC bias and very long read length but has low accuracy and modest throughput. De novo sequence assembly has constraints of short-read sequence data. The preparation of a sequence-ready library is considerably constrained by the lengthy and labor-intensive library construction process, which includes numerous phases, such as gel-based size selection. Enormous amounts of high-throughput sequential information generated by NGS technology have some difficulties in terms of storage, quality assurance, transmission of data, laboratory information management, and informatics processes to align or assemble read data.

The hybrid-sequencing approach was recently created to rectify faults in PacBio's long reads using the high throughput and precision short-read data produced by the SGS system. For understanding the whole expression of gene isoforms, hybrid sequencing techniques are used. The previously discussed upgrades in reading length and the capabilities of such systems to generate information through both edges of each library segment have improved the accuracy of reading location on intricate reference genomes and boosted the number and categories of versions that can be discovered after this step using the computational analysis methodology. Even with the now-relatively sophisticated state of data analysis following short-read alignment, the complexity of many genomes results in large false-positive rates when detecting specific forms of variation, such as structural modifications (Mardis, 2017).

2.8 FUTURE PROSPECTS

The future of NGS in breeding programs is the discovery of novel allelic variations in different crops' genomes. NGS technologies have enabled researchers to access the genome sequences of several important crops, opening the door for genome editing techniques. Such techniques enable site-directed mutagenesis to progress in a profitable way by means of modifying a specific locus. For the same purpose, sequence-specific nucleases such as zinc finger nucleases (ZFN), transcription activator-like effector nucleases (TALEN), and the CRISPR/Cas9 system can be used (Vlk et al., 2017). Approaches to create mutants and characterize them using refined NGS pipelines and various irradiations could appear more frequently in upcoming investigations. Combining irradiation facilities such as heavy ion beams, cosmic rays, and so on with DNA sequencing technology will increase mutagenesis efficiency and optimize the utilization of generated genetic resources for plant breeding and functional genomics research (Jo et al., 2019; Ståhlberg et al., 2016; Ståhlberg et al., 2017)

The world's population is growing at such a rapid pace that nature is unable to manage and support itself on its own. Feeding and maintaining a healthy lifestyle have become difficult, mainly in developing nations. Long-term usage of synthetic pharmaceuticals has resulted in major health issues in both underdeveloped and industrialized countries. The inclusion of plant-based goods is extremely important. Phytochemicals are promising materials for maintaining long-term and excellent health. Many plants flourish in the wild and are consequently utilized despite their inability to reproduce within a short amount of time. Due to this, they are becoming endangered, and their population is gradually declining, putting them in the category of threatened plants. Cultivation and certain other conventional breeding procedures are no longer beneficial in addressing the ignored state of these therapeutic plants. Thus, we can depend on biotechnology for great assistance, as well as NGS to recognize and categorize the wild population of therapeutic plants. Based upon surveys undertaken by many researchers, for a considerable proportion of the global population, herbal medicines continue to be a major source of healthcare. Pharmaceutical and biotechnology firms require these plants as key active components or as nutritional supplements. In general, two components are involved: (1) the biosynthetic pathway, as well as pathway control of important bioactive compounds, and (2) the mechanism of action of phytochemicals. These obstacles could be solved by means of devising computational metabolomics methodologies, recreating the de novo route, as well as studying possible variations.

Recent advancements have converged around a trend in the development of NGS-based technologies, such as creative modification of already-existing locus-specific techniques that now enable genome-wide reading. Growing interest in medicinal plants and technical approaches to better understand the biological mechanisms involved in the production of bioactive molecules have increased the supply of instruments. Genome elucidation provides knowledge on genes associated with secondary metabolite biosynthesis, metabolism, and transport. NGS data can be utilized to enhance our understanding of interactions among nuclear and organelle biosynthetic pathways. A full description of the plastid genome makes genetic modification of high-yielding designer therapeutic plants possible. NGS creates a snapshot of the genome that has the potential to find common or obscure gene groups that can be triggered or altered to make better phytopharmaceuticals. It will assist in the exploration of new bioactive chemicals that were not previously known.

NGS provides support units for the production of bioactive compounds in a more efficient and controlled manner. These genes can be modified to generate more or less of a certain bioactive compound, or they can be cloned for heterologous generation of phytochemicals in other species. Medicinal plants have come a long way due to NGS, although they still have a long way to go before they are faultless. Simply sequencing as well as assembly is insufficient to produce secure and efficient phytomedicines. To uncover the complete biosynthetic process, each finding must be linked to other cutting-edge phytochemical, biotechnological, metabolomic, pharmacological, and bioinformatics techniques (Sumner et al., 2015). To obtain additional information on biological processes, genome and transcriptome analyses should be combined. These integrated techniques have the potential to establish new phytochemical genetics foci (Zheng and Chan 2002). As a result, technologies based on NGS altered the field of study of medicinal plants as well as associated disciplines. More pathways of plant biosynthesis mechanisms will open up as discoveries and improvements continue, and we can expect more improved and functionally tailored phytomedicines. The future may offer novel and innovative technologies for the creation of sequencing data. This research and advancement area, in particular, encourages significant innovation as well as financial interest.

2.9 CONCLUSION

Therefore, the protection of endangered therapeutic plants is critical for the coming decade's viability of biodiversity. Bioinformatics and the use of computational and statistical techniques play a significant role in explaining different evolutionary processes of organisms at the molecular level in this era of genomics. The cost-effectiveness and viability of genomic techniques in conservation research and management must be considered. Traditional genetic markers, such as microsatellites, that have previously been created will often be the most cost-effective and efficient answer to a specific conservation genetics concern. This might be the case for conservation initiatives in more remote areas where the use of genetic technology is presently not viable. Extending the use of genomic research to a wider number and diversity of species is critical for biodiversity classification and conservation, as well as assessing genetic factors of population viability. The proper implementation of genomic technologies in the conservation context, on the other hand, should be driven by the required resolving power rather than the novelty of the approaches. Finally, these technologies can give critical indications for conservation strategy implementation. Despite the many technological hurdles that remain in the application of bioinformatics as a conservation technique, all stakeholders must actively participate in making this endeavor a reality.

REFERENCES

Abera B (2014) Medicinal plants used in traditional medicine by Oromo people, Ghimbi District, Southwest Ethiopia. *Journal of Ethnobiology and Ethnomedicine* 10: pp 1–15.

Ahmad J, Malik AA, Shakya L (2013) Urban development: a threat to wild species of medicinal and aromatic plants. *Middle-East Journal of Scientific Research* 13(7): pp 947–951.

Ahmadian A, Ehn M, Hober S (2006) Pyrosequencing: history, biochemistry and future. *Clinica Chimica Acta* 363(1–2): pp 83–94.

Alkan C, Sajjadian S, Eichler EE (2011) Limitations of next-generation genome sequence assembly. *Nature Methods* 8(1): pp 61–65.

Archer J, Baillie G, Watson SJ, Kellam P, Rambaut A, Robertson DL (2012) Analysis of high-depth sequence data for studying viral diversity: a comparison of next generation sequencing platforms using Segminator II. *BMC Bioinformatics* 13: pp 1–11.

Balaji R, Ravichandiran K, Tanuja, Parani M (2021) The complete chloroplast genome of *Ocimum gratissimum* from India–a medicinal plant in the Lamiaceae. *Mitochondrial DNA Part B* 6(3): pp 948–950.

Barba M, Czosnek H, Hadidi A (2014) Historical perspective, development and applications of next-generation sequencing in plant virology. *Viruses* 6(1): pp 106–136.

Behjati S, Tarpey PS (2013) What is next generation sequencing? *Archives of Disease in Childhood-Education and Practice* 98(6): pp 236–238.

Bodeker G, van 't Klooster C, Weisbord E (2014) *Prunus africana* (Hook. f.) Kalkman: the overexploitation of a medicinal plant species and its legal context. *The Journal of Alternative and Complementary Medicine* 20(11): pp 810–822.

Borah R, Bhattacharjee A, Rao SR, Kumar V, Sharma P, Upadhaya K, Choudhury H (2021) Genetic diversity and population structure assessment using molecular markers and SPAR approach in *Illicium griffithii*, a medicinally important endangered species of Northeast India. *Journal of Genetic Engineering and Biotechnology* 19(1): p 118.

ten Bosch JR, Grody WW (2008) Keeping up with the next generation: massively parallel sequencing in clinical diagnostics. *The Journal of Molecular Diagnostics* 10(6): pp 484–492.

Bragg LM, Stone G, Butler M.K, Hugenholtz P, Tyson GW (2013) Shining a light on dark sequencing: characterizing errors in Ion Torrent PGM data. *PLoS Computational Biology* 9(4): p. e1003031.

Carneiro MO, Russ C, Ross MG, Gabriel SB, Nusbaum C, DePristo MA (2012) Pacific biosciences sequencing technology for genotyping and variation discovery in human data. *BMC Genomics* 13(1): pp 1–7.

Caruccio N (2011) Preparation of next-generation sequencing libraries using Nextera™ technology: simultaneous DNA fragmentation and adaptor tagging by in vitro transposition. *High-Throughput Next Generation Sequencing: Methods and Applications* 733: pp 241–255.

Chaudhary S, Sharma PC (2016) Next generation sequencing-based exploration of genomes and transcriptomes of medicinal plants. *Indian Journal of Plant Physiology* 21: pp 489–503.

Chen J, Dong X, Li Q, Zhou X, Gao S, Chen R, Sun L, Zhang L, Chen W (2013) Biosynthesis of the active compounds of *Isatis indigotica* based on transcriptome sequencing and metabolites profiling. *BMC Genomics* 14(1): pp 1–13.

Chi KR (2008) The year of sequencing: in 2007, the next-generation sequencing technologies have come into their own with an impressive array of successful applications. *Nature Methods* 5(1): pp 11–15.

Desai AN, Jere A (2012) Next-generation sequencing: ready for the clinics? *Clinical Genetics* 81(6): pp 503–510.

Edwards D, Batley J (2010) Plant genome sequencing: applications for crop improvement. *Plant Biotechnology Journal* 8(1): pp 2–9.

Egan AN, Schlueter J, Spooner DM (2012) Applications of next-generation sequencing in plant biology. *American Journal of Botany* 99(2): pp 175–185.

Eisenstein M (2012) The battle for sequencing supremacy: which companies are leading the pack in terms of market share of the sequencing sector? Michael Eisenstein reports. *Nature Biotechnology* 30(11): pp 1023–1027.

Erlich H (2012) HLA DNA typing: past, present, and future. *Tissue Antigens* 80(1): pp 1–11.

Erwin KL (2009) Wetlands and global climate change: the role of wetland restoration in a changing world. *Wetlands Ecology and Management* 17(1): pp 71–84.

Fakruddin MD, Chowdhury A (2012) Pyrosequencing an alternative to traditional Sanger sequencing. *American Journal of Biochemistry and Biotechnology* 8(1): pp 14–20.

Gabriel C, Fürst D, Faé I, Wenda S, Zollikofer C, Mytilineos J, Fischer GF (2014) HLA typing by next-generation sequencing–getting closer to reality. *Tissue Antigens* 83(2): pp 65–75.

Gandhi SG, Mahajan V, Bedi YS (2015) Changing trends in biotechnology of secondary metabolism in medicinal and aromatic plants. *Planta* 241: pp 303–317.

Ganie AH, Tali BA, Khuroo AA, Reshi ZA, Nawchoo IA (2019) Impact assessment of anthropogenic threats to high-valued medicinal plants of Kashmir Himalaya, India. *Journal for Nature Conservation* 50: p 125715.

Ghazal H, Adam Y, Idrissi Azami A, Sehli S, Nyarko HN, Chaouni B, Olasehinde G, Isewon I, Adebiyi M, Ajani O, Matovu E (2021) Plant genomics in Africa: present and prospects. *The Plant Journal* 107(1): pp 21–36.

Grada A, Weinbrecht K (2013) Next-generation sequencing: methodology and application. *The Journal of Investigative Dermatology* 133(8): p e11.

Green P (2002) Whole-genome disassembly. *Proceedings of the National Academy of Sciences* 99(7): pp 4143–4144.

Gullapalli RR, Desai KV, Santana-Santos L, Kant JA, Becich MJ (2012) Next generation sequencing in clinical medicine: challenges and lessons for pathology and biomedical informatics. *Journal of Pathology Informatics* 3(1): p 40.

Hamilton JP, Robin Buell C (2012) Advances in plant genome sequencing. *The Plant Journal* 70(1): pp 177–190.

Hansen BK, Bekkevold D, Clausen LW, Nielsen EE (2018) The sceptical optimist: challenges and perspectives for the application of environmental DNA in marine fisheries. *Fish and Fisheries* 19(5): pp 751–768.

Henry RJ (2012) Next-generation sequencing for understanding and accelerating crop domestication. *Briefings in Functional Genomics* 11(1): pp 51–56.

Hirsch CN, Robin Buell C (2013) Tapping the promise of genomics in species with complex, nonmodel genomes. *Annual Review of Plant Biology* 64: pp 89–110.

Hu L, Hao C, Fan R, Wu B, Tan L, Wu H (2015) De novo assembly and characterization of fruit transcriptome in black pepper (Piper nigrum). *PLoS One* 10(6): p. e0129822.

Hua ZS, Han YJ, Chen LX, Liu J, Hu M, Li SJ, Kuang JL, Chain PS, Huang LN, Shu WS (2015) Ecological roles of dominant and rare prokaryotes in acid mine drainage revealed by metagenomics and metatranscriptomics. *The ISME Journal* 9(6): pp 1280–1294.

Huang Y, Zhang JL, Yu XL, Xu TS, Wang ZB, Cheng XC (2013) Molecular functions of small regulatory non coding RNA. *Biochemistry (Moscow)* 78(3): pp 221–230.

Jo YD, Kim JB (2019) Frequency and spectrum of radiation-induced mutations revealed by whole-genome sequencing analyses of plants. *Quantum Beam Science* 3(2): p 7.

Kalmankar NV, Venkatesan R, Balaram P, Sowdhamini R (2020) Transcriptomic profiling of the medicinal plant Clitoria ternatea: identification of potential genes in cyclotide biosynthesis. *Scientific Reports* 10(1): pp 1–20.

Kapoor B, Kumar A, Kumar P (2021) Transcriptome repository of North-Western Himalayan endangered medicinal herbs: a paramount approach illuminating molecular perspective of phytoactive molecules and secondary metabolism. *Molecular Genetics and Genomics* 296(6): pp 1–26.

Kircher M, Kelso J (2010) High-throughput DNA sequencing–concepts and limitations. *Bioessays* 32(6): pp 524–536.

Klumpp J, Fouts DE, Sozhamannan S (2012) Next generation sequencing technologies and the changing landscape of phage genomics. *Bacteriophage* 2(3): pp 190–199.

Kotwal S, Kaul S, Sharma P, Gupta M, Shankar R, Jain M, Dhar MK (2016) De novo transcriptome analysis of medicinally important Plantago ovata using RNA-Seq. *PloS One* 11(3): p e0150273.

Kulski JK (2016) Next-generation sequencing—an overview of the history, tools, and "Omic" applications. *Next Generation Sequencing-Advances, Applications and Challenges* 10: p 61964.

Kumar S, Yadav M, Yadav A, Yadav JP (2017) Impact of spatial and climatic conditions on phytochemical diversity and in vitro antioxidant activity of Indian *Aloe vera* (L.) Burm. f. *South African Journal of Botany* 111: pp 50–59.

Le Nguyen K, Grondin A, Courtois B, Gantet P (2019) Next-generation sequencing accelerates crop gene discovery. *Trends in Plant Science* 24(3): pp 263–274.

Li MR, Shi FX, Zhou YX, Li YL, Wang XF, Zhang C, Wang XT, Liu B, Xiao HX, Li LF (2015) Genetic and epigenetic diversities shed light on domestication of cultivated ginseng (Panax ginseng). *Molecular Plant* 8(11): pp 1612–1622.

Lin W, Sun F, Zhang Y, Xu X, Lu X, Li L, Xu R (2019) Comparative transcriptome and metabolite profiling of four tissues from *Alisma orientale* (Sam.) Juzep reveals its inflorescence developmental and medicinal characteristics. *Scientific Reports* 9(1): pp 1–12.

Liu Y, Wang Y, Guo F, Zhan L, Mohr T, Cheng P, Huo N, Gu R, Pei D, Sun J, Tang L (2017) Deep sequencing and transcriptome analyses to identify genes involved in secoiridoid biosynthesis in the Tibetan medicinal plant *Swertia mussotii. Scientific Reports* 7(1): pp 1–14.

Lodge DM, Turner CR, Jerde CL, Barnes MA, Chadderton L, Egan SP, Feder JL, Mahon AR, Pfrender ME (2012) Conservation in a cup of water: estimating biodiversity and population abundance from environmental DNA. *Molecular Ecology* 21(11): pp 2555–2558.

Lohse M, Drechsel O, Kahlau S, Bock R (2013) Organellar Genome DRAW—a suite of tools for generating physical maps of plastid and mitochondrial genomes and visualizing expression data sets. *Nucleic acids research* 41(W1): pp W575–W581.

Luo H, Li Q, Flower A, Lewith G, Liu J (2012) Comparison of effectiveness and safety between granules and decoction of Chinese herbal medicine: a systematic review of randomized clinical trials. *Journal of Ethnopharmacology* 140(3): pp 555–567.

Ma L, Dong C, Song C, Wang X, Zheng X, Niu Y, Chen S, Feng W (2021) De novo genome assembly of the potent medicinal plant *Rehmannia glutinosa* using nanopore technology. *Computational and Structural Biotechnology Journal* 19: pp 3954–3963.

Mak SST, Gopalakrishnan S, Carøe C, Geng C, Liu S, Sinding MHS, Kuderna LF, Zhang W, Fu S, Vieira FG, Germonpré M (2017) Comparative performance of the BGISEQ-500 vs Illumina HiSeq2500 sequencing platforms for palaeogenomic sequencing. *Gigascience* 6(8): p gix049.

Manish M (2013) Current status of endangered medicinal plant *Hedychium coronarium* and causes of population decline in the natural forests of Anuppur and Dindori districts of Madhya Pradesh, India. *International Research Journal of Biological Sciences* 2(3): pp 1–6.

Mardis ER (2017) DNA sequencing technologies: 2006–2016. *Nature Protocols* 12(2): pp 213–218.

Metzker ML (2010) Sequencing technologies—the next generation. *Nature Reviews Genetics* 11(1): pp 31–46.

Monson-Miller J, Sanchez-Mendez DC, Fass J, Henry IM, Tai TH, Comai L (2012) Reference genome-independent assessment of mutation density using restriction enzyme-phased sequencing. *BMC Genomics* 13(1): pp 1–15.

Muthu C, Ayyanar M, Raja N, Ignacimuthu S (2006) Medicinal plants used by traditional healers in Kancheepuram District of Tamil Nadu, India. *Journal of Ethnobiology and Ethnomedicine* 2(1): pp 1–10.

Niazian M (2019) Application of genetics and biotechnology for improving medicinal plants. *Planta* 249: pp 953–973.

Ninomiya M, Ueno Y, Shimosegawa T (2014) Application of deep sequence technology in hepatology. *Hepatology Research* 44(2): pp 141–148.

Nocker A, Burr M, Camper AK (2007) Genotypic microbial community profiling: a critical technical review. *Microbial Ecology* 54: pp 276–289.

Nogueira M, Enfissi EM, Almeida J, Fraser PD (2018) Creating plant molecular factories for industrial and nutritional isoprenoid production. *Current Opinion in Biotechnology* 49: pp 80–87.

Ong M, Carreira S, Goodall J, Mateo J, Figueiredo I, Rodrigues DN, Perkins G, Seed G, Yap TA, Attard G, De Bono JS (2014) Validation and utilisation of high-coverage next-generation sequencing to deliver the pharmacological audit trail. *British Journal of Cancer* 111(5): pp 828–836.

Petrovska BB (2012) Historical review of medicinal plants' usage. *Pharmacognosy Reviews* 6(11): p 1.

Quail MA, Smith M, Coupland P, Otto TD, Harris SR, Connor TR, Bertoni A, Swerdlow HP, Gu Y (2012) A tale of three next generation sequencing platforms: comparison of Ion Torrent, Pacific Biosciences and Illumina MiSeq sequencers. *BMC Genomics* 13(1): pp 1–13.

Rai A, Saito K, Yamazaki M (2017) Integrated omics analysis of specialized metabolism in medicinal plants. *The Plant Journal* 90(4): pp 764–787.

Roy NS, Choi IY, Um T, Jeon MJ, Kim BY, Kim YD, Yu JK, Kim S, Kim NS (2021) Gene expression and isoform identification of PacBio full-length cDNA sequences for berberine biosynthesis in *Berberis koreana*. *Plants* 10(7): p 1314.

Schadt EE, Turner S, Kasarskis A (2010) A window into third-generation sequencing. *Human Molecular Genetics* 19(R2): pp R227–R240.

Schmitt MW, Kennedy SR, Salk JJ, Fox EJ, Hiatt JB, Loeb LA (2012) Detection of ultra-rare mutations by next-generation sequencing. *Proceedings of the National Academy of Sciences* 109(36): pp 14508–14513.

Sharma S, Shrivastava N (2016) Renaissance in phytomedicines: promising implications of NGS technologies. *Planta* 244: pp 19–38.

Shendure JA, Porreca GJ, Church GM, Gardner AF, Hendrickson CL, Kieleczawa J, Slatko BE (2011) Overview of DNA sequencing strategies. *Current Protocols in Molecular Biology* 96(1): p 7-1.

Ståhlberg A, Krzyzanowski PM, Egyud M, Filges S, Stein L, Godfrey TE (2017) Simple multiplexed PCR-based barcoding of DNA for ultrasensitive mutation detection by next-generation sequencing. *Nature Protocols* 12(4): pp 664–682.

Ståhlberg A, Krzyzanowski PM, Jackson JB, Egyud M, Stein L, Godfrey TE (2016) Simple, multiplexed, PCR-based barcoding of DNA enables sensitive mutation detection in liquid biopsies using sequencing. *Nucleic Acids Research* 44(11): pp e105–e105.

Su Z, Ning B, Fang H, Hong H, Perkins R, Tong W, Shi L (2011) Next-generation sequencing and its applications in molecular diagnostics. *Expert Review of Molecular Diagnostics* 11(3): pp 333–343.

Sumner LW, Lei Z, Nikolau BJ, Saito K (2015) Modern plant metabolomics: advanced natural product gene discoveries, improved technologies, and future prospects. *Natural Product Reports* 32(2): pp 212–229.

Tripathi R, Sharma P, Chakraborty P, Varadwaj PK (2016) Next-generation sequencing revolution through big data analytics. *Frontiers in Life Science* 9(2): pp 119–149.

Van Dijk EL, Auger H, Jaszczyszyn Y, Thermes C (2014) Ten years of next-generation sequencing technology. *Trends in Genetics* 30(9): pp 418–426.

Vlk D, Řepková J (2017) Application of next-generation sequencing in plant breeding. *Czech Journal of Genetics and Plant Breeding* 53(3): pp 89–96.

Voelkerding KV, Dames S, Durtschi JD (2010) Next generation sequencing for clinical diagnostics-principles and application to targeted resequencing for hypertrophic cardiomyopathy: a paper from the 2009 William Beaumont Hospital Symposium on Molecular Pathology. *The Journal of Molecular Diagnostics* 12(5): pp 539–551.

Voelkerding KV, Dames SA, Durtschi JD (2009) Next-generation sequencing: from basic research to diagnostics. *Clinical Chemistry* 55(4): pp 641–658.

Wang M (2021) Next-generation sequencing (NGS). *Clinical Molecular Diagnostics* pp 305–327.

Wang S, Chen J, Ma J, Jin J, Chen L, Yao M. (2020) Novel insight into theacrine metabolism revealed by transcriptome analysis in bitter tea (Kucha, *Camellia sinensis*). *Scientific Reports* 10(1): p 6286.

Weisenfeld NI, Kumar V, Shah P, Church DM, Jaffe DB (2017) Direct determination of diploid genome sequences. *Genome Research* 27(5): pp 757–767.

Wommack KE, Bhavsar J, Ravel J (2008) Metagenomics: read length matters. *Applied and Environmental Microbiology* 74(5): pp 1453–1463.

Xiang LX, He D, Dong WR, Zhang YW, Shao JZ (2010) Deep sequencing-based transcriptome profiling analysis of bacteria-challenged *Lateolabrax japonicus* reveals insight into the immune-relevant genes in marine fish. *BMC Genomics* 11(1): p 472.

Yang L, Ding G, Lin H, Cheng H, Kong Y, Wei Y, Fang X, Liu R, Wang L, Chen X, Yang C (2013) Transcriptome analysis of medicinal plant *Salvia miltiorrhiza* and identification of genes related to tanshinone biosynthesis. *PLoS One* 8(11): p e80464.

Yuan Y, Bayer PE, Batley J, Edwards D (2017) Improvements in genomic technologies: application to crop genomics. *Trends in Biotechnology* 35(6): pp 547–558.

Yun YE, Yu JN, Nam GH, Ryu SA, Kim S, Oh K, Lim CE (2015) Next-generation sequencing identification and characterization of microsatellite markers in *Aconitum austrokoreense* Koidz., an endemic and endangered medicinal plant of Korea. *Genetics and Molecular Research* 14(2): pp 4812–4817.

Zaura E, Pappalardo VY, Buijs MJ, Volgenant CM, Brandt BW (2021) Optimizing the quality of clinical studies on oral microbiome: a practical guide for planning, performing, and reporting. *Periodontology* 200085(1): pp 210–236.

Zhang D, Li W, Chen ZJ, Wei FG, Liu YL, Gao LZ (2020) SMRT- and Illumina-based RNA-seq analyses unveil the ginsinoside biosynthesis and transcriptomic complexity in *Panax notoginseng*. *Scientific Reports* 10(1): p 15310.

Zhang Z, Schwartz S, Wagner L, Miller W (2000) A greedy algorithm for aligning DNA sequences. *Journal of Computational Biology* 7(1–2): pp 203–214.

Zheng XF, Chan TF (2002) Chemical genomics: a systematic approach in biological research and drug discovery. *Current Issues in Molecular Biology* 4(2): pp 33–43.

Zhong F, Huang L, Qi L, Ma Y, Yan Z (2020) Full-length transcriptome analysis of *Coptis deltoidea* and identification of putative genes involved in benzylisoquinoline alkaloids biosynthesis based on combined sequencing platforms. *Plant Molecular Biology* 102: pp 477–499.

Zhou X, Ren L, Li Y, Zhang M, Yu Y, Yu J (2010) The next-generation sequencing technology: a technology review and future perspective. *Science China Life Sciences* 53: pp 44–57.

Zhu Z, Jenkins G, Zhang W, Zhang M, Guan Z, Yang CJ (2012) Single-molecule emulsion PCR in microfluidic droplets. *Analytical and Bioanalytical Chemistry* 403: pp 2127–2143.

Zook JM, Catoe D, McDaniel J, Vang L, Spies N, Sidow A, Weng Z, Liu Y, Mason CE, Alexander N, Henaff E (2016) Extensive sequencing of seven human genomes to characterize benchmark reference materials. *Scientific Data* 3(1): pp 1–26.

3 Genomic Tools for Animal Identification and Characterization

Asha Poonia[1] *Radhika Sharma*[2], *Seema Rani*[3]*
[1]Faculty of Life Sciences, Chaudhary BansiLal University, Bhiwani, India
[2]Department of Biology and Environmental Sciences, College of Basic Sciences, CSK Himachal Pradesh Krishi Vishvidayala, Palampur, H.P.
[3]Department of Zoology, Faculty of Life Sciences, Chaudhary BansiLal University, Bhiwani, India
*Corresponding Author: Dr. Seema Rani, Department of Zoology, Faculty of Life Sciences, Chaudhary Bansi Lal University, Bhiwani, India,

3.1 INTRODUCTION TO GENOMIC TOOLS

Genomic tools encompass a range of cutting-edge techniques that leverage the information encoded within an organism's DNA. With the introduction of high-throughput sequencing technologies, researchers can now decipher the genetic blueprints of animals with unprecedented speed and accuracy. This ability has not only transformed our capacity to identify and distinguish between species but has also provided a comprehensive framework for characterizing genetic diversity, understanding evolutionary relationships, and unraveling the genetic basis of various traits.Genome analysis has made use of a wide range of technologies and instruments. High-throughput technologies, which range from conventional real-time polymerase chain reaction to more complex systems, such as microarray, Sanger sequencing, and next-generation sequencing, have recently undergone rapid revolutionary advancements. These techniques allow for precise and thorough functional analysis in multiple fields of study, including genomics, epigenomics, proteomics, and interactomics. This is necessary to close knowledge gaps regarding dynamic biological processes at the cellular and organism levels (Gasperskaja & Kucinskas, 2017). In the rapidly evolving landscape of modern biology, the integration of genomics has revolutionized our understanding of diverse life forms. This integration is particularly pronounced in the field of animal identification and characterization, where traditional methods have often been limited in their scope and precision. The emergence of genomic tools has ushered in a new era of research, enabling scientists to delve deeper into the intricacies of animal species, populations, and individuals (Figure 3.1).

3.2 SANGER SEQUENCING

Sanger sequencing, also known as chain termination sequencing, is a widely used method for determining the nucleotide sequence of DNA. Developed by Frederick Sanger in the late 1970s, this sequencing technique was a groundbreaking advancement that laid the foundation for modern genomics. Sanger sequencing played a crucial role in many significant scientific discoveries, including the sequencing of the human genome (Figure 3.2).

DOI: 10.1201/9781032697550-3

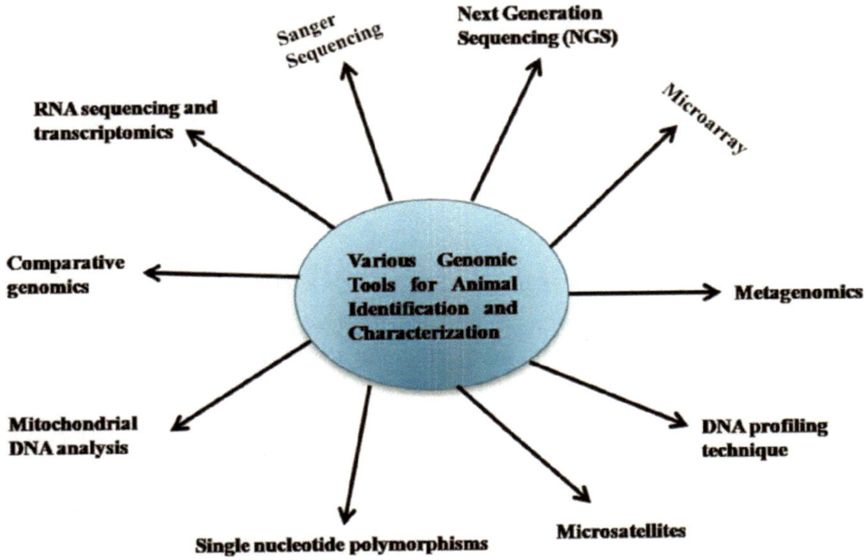

FIGURE 3.1 Various tools used for animal identification and characterization.

The advantages of Sanger sequencing include the following:

- Sanger sequencing is considered the gold standard for accuracy in DNA sequencing.
- It is well-suited for sequencing shorter DNA fragments or specific regions of interest.
- The fluorescent labeling of ddNTPs allows for automated and high-throughput sequencing.

3.2.1 Applications in Animal Identification and Characterization

DNA barcoding. Sanger sequencing is a fundamental tool for DNA barcoding, a method that uses short standardized gene sequences to identify species. The cytochrome c oxidase subunit I (COI) gene is a commonly used marker for DNA barcoding in animals. By sequencing the COI gene from an unknown sample and comparing it to a reference database, researchers can accurately identify the species.

Species differentiation. In situations where closely related species need to be distinguished, Sanger sequencing can be used to analyze specific genetic markers. For example, in cryptic species complexes (species that are morphologically similar but genetically distinct), Sanger sequencing of targeted genes can reveal subtle genetic differences that aid in accurate identification.

Population genetics. Sanger sequencing has been used to study genetic diversity within and among populations of animal species. By sequencing specific genomic regions, such as microsatellites or mitochondrial DNA segments, researchers can gain insights into the genetic structure of populations and identify unique genetic signatures associated with specific regions.

Conservation genetics. Sanger sequencing plays a role in conservation efforts by identifying and characterizing genetic diversity within endangered or threatened species. This information helps guide conservation strategies and management plans to ensure the survival of these species.

Forensic identification. Sanger sequencing can contribute to wildlife forensics, where it is used to identify animal species from trace samples. For example, Sanger sequencing of short DNA fragments recovered from animal products can provide evidence of species origin in cases of illegal wildlife trade.

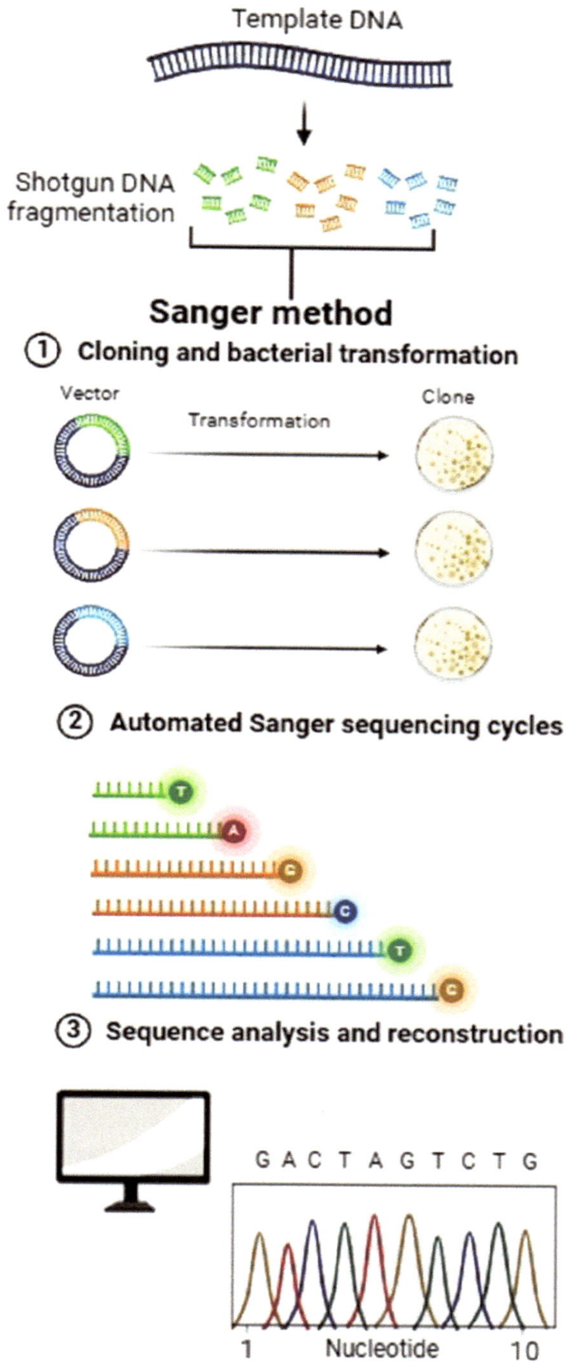

FIGURE 3.2 Process of Sanger sequencing.

Hybridization and introgression. Sanger sequencing can be used to determine if hybridization (crossbreeding between different species) or introgression (relocation of genes between species due to hybridization) has occurred. By analyzing specific genetic markers, researchers can identify signs of genetic mixing between species.

Molecular taxonomy. Sanger sequencing has been instrumental in revising and refining the classification of species based on genetic data. DNA sequences obtained through Sanger sequencing can help clarify evolutionary relationships and validate species assignments.

Zoo and aquarium management. Sanger sequencing can aid in confirming the identity of animals in captivity, ensuring that the correct species are being bred and conserved in zoo and aquarium settings.

3.3 NEXT-GENERATION SEQUENCING (NGS)

Advancement in DNA sequencing technology, known as Nnxt-generation sequencing (NGS), has hundreds of publications to credit its use and efficacy as a method. NGS has proven itself as a reliable method and a novel route for illness diagnosis and molecular characterization. It has unpredictably increased speed and reduced its per-sample cost. NGS can be used to generate DNA barcodes, which are short genetic sequences (typically a region of the mitochondrial genome) unique to each species. These barcodes can help identify and differentiate species quickly and accurately. NGS enables the analysis of entire communities of organisms present in a given sample, such as soil, water, or gut contents. This can aid in identifying animals indirectly through their DNA or RNA, even if they are not physically present in the sample. It is useful for studying ecosystems and tracking invasive species. NGS can provide an understanding of the genetic diversity and population structure of animal species. By sequencing multiple individuals within a population, researchers can assess genetic variation, gene flow, and population dynamics. This information can be valuable for conservation efforts and understanding species evolutionary history. NGS data can be used to reconstruct the evolutionary relationships among different animal species or populations. This can help clarify taxonomic uncertainties and provide a more accurate understanding of the evolutionary history of animals. NGS can generate vast amounts of genomic data that can be stored in databases and used as a resource for future studies on animal identification, evolution, and ecology.

For this procedure, linked adaptors are used with extracted DNA material. The created fragment library is clonally amplified, and the fragments are then sequenced in a massively parallel fashion (Figure 3.3).

3.3.1 Applications in Animal Identification and Characterization

NGS technologies are being used for whole genome sequencing (WGS), investigation of genome diversity including antimicrobial resistance studies and metagenomics, epigenetics (changes in DNA methylation, histone modifications, nucleosome positioning, and noncoding RNA), different protein interaction studies, and determination of regulating protein-binding sites (de Vries et al., 2021; Barzon et al., 2011). NGS has developed a powerful genetic tool for animal identification, revolutionizing the field of wildlife conservation and livestock management. NGS allows researchers and breeders to decipher an animal's entire genetic code quickly and cost-effectively, providing a wealth of information for identification purposes. By analyzing specific genetic markers or whole genomes, NGS can determine an animal's unique genetic fingerprint, facilitating accurate species identification, individual tracking, and pedigree verification. In wildlife conservation, NGS helps combat illegal wildlife trade by identifying the origin of confiscated animal products, aiding in the prosecution of poachers and traffickers. In agriculture, NGS assists in parentage verification, breeding program optimization, and disease resistance selection, ensuring the integrity and quality of livestock populations.

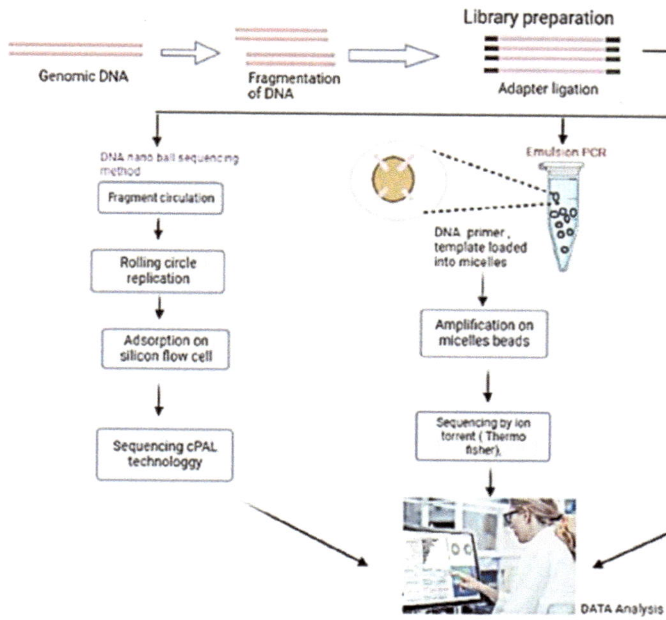

FIGURE 3.3 The stepwise process of next-generation sequencing.

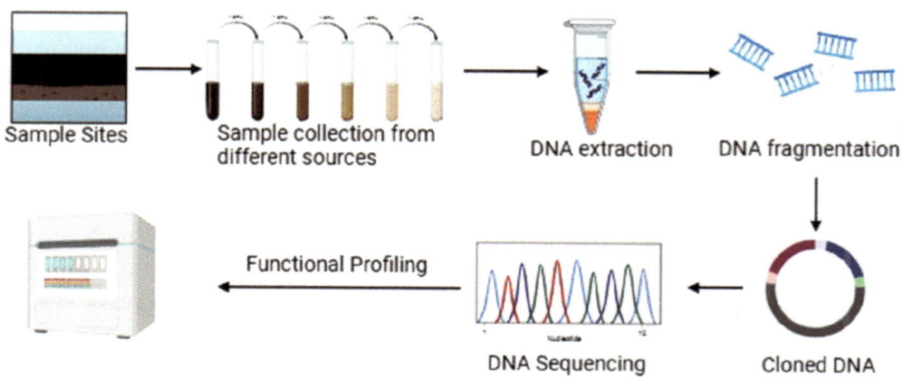

FIGURE 3.4 Diagram showing steps of metagenomics.

3.4 METAGENOMICS

The study of genetic material directly extracted from environmental or clinical samples using a technique called sequencing is known as metagenomics. The vast field may alternatively be known as microbiomics, community genomics, environmental genomics, or ecogenomics. Reconstructing the metabolism of the organisms that make up the community and predicting their functional roles in the environment are the primary goals of sequencing-based metagenomics (Figure 3.4). Using a variety of DNA sequences, co-expression networks can be created by combining large-scale genomic data analysis with gene expression analysis (transcriptome) in order to further uncover the genes linked to underlying connections between genes and phenotypes (Cavalieri & Grosu, 2004; Ferrara et al., 2008). It plays an important role in the study of population structure, identifying the diversity, and examining and separating our target genes from the members that haven't been cultured.

3.4.1 Applications in Animal Identification and Characterization

Genome recovery from metagenomes. As the number of sequences rises per sample read, it becomes possible to recover a genome from metagenomics in complex communities. For example, the study of archaea by metagenomics revealed that many eukaryotic characteristics are present in the cellular structure of archaea, which supports the development of eukaryotes from archaea (Spang et al., 2015; Neslihan et al., 2021). A key advantage of genome-resolved metagenomics, which will continue to open the door for new understandings of microbial ecology, is the explicit coupling of microbial activity to taxonomy.

Viral metagenomics. The development of meta-omic methods has made it possible to examine untamed viruses and has completely altered how we think about them as key players in many biogeochemical cycles on Earth (Neslihan et al., 2021). In addition to exposing viral expression levels and virus–host interactions, this enables the characterization of proviruses, viral episomal elements, and virions (Martinez-Hernandez et al., 2017).

3.5 MICROARRAYS

Microarrays are powerful tools used in genomics and molecular biology to simultaneously analyze the expression levels of thousands of genes or genetic variations in a single experiment. They enable researchers to study gene expression, DNA methylation, protein interactions, and more on a large scale. Microarrays have been widely used for various applications, including animal identification and characterization.

A microarray is a small, solid support (usually a glass slide or silicon chip) onto which numerous DNA or RNA probes are immobilized in a grid-like pattern. The biological material being analyzed, such as DNA, RNA, or proteins, is labeled with a fluorescent dye. When applied to the microarray, the labeled material binds to its complementary probe on the array, and the fluorescence intensity at each spot corresponds to the abundance of the target molecule (Figure 3.5).

3.5.1 Applications in Animal Identification and Characterization

Microarrays can be designed with probes specific to certain genetic markers or regions that are unique to different animal species. By hybridizing the DNA of an unknown sample with the microarray, researchers can identify the species based on the pattern of hybridization. This approach is particularly useful for identifying closely related species or for distinguishing between cryptic species. Microarrays are used to analyze gene expression patterns in different animal species or populations. By comparing the expression profiles of genes across samples, researchers can identify genes that are upregulated or downregulated in specific conditions, tissues, or developmental stages. This can help differentiate between species or reveal unique traits. Microarrays can help identify hybrid individuals resulting from interbreeding between different species. By analyzing the hybrid's genetic material against the microarray, researchers can identify the presence of genetic markers from both parent species, providing evidence of hybridization.

Also, microarrays designed with known single nucleotide polymorphisms (SNPs) can be used for genotyping animals. By hybridizing the animal's DNA with the microarray, researchers can detect specific SNPs and assess genetic variations among individuals or populations. This information is valuable for understanding genetic diversity and relationships. Microarrays enable the comparison of gene expression profiles between different animal species or populations. This can shed light on evolutionary relationships, adaptive traits, and conserved or divergent genetic pathways. By comparing gene expression patterns in animals with varying traits, researchers can pinpoint genes

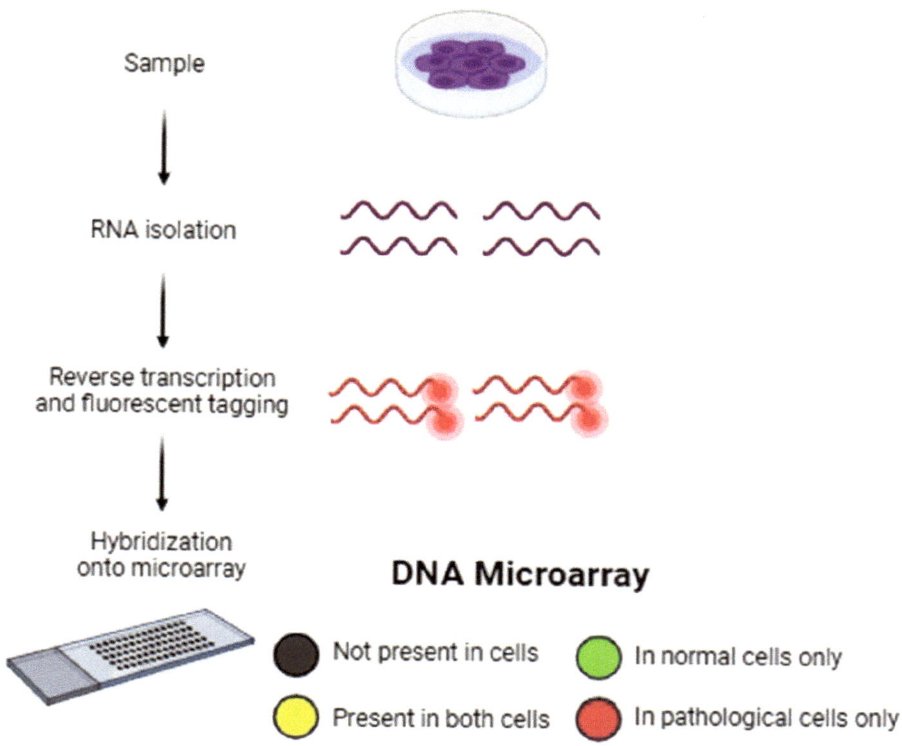

FIGURE 3.5 Diagram showing steps of microarray.

linked to particular characteristics. Microarrays can identify unique genomic signatures or genetic markers associated with certain animal traits, behaviors, or adaptations. These signatures can aid in the identification and classification of animals based on their genetic makeup.

Microarrays can also analyze DNA methylation patterns, which play a role in gene regulation. Differences in DNA methylation between species or populations can contribute to phenotypic variations. Microarrays still offer a cost-effective and targeted approach for studying specific genetic markers and gene expression patterns, making them relevant in various aspects of animal identification and research.

3.6 RNA SEQUENCING AND TRANSCRIPTOMICS

The technique known as RNA sequencing (RNA-Seq) and transcriptomics is currently widely used to examine gene expression and discover new RNA species. It makes use of NGS to determine the presence and amount of RNA in a biological sample, serving as a combined picture of the dynamic pool of RNAs present in cells, also known as the transcriptome (Chu & Corey 2012; Wang et al., 2009). Using specialized techniques for cDNA library production, aspects of RNA biosynthesis and metabolism can be investigated (Hrdlickova et al., 2017). RNA-Seq technology can be simply applied to species with high-quality reference genomes to gather abundant valuable information (Figure 3.6).

However, in many wildlife and aquaculture species, RNA-Seq analysis is difficult and requires more resources for the de novo association of the genes due to the lack of an annotated reference sequence and the restricted EST and mRNA characterization (Wickramasinghe et al., 2014). It also analyzes the biological purpose of a gene and uncovers the activity and mechanism of the molecular

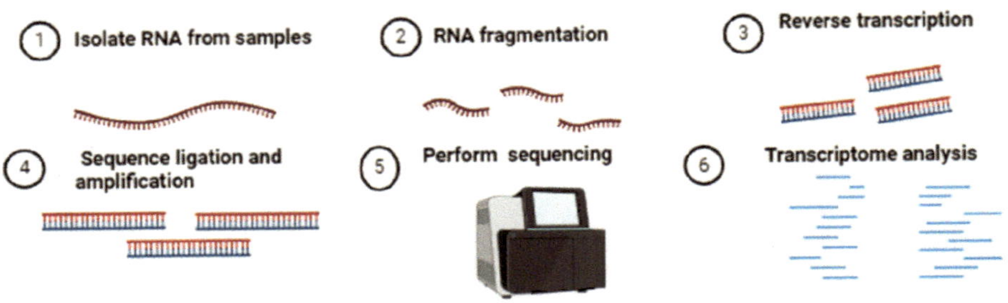

FIGURE 3.6 Mechanism of RNA sequencing.

structure (Zhao et al., 2017). The simple characterization of a specie's transcriptome is one of the most fundamental and often used applications.

3.7 DNA PROFILING

DNA fingerprinting, also known as forensic genetics or DNA profiling, is a method used to help identify people or items based on their unique DNA profiles (Figure 3.7). This technique is commonly used in paternity testing, immigration eligibility, and in genealogical and medical research (Petersen, 2012). The life sciences underwent a significant transformation as a result of the further development of this technology, particularly in the fields of ecology, evolutionary biology, and forensics (Chambers et al., 2014). In the disciplines of botany, zoology, and agriculture, DNA fingerprinting has also been applied to the study of plant and animal populations (Chambers et al., 2014). Multilocus profiles have also been used to find species hybrids, and DNA profiling has given zoologists a real-world tool to test the predictions of kin selection models (Hamilton, 1964; Ma & Lambert, 1997).

3.7.1 Applications in Animal Identification and Characterization

The investigation of bird mating systems was one of the first and most influential fields where minisatellite DNA fingerprinting technology was used (Burke & Bruford, 1987; Wetton et al., 1987). This

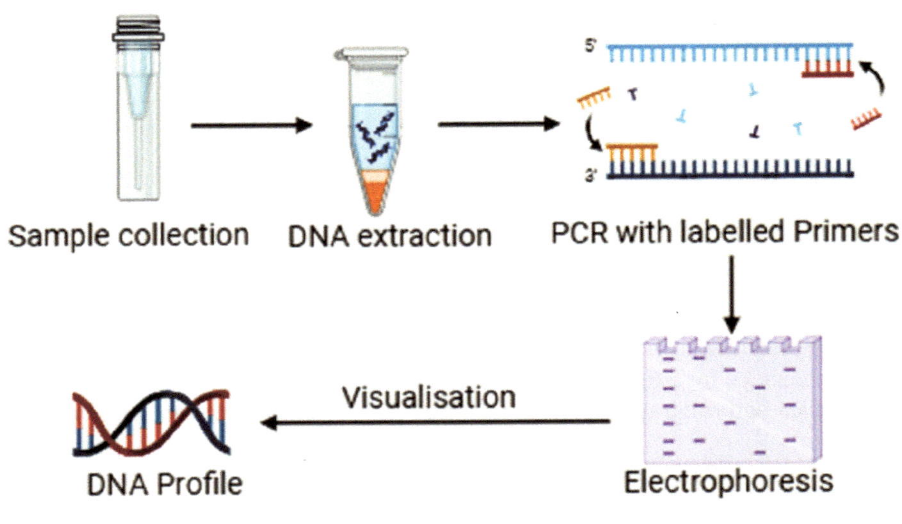

FIGURE 3.7 DNA profiling technique.

technology is also used to study population variation in other wild animals such as the California Channel Island fox and humpback whale (Baker et al., 1992; Baker et al., 1993). Additional minisatellite analysis uses were attempted, with varying degrees of success, such as stock assignment and trait mapping (Chambers et al., 2014). Several years of research on the mutational dynamics of minisatellites in primates and rodents revealed a variety of fascinating features (Gray & Jeffreys, 1991; Kelly et al., 1991). The Jeffreys's research team also demonstrated that captive breeding programs for uncommon and endangered species, such as the Waldrapp ibis, could benefit from DNA fingerprinting (Signer et al., 1994). Assays for sex testing, forensic wildlife services, and tests to prevent the illegal trade in endangered species and animal smuggling were all developed directly from DNA fingerprinting techniques (Ma & Lambert, 1997; Chambers et al., 2014).

3.8 MICROSATELLITES

Microsatellites are short DNA sequences consisting of tandemly repeating mono-, di-, tri-, and tetranucleotide units in the genomes of eukaryotes, prokaryotes, and viruses as they are pervasive and exceedingly abundant (Lei et al., 2021). Due to their functions in the organization of gene expression, chromosome structure, cell cycle dynamics, DNA recombination, and replication, microsatellites have undoubtedly garnered a lot of attention (Chistiakov et al., 2006). Microsatellites are more common in some parts of the genome than others, and their distribution in non-coding and coding regions varies greatly. Microsatellite repeats are excessive in non-coding areas of eukaryotic organisms but relatively infrequent in coding regions, with higher plants having between 7% and 10% of them and vertebrates having between 9% and 15% (Metzgar et al., 2000; Varshney et al., 2002) (Figure 3.8).

3.8.1 Applications in Animal Identification and Characterization

Microsatellites are the most widely used and flexible genetic markers with varied applications in population genetics, evolutionary biology, and conservation biology. These markers are codominant

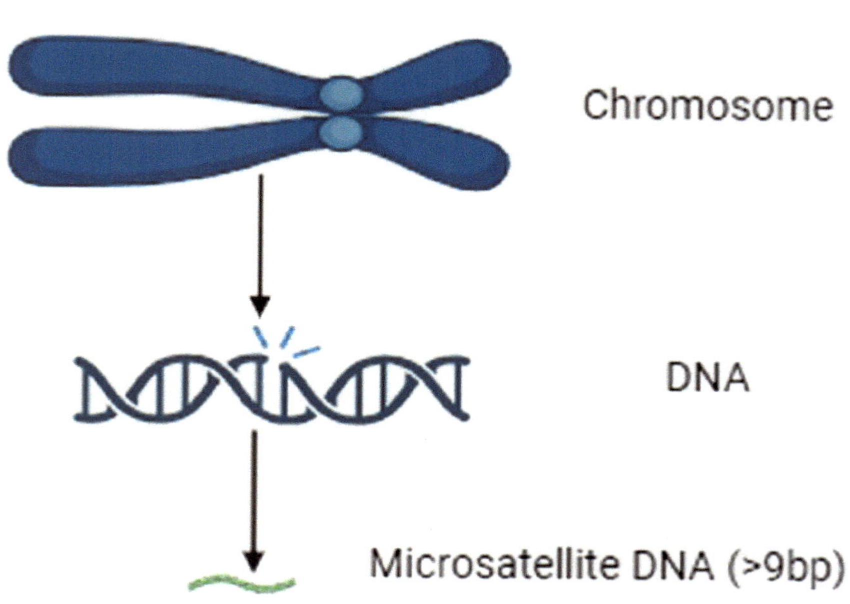

FIGURE 3.8 Microsatellite DNA.

in nature, stable, highly polymorphic, easily typed, and very simple to analyze, making them excellent candidates for studying population structure, performing pedigree analyses, and identifying variations between closely related species (Mahmoud et al., 2020; Abdul-Muneer, 2014). Microsatellite markers are useful genome structures, which are valuable in strain typing and identification of genomes in pathogen. Numerous studies have shown that the hotspots of phenotypic features may be related to the distribution of microsatellites (Gylfe et al., 2014). The distribution of repeat types (mono- to hexanucleotide) differs among various species, and it has been hypothesized that this variance is related to the interaction of mutation and varied selection pressure (Li et al., 2002). It has also been studied that microsatellite markers can be used to analyze Algerian cattle breeds' genetic diversity and population dynamics (Rahal et al., 2021).

3.9 SINGLE NUCLEOTIDE POLYMORPHISMS (SNPS)

A possible method for establishing an animal's genetic ancestry and identification is the use of DNA markers. A single base change in a DNA sequence constitutes an SNP marker (Beuzen et al., 2000). SNPs are the best genetic variation resource for population studies and genome mapping since they account for more than 90% of all individual differences (Frohlich et al., 2004).

3.9.1 Applications in Animal Identification and Characterization

SNPs have been extensively utilized to select livestock animals for desired qualities (Matukumalli et al., 2009; Surya et al., 2018). Structural nucleotide variant arrays can be used to characterize the genetic variants driving complex phenomic phenomena, including animal reproduction, health, and productivity. These traits are the result of the interaction polygenic and genomic additive effects. The success of any genetic improvement program for livestock species depends on the cumulative metric based on each individual's genomic information (Barker, 1994; Notter, 1999). Studies of the genetic makeup of endemic species with local distribution can potentially benefit from SNPs (Wenne, 2023). These have also been used to identify species and their hybrids in their natural habitats, to explore the genetic effects of restocking for conservation, and to examine the detrimental effects of fish accidentally escaping from aquaculture on natural populations (Wenne, 2023). These have proved helpful in assessments of the genetic makeup of marine species with great dispersion potential, sedentary freshwater populations, and benthic species, as well as natural and wild populations (Wenne, 2023).

3.10 MITOCHONDRIAL DNA ANALYSIS

Mitochondrial DNA (mtDNA or mDNA) is the DNA located in mitochondria, cell organelles present in eukaryotic cells that convert food energy into chemical energy, for example, adenosine triphosphate (ATP). Animals have strong conservation of the mitochondrial gene content, which has no introns, very few duplications, and very short intergenic regions (Gissi et al., 2008). mtDNA is present in the cell in several copies and is amplified very easily in experiments. These features make it a perfect genetic marker, according to early investigations (Avise, 2004). Maternal inheritance and the absence of recombination are very desirable characteristics for a tool in the reconstruction of phylogenetic histories. Additionally, the presence of both conserved and less conserved sections inside a single molecule, as well as the higher mutation rate compared to nuclear DNA, make mtDNA suited for comparisons between members of the same population and species that are distantly related (Ladoukakis & Zouros, 2017).

3.10.1 Applications in Animal Identification and Characterization

An examination of the mitochondrial genomic sequences of 11 animal species using bioinformatics revealed inter- and intraspecific differences in mtDNA. The mitochondrial 12S and 16S ribosomal

RNA (rRNA) genes of these species have a few highly conserved regions. Numerous studies have examined the evolutionary links both within and across species using analysis of mtDNA sequencing data. It provides an especially abundant supply of markers for the investigation of closely related species (Daniel & Brenden, 2005). mtDNA surveys with appropriately chosen samples can be an effective way to monitor gene flow, identify hybrid zones, gauge the degree of reproductive isolation, determine species boundaries, and reveal underlying historical patterns of population structure (Daniel & Brenden, 2005). In phylogenetic investigations to ascertain ciliate relationships and species delimitation, mtDNA has shown to be incredibly helpful (Zhang et al., 2019).

In order to correlate the evolutionary links of genetic lines with their geographic distribution and better understand the processes that shaped the current distribution of genes, populations, and species, mtDNA sequences can also be employed in phylogeographic investigations (Kowalczyk et al., 2021). The manner in which the mitochondrial genome is organized makes it is easy to create universal primers for a wide variety of species in conserved regions, and the variation in the fragments flanking them can be used to identify species in biological samples from unknown provenance (Kowalczyk et al., 2021).

The existence of species-specific sequences in mtDNA has made species identification based on polymorphism in mtDNA a prominent trend in wildlife forensic genetics. The cytochrome oxidase I and cytochrome b gene sequences are most frequently utilized in forensics to identify species. mtDNA analyses were quickly embraced for the purposes of forensic genetics dealing with crimes against protected species (wildlife forensics), as well as in breeding and companion animals. The illegal killing of wild animals, poaching, illegal farm animal slaughter, hunting, and illegal trade of protected species are just a few examples of crimes against animals where mtDNA analysis can be used to identify the species and trace the geographic origin of a sample (Kowalczyk et al., 2021; Lorenzini & Garofalo 2021).

3.11 COMPARATIVE GENOMICS FOR SPECIES IDENTIFICATION AND DIFFERENTIATION

Comparative genomics is the practice of directly comparing two organisms' complete genetic compositions to better understand how species have evolved and to determine the functions of genes and noncoding regions in genomes. In addition to showing how different creatures have evolved, comparative genomics may also show the differences and similarity in species and also disclose the differences between other organisms and humans (Sivashankari & Shanmughavel, 2007). Genome correspondence, the first stage in comparative genomics, is the process of determining the appropriate chromosomal segment and functional component for each species under study.

3.11.1 Applications in Animal Identification and Characterization

Comparative genomics can help with gene identification once genome correspondence has been established. Based on the nucleotide conservation patterns of genuine genes over evolutionary time, comparative genomics can identify them.

Comparative genomics is a comprehensive, large-scale technique that examines the biology of each individual genome while identifying similarities and differences between two or more genomes. Whole genomes or syntenic sections of various species, subspecies, or strains of the same species can be studied using comparative genomics (Liping et al., 2002).

The development of full genome sequencing technology has made it possible to examine vibrios genetic traits at the genomic level (Lin et al., 2018).

3.12 ETHICAL AND SOCIETAL CONSIDERATIONS

The use of genomic tools greatly helps scientists to accelerate conservation efforts by accurate species identification and to study genetic diversity among and between subspecies and higher taxonomic categories (Lin et al., 2018). However, using genomic tools in animal identification and characterization raises several ethical considerations (National Academies of Sciences Engineering and Medicine (U.S.), 2017) such as the number of samples collected, especially in invasive sampling. Another concern is higher human interference, especially if the species has a very small population or is endemic.

Keeping this in view, a few precautions need to be taken using genomic tools, such as using non-invasive methods whenever possible to minimize stress and harm to the animals, anonymizing data to protect the privacy of animal groups, respecting indigenous perspectives, and engaging in ethical consultations when conducting research on animals that are culturally significant to indigenous communities. Another important factor includes ensuring that research is not used to exploit animals or enable activities such as illegal wildlife trafficking (Baruah et al., 2013). Equitable sharing of benefits with the communities or countries from which the genetic samples were collected, using genomic research as a tool for education, and raising awareness about the importance of biodiversity and conservation should also be adopted. It becomes essential that researchers adhere to ethical guidelines and regulatory frameworks, monitor the environmental impact of genetic research over time, and adapt strategies accordingly. The obtained genomic data should be shared transparently and ethically with policymakers, conservationists, and other stakeholders in an accurate and unbiased manner, as it can be misused for purposes that harm animals, such as genetic modification or exploitation.

3.13 FUTURE DIRECTIONS AND TECHNOLOGICAL ADVANCES

Recent advances in genome editing techniques, such as CRISPR/Cas9, have opened up exciting possibilities for genetic engineering. These methods enable us to delete, add, or modify alleles at specific locations in the genome of cells, and when applied to germline cells, these changes become permanent and heritable. The potential of these advances has garnered significant attention in their application to accelerate genetic improvements in breeding programs (Jenko et al., 2015) including "double muscling" in cows, sheep, and pigs (He et al., 2015), hornless cattle (Tan et al., 2013), and resistance to viral diseases like porcine reproductive respiratory syndrome and African swine fever (Lillico et al., 2013; Whitworth et al., 2017). Beyond selection, genome editing can be used to reintroduce genetic diversity that has disappeared from a population or even to artificially introduce desirable variations as used by Shen et al. (2017) in crop breeding, revolutionizing the agricultural sector.

In addition to genome editing, other technological advancements are transforming the field of genetic engineering. Nanopore sequencing, for instance, is enabling real-time sequencing of DNA and RNA, which is invaluable for fieldwork and rapid species identification. Single-cell genomics and environmental DNA (eDNA) analysis hold tremendous potential for non-invasive population monitoring. Machine learning and bioinformatics are improving the speed and accuracy of data analysis, facilitating the creation of reference databases crucial for precise species identification. Long-read sequencing is enhancing our ability to characterize complex animal genomes, particularly those with repetitive elements. These technologies are expected to become even more accessible and user-friendly in the future.

Genome editing has the potential to revolutionize how we approach genetic diversity improvement in species. However, it's important to note that this technology still presents challenges, and there is uncertainty about its future acceptance in animal breeding. A crucial aspect to address is the ethical debate on when and why we should apply genome editing in livestock breeding. If accepted, we must also carefully consider the potential costs and benefits (Bastiaansen et al., 2018). This reflects the need for a comprehensive discussion on the future of genetic engineering and its implications for agriculture and beyond.

REFERENCES

Abdul-Muneer, P. M. 2014. Application of microsatellite markers in conservation genetics and fisheries management: recent advances in population structure analysis and conservation strategies. *Genet Res Int* 2014:691759.

Avise, J. C. 2004. *Molecular Markers, Natural History, and Evolution*. 2nd Edition. Sunderland: Sinauer AssociatesInc. Publishers.

Baker, C. S., Gilbert, D. A., Weinrich, M. T., et al. 1993. Population characteristics of DNA fingerprints in humpback whales (*Megaptera novaeangliae*). *J Hered* 84:281–290.

Baker, C. S., MacCarthy, M., Smith, P. J., et al. 1992. DNA fingerprints of orange roughy, *Hoplostethus atlanticus*: a population comparison. *Mar Biol* 113:561–567.

Barker, J. 1994. A global protocol for determining genetic distances among domestic livestock breeds. *In*: Proceedings *of the 5th World Congress on Genetics Applied to Livestock Production*, Guelph, Ontario, Canada, 501–508.

Baruah, C., Bhuyan, S. K., Devi, P., et al. 2013. Emerging trends in wildlife trade in Northeast India: a serious concern. *Asian J Conserv Biol* 2(2):86–87.

Barzon, L., Lavezzo, E., Militello, V., et al. 2011. Applications of next-generation sequencing technologies to diagnostic virology. *Int J Mol Sci* 12(11):7861–7884.

Bastiaansen, J. W., Bovenhuis, H., Groenen, M. A., et al. 2018. The impact of genome editing on the introduction of monogenic traits in livestock. *Genet Sel Evol* 50:18.

Beuzen, N. D., Stear, M. J., Chang, K. C. 2000. Molecular markers and their use in animal breeding. *Vet J* 160:42–52.

Burke, T., Bruford, M. 1987. DNA fingerprinting in birds. *Nat* 327:149–152.

Cavalieri, D., Grosu, P. 2004. Integrating whole-genome expression results into metabolic networks with Pathway Processor. *Curr Protoc Bioinformat* 7:7.

Chambers, G. K., Curtis, C., Millar, C. D., et al. 2014. DNA fingerprinting in zoology: past, present, future. *Investig Genet* 5(1):3.

Chistiakov, D. A., Hellemans, B., Volckaert, F. A. M. 2006. Microsatellites and their genomic distribution, evolution, function and applications: a review with special reference to fish genetics. *Aquac* 255(1):1–29.

Chu, Y., Corey, D. R. 2012. RNA sequencing: platform selection, experimental design, and data interpretation. *Nucleic Acid Ther* 22(4):271–274.

Daniel, R., Brenden, S. H. 2005. Between two extremes: mitochondrial DNA is neither the panacea nor the nemesis of phylogenetic and taxonomic inference. *Syst Biol* 54 (6):952–961.

De Vries, J. J. C., Brown, J. R., Couto, N., et al. 2021. Recommendations for the introduction of metagenomic next-generation sequencing in clinical virology, part II: bioinformatic analysis and reporting. *J Clin Virol* 138:104812.

Ferrara, C. T., Wang, P., Neto, E. C., et al. 2008. Genetic networks of liver metabolism revealed by integration of metabolic and transcriptional profiling. *PLoS Genet* 3:e1000034.

Frohlich, T., Kirschbaum, T., Thoenes, U., et al. 2004. The light typer instrument: high throughput genotyping of single nucleotide polymorphism. *Biochemica* 2:9–11.

Gasperskaja, E., Kucinskas, V. 2017. The most common technologies and tools for functional genome analysis. *Acta Med Litu* 24(1):1–11.

Gissi, C., Iannelli, F., Pesole, G. 2008. Evolution of the mitochondrial genome of Metazoa as exemplified by comparison of congeneric species. *Heredity* 101:301–320.

Gray, I. C., Jeffreys, A. J. 1991. Evolutionary transience of hypervariable minisatellites in man and the primates. *Proc Biol Sci* 243:241–253.

Gylfe, A. E., Tuupanen, S., Hanninen, U., et al. 2014. Novel candidate oncogenes with mutation hot spots in microsatellite unstable colorectal cancer. *Cancer Res* 74(19):5193.

Hamilton, W. D. 1964. The genetical evolution of social behaviour, I and II. *J Theor Biol* 7:1–52.

Hao, Y., Cui, Y., Gu, X. 2016. Genome-wide DNA methylation profiles changes associated with constant heat stress in pigs as measured by bisulfite sequencing. *Sci Rep* 6:27507.

He, Z., Proudfoot, C., Mileham, A. J., et al. 2015. Highly efficient targeted chromosome deletions using CRISPR/Cas9. *Biotechnol Bioeng* 112:1060–1064.

Hrdlickova, R., Toloue, M., Tian, B. 2017. RNA-Seq methods for transcriptome analysis. *Wiley Interdiscip Rev RNA* 8(1):10.

Jenko, J., Gorjanc, G., Cleveland, M. A., et al. 2015. Potential of promotion of alleles by genome editing to improve quantitative traits in livestock breeding programs. *Genet Sel Evol* 47:55.

Kelly, R., Gibbs, M. A., Collick, A., et al. 1991. Spontaneous mutation at the hypervariable mouse minisatellite locus Ms6-hm: flanking DNA sequence and analysis of germline and early somatic mutation events. *Proc Biol Sci* 245:235–245.

Kowalczyk, M., Staniszewski, A., Kaminska, K., et al. 2021.Advantages, possibilities, and limitations of mitochondrial DNA analysis in molecular identification. *Folia Biologica (Krakow)* 69:101–111.

Ladoukakis, E. D., Zouros, E. 2017. Evolution and inheritance of animal mitochondrial DNA: rules and exceptions. *J Biol Res (Thessalon)* 24:2.

Lei, Y., Zhou, Y., Price, M., et al. 2021. Genome-wide characterization of microsatellite DNA in fishes: survey and analysis of their abundance and frequency in genome-specific regions. *BMC Genomics* 22:421.

Li, Y. C., Korol, A. B., Fahima, T., et al. 2002. Microsatellites: genomic distribution, putative functions and mutational mechanisms: a review. *Mol Ecol* 11:2453–2465.

Lillico, S. G., Proudfoot, C., Carlson, D. F., et al. 2013. Live pigs produced from genome edited zygotes. *Sci Rep* 3:2847.

Lin, H., Yu, M., Wang, X., et al. 2018. Comparative genomic analysis reveals the evolution and environmental adaptation strategies of vibrios. *BMC Genomics* 19:135.

Liping, W., Yueyi, L., Inna, D., et al. 2002. Comparative genomics approaches to study organism similarities and differences. *J Biomed Inform* 35(2):142–150.

Lorenzini, R., Garofalo l. 2021. Wildlife Forensics: DNA analysis in wildlife forensic investigations. In *Forensic DNA Analysis: Technological Development and Innovative Applications*, Pilli, E., Berti, A. (eds), New Jersey, USA: Apple Academic Press, 357–384.

Ma, W., Lambert, D. M. 1997. Minisatellite DNA markers reveal hybridisation between the endangered black robin and tomtits. *Electrophoresis* 18:1682–1687.

Mahmoud A. H., Farah M. A., Rady, A., et al. 2020. Molecular characterization of goats from Saudi Arabia using microsatellite markers. *J King Saud University Sci* 32(2):1681–1686.

Martinez-Hernandez ,F., Fornas, O., Gomez M. L., et al. 2017. Single-virus genomics reveals hidden cosmopolitan and abundant viruses. *Nat Commun*, 8:1–13.

Matukumalli, L. K., Lawley, C. T., Schnabel, R. D., et al. 2009. Development and characterization of a high density SNP genotyping assay for cattle. *PLoS One* 4:e5350.

Metzgar, D., Bytof, J., Wills, C. 2000. Selection against frameshift mutations limits microsatellite expansion in coding DNA. *Genome Res* 10(1):72–80.

Neslihan, T., Anniek, E. E. d. J., Yaoming, L., Gareth, T., et al. 2021. Metagenomic tools in microbial ecology research. *Curr Opin Biotechnol* 67:184–191.

Notter, D. R. 1999. The importance of genetic diversity in livestock populations of the future. *J Anim Sci* 77:61–69.

Petersen, K. J.2012. *Handbook of Surveillance Technologies*. 3rd ed. Boca Raton, FL: CRC Press: 815.

Rahal, O., Aissaoui, C., Ata, N., et al. 2021. Genetic characterization of four Algerian cattle breeds using microsatellite markers. *Anim Biotechnol* 32(6):699–707.

Shen, L., Hua, Y., Fu, Y., et al. 2017. Rapid generation of genetic diversity by multiplex CRISPR/Cas9 genome editing in rice. *Sci. China Life Sci* 60:506–515.

Signer, E. N., Schmidt, C. R., Jeffreys, A. J. 1994. DNA variability and parentage testing in captive Waldrapp ibises. *Mol Ecol* 3:291–300.

Sivashankari, S., Shanmughavel, P. 2007. Comparative genomics: a perspective. *Bioinform* 1(9):376–378.

Spang A., Saw J. H., Jorgensen S. L., et al. 2015. Complex archaea that bridge the gap between prokaryotes and eukaryotes. *Nat* 521:173–179.

Surya, T., Vineeth, M., Sivalingam, J., et al. 2018. Genomewide identifcation and annotation of SNPs in Bubalus bubalis. *Genomics* 111(6):1695–1698.

Tan, W., Carlson, D. F., Lancto, C. A., et al. 2013. Efficient nonmeiotic allele introgression in livestock using custom endonucleases. *Proc Natl Acad Sci USA* 110:16526–16531.

Varshney, R. K., Thiel, T., Stein, N., et al. 2002. In silico analysis on frequency and distribution of microsatellites in ESTs of some cereal species. *Cell Mol Biol Lett* 7(2A):537–546.

Wang, Z., Gerstein, M., Snyder, M. 2009. RNA-Seq: a revolutionary tool for transcriptomics. *Nat Rev Genet* 10(1):57–63.

Wenne, R. 2023. Single nucleotide polymorphism markers with applications in conservation and exploitation of aquatic natural populations. *Animals (Basel)* 13(6):1089.

Wetton, J. H., Carter, R. E., Parkin, D. T., et al. 1987. Demographic study of a wild house sparrow population by DNA fingerprinting. *Nat* 327:147–149.

Whitworth, K. M., Benne, J. A., Spate, L. D., et al. 2017. Zygote injection of CRISPR/Cas9 RNA successfully modifies the target gene without delaying blastocyst development or altering the sex ratio in pigs. *Transgenic Res* 26:97–107.

Wickramasinghe, S., Canovas, A., Rincon, G., et al. 2014. RNA-Sequencing: a tool to explore new frontiers in animal genetics. *Livest Sci* 166:206–216.

Zhang, T., Fan, X., Gao, F., et al. 2019. Further analyses on the phylogeny of the subclass Scuticociliatia (Protozoa, Ciliophora) based on both nuclear and mitochondrial data. *Mol Phylogenet Evol* 139:106565.

Zhao, F. R., Xie, Y. L., Liu, Z. Z., et al. 2017. Transcriptomic analysis of porcine PBMCs in response to FMDV infection. *Acta Trop* 173:69–75.

4 DNA Barcoding for Plant Identification and Characterization

Neetika Kimta[1], *Nidhi Bhardwaj*[1], *Sunil Puri*[1]

[1]School of Biological and Environmental Sciences, Shoolini University of Biotechnology and Management Sciences, Solan, India

4.1 INTRODUCTION

In today's global consumer market, the usage and sale of medicinal plants and herbal supplements have come to depend more and more on the authentication and verification of all products. Over the last ten years, single-gene, combined-gene, genome, and, most recently, metabarcoding DNA barcode technologies have all been developed and are now globally used to track the usage, commercialization, and authenticity of medicinal plants worldwide. Furthermore, in their usage and commercialization process, the vitality of human health is the primary aspect that has majorly benefited from the applications enclosed by DNA barcode technology (Yu et al., 2021). In general, the DNA barcoding approach was first suggested as a method in 2003 by Hebert et al. to enable quick, reliable, and automatic identification of species using a standard, small DNA sequence or DNA barcode (Hebert et al., 2003; Hebert and Gregory, 2005). In actuality, a very small sample of tissue from an unknown organism can be used to create a sequence from such a small standardized region of the DNA, which can then be compared to a reference sequence library of known species (Hebert et al., 2003; Casiraghi et al., 2010). Traditionally, identifying medicinal plants includes chemical profiling (TLC, HPLC-MS, and HPLC-UV); microscopic and macroscopic techniques that use shape, color, and texture; and organoleptic techniques that use senses, taste, sight, smell, and touch, which necessitate the employment of experienced personnel for both macroscopic and microscopic investigations. Even though storage and physiological circumstances can have an impact on the chemical profiles or markers in the latter approach, DNA as a macromolecule is unaffected by environmental influences, as opposed to RNA, and confirmation at the DNA level offers more consistency. Consequently, the creation of markers (DNA-based) is crucial for the verification of medicinal plant authenticity, according to Techen et al. (2014). They also underlined that species borders, species delimitation, and the documentation of novel species, can all be achieved by DNA barcoding (Hebert et al., 2003; Hebert et al., 2004; Techen et al., 2014). In fact, it is known that medicinal plants belong to 219 distinct families, out of which DNA barcodes have been used on only 142 families (64.8%), including 832 genera, but there are no published records of sequence data of DNA barcodes for the leftover 77 families of medicinal plants. However, for angiosperm evolutionary lineages, the percentage of families having records of medicinal plants ranges from 43.4% to 54.6% in monocots to basal angiosperms, which is significant in DNA barcoding, and from 33.3% to 78.1% for monocots to basal eudicots (Yu et al., 2021). In terms of their other scientific importance, barcoding is essentially a taxonomic technique that identifies a brief genetic signature from a specific organism and connects it to a specific species (Figure 4.1). The technique's main objective is to identify samples of species (unknown) and link them to established classification schemes (DeSalle, 2006; Sarwat and Yamdagni, 2016). For

DOI: 10.1201/9781032697550-4

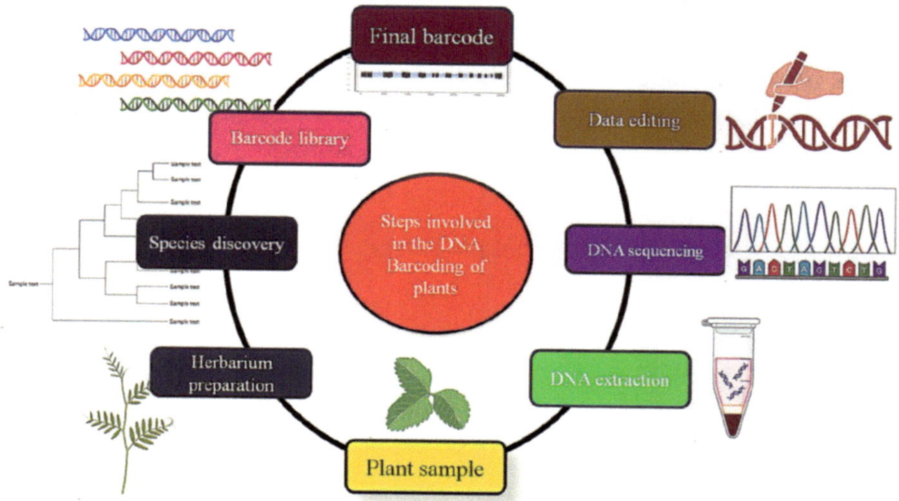

FIGURE 4.1 Steps involved in the generation of a DNA barcode.

optimal DNA barcoding, some qualities must be met such as (i) individuals belonging to the same species should have roughly identical gene sequences, but those of different species should differ; (ii) it needs to be standardized using the same region of the DNA for all taxonomic groups; (iii) the targeted region of DNA must contain all the information needed to classify species that are unknown or have not yet been barcoded into the appropriate taxonomic groups; (iv) it needs to be incredibly durable; (v) both the sequencing and the amplification of DNA should be extremely dependable; and (vi) the regions ought to be sufficiently brief to provide "minimal amplification" of the target region of the DNA (Newmaster et al., 2006). For animal identification, a section of the mitochondrial cox I "cytochrome c oxidase subunit 1" was primarily used as a DNA barcode sequence (Hebert et al., 2003), but this region is extremely invariant and inappropriate for use as a DNA barcode for all terrestrial plants. To overcome this, several conserved regions of plastid have been postulated in place of mitochondrial cox I for their (plant) identification, such as "matK, rpoB, rpoC1, rbcL genes, atpF-atpH, trnH-psbA, psbK-psbI (intergenic plastidial spacers), and nuclear internal transcribed spacers" (Fazekas et al., 2008). Chen and colleagues (2010) examined over 6600 plant samples from 4800 different species over 753 different genera by utilizing the genomic areas (psbA-trnH, matK, rbcL, rpoC1, ycf5, ITS, and ITS2) as mentioned in Table 4.1. Their findings indicated that the best area for DNA barcoding applications is the nuclear ribosomal DNA's second internal transcribed spacer (ITS2), as it constitutes 92.7% success rate in the identification at species level. In the literature, there are a total of 26 references to the ITS area: 21 references to psbA-trnH, 19 references to matK, and 14 references to rbcL. These were the most often discussed barcoding regions, while ITS2 with 9 references, 6 references of rpoC1, 4 references of rpoB, and 3 references of trnL-F showed additional genomic areas barcoded (Techen et al., 2014). As there were various DNA barcodes in the literature, Selvaraj et al. (2012) examined several genomic barcode areas and reached a similar conclusion: after all samples were examined, ITS or ITS2 had the highest discrimination rate. But, on the other hand, Asahina et al. (2010) gave a contradictory statement that matK and psbA-trntrnH had the greatest discrimination rates. Similarly, Patwardhan et al. (2014) and Guo et al. (2016) also stated that the matK gene is frequently used for phylogenetic and molecular plant identification because of its high rate of evolution and very varied sequence. The matK gene has superior individual differentiation and success rate in barcoding compared to rbcL according to Biswas et al. (2019), can be used to classify a species, and is easily amplified to create a decent nucleotide sequence according to Hollingsworth et al. (2011), in contrast to Estevez

TABLE 4.1
Overview of Analytical Material and Genomic Regions Analyzed for Medicinal Plant Identification

Analysed Material	Genomic Regions Analyzed					References
Angelica spp.	matK	rbcL	ITS	ITS2	psbA-trnH	He et al., 2012
Rhododendron spp.	matK	rbcL	ITS	ITS2	psbA-trnH	Liu et al., 2012b
Rubus spp.	ITS	psbA-trnH	trnL-F	SCAR	SNP	Yang et al., 2012a
Citrus spp.	ITS	SNP				Wang et al., 2012a
Glycerrhiza spp.	SCAR					Techen et al., 2014
Phyllanthus spp.	ISSR					Senapati et al., 2011
Dendrobium spp.	ITS					Yuan et al., 2011
Aconitum	PsbA-trnH					He et al., 2010
Angelica anomala	matK	ITS	ITS2	rbcL,	psbA-trnH	He et al., 2012
Mentha canadensis	ITS2					Pang et al., 2012a, 2012b
Rhododendron spp.	PsbA-trnH					Liu et al., 2012a
Ginkgo biloba	Short region in rbcL					Liu et al., 2018
Saposhnikoviadivaricata	ITS2					Zhang et al., 2016
Phyllanthus amarus	trnL					Buddhachat et al., 2015
Serenoa repens	Short region in rbcL, matK					Little and Jeanson, 2013

et al. (2015), who state that there is sufficient variation in rbcL to differentiate the majority of terrestrial plant species. Numerous research teams have suggested various combinations of these loci to improve species identification accuracy, particularly for the essential groups (Starr et al., 2009). Nonetheless, disagreements persist about the designation and quantity of the most appropriate areas for plant DNA barcoding (Starr et al., 2009).

4.1.1 DNA Barcoding Identification Methodology

4.1.1.1 Experimental Procedure

The steps involved in the identification process of DNA barcoding technology are as follows (Figure 4.2): (1) collection of samples, which includes fresh, mixed, dry, and processed samples; (2) extraction of DNA (e.g., sodium dodecyl sulfate [SDS] method, cetyl trimethylammonium bromide [CTAB] method, polyvinylpyrrolidone [PVP], and phenol-chloroform method); (3) PCR (polymerase chain reaction) amplification, which involves three major points in a cycle: denaturation, annealing, and extension; (4) sequencing of DNA, which includes the first-generation Sanger dideoxy DNA sequencing and progresses to next-generation sequencing (NGS) by synthesis and

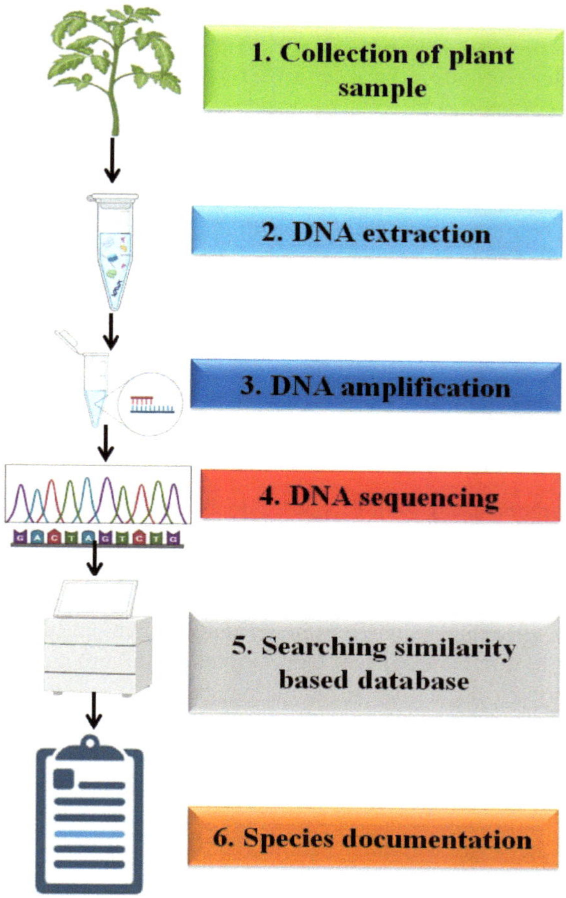

FIGURE 4.2 Experimental protocol of DNA barcoding.

third-generation sequencing by single molecule; (5) using software tools like BLAST for similarity-based database search; and (6) documentation of species (Yu et al., 2021).

4.1.1.2 Bioinformatics Analysis

The most crucial aspect of DNA barcoding technology is sequence analysis. Numerous analytical techniques, including genetic similarity analysis, the genetic distance method, and phylogenetic analysis, are included in DNA barcodes. Numerous molecular software tools were employed, including MEGA 7.0 (Kumar et al., 2016), TaxonDNA (Meier et al., 2006), and ABGD (Puillandre et al., 2012).

4.2 TYPES OF DNA BARCODES

Recent advances in molecular genetics have provided workforces with a variety of novel approaches for dependable and simple plant species identification. This tremendous advancement in taxonomy and biotechnology has had a significant impact on the creation of alternative potent technology. Hebert et al. (2003) suggested a new tool called "DNA barcoding," which is a helpful addition to the taxonomic toolbox. They promoted the utilization of DNA barcodes, or short DNA sequences from a specific section of the genome, for biological identification. It entails the use of regular DNA locus sequencing as a technique for species identification. A perfect DNA barcode should be easy to obtain using just one pair of primers, enable bidirectional sequencing, and efficiently give good discrimination between different types of DNA (DeSalle, 2006; Rubinoff et al., 2006). DNA barcoding has gone from science fiction to reality in a few years (Frézal and Leblois, 2008). A marker must have high interspecific and low intraspecific variability in order to have effective discriminating power. The "DNA barcoding gap" states the discrepancy within inter- and intraspecific distances. The DNA barcoding revolution is rooted in the automation and standardization of taxonomic approaches (Casiraghi et al., 2010). In the postgenomics era, several researchers have proposed DNA barcoding as a combined technique with traditional taxonomy to identify and authenticate species (Sahare and Srinivasu, 2012; Vohra and Khera, 2013). Barcodes from numerous locations have also been efficaciously combined. Several ways are discussed here to familiarize academics with current trends in addressing the problems of DNA barcoding in plants.

4.2.1 Single-Locus DNA Barcodes

The matK and rbcL markers, along with ITS, were declared to be the core sequences of plant DNA barcodes and trnH-psbA as complementary sequences at the third World DNA Barcode Conference in 2009 (CBOL Plant Working Group, 2009). Chen and associates (2010) projected the ITS2 section as the principal DNA barcode and trnH-psbA as a complementary sequence for identifying medicinal plant species after thorough experiments and verification. In addition to rbcL and matK, the chloroplast intergenic trnH-psbA spacer (approximately 450 bp) has recently emerged as a useful tool in low taxonomic level plant DNA-barcoding studies (Yao et al., 2009; Pang et al., 2012c). The fundamental problem with the locus is the high frequency of mononucleotide repetitions, which result in unidirectional readings and make recovery of bidirectional sequences difficult (Devey et al., 2009; Whitlock et al., 2010). It is difficult to use trnH-psbA as a single barcode due to its intricate design (Hao et al., 2010; Štorchová and Olson, 2007). In contrast, Pang et al. (2012c) provided a summary of the comprehensive assessment of the effectiveness of trnH-psbA and its combinations. In addition to the other successful loci, it has been reported that the variable region of the 5S rDNA IGS, plastid protein coding (rpoB, rpoC1), plastid IGSs (atpF-H, psbK-I), and low copy number genes are useful for plant material identification (Chen et al., 2008; Law et al., 2011; Pillon et al., 2013). These genes are being tested for their identification success in different families.

4.2.1.1 matK

Phylogenetic relationships at superficial to deep taxonomic levels have been resolved using molecular data from matK. matK is one of the plastid genome's most quickly developing coding areas and is thought to be the closest plant counterpart of the animal barcode, cytochrome oxidase 1 (COI), a mitochondrial gene (Hollingsworth et al., 2011). Among all the plastid regions used in plant systematics, matK stands out because of its faster pace of development (Wicke and Quandt, 2009). Lahaye et al. (2008) examined 1084 plant species (almost 96% of which were orchids) and discovered that the plastid matK gene fragment could function as a universal flowering plant DNA barcode. matK has a strong evolutionary rate, an appropriate length, clear interspecific divergence, and a low transition/transversion rate (Min & Hickey, 2007). Unfortunately, matK is difficult to amplify universally using the primer sets that are currently available. The CBOL Plant Working Group (2009) discovered that amplifying angiosperm DNA with a single primer pair had almost a 90% success rate. Even with several primer sets, however, success was restricted in gymnosperms (83%) and significantly poorer in cryptogams (10%). In different taxonomic families, different primer combinations were required (Chase et al., 2007; Hollingsworth, 2008). Lahaye et al. (2008) used specific primers to successfully amplify the matK gene in 1667 samples of angiosperm plants (Cuénoud et al., 2002). The differences in discriminating rates between taxonomic groups present another challenge.

According to Kress and Erickson (2007), matK can identify over 90% of *Orchidaceae* species, but only 49% of nutmeg species (Newmaster et al., 2008). Fazekas et al. (2008) were only successful 56% of the time while attempting to identify 92 species from 32 genera using the matK barcode. These results demonstrate that the matK barcode alone is insufficient as a universal barcode.

4.2.1.2 rbcL

In contrast to matK, a large subunit of RUBISCO, encoded by the highly conserved chloroplast gene rbcL, has been extensively sequenced from a variety of plant taxa at different levels (Mishra et al., 2016). With the exception of the last 30 base pairs, this single-copy gene is about 1430 base pairs long and evolves at a rather conservative rate. However, it is evident that rbcL regularly fails to clarify phylogenetic relationships below the family level (Mishra et al., 2016). As a result, there is interest in discovering other valuable DNA regions that change more quickly than rbcL to aid in lower-level phylogenetic reconstruction. With over 50,000 sequences accessible in GenBank, rbcL is commonly used in phylogenetic studies. This gene has the benefit of being an excellent location for DNA barcoding for plants at the family and genus levels, and it is also easy to amplify, sequence, and align in most terrestrial plants. In contrast, rbcL sequences exhibit slow evolution, and among plastid genes in flowering plants, this locus has the least divergence (Kress et al., 2005). It is inappropriate at the species level due to its weak discriminatory strength (Fazekas et al., 2008; Lahaye et al., 2008; CBOL Plant Working Group, 2009; Chen et al., 2010). The length of the gene may also be a problem because four primers may be required for double-stranded sequencing of the entire gene sequence. Notwithstanding these drawbacks, rbcL was recommended as one of the most promising candidate plant barcodes because of its ease of gene sequence recovery, abundance of readily available data, and respectable but not remarkable discriminatory power (CBOL Plant Working Group, 2009; Hollingsworth et al., 2011).

4.2.1.3 trnH-psbA

The most widely used plastic barcode at the time is trnH-psbA. A single primer pair should be able to amplify almost all angiosperms due to the presence of highly conserved coding areas on both sides, making it possible to design universal primers (Shaw et al., 2005). The highest rates of insertion/deletion and sequence diversity are found in the non-coding intergenic region (Kress & Erickson, 2007). According to Kress and Erickson (2007) and Shaw et al. (2005), these characteristics make trnH-psbA a highly effective plant barcode for species discrimination. In addition, extensive barcoding studies have shown that nearly all species in several land plant groupings, such as *Hydrocotyle*, *Dendrobium*, and *Pteridophytes*, were recognized by the trnH-psbA region. Due to its complex molecular evolution, considerable length variation, and high rates of insertion/deletion in bigger angiosperm families (Chang et al., 2006; Chase et al., 2007), alignment of the trnH-psbA spacer can be quite perplexing. Duplicate loci and a pseudogene are the cause of the extremely short trnH-psbA sequence, which is fewer than 300 base pairs (bp) in other groups (Kress et al., 2005) and less than 100 bp in bryophytes (Stech & Quandt, 2010). The prevalence of inversions in some plant lineages is a major problem with utilizing trnH-psbA as a standard barcode since it can result in greatly inflated estimates of genetic divergence and erroneous phylogenetic assignment (Whitlock et al., 2010).

4.2.1.4 ITS

The nuclear ribosomal cistron's internal transcribed spacer (ITS) region (18S–4.8S26S) has been sequenced the most frequently in plants and has the best-characterized barcode gap between intra- and interspecific variations. The two spacers in this area, ITS1 and ITS2 (each 300 bp), are located next to the 4.8S locus. They differ more from each other than the rRNA genes they separate, and they have enough phylogenetic signal to distinguish between different plants and animals (Chen et al., 2010; Yao et al., 2010). High levels of interspecific divergence make the ITS spacer a powerful phylogenetic marker at the species level (Álvarez & Wendel, 2003). ITS has been suggested as a plant barcode (Sass et al., 2007) due to its higher discriminatory power over plastid areas at low taxonomic levels, especially in parasitic plants that provide less precision than plastid barcodes (Hollingsworth et al., 2011). ITS2 is thought to be more amenable to amplification and sequencing than ITS1 due to its shorter target region (Gao et al., 2010a; Han et al., 2013). Following the endorsement of the nuclear ITS region as the standard barcode by Li et al. (2011), it was thought to be one of the primary barcodes for seed plants. The issue of paralogy, which results from the prevalence of divergent copies among individuals and can cause sample misidentification, is one of the major limitations associated with the area (Bailey et al., 2003). The ITS sequences of fungi and plants are quite similar. Because of this, the primers used for sequencing and amplification are almost identical, and fungal DNA is frequently amplified by accident, especially in plants that have fungal endophytes. Misidentification of samples may result from this. Because of this, amplification and sequencing have proven challenging for a range of samples, irrespective of the quantity of primer sets accessible for this particular barcode region (Gonzalez et al., 2009). Although the ITS2 barcode offers a number of advantages over other putative loci, such as the ITS, this area has received little attention from researchers. A major concern that could produce false or misleading results is the existence of many copies of the genome with notable within-species and even within-individual sequence differentiation (Table 4.2) (Yamaguchi et al., 2006).

4.2.2 Multiple Locus DNA Barcodes

The most recent option that proved to be beneficial was the combination of barcodes, which arose after early in silico and laboratory-based evaluation. Several researchers employed a combination of markers to identify therapeutic plants since single-locus marker sequences may not always provide enough information for low-level identification. matK, rbcL, trnH-psbA, and ITS are the most common combinations of DNA markers (Fu et al., 2011; Newmaster et al., 2013; Purushothaman

TABLE 4.2
Specific Loci as Key Tool for Identification of Plants at Family Level

Loci	Family	References
atpF-atp + psbK-psbI + trnH-psbA	Orchidaceae	Kim et al., 2014
atpF-atpH	Lemnaceae	Wang et al., 2010
ITS	Nyssaceae	Wang et al., 2012b
ITS2	Asteraceae, Fabaceae, Rosaceae, Zingiberace, Araliaceaes, Leguminaceae, Apiaceae, Rutaceae	Gao et al., 2010b; Gao et al., 2010a; Pang et al., 2011; Shi et al., 2011, Liu et al., 2012b, Madesis et al., 2012; Liu et al., 2014; Luo et al., 2010
matK	Juglandaceae, Lamiaceae, Zingiberaceae, Combretaceae	Xiang et al., 2011; Theodoridis et al., 2012; Vinitha et al., 2014; Gere et al., 2013
matK + rbcL + trnH-psbA	Arecaceae	Yang et al. 2012b
rbcL	Zingiberaceae	Vinitha et al., 2014
rbcLa + matK + trnH-psbA	Combretaceae	Gere et al., 2013
trnH-psbA	Lamiaceae, Lauraceae, Polygonaceae, Orchidaceae, Combretaceae	Theodoridis et al., 2012; Liu et al., 2012c; Song et al., 2009; Kim et al., 2014; Gere et al., 2013
trnL	Leguminaceae	Madesis et al., 2012

et al., 2014). The CBOL Plant Working Group assessed the viability of several top candidate markers before proposing the two-locus combination of matK and rbcL as the core plant barcode (Newmaster et al., 2008; CBOL Plant Working Group, 2009). Further research is necessary, nevertheless, due to matK's primer universality and rbcL's lack of selective power (Hollingsworth et al., 2011). Nithaniyal et al. (2014) have found that when two loci are present, the conserved coding locus will produce deep phylogenetic branches when it aligns well with the taxa of a community sample, while the hypervariable portion of the DNA barcode will easily align in the subclades of closely related species. The two-loci barcode was declared the standard barcode for land plants in order to save money by using a three-loci combination for large datasets. However, in certain circumstances, using a three-loci combination did not enhance discrimination beyond the best-performing two-loci barcodes. Thus, a tiered approach that offers the greatest diversity from two-barcode regions for identifying closely related species can circumvent the problem of noncoding area alignment (Newmaster et al., 2013; Purushothaman et al., 2014). Newmaster et al. (2006) determined that rbcL is a feasible standard first-tier core coding region based on the most well-characterized plastid coding region in the gene bank. It has been demonstrated to resolve about 85% of instances in congeneric species due to its universality and ease of amplification and alignment. Rather than a lack of diversity, several scientists hypothesized that inconsistencies between the plastid gene tree and species boundaries were the cause of multiple-locus barcodes' inability to improve species recognition (Fazekas et al., 2009; Hollingsworth et al., 2011). Thus, prospective locus combinations are unable to compensate for the inherent limitations of plant DNA barcoding.

4.2.3 Genome-Based DNA Barcodes

The chloroplast (cp) genome, which has more genetic material for species identification than any commonly used single-locus marker, houses all of the DNA sequences found in a plastid. On October 27, 2019, the chloroplast genomes of 3452 plants were made available on NCBI. The use of the entire plastid genome for DNA barcoding was suggested by Govindaraghavan and Li (Govindaraghavan et al., 2012; Li et al., 2015). According to Sucher and Carles (2008), genetic fingerprinting is useful for identifying sample homogeneity and the presence of adulterants in herbal supplements. It can also be used to distinguish between individuals, species, and populations. Yang used the chloroplast genome to identify the species of *Datura stramonium* (Yang et al., 2014). He examined and sequenced the *Lonicera japonica* whole chloroplast genome. According to He et al. (2017), this study identified unique characteristics of the *L. japonica* cp genome and provided helpful data for the phylogenetic classification and particular barcoding of this therapeutic plant. Zhou et al. analyzed the molecular structures and evolutionary relationships of the chloroplast genomes of *Papaver orientale* and *Papaver rhoeas*. They concluded that the Papaveraceae, a significant family of medicinal plants, may have its phylogenetic positions and relationships resolved by using the chloroplast genome (Zhou et al., 2018).

4.2.4 Super DNA Barcodes

A new technique is needed to identify closely related plant species since single-locus DNA barcodes have intrinsic limitations (Heinze, 2007). Recent research has demonstrated that the full cp genome may be utilized as a plant barcode and has as much variety as the CO1 locus in mammals (Kane & Cronk, 2008). A conserved sequence spanning 110–160 kbp is present throughout the entire cp genome, which is significantly longer than commonly used DNA barcodes and offers more variation for differentiating closely related plants. The cp genome has been used as a flexible phylogenetic instrument. In plant phylogenetic, phylogeographic, and population genetic studies, it has been suggested as a species-level DNA barcode since it can significantly increase resolution at lower taxonomic levels, supporting the recovery of lineages as monophyletic (Parks et al., 2009). The analysis of this super-barcode also removes the sequence retrieval problems that are frequently encountered

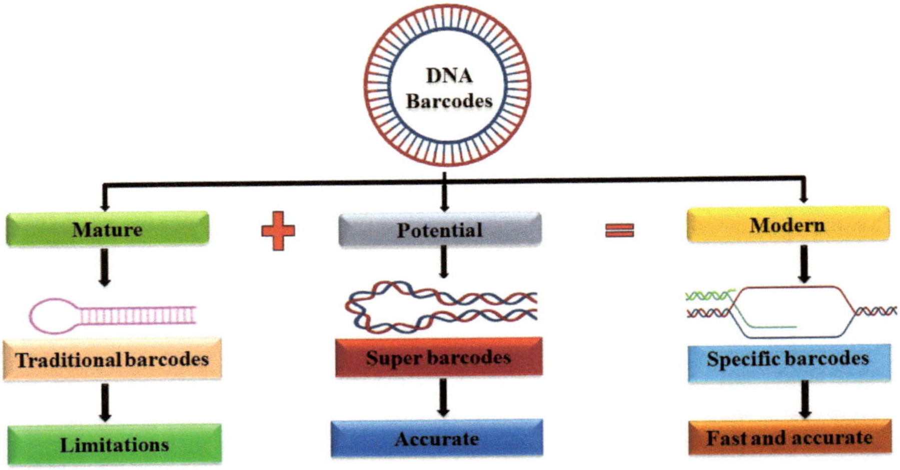

FIGURE 4.3 Schematic representation of plant barcoding developments.

in standard barcoding studies. As a genome-based barcode, the cp genome is more suitable than the nuclear genome due to its smaller size, and higher interspecific and lower intraspecific divergence. According to Hebert et al. (2004), one of the most fundamental tests for species identification using barcoding techniques is the determination of whether a gene is present in two different species. This is due to the fact that super-barcoding is superior to standard barcoding in terms of its ability to identify gene loss and establish gene order (Luo et al., 2009). Establishing a comprehensive cp genome database, cutting sequencing costs, and obtaining greater quality and quantity of DNA are the main obstacles to super-barcoding (Kane et al., 2012) (Figure 4.3).

4.2.5 Real-Time DNA Barcodes

Bar-HRM, or high-resolution melting (HRM) analysis plus DNA barcoding, is a commonly used technique to identify contamination in herbal blends. Without the need for sequencing or hybridization procedures, it is a unique DNA-based method for genotyping and fingerprinting that differentiates between DNA sequence variants based on the thermal denaturation of the amplicons (Wittwer, 2009). The amplicon is analyzed in a real-time PCR instrument using fluorescence monitoring of the dsDNA melting curve produced by the release of the intercalating dye SYBR Green I. Compared to other methods for researching genetic diversity, HRM analysis has a low reaction cost, is conducted in a closed-tube system, and does not require manual post-PCR processing. According to Jiang et al. (2014), these benefits have led to its widespread usage in biological diagnostics, food authenticity, and disease detection. Kalivas et al. (2014) reported another successful use of HRM in species identification, wherein seven Greek *Sideritis* taxa were taxonomically identified and their authentication was carried out using the nuclear ITS2 DNA barcoding sequence. Consequently, a large variety of medicinal plants may benefit from the approach, especially those belonging to closely related species that are utilized in the herbal industry. HRM has also been demonstrated to reliably identify products in the food industry when conjugated with particular barcode sections, such as mitochondrial DNA, 16S rDNA regions, and microsatellite markers (Bosmali et al., 2012; Ganopoulos et al., 2011)

4.3 FUTURE CHALLENGES AND RECOMMENDED SOLUTIONS FOR DNA BARCODING

Over the next ten years, establishing the worldwide plant's DNA barcode collection will be one of the main difficulties because of the vouchered number collection, tagging of individual trees that botanists may revisit if needed, and well-verified identifications, which constitute excellent sources for developing the library of plants' DNA barcodes. There are two further ways to expand the plant library: the lineage-based approach and the floristic approach. As part of fundamental molecular phylogenetic detection, in which the markers (DNA barcodes) are used to determine the relationships at the evolutionary level, or as testing for sequencing accomplishments using the standard barcodes (Chen et al., 2010, 2014; Wang et al., 2017), individual taxonomists are also creating DNA barcodes for particular groups of plants. All of these "DNA barcodes" are contributing to the collection of sequences that go along with the conventional DNA barcode marker, even though many of them might not be given the official GenBank DNA barcode description.

The second future challenge is identifying the percentage of distinct species, which is the main obstacle to a plant's DNA barcoding process. In plant barcoding, the requirements of employing numerous loci or a multitiered system result in increased time for preparation, sample handling, and expenses. For example, the non-coding plastid regions trnH-psbA, atpF-atpH, and psbK-psbI, and the trnL intron, which has been recommended for circumstances involving severely damaged tissue, will be added to the core rbcL-matK barcode in separate projects as needed in the short term, when more universality and resolution are the basic needs (Taberlet et al., 2007). In the taxonomic groupings where "direct sequencing" of this locus is feasible, the continuously changing ITS of nuclear ribosomal DNA also serves as a helpful supplemental barcode (Feliner et al., 2007). Beyond these currently accessible supplemental barcodes, opportunities to target the nuclear genome as a basis for enlightening characters will increase significantly due to currently adopted improvements in sequencing technologies and the concurrent collection of transcriptomic and genomic sequence information from plants. It is highly likely that future developments will lead to further refinement of plant DNA barcoding techniques (Fazekas et al., 2009). To facilitate large-scale or extensive sequencing and the creation of an infrastructure of plant barcoding at the international level, agreeing on a uniform set of loci is a crucial first step in the process of plant barcoding. Thus, the widespread community consensus sequence of rbcL and matK (i.e., standard 2-locus barcode) is a crucial first step toward creating a centralized database of plant barcodes as a tool for taxonomy, conservation, and the myriad other uses for plant material identification (Newmaster et al., 2009). The development of high-throughput sequencing technology has also helped to address several drawbacks of classical plant DNA barcoding, which has accelerated the field of plant genomics, particularly chloroplast genomics. It has also recently been demonstrated that whole chloroplast genomes may effectively distinguish closely related species. In addition to being useful for studying large-scale biodiversity patterns and the genesis, identification, and evolution of related species, DNA barcoding technology will also make it easier to preserve and use biological resources. Moreover, a quicker and economically cheap solution for large-scale/extensive multilocus plant DNA barcoding is provided by the use of alternative sequencing technologies, such as the application of microfluidic PCR-based targets (Das, 2019), which is followed by sequencing based on dideoxy chain termination (Sanger). Alternatively, big genomes (i.e., genome skimming) or entire genomes (i.e., organellar or otherwise genome skimming) can be partially sequenced using NGS technologies, which will reduce sequencing costs in the future (Li et al., 2015; Coissac et al., 2016).

The effective use of molecular techniques depends on the separation of pure, high-molecular-weight DNA. This might be difficult since processed medicinal plant material frequently has severely damaged DNA or large concentrations of secondary metabolites (including alkaloids and flavonoids), polysaccharides, and polyphenols. For the purpose of producing high-quality DNA for later uses from powdered plant materials having medicinal properties, herbarium specimens, pills, capsules, and tinctures, a variety of commercially available kits and modified protocols of traditional procedures exist (Kazi et al., 2013). Shorter pieces of herbarium DNA are simpler to amplify, according to Sarkinen et al.'s substantial negative association between amplicon size and PCR performance. Only 10% of the examined samples showed successful amplification of the rbcL (670 bp), 24% of the medium-sized LEAFY intron 3 with 260 bp exhibited successful amplification, and 78% of the short genomic region trnL P6 loop with 10–143 bp showed successful amplification. Therefore, it is mostly advised to create conditions that allow for the production of very short amplicons that are simpler to amplify and/or to carry out DNA "repair reactions" for severely damaged DNA (Särkinen et al., 2012; Pusch et al., 1998), both of which are offered as commercial kits. Given that the NGS process needs fragmented or ligated DNA as the starting material, Särkinen et al. predict that fragmented DNA from medicinal plants will be much less of an issue in the near future.

At present, the current improvement in demand for herbal or natural products is due to the continuous loss of safety and quality of these plant products. For this reason, DNA barcoding technology has been used efficiently in the identification or authentication of medicinal plant-derived herbal supplements that promise the protection of their respective users. According to a literature survey, a minimum of 17 molecular markers have been recorded for the identification and authentication of herbal beneficial plants, but despite all this progress, we have not yet found a possible or common barcode for all groups of plants. The reason or limitation behind this problem (not having a common barcode) is the state of DNA in the tested samples of herbal products because, in most cases, tested samples contained degraded DNA with some additive contamination. For this situation, a possible solution worth mentioning here is the use of DNA mini-barcodes (<200 bp), although this technique may limit the use of polymorphic regions. The identification of several species of medicinal plants in a single sample is another drawback of the DNA barcode method; nevertheless, in these situations, methods like metabarcoding and mass sequencing should be applied to get beyond these restrictions (Gesto-Borroto et al., 2021).

DNA barcoding of plant material in combination with conservative plant traits (such as flowering phenology) is crucial for comparing natural reserves or community evolutions from forest dynamics plots. It also allows for a deeper exploration of the effects of individual plants' assembly patterns in communities, the divergence of distinct groups along an environmental gradient, and mostly the non-random conditions within communities. As a result, some potential paths are as follows: (i) putting forth a "purpose-driven barcode" (such as mini-barcode and metabarcoding) suitable for multilevel uses such as biomolecule identification, community phylogeny reconstruction, biodiversity of environmental detection, and ecological network structure exploration; (ii) establishing libraries of DNA barcode sequences of all economically valuable tree species for the identification of endangered plant taxa against illegal international trade at a national level; (iii) adoption of new consolidative sequencing strategies like genome skimming to create mega-phylogenies in the face of the post-genomic era; and (iv) creating intelligent identification systems for land-growing plants that integrate environmental, morphological, and genetic information, which will make plant identification more accurate and convenient at the DNA level.

4.4 CONCLUSION AND FUTURE PERSPECTIVE

Although it hasn't been used to its full potential yet, the identification of plant species in herbal medicine by DNA barcoding seems promising. This chapter gives a brief history of DNA barcoding in medicinal plants, a rundown of the various varieties of DNA barcoding, and an overview of some popular and cutting-edge technologies. As a widely accepted technique, DNA barcoding has been crucial to the regulation of the pharmaceutical industry, the classification of therapeutic plants, and the identification of adulterants and replacements. With the present technological instruments available, the majority of plant raw materials and unfinished items available on the medicinal materials market can be successfully identified and employed. Creating a single sequence that can be used to identify every species is the aim of DNA barcoding. Investigations have revealed that multilocus markers, which were previously believed to be more effective in identifying species, are likewise inadequate for identifying all plants. Particularly for closely related taxa, whole-plastid-based barcodes have shown great promise in species discrimination. Plant identification may adopt these super-barcodes as the standard method when sequencing technology advances. Finding a universal identifier is not the ultimate goal of DNA barcoding; rather, it is to discriminate between species. Thanks to an international, publicly accessible database of reference barcode sequences produced via barcoding, non-taxonomists are able to identify specimens. There isn't a single-locus barcode that can be used as a universal DNA barcode for plants due to limitations like low PCR efficiency, gene loss, and inadequate variation. A multilocus methodology is a much better idea and is generally accepted as a workable method for barcoding terrestrial plants if discrimination success is necessary. Combining each locus's universality, discriminatory power, and amplification success will produce high discrimination-oriented results. The certification of effective medicinal plants for herbal medication formulations is being made possible by the latest developments in plant DNA barcoding with NGS and HRM techniques. In the past, the selection of markers and the building of sequence databases constituted the main areas of concentration for DNA barcoding technology development. While there is disagreement among scientists over universal plant DNA barcodes, the four standard markers—matK, ITS, rbcL, and trnH-psbA—are used in most applications. With continued advancement and full technological application, the super-barcode and metabarcoding become members of the extended family of barcodes, similar to the DNA barcode. Thus, DNA barcoding needs to cooperate with other suitable molecular biology and analytical chemistry technologies if it is to be successfully used for authentication purposes by the herbal industry. This would require substantial application of systems biology components such as proteomics (specific proteome) and transcriptomics (specific expression subset analysis [SESA]), supplemented by need-based use of these components (genomics, DNA barcoding) and metabolomics (for active secondary metabolites).

REFERENCES

Álvarez IJ, Wendel JF. Ribosomal ITS sequences and plant phylogenetic inference. *Molecular Phylogenetics and Evolution*. 2003 Dec 1;29(3):417–34.

Asahina H, Shinozaki J, Masuda K, Morimitsu Y, Satake M. Identification of medicinal Dendrobium species by phylogenetic analyses using matK and rbcL sequences. *Journal of Natural Medicines*. 2010 Apr;64:133–8.

Bailey CD, Carr TG, Harris SA, Hughes CE. Characterization of angiosperm nrDNA polymorphism, paralogy, and pseudogenes. *Molecular Phylogenetics and Evolution*. 2003 Dec 1;29(3):435–54.

Biswas R, Panja AS, Bandopadhyay R. In silico analyses of burial codon bias among the species of Dipterocarpaceae through molecular and phylogenetic data. *Evolutionary Bioinformatics*. 2019 Mar;15:1176934319834888.

Bosmali I, Ganopoulos I, Madesis P, Tsaftaris A. Microsatellite and DNA-barcode regions typing combined with high resolution melting (HRM) analysis for food forensic uses: A case study on lentils (Lens culinaris). *Food Research International*. 2012 Apr 1;46(1):141–7.

Buddhachat K, Osathanunkul M, Madesis P, Chomdej S, Ongchai S. Authenticity analyses of *Phyllanthus amarus* using barcoding coupled with HRM analysis to control its quality for medicinal plant product. *Gene*. 2015 Nov 15;573(1):84–90.

Casiraghi M, Labra M, Ferri E, Galimberti A, De Mattia F. DNA barcoding: a six-question tour to improve users' awareness about the method. *Briefings in Bioinformatics*. 2010 Jul 1;11(4):440–53.

CBOL Plant Working Group. A DNA barcode for land plants. *Proceedings of the National Academy of Sciences of the United States of America*. 2009;106(12794):7.

Chang CC, Lin HC, Lin IP, Chow TY, Chen HH, Chen WH, Cheng CH, Lin CY, Liu SM, Chang CC, Chaw SM. The chloroplast genome of Phalaenopsis aphrodite (Orchidaceae): comparative analysis of evolutionary rate with that of grasses and its phylogenetic implications. *Molecular Biology and Evolution*. 2006 Feb 1;23(2):279–91.

Chase MW, Cowan RS, Hollingsworth PM, Van Den Berg C, Madriñán S, Petersen G, Seberg O, Jørgsensen T, Cameron KM, Carine M, Pedersen N. A proposal for a standardised protocol to barcode all land plants. Taxon. 2007 May;56(2):295–9.

Chen F, Chan HY, Wong KL, Wang J, Yu MT, But PP, Shaw PC. Authentication of Saussurealappa, an endangered medicinal material, by ITS DNA and 5S rRNA sequencing. *Planta Medica.* 2008 Jun;74(08):889–92.

Chen J, Zhao J, Erickson DL, Xia N, Kress WJ. 2014. Testing DNA barcodes in closely related species of Curcuma (Zingiberaceae) from Myanmar and China. Molecular Ecology Resources 15: 337–348.

Chen S, Yao H, Han J, Liu C, Song J, Shi L, Zhu Y, Ma X, Gao T, Pang X, Luo K. Validation of the ITS2 region as a novel DNA barcode for identifying medicinal plant species. *PloS One.* 2010 Jan 7;5(1):e8613.

China Plant BOL Group 1, Li DZ, Gao LM, Li HT, Wang H, Ge XJ, Liu JQ, Chen ZD, Zhou SL, Chen SL, Yang JB. Comparative analysis of a large dataset indicates that internal transcribed spacer (ITS) should be incorporated into the core barcode for seed plants. *Proceedings of the National Academy of Sciences.* 2011 Dec 6;108(49):19641–6.

Coissac E, Hollingsworth PM, Lavergne S, Taberlet P. From barcodes to genomes: extending the concept of DNA barcoding. Wiley Online Library. 2016.

Cuénoud P, Savolainen V, Chatrou LW, Powell M, Grayer RJ, Chase MW. Molecular phylogenetics of Caryophyllales based on nuclear 18S rDNA and plastid rbcL, atpB, and matK DNA sequences. *American Journal of Botany.* 2002 Jan;89(1):132–44.

Das K. Authentic identification and new drug discovery from natural plant-based constituents through DNA barcoding: a challenging task to the researchers. *Ann. Phytomedicine.* 2019;8:19–27.

DeSalle RO. Species discovery versus species identification in DNA barcoding efforts: response to Rubinoff. *Conservation Biology.* 2006 Oct 1;20(5):1545–7.

Devey DS, Chase MW, Clarkson JJ. A stuttering start to plant DNA barcoding: microsatellites present a previously overlooked problem in non-coding plastid regions. Taxon. 2009 Feb;58(1):7–14.

Estevez I, Mensah S, Tian R, Fein M. DNA barcoding and phylogenetic analysis of plant species: Urban barcode project of New York City. *International Journal of Science Research Science Engineering Technology.* 2015;1(4):347–52.

Fazekas AJ, Burgess KS, Kesanakurti PR, Graham SW, Newmaster SG, Husband BC, Percy DM, Hajibabaei M, Barrett SC. Multiple multilocus DNA barcodes from the plastid genome discriminate plant species equally well. *PLoS One.* 2008 Jul 30;3(7):e2802.

Fazekas AJ, Kesanakurti PR, Burgess KS, Percy DM, Graham SW, Barrett SC, Newmaster SG, Hajibabaei M, Husband BC. Are plant species inherently harder to discriminate than animal species using DNA barcoding markers? *Molecular Ecology Resources.* 2009 May;9:130–9.

Feliner GN, Rosselló JA. Better the devil you know? Guidelines for insightful utilization of nrDNA ITS in species-level evolutionary studies in plants. *Molecular Phylogenetics and Evolution.* 2007 Aug 1;44(2):911–9.

Frézal L, Leblois R. Four years of DNA barcoding: current advances and prospects. *Infection, Genetics and Evolution.* 2008 Sep 1;8(5):727–36.

Fu YM, JIiang WM, Fu CX. Identification of species within Tetrastigma (Miq.) Planch.(Vitaceae) based on DNA barcoding techniques. *Journal of Systematics and Evolution.* 2011 May;49(3):237–44.

Ganopoulos I, Argiriou A, Tsaftaris A. Adulterations in Basmati rice detected quantitatively by combined use of microsatellite and fragrance typing with high resolution melting (HRM) analysis. *Food Chemistry.* 2011 Nov 15;129(2):652–9.

Gao T, Yao H, Song J, Liu C, Zhu Y, Ma X, Pang X, Xu H, Chen S. Identification of medicinal plants in the family Fabaceae using a potential DNA barcode ITS2. *Journal of Ethnopharmacology.* 2010a Jul 6;130(1):116–21.

Gao T, Yao H, Song J, Zhu Y, Liu C, Chen S. Evaluating the feasibility of using candidate DNA barcodes in discriminating species of the large Asteraceae family. *BMC Evolutionary Biology.* 2010b Dec;10:1–7.

Gere J, Yessoufou K, Daru BH, Mankga LT, Maurin O, van der Bank M. Incorporating trnH-psbA to the core DNA barcodes improves significantly species discrimination within southern African Combretaceae. *ZooKeys.* 2013;2013(365):129.

Gesto-Borroto R, Medina-Jimenez K, Lorence A, Villarreal ML. Application of DNA barcoding for quality control of herbal drugs and their phytopharmaceuticals. *Revista Brasileira de Farmacognosia.* 2021 Apr;31:127–41.

Gonzalez MA, Baraloto C, Engel J, Mori SA, Pétronelli P, Riéra B, Roger A, Thébaud C, Chave J. Identification of Amazonian trees with DNA barcodes. *PLoS One.* 2009 Oct 16;4(10):e7483.

Govindaraghavan S, Hennell JR, Sucher NJ. From classical taxonomy to genome and metabolome: towards comprehensive quality standards for medicinal herb raw materials and extracts. *Fitoterapia*. 2012 Sep 1;83(6):979–88.

Guo YY, Huang LQ, Liu ZJ, Wang XQ. Promise and challenge of DNA barcoding in Venus Slipper (Paphiopedilum). *PloS One*. 2016 Jan 11;11(1):e0146880.

Han J, Zhu Y, Chen X, Liao B, Yao H, Song J, Chen S, Meng F. The short ITS2 sequence serves as an efficient taxonomic sequence tag in comparison with the full-length ITS. *BioMed Research International*. 2013 Jan 1;2013:741476–83.

Hao DC, Chen SL, Xiao PG. Sequence characteristics and divergent evolution of the chloroplast psbA-trnH noncoding region in gymnosperms. *Journal of Applied Genetics*. 2010Sep;51:259–73.

He J, Wong KL, Shaw PC, Wang H, Li DZ. Identification of the medicinal plants in Aconitum L. by DNA barcoding technique. *Planta Medica*. 2010 Oct;76(14):1622–8.

He L, Qian J, Li X, Sun Z, Xu X, Chen S. Complete chloroplast genome of medicinal plant Lonicera japonica: genome rearrangement, intron gain and loss, and implications for phylogenetic studies. *Molecules*. 2017 Feb 7;22(2):249.

He Y, Hou P, Fan G, Song Z, Arain S, Shu H, Tang C, Yue Q, Zhang Y. Authentication of Angelica anomalaAvé-Lall cultivars through DNA barcodes. *Mitochondrial DNA*. 2012 Apr 1;23(2):100–4.

Hebert PD, Cywinska A, Ball SL, DeWaard JR. Biological identifications through DNA barcodes. *Proceedings of the Royal Society of London. Series B: Biological Sciences*. 2003 Feb 7;270(1512):313–21.

Hebert PD, Gregory TR. The promise of DNA barcoding for taxonomy. *Systematic Biology*. 2005 Oct 1;54(5):852–9.

Hebert PD, Penton EH, Burns JM, Janzen DH, Hallwachs W. Ten species in one: DNA barcoding reveals cryptic species in the neotropical skipper butterfly Astraptesfulgerator. *Proceedings of the National Academy of Sciences*. 2004 Oct 12;101(41):14812-7.

Heinze B. A database of PCR primers for the chloroplast genomes of higher plants. *Plant Methods*. 2007 Dec;3(1):1–7.

Hollingsworth PM, Graham SW, Little DP. Choosing and using a plant DNA barcode. *PLoS One*. 2011 May 26;6(5):e19254.

Hollingsworth PM. Progress and outstanding questions. *Heredity*. 2008;101:1–2.

Jiang C, Cao L, Yuan Y, Chen M, Jin Y, Huang L. Barcoding melting curve analysis for rapid, sensitive, and discriminating authentication of saffron (*Crocus sativus* L.) from its adulterants. *BioMed Research International*. 2014 Oct 30;2014:809037–47.

Kalivas A, Ganopoulos I, Xanthopoulou A, Chatzopoulou P, Tsaftaris A, Madesis P. DNA barcode ITS2 coupled with high resolution melting (HRM) analysis for taxonomic identification of Sideritis species growing in Greece. *Molecular Biology Reports*. 2014 Aug;41:5147–54.

Kane N, Sveinsson S, Dempewolf H, Yang JY, Zhang D, Engels JM, Cronk Q. Ultra-barcoding in cacao (Theobroma spp.; Malvaceae) using whole chloroplast genomes and nuclear ribosomal DNA. *American Journal of Botany*. 2012 Feb;99(2):320–9.

Kane NC, Cronk Q. Botany without borders: barcoding in focus. *Molecular Ecology*. 2008 Dec;17(24):5175–6.

Kazi T, Hussain N, Bremner P, Slater A, Howard C. The application of a DNA-based identification technique to over-the-counter herbal medicines. *Fitoterapia*. 2013 Jun 1;87:27–30.

Kim HM, Oh SH, Bhandari GS, Kim CS, Park CW. DNA barcoding of Orchidaceae in Korea. *Molecular Ecology Resources*. 2014 May;14(3):499–507.

Kress WJ, Erickson DL. A two-locus global DNA barcode for land plants: the coding rbcL gene complements the non-coding trnH-psbA spacer region. *PLoS One*. 2007 Jun 6;2(6):e508.

Kress WJ, Wurdack KJ, Zimmer EA, Weigt LA, Janzen DH. Use of DNA barcodes to identify flowering plants. *Proceedings of the National Academy of Sciences*. 2005 Jun 7;102(23):8369–74.

Kumar S, Stecher G, Tamura K. MEGA7: molecular evolutionary genetics analysis version 7.0 for bigger datasets. *Molecular Biology and Evolution*. 2016 Jul 1;33(7):1870–4.

Lahaye R, Van der Bank M, Bogarin D, Warner J, Pupulin F, Gigot G, Maurin O, Duthoit S, Barraclough TG, Savolainen V. DNA barcoding the floras of biodiversity hotspots. *Proceedings of the National Academy of Sciences*. 2008 Feb 26;105(8):2923–8.

Law SK, Simmons MP, Techen N, Khan IA, He MF, Shaw PC, But PP. Molecular analyses of the Chinese herb Leigongteng (*Tripterygium wilfordii* Hook. f.). *Phytochemistry*. 2011 Jan 1;72(1):21–6.

Li X, Yang Y, Henry RJ, Rossetto M, Wang Y, Chen S. Plant DNA barcoding: from gene to genome. *Biological Reviews*. 2015 Feb;90(1):157–166.

Little DP, Jeanson ML. DNA barcode authentication of saw palmetto herbal dietary supplements. *Scientific Reports*. 2013 Dec 17;3(1):3518.

Liu J, Shi L, Han J, Li G, Lu H, Hou J, Zhou X, Meng F, Downie SR. Identification of species in the angiosperm family Apiaceae using DNA barcodes. *Molecular Ecology Resources*. 2014 Nov;14(6):1231–8.

Liu Y, Wang XY, Wei XM, Gao ZT, Han JP. Rapid authentication of Ginkgo biloba herbal products using the recombinase polymerase amplification assay. *Scientific Reports*. 2018 May 22;8(1):8002.

Liu Y, Zhang L, Liu Z, Luo K, Chen S, Chen K. Species identification of Rhododendron (Ericaceae) using the chloroplast deoxyribonucleic acid PsbA-trnH genetic marker. *Pharmacognosy Magazine*. 2012b Jan;8(29):29.

Liu Z, Chen SL, Song JY, Zhang SJ, Chen KL. Application of deoxyribonucleic acid barcoding in Lauraceae plants. *Pharmacognosy Magazine*. 2012a Jan;8(29):4.

Liu Z, Zeng X, Yang D, Chu G, Yuan Z, Chen S. Applying DNA barcodes for identification of plant species in the family Araliaceae. *Gene*. 2012c May 10;499(1):76–80.

Luo H, Sun Z, Arndt W, Shi J, Friedman R, Tang J. Gene order phylogeny and the evolution of methanogens. *PLoS One*. 2009 Jun 29;4(6):e6069.

Luo K, Chen S, Chen K, Song J, Yao H, Ma X, Zhu Y, Pang X, Yu H, Li X, Liu Z. Assessment of candidate plant DNA barcodes using the Rutaceae family. *Science China Life Sciences*. 2010 Jun;53:701–8.

Madesis P, Ganopoulos I, Ralli P, Tsaftaris A. Barcoding the major Mediterranean leguminous crops by combining universal chloroplast and nuclear DNA sequence targets. *Genetics and Molecular Research*. 2012 Aug 16;11(3):2548–58.

Meier R, Shiyang K, Vaidya G, Ng PK. DNA barcoding and taxonomy in Diptera: a tale of high intraspecific variability and low identification success. *Systematic Biology*. 2006 Oct 1;55(5):715–28.

Min XJ, Hickey DA. Assessing the effect of varying sequence length on DNA barcoding of fungi. *Molecular Ecology Notes*. 2007 May;7(3):365–73.

Mishra P, Kumar A, Nagireddy A, Mani DN, Shukla AK, Tiwari R, Sundaresan V. DNA barcoding: an efficient tool to overcome authentication challenges in the herbal market. *Plant Biotechnology Journal*. 2016 Jan;14(1):8–21.

Newmaster SG, Fazekas AJ, Ragupathy SD. DNA barcoding in land plants: evaluation of rbcL in a multigene tiered approach. *Botany*. 2006 Mar;84(3):335–41.

Newmaster SG, Fazekas AJ, Steeves RA, Janovec J. Testing candidate plant barcode regions in the Myristicaceae. *Molecular Ecology Resources*. 2008 May;8(3):480–90.

Newmaster SG, Grguric M, Shanmughanandhan D, Ramalingam S, Ragupathy S. DNA barcoding detects contamination and substitution in North American herbal products. *BMC Medicine*. 2013 Dec;11:1–3.

Newmaster SG, Ragupathy S, Janovec J. A botanical renaissance: state-of-the-art DNA bar coding facilitates an automated identification technology system for plants. *International Journal of Computer Applications in Technology*. 2009 Jan 1;35(1):50–60.

Nithaniyal S, Newmaster SG, Ragupathy S, Krishnamoorthy D, Vassou SL, Parani M. DNA barcode authentication of wood samples of threatened and commercial timber trees within the tropical dry evergreen forest of India. *PLoS One*. 2014 Sep 26;9(9):e107669.

Pang X, Liu C, Shi L, Liu R, Liang D, Li H, Cherny SS, Chen S. Utility of the trnH–psbA intergenic spacer region and its combinations as plant DNA barcodes: a meta-analysis. *PLoS One*. 2012c Nov 14;7(11):e48833.

Pang X, Song J, Chen S. Identification of Junci Medulla using DNA barcoding technique. *Zhongguo Zhong yao za zhi= ZhongguoZhongyaoZazhi= China Journal of Chinese Materia Medica*. 2012a Apr 1;37(8):1097–9.

Pang X, Song J, Zhu Y, Xu H, Huang L, Chen S. Applying plant DNA barcodes for Rosaceae species identification. *Cladistics*. 2011 Apr;27(2):165–70.

Pang X, Xu HB, Han J, Song J. Applying DNA barcoding technique to identify menthaehaplocalycisherba. *Zhongguo Zhong yao za zhi= ZhongguoZhongyaoZazhi= China Journal of Chinese Materia Medica*. 2012b Apr 1;37(8):1114–7.

Parks M, Cronn R, Liston A. Increasing phylogenetic resolution at low taxonomic levels using massively parallel sequencing of chloroplast genomes. *BMC Biology*. 2009 Dec;7(1):1–7.

Patwardhan A, Ray S, Roy A. Molecular markers in phylogenetic studies: a review. *Journal of Phylogenetics & Evolutionary Biology*. 2014;2(2):131.

Pillon Y, Johansen J, Sakishima T, Chamala S, Barbazuk WB, Roalson EH, Price DK, Stacy EA. Potential use of low-copy nuclear genes in DNA barcoding: a comparison with plastid genes in two Hawaiian plant radiations. *BMC Evolutionary Biology*. 2013 Dec;13:1–1.

Puillandre N, Lambert A, Brouillet S, Achaz GJ. ABGD, Automatic Barcode Gap Discovery for primary species delimitation. *Molecular Ecology*. 2012 Apr;21(8):1864–77.

Purushothaman N, Newmaster SG, Ragupathy S, Stalin N, Suresh D, Arunraj DR, Gnanasekaran G, Vassou SL, Narasimhan D, Parani M. A tiered barcode authentication tool to differentiate medicinal Cassia species in India. *Genetics and Molecular Research.* 2014 Apr 16;13(2):2959–68.

Pusch CM, Giddings I, Scholz M. Repair of degraded duplex DNA from prehistoric samples using Escherichia coli DNA polymerase I and T4 DNA ligase. *Nucleic Acids Research.* 1998 Feb 1;26(3):3189–94.

Rubinoff D, Cameron S, Will K. Are plant DNA barcodes a search for the Holy Grail? *Trends in Ecology & Evolution.* 2006 Jan 1;21(1):1–2.

Sahare P, Srinivasu T. Barcoding for authentic identification of medicinal plants. *International Journal of Engineering Science (IJES).* 2012;1(2):33–6.

Särkinen T, Staats M, Richardson JE, Cowan RS, Bakker FT. How to open the treasure chest? *Optimising DNA Extraction from Herbarium Specimens.* 2012;1–9:e43808.

Sarwat M, Yamdagni MM. DNA barcoding, microarrays and next generation sequencing: recent tools for genetic diversity estimation and authentication of medicinal plants. *Critical Reviews in Biotechnology.* 2016 Mar 3;36(2):191–203.

Sass C, Little DP, Stevenson DW, Specht CD. DNA barcoding in the cycadales: testing the potential of proposed barcoding markers for species identification of cycads. *PLoS One.* 2007 Nov 7;2(11):e1154.

Selvaraj D, Shanmughanandhan D, Sarma RK, Joseph JC, Srinivasan RV, Ramalingam S. DNA barcode ITS effectively distinguishes the medicinal plant Boerhaviadiffusa from its adulterants. *Genomics, Proteomics & Bioinformatics.* 2012 Dec 1;10(6):364–7.

Senapati SK, Aparajita S, Rout GR. Identification of species-diagnostic inter simple sequence repeat markers for ten Phyllanthus species. *Zeitschrift für Naturforschung C.* 2011 Apr 1;66(3–4):167–72.

Shaw J, Lickey EB, Beck JT, Farmer SB, Liu W, Miller J, Siripun KC, Winder CT, Schilling EE, Small RL. The tortoise and the hare II: relative utility of 21 noncoding chloroplast DNA sequences for phylogenetic analysis. *American Journal of Botany.* 2005 Jan;92(1):142–66.

Shi LC, Zhang J, Han JP, Song JY, Yao H, Zhu YJ, Li JC, Wang ZZ, Xiao W, Lin YL, Xie CX. Testing the potential of proposed DNA barcodes for species identification of Zingiberaceae. *Journal of Systematics and Evolution.* 2011 May;49(3):261–6.

Song J, Yao H, Li Y, Li X, Lin Y, Liu C, Han J, Xie C, Chen S. Authentication of the family Polygonaceae in Chinese pharmacopoeia by DNA barcoding technique. *Journal of Ethnopharmacology.* 2009 Jul 30;124(3):434–9.

Starr JR, Naczi RF, Chouinard BN. Plant DNA barcodes and species resolution in sedges (Carex, Cyperaceae). *Molecular Ecology Resources.* 2009 May;9:151–63.

Stech M, Quandt D. 20,000 species and five key markers: the status of molecular bryophyte phylogenetics. *Phytotaxa.* 2010;9(1):196–228.

Štorchová H, Olson MS. The architecture of the chloroplast psb A-trn H non-coding region in angiosperms. *Plant Systematics and Evolution.* 2007 Oct;268:235–56.

Sucher NJ, Carles MC. Genome-based approaches to the authentication of medicinal plants. *Planta Medica.* 2008 May;74(06):603–23.

Taberlet P, Coissac E, Pompanon F, Gielly L, Miquel C, Valentini A, Vermat T, Corthier G, Brochmann C, Willerslev E. Power and limitations of the chloroplast trn L (UAA) intron for plant DNA barcoding. *Nucleic Acids Research.* 2007 Feb 1;35(3):e14.

Techen N, Parveen I, Pan Z, Khan IA. DNA barcoding of medicinal plant material for identification. *Current Opinion in Biotechnology.* 2014 Feb 1;25:103–10.

Theodoridis S, Stefanaki A, Tezcan M, Aki C, Kokkini S, Vlachonasios KE. DNA barcoding in native plants of the Labiatae (Lamiaceae) family from Chios Island (Greece) and the adjacent Çeşme-Karaburun Peninsula (Turkey). *Molecular Ecology Resources.* 2012 Jul;12(4):620–33.

Vinitha MR, Kumar US, Aishwarya K, Sabu M, Thomas G. Prospects for discriminating Zingiberaceae species in India using DNA barcodes. *Journal of Integrative Plant Biology.* 2014 Aug;56(8):760–73.

Vohra P, Khera KS. DNA barcoding: current advances and future prospects-a review. *Asian Journal of Biological and Life Sciences.* 2013;2(3):185–9.

Wang D-Y, Wang Q, Wang Y-L, Xiang X-G, Huang L-Q, Jin X-J. 2017. Evaluation of DNA barcodes in Codonopsis (Campanulaceae) and in some large angiosperm plant genera. *PLoS One.* 2017;12:e0170286.

Wang H, Kim MK, Kim YJ, Lee HN, Jin H, Chen J, Yang DC. Molecular authentication of the oriental medicines Pericarpium Citri Reticulatae and Citri Unshius Pericarpium using SNP markers. *Gene.* 2012a Feb 15;494(1):92–4.

Wang N, Jacques F, Milne RI, Zhang CQ, Yang JB. DNA barcoding of Nyssaceae (Cornales) and taxonomic issues. *Botanical Studies.* 2012b Apr 1;53(2):265–74.

Wang W, Wu Y, Yan Y, Ermakova M, Kerstetter R, Messing J. DNA barcoding of the Lemnaceae, a family of aquatic monocots. *BMC Plant Biology*. 2010 Dec;10:1–1.

Whitlock BA, Hale AM, Groff PA. Intraspecific inversions pose a challenge for the trnH-psbA plant DNA barcode. *PLoS One*. 2010 Jul 13;5(7):e11533.

Wicke S, Quandt D. Universal primers for the amplification of the plastid trnK/matK region in land plants. *Anales del Jardín Botánico de Madrid*. 2009;66(2):285–8).

Wittwer CT. High-resolution DNA melting analysis: advancements and limitations. *Human Mutation*. 2009 Jun;30(6):857–9.

Xiang XG, Zhang JB, Lu AM, Li RQ. Molecular identification of species in Juglandaceae: a tiered method. Journal of Systematics and Evolution. 2011 May;49(3):252–60.

Yamaguchi A, Kawamura H, Horiguchi T. A further phylogenetic study of the heterotrophic dinoflagellate genus, Protoperidinium (Dinophyceae) based on small and large subunit ribosomal RNA gene sequences. *Phycological Research*. 2006 Dec;54(4):317–29.

Yang HQ, Dong YR, Gu ZJ, Liang N, Yang JB. A Preliminary Assessment of matK, rbcL and trnH—psbA as DNA Barcodes for Calamus (Arecaceae) Species in China with a note on ITS. In *Annales Botanici Fennici,* 2012a Jan 1 (pp. 319–330). Finnish Zoological and Botanical Publishing Board.

Yang JY, Jang SY, Kim HK, Park SJ. Development of a molecular marker to discriminate Korean Rubus species medicinal plants based on the nuclear ribosomal DNA internal transcribed spacer and chloroplast trnL-F intergenic region sequences. *Journal of the Korean Society for Applied Biological Chemistry*. 2012b Apr;55:281–9.

Yang Y, Dang YY, Li Q, Lu JJ, Li XW, Wang YT. Complete chloroplast genome sequence of poisonous and medicinal plant *Datura stramonium*: organizations and implications for genetic engineering. *PLoS One*. 2014;9:e110656.

Yao H, Song J, Liu C, Luo K, Han J, Li Y, Pang X, Xu H, Zhu Y, Xiao P, Chen S. Use of ITS2 region as the universal DNA barcode for plants and animals. *PLoS One*. 2010 Oct 1;5(10):e13102.

Yao H, Song JY, Ma XY, Liu C, Li Y, Xu HX, Han JP, Duan LS, Chen SL. Identification of Dendrobium species by a candidate DNA barcode sequence: the chloroplast psbA-trnH intergenic region. *Planta Medica*. 2009 May;75(06):667–9.

Yu J, Wu XI, Liu C, Newmaster S, Ragupathy S, Kress WJ. Progress in the use of DNA barcodes in the identification and classification of medicinal plants. *Ecotoxicology and Environmental Safety*. 2021 Jan 15;208:111691.

Yuan YH, Hou BW, Xu HJ, Luo J, Ding XY. Identification of the geographic origin of Dendrobium thyrsiflorum on Chinese herbal medicine market using trinucleotide microsatellite markers. *Biological and Pharmaceutical Bulletin*. 2011 Dec 1;34(12):1794–800.

Zhang W, Yang S, Zhao H, Huang L. Using the ITS2 sequence-structure as a DNA mini-barcode: A case study in authenticating the traditional medicine "Fang Feng". *Biochemical Systematics and Ecology*. 2016 Dec 1;69:188–94.

Zhou J, Cui Y, Chen X, Li Y, Xu Z, Duan B, Li Y, Song J, Yao H. Complete chloroplast genomes of Papaver rhoeas and Papaver orientale: molecular structures, comparative analysis, and phylogenetic analysis. *Molecules*. 2018 Feb 16;23(2):437.

5 The Science of Omics
A Recent Approach for Screening and Enhancement of Bioactive Compounds in Medicinal Plants

Kawalpreet Kaur[1], *Varruchi Sharma*[1],
Bhavna Kabila[2], *Malkiat Chand Sidhu*[3]
[1]SGGS College Sector 26, Chandigarh
[2]K.M.V College, Jalandhar, Punjab
[3]Department of Botany, Panjab University Chandigarh

5.1 INTRODUCTION

The rising global human population has spurred the need for medicinal compounds to cultivate additional and alternative metabolites for medicinal preparation. According to the World Health Organization (WHO), around 80% of the world's population predominantly depends on plant-based remedies. These plant supplements have found their therapeutic value in different fields ranging from wellness to fitness and extending to neutraceuticals. The utilization of drugs such as antitumor (camptothecin and taxol), antimalarial (artemisinin and quinine), antigout (colchicine), analgesics (codeine and morphine), antidiabetic (allicin), and for brain functions (caffeine and nicotine) are some of the recognized drugs (Chakraborty, 2018). Omics can be classified as genomics, metabolomics, proteomics, and transcriptomics (Figure 5.1). Despite the increasing demand for these plant-based drugs, many important parameters like the genetic background, quality of the secondary metabolites, and growth parameters need further exploration. However, recent omics studies have brought some advances in this field of medicinal plants. These comprehensive genomic analyses have contributed to the increasingly intricate understanding of medicinal plants, offering valuable insights for various applications. They have facilitated the identification of molecular alterations related to secondary metabolites, biological pathways, and plant-based drug discoveries. Recent case studies on medicinal plant genomes have been summarized to exemplify the expansion of knowledge resulting from the paradigm shift toward omics-based approaches.

5.2 GENOMICS

Genomics deals with the understanding of a complete set of genes in an organism and their interactions with each other and with the environment. It can be divided into structural and functional genomic studies. The genotypic and phenotypic characterization of medicinal plants is a prerequisite to designing breeding experiments and enhancing the production of desired secondary metabolites. Experiments on DNA sequencing and barcoding have facilitated genome advancement and functional research (Shendure *et al.* 2017). Next-generation sequencing associated with massively parallel sequencing permits plant genomics to understand the DNA or RNA sequencing for genetic variations. These initial plant genome studies have described the role of hundreds of genes involved

DOI: 10.1201/9781032697550-5

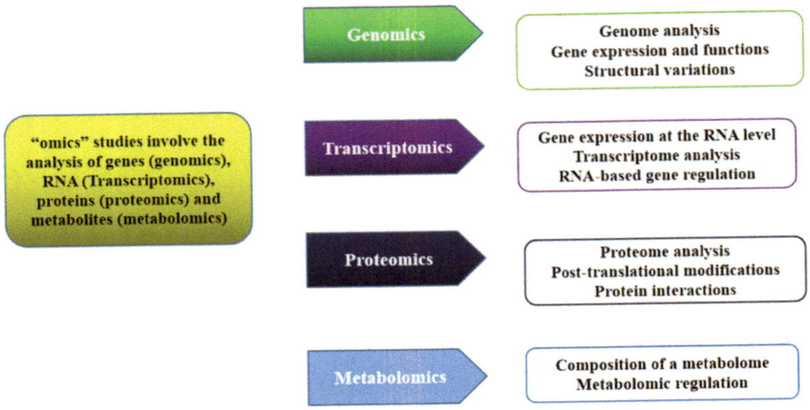

FIGURE 5.1 Omics classified as genomics, metabolomics, proteomics, and transcriptomics.

in marker-based breeding and the evolutionary history of plants. To meet the escalating demand for plant-based medicines in a rapidly growing population, it is imperative to introduce stress-tolerant varieties exhibiting high yields coupled with enhanced secondary metabolite contents. The genomics study of medicinal plants aims to elucidate and regulate the molecular mechanism for the production of secondary metabolites.

As genome sequencing technology has become more affordable (Cheng *et al.* 2021), significant strides have been made in unraveling the genetic code of various organisms. Notably, the genome of the medicinal plant (*Ricinus communis*) was first reported in 2010 (Chan *et al.* 2010) marking a crucial milestone in understanding its genetic makeup and potential applications. Since then, advancements in sequencing technologies have accelerated, paving the way for comprehensive analyses and insights into the genomes of diverse plant species. Further, Cheng *et al.* (2021) reported the genome deciphering of five medicinal plants with large-sized genomes: *Allium sativum*, *Aloe vera*, *Paeonia suffruticosa*, *Ginkgo biloba*, and *Taxus chinensis*. Concurrently, studies focusing on inter- and intragenomic comparisons have been conducted in genera like *Cannabis*, *Erythroxylum*, *Panax*, *Papaver*, and *Perilla* (Sun *et al.* 2022). Most of the genomes have been sequenced using Sanger technology, which provides long, high-quality ~1000 base pair reads. While Sanger sequencing demands a cloning step and is time-consuming with a relatively high cost, the final results are refined and novel (Michael and Jackson, 2013).

Azadirachta indica is being used in approximately 700 herbal preparations in India (Brahmachari, 2004). However, a few studies have attempted to generate genomic resources for the Neem tree (Rajakani *et al.* 2014; Krishnan *et al.* 2012). According to Kuravadi *et al.* (2015) 44,495 genes are predicted in the Neem genome, of which 32,278 were expressed in Neem tissues. While the biochemical constituents of Neem have been widely investigated, its genetic, molecular, and genomic resources are scarce (Dave *et al.* 2023).

The genus *Ocimum* possesses wide genetic diversity at intra- and interspecies levels (Upadhyay *et al.* 2015). All the species of *Ocimum* are medicinally important. Unraveling the whole genome sequencing of *O. sanctum* (Rastogi *et al.* 2015) provided an opportunity for the designing of new DNA markers. Further advancements in high-throughput sequencing and bioinformatics tools have charged ongoing research, enabling scientists to delve deeper into the intricate details of plant genomes and unlocking new dimensions in the study of plant biology.

5.3 METABOLOMICS

Medicinal plants synthesize a huge repertoire of secondary metabolites when they are confronted with complex abiotic and biotic stresses. These secondary metabolites, such as terpenes, phenolic compounds, and alkaloids, are classified based on their biosynthetic origin. Different classes of these compounds are associated with a narrow set of species within a phylogenetic group and constitute the bioactive compounds in several medicinal, aromatic, colorant, and spice plants and/or functional foods. A wide array of monoterpenoid indole alkaloids have been reported from the families Apocynaceae, Loganiaceae, and Rubiaceae (Facchini *et al.* 2012). Metabolic profiling technology (metabolomics), also called phytochemical genomics, has revolutionized analytical techniques. Metabolomics studies primarily involve targeted, widely targeted, untargeted, and spatial metabolome studies of secondary metabolites. It has empowered plant biologists to uncover the elusive biosynthetic pathways responsible for the production of bioactive compounds in plants. The pioneering work was carried out on a model plant (*Arabidopsis*) with known genome sequences and later it was extended to crops and other medicinal plants, where no genome sequences were readily available (Saito, 2013). Isotopic labeling and tracer studies are crucial to understanding the precursor compounds and their resulting metabolites. These approaches can play an important role in the pathway design of various substances. They have been essential in unraveling both primary and secondary metabolic routes.

Members of the genus *Ocimum* are among the most revered, having anti-inflammatory, antimicrobial, and cardioprotective properties. Metabolomics analyses have revealed the presence of approximately 100 non-volatile and 134 volatile compounds. Further, differential analysis has suggested compounds such as eugenol, γ-terpinene, germacrene D, and malic acid in these species. According to Du *et al.* (2023), the quality and flavor of metabolites can be improved through controlled hybridization of *Ocimum* species. The study by Kuravadi *et al.* (2015) has provided a quantity of azadirachtin, nimbin, and salanin (Neem metabolites), thus facilitating the advancement of fundamental research in Neem to understand the biochemical pathways.

Only the species of the genus *Papaver* have the potential to synthesize morphinan, an alkaloid. Metabolic engineering has increased its production both by overexpression and suppression of the gene encoding the morphinan pathway enzyme salutaridinol 7-O-acetyltransferase (SalAT) and subsequently resulted in the accumulation of this alkaloid (Allen *et al.* 2008).

Taxol (paclitaxel), an effective anticancer drug, is obtained from *Taxus brevifolia*. Attempts are in progress to make it cost-effective. Ajikumar *et al.* (2010) reported a metabolic pathway that has successfully increased the production of taxadiene. This is the first committed taxol intermediate, approximately 15,000-fold in an engineered *Escherichia coli* strain. The metabolic pathway has been partitioned into two modules: a native upstream methylerythritol–phosphate pathway forming isopentenyl pyrophosphate and a secondary heterologous downstream terpenoids-forming pathway.

Rauvolfia serpentina (family Apocynaceae) can synthesize an important monoterpenoid indole alkaloid, including ajmaline, which is effective in the treatment of cardiac arrhythmias (Brugada *et al.* 2003) and reserpine, an antihypertensive drug (Achor *et al.* 1955). Cázares-Flores *et al.* (2016) reported molecular cloning and functional expression of two *Rauvolfia serpentina* DNAs in *Escherichia coli* that are involved in indole and side-chain N-methylation of ajmaline. Recombinant proteins have shown remarkable substrate specificity for molecules with an ajmalan-type backbone and strict region-specific N-methylation. These N-methyltransferase gene transcripts and enzyme activity were enriched in *Rauvolfia serpentina* roots, resulting in the accumulation of ajmaline alkaloids. This study has elucidated the final step in the ajmaline biosynthetic pathway.

Etoposide, a topoisomerase (anti-cancerous), is obtained from the slow-growing *Podophyllum hexandrum*. Lau and Sattely (2015) identified different pathways for the production of etoposide aglycone in *Nicotiana benthamiana* (tobacco) and circumvented the need for the cultivation of *Podophyllum hexandrum*. Iridoids are a large family of distinctive bicyclic monoterpenes with anti-cancer, anti-inflammatory, antifungal, and antibacterial properties. To harness their biotechnological potential, Geu-Flores *et al.* (2012) reported iridoid synthase, a plant-derived enzyme capable of producing the iridoid ring scaffold, as evidenced by biochemical assays, co-expression analysis, and gene silencing studies.

The development of novel pathways can profoundly influence the detection of unexplored molecules of medicinal importance. Thus, the omics data contribute significantly to the discovery of new pathways and facilitate the identification of genes with novel proteins across various plant species. These investigations are poised to unveil a spectrum of biologically active secondary metabolites, thereby nurturing and conserving the rich biodiversity of plants for potential breakthroughs in drug discovery.

5.4 TRANSCRIPTOMICS

Transcriptomics is a highly sensitive and high-resolution technique that involves the study of RNA transcripts generated by an organism's genome. It encompasses investigations of bulk RNA sequencing (RNA-seq), single-cell RNA-seq, spatial RNA-seq, degradome, and non-coding RNAs (Zhou *et al.* 2022). This method can be classified into two categories: microarray studies, which rely on hybridization to detect the expression levels of known genes, and RNA sequencing, which utilizes sequencing methods to capture the expression information of all genes. The advent of next-generation sequencing (NGS) techniques has further intensified transcriptomic research (Mir *et al.* 2022). It is a novel technology used for DNA/RNA sequencing and to detect the variations occurring due to mutations. It also provides better sequencing results of genes or of the whole genome within a short time. The functional genes and regulatory mechanisms in medicinal plants can be fruitfully examined at the transcriptome level. It can subsequently be utilized to improve the hybridization and cultivation mechanisms to increase the production of secondary metabolites. RNA sequencing-based investigation provides more precise results to characterize the transcriptome compared to microarray. It further identifies the presence and abundance of RNA transcripts in an unbiased and high-throughput manner.

Transcriptome analysis can be effectively used to analyze unexplored medicinal plant species. The gene expression patterns obtained reflect alterations in the accumulation of secondary metabolites in medicinal plants. The same patterns are exhibited under different growth conditions, and at various ages and growth stages, even if they belong to the same species. Thus, transcriptome level analysis is more suitable for identifying genes and their expression related to medicinal compounds in plants (Wang *et al.* 2009). According to Rastogi *et al.* (2014), transcriptome analysis has revealed the presence of essential oil compositions in *Ocimum basilicum* and *O. sanctum*. The transcriptome database has suggested the genes encoding pathway enzymes, which have a significant association with non-volatile substances, including pentacyclic triterpenes and rosmarinic acid. This study has characterized genes associated with secondary metabolism, their control, and the breeding of chemotypes. According to Shah *et al.* (2020), *Lantana camara* transcriptome sequencing of leaves and roots has indicated the presence of 72,877 and 513,985 unigenes from the leaves and roots, respectively.

Through microarray analysis, the roles of specific messenger RNAs (mRNAs) and microRNAs (miRNAs) in stress conditions can be determined. These miRNAs serve as potent regulators of gene expression at both transcriptional and post-transcriptional levels. Prakash *et al.* (2016) investigated 15 conserved microRNAs in *Rauvolfia serpentina* for identification and characterization. Gene ontology annotation suggested that these targets include transcription factors involved in diverse metabolic processes. Some of these miRNAs were expected to target genes of pharmaceutically important secondary metabolic pathways, such as alkaloids and anthocyanin biosynthesis. These findings have contributed to a better understanding of miRNA-mediated gene regulatory processes in the production of secondary metabolites.

RNA-seq provides insight into the spatiotemporal expression of genes related to the production of therapeutically important secondary metabolites in medicinal plants. Wang *et al.* (2021) studied the candidate genes involved in the biosynthesis of stilbenes, flavonoids, and anthraquinones by analyzing the transcriptomes of *Polygonum cuspidatum.* The findings revealed that phenylpropanoid pathway genes were predominantly expressed in the root and stem, while the methylerythritol 4-phosphate and isochorismate pathways related to anthraquinone biosynthesis were predominantly expressed in the leaf. The expression patterns of synthase genes were in accordance with the metabolite profiles in various tissues of *P. cuspidatum.*

A third-generation single-molecule sequencing stage is nanopore sequencing. The application of long-read sequencing in *Moringa oliefera*, a well-known nutraceutical, has wider utilization in the pharma industry (Deepa *et al.* 2021). Li *et al.* (2017) reconstructed the full-length transcriptomes of *Astragalus membranaceus* using PacBio Iso-Seq long reads. They identified 27,975 and 22,343 full-length unique transcript models in the leaf and root tissues, respectively.

Single-cell RNA sequencing enables the examination of gene expression with a resolution specific to individual cells (Chen *et al.* 2019). Novel sets of genes related to the biosynthesis of monoterpene indole alkaloids were identified in *Catharanthus roseus*, which is a prominent source of potent anticancer drugs (Li *et al.* 2022).

Spatial transcriptome sequencing (spRNA-seq) at the cellular level presents significant opportunities in natural drug research. It enables the exploration of vast resources of medicinal plant tissue specimens, offering insights into the production mechanisms of secondary metabolites and the discovery of potential therapeutic targets. Liu *et al.* (2020) selected the reference genes from *Glycyrrhiza uralensis* and *G. inflat*e species on the basis of their spatiotemporal expression in several tissues and at different developmental stages. The results have indicated that the transcriptome is a rich reservoir for selecting stably expressed candidate reference genes.

Degradome and non-coding RNA sequencing provide information on RNA degradation. High-throughput sequencing was employed to explore microRNAs and the degradome in *Ginkgo biloba*, revealing 3314 microRNAs, including 1197 conserved and 2117 novel microRNAs (Ye *et al.* 2020). The genes involved in the production of different flavonoids in *Ginkgo* leaves were investigated by Wu *et al.* (2018).

5.5 PROTEOMICS

Plants utilize their metabolic systems to biosynthesize a wide variety of secondary metabolites, which are fundamentally not required for the maintenance of plant life processes. However, these are important as food additives, flavors, nutraceuticals, and in the pharmaceutical industry. The production, composition, and accumulation of these compounds depend on plant genetics and an array of environmental factors, especially stresses. The proteomics-based medicinal plant research with special reference to the pharmacological actions is applicable in a proteomic approach to indigenous plants with limited sequence data. Proteomics studies deal with the structure, functions, and interactions of protein molecules. Its application has proven to be integral in the advancements of medicinal plant biology, not only elucidating physiological changes at a systemic level but also uncovering crucial enzymes involved in the synthesis of valuable compounds. Utilizing proteomic methodologies at diverse levels, the active compounds in *Centella asiatica*, identified as saponins, underwent modifications through multiple glycosylation processes, resulting in the identification of numerous novel proteins (Gohil *et al.* 2010).

Cancer patients undergoing conventional chemotherapeutic treatments often face difficulties as these drugs tend to target non-specific cells and lead to immuno- or myelosuppression (decreased production of blood cells), mucositis (inflammation of the lining of the digestive tract), and alopecia (hair loss), hence affecting the quality of life quality of cancer patients (Cassidy *et al.* 2015; Kaur and Kaur (2021). Immunotherapy and nanotherapy are promising, although these are still in their early phases. Therefore, there has been an upsurge in the field of modern medicines to identify plant-based anticancer compounds.

Scutellarin, a drug obtained from *Scutellaria laterifolia*, increases the sensitivity of prostate cancer cells to cisplatin. It suppressed cell proliferation by promoting G2/M arrest and thereby inducing apoptosis. The anticolorectal cancer effects of scutellarin have been described through genomic and proteomic analysis (Xiong *et al.* 2020). Efferth (2017) reviewed the efficacy of Artemisinin, i.e., 1, 2,-trioxane from *Artemisia annua* (Asteraceae), as an anticancer drug. Its antitumor activity regulates oxidative stress by managing the formation of reactive oxygen species. It has also demonstrated its impact on apoptosis and anticancer activity. The microarray-based mRNA expression of various DNA damage response and repair genes in the NCI panel of tumor cell lines.

Proteomic analysis of mung seed (*Vigna radiata*) coat extract has suggested the abundance of 190 proteins changed in lipopolysaccharide-stimulated RAW264.7 cells. The immunomodulatory effect has been attributed to the antioxidant properties of flavonoids (Hashiguchi *et al.* 2017). Zhu *et al.* (2012) reported that mung bean seed coat is protective against lethal sepsis by stimulating autophagic HMGB1 degradation through nucleosomal protein, which is a mediator of lethal systemic inflammation.

Cannabinoids found in *Cannabis sativa* demonstrate efficacy in addressing chronic pain. Raharjo *et al.* (2004) suggested that the genes responsible for the biosynthesis of cannabinoids overlapped with the primary metabolic proteins. According to Jenkins and Orsburn (2020), *Cannabis* contains numerous post-translationally modified proteins that are involved in the production of *Cannabis* molecules of commercial and medical interest. Ceccacci *et al.* (2024) proposed that the anticancer activity of cannabidiol is related to an alteration of spliceosome functionality mediated by the interaction of natural products with the EFTUD2 factor.

Proteomic studies also play a significant role in understanding diabetes, a chronic metabolic disorder that causes postprandial hyperglycemia due to a deficiency of insulin or its malfunction. The link between proteomics and diabetes lies in unraveling the intricate molecular mechanisms and identifying potential therapeutic targets through the study of protein expression patterns and modifications. Aryal *et al.* (2021) demonstrated that flavonoids and polyphenols present in the bark of *Acacia catechu* substantially impede carbohydrate hydrolases. The study also revealed that the ethanol extract exhibited the most pronounced inhibition of α-amylase and α-glucosidase. In a separate investigation, Ahmad *et al.* (2021) highlighted the remarkable antioxidant capabilities of *Aegle marmelos*. This study found significant inhibition of the enzymes α-amylase and α-glucosidase in relation to the observed antioxidant potential.

5.6 CONCLUSION

The study explores the use of omics technology in plant-based medicines to enhance the production and quality of bioactive compounds. Omics provides a potential solution to the complexity of drug development by analyzing genes, transcripts, and proteins in detail. Understanding the genetic blueprints and their intricate pathways can lead to novel interventions and an improved understanding of biological processes. The study highlights the importance of stress-tolerant varieties with high yields and enhanced secondary metabolite contents in meeting the growing demand for plant-based medicinal preparations. Advancements in genome sequencing technology have led to significant advancements in understanding the genetic code of various organisms. Comprehensive genomic analyses contribute to an increasingly intricate understanding of medicinal plants, offering valuable insights for various applications. Metabolomics studies, also known as phytochemical genomics, have revolutionized analytical techniques, enabling plant biologists to uncover the elusive biosynthetic pathways responsible for the production of bioactive compounds in plants. Despite strides in omics technologies, further research is required to instigate significant advancements in the realms of medicinal plants as well as in the production and manipulation of secondary metabolites. In addition, the development of high-quality and cost-effective sequencing of complete individual genomes and transcriptomes can also accelerate omics studies.

REFERENCES

Ahmad, W., Amir, M., Ahmad, A., Ali, A., Ali, A., Wahab, S., *et al.* (2021). *Aegle marmelos* leaf extract phytochemical analysis, cytotoxicity, in-vitro antioxidant and antidiabetic activities. *Plants*, 10(12): 1–13.

Ajikumar, P. K., Xiao, W. H., Tyo, K. E., Wang, Y., Simeon, F., Leonard, E., *et al.* (2010). Isoprenoid pathway optimization for taxol precursor overproduction in *Escherichia coli*. *Science*, 330(6000): 70–74.

Allen, R. S., Miller, J. A., Chitty, J. A., Fist, A. J., Gerlach, W. L., and Larkin, P. J. (2008). Metabolic engineering of morphinan alkaloids by over-expression and RNAi suppression of salutaridinol 7-O-acetyltransferase in opium poppy. *Plant Biotechnology Journal*, 6(1): 22–30.

Aryal, B., Adhikari, B., Aryal, N., Bhattarai, B. R., Khadayat, K., and Parajuli, N. (2021). LC-HRMS profiling and antidiabetic, antioxidant, and antibacterial activities of *Acacia catechu* (Lf) Willd. *BioMed Research International*, 2021: 1–16.

Brahmachari, G. (2004). Neem—an omnipotent plant: a retrospection. *Chembiochem,* 2004 Apr 2; 5(4): 408–21.

Brugada, J., Brugada, P., and Brugada, R. (2003). The ajmaline challenge in Brugada syndrome. *European Heart Journal*, 24: 1085–1086.

Cassidy, J., Bissett, D., OBE, R. A., Payne, M., and Morris-Stiff, G., editors. (2015). Oxford Handbook of Oncology. Oxford: Oxford University Press.

Cazares-Flores, P., Levac, D., and De Luca, V. (2016). *Rauvolfia serpentina* N-methyltransferases involved in ajmaline and Nβ-methylajmaline biosynthesis belong to a gene family derived from γ-tocopherol C-methyltransferase. *The Plant Journal*, 87(4): 335–342.

Ceccacci, S., Corsi, L., Spinelli, L., Caroli, C., Marani, M., Anceschi, L., *et al.* (2024). A label free chemoproteomic-based platform to disclose cannabidiol molecular mechanism of action on chronic myelogenous leukemia cancer cells. *Heliyon*, 10(01): 1–8.

Chakraborty, P. (2018). Herbal genomics as tools for dissecting new metabolic pathways of unexplored medicinal plants and drug discovery. *Biochimie Open*, 6: 9–16.

Chan, A. P., Crabtree, J., Zhao, Q., Lorenzi, H., Orvis, J., Puiu, D., *et al.* (2010). Draft genome sequence of the ricin-producing oilseed castor bean. *Nature Biotechnology*, 28(9): 951–956.

Chen, G., Ning, B., and Shi, T. (2019). Single-cell RNA-seq technologies and related computational data analysis. *Frontiers in Genetics*, 10: 1–13.

Cheng, Q. Q., Ouyang, Y., Tang, Z. Y., Lao, C. C., Zhang, Y. Y., Cheng, C. S., and Zhou, H. (2021). Review on the development and applications of medicinal plant genomes. *Frontiers in Plant Science*, 12: 1–22.

Dave, N., Iqbal, A., Patel, M., Kant, T., Yadav, V. K., Sahoo, D. K., and Patel, A. (2023). Deciphering the key pathway for triterpenoid biosynthesis in *Azadirachta indica* A. Juss.: a comprehensive review of omics studies in nature's pharmacy. *Frontiers in Plant Science*, 14: 1–11.

De Costa, F., Barber, C. J. S., Kim, Y. B., Reed, D. W., Zhang, H., Fett-Neto, A. G., *et al.* (2017). Molecular cloning of an ester-forming triterpenoid: UDP-glucose 28-O-glucosyltransferase involved in saponin biosynthesis from the medicinal plant *Centella asiatica*. *Plant Science*, 262: 9–17.

Deepa, J., Kambale, R., Bharathi, A., Williams, M., Boopathi, N. M. and Raveendran, M. (2021). *Moringa Functional Genomics: Implications of Long Read Sequencing Technologies.* In: Boopathi, N.M., Raveendran, M., Kole, C. (eds) The Moringa Genome. Compendium of Plant Genomes, pp. 133–143.

Du, P., Yuan, H., Chen, Y., Zhou, H., Zhang, Y., Huang, M., *et al.* (2023). Identification of key aromatic compounds in Basil (*Ocimum* L.) using sensory evaluation, metabolomics and volatilomics analysis. *Metabolites*, 13(1): 1–12.

Efferth, T. (2017). From ancient herb to modern drug: *Artemisia annua* and artemisinin for cancer therapy. *Seminars Cancer Biology*, 46: 65–83.

Facchini, P. J., Bohlmann, J., Covello, P. S., De Luca, V., Mahadevan, R., Page, J. E., *et al.* (2012). Synthetic biosystems for the production of high-value plant metabolites. *Trends in Biotechnology*, 30(3): 127–131.

Geu-Flores, F., Sherden, N. H., Courdavault, V., Burlat, V., Glenn, W. S., Wu, C., *et al.* (2012). An alternative route to cyclic terpenes by reductive cyclization in iridoid biosynthesis. *Nature*, 492(7427): 138–142.

Gohil, K. J., Patel, J. A., and Gajjar, A. K. (2010). Pharmacological review on *Centella asiatica*: A potential herbal cure all. *Indian Journal of Pharmaceutical Sciences*, 72(5): 546–556.

Guo, C., Wang, Y., Yang, A., He, J., Xiao, C., Lv, S., *et al.* (2020). The *Coix* genome provides insights into Panicoideae evolution and papery hull domestication. *Molecular Plant*, 13: 309–320.

Jenkins, C., and Orsburn, B. (2020). The cannabis proteome draft map project. *International Journal of Molecular Science*, 21(3): 1–16.

Kaur, K., and Kaur, S. (2021). Spirulina-A wonder nutraceutical against cancer: a review. *Plant Archives*, 21(1): 1333–1338.

Krishnan, N. M., Pattnaik, S., Jain, P., Gaur, P., Choudhary, R., Vaidyanathan, S., *et al.* (2012). A draft of the genome and four transcriptomes of a medicinal and pesticidal angiosperm *Azadirachta indica. BMC Genomics,* 13: 1–13.

Kuravadi, N. A., Yenagi, V., Rangiah, K., Mahesh, H. B., Rajamani, A., Shirke, M. D., *et al.* (2015). Comprehensive analyses of genomes, transcriptomes and metabolites of neem tree. *PeerJ*, 3: 1–25.

Lau, W., and Sattely, E. S. (2015). Six enzymes from mayapple that complete the biosynthetic pathway to the etoposide aglycone. *Science*, 349(6253): 1224–1228.

Li, C., Wood, J. C., Vu, A. H., Hamilton, J. P., Lopez, C. E. R., Payne, R. M., *et al.* (2022). Single-cell multi-omics enabled discovery of alkaloid biosynthetic pathway genes in the medical plant *Catharanthus roseus. bioRxiv*, 1–40.

Li, J., Harata-Lee, Y., Denton, M. D., Feng, Q., Rathjen, J. R., Qu, Z., and Adelson, D. L. (2017). Long read reference genome-free reconstruction of a full-length transcriptome from *Astragalus membranaceus* reveals transcript variants involved in bioactive compound biosynthesis. *Cell Discovery*, 3(1): 1–13.

Liu, Y., Enninful, A., Deng, Y. and Fan, R. (2020). Spatial transcriptome sequencing of FFPE tissues at the cellular level. *bioRxiv*, 1–38.

Michael, T. P., and Jackson, S. (2013). The first 50 plant genomes. *Plant Genome*, 6(2): 1–7.

Mir, R. R., Rustgi, S., Zhang, Y. M., and Xu, C. (2022). Multi-faceted approaches for breeding nutrient-dense, disease-resistant and climate-resilient crop varieties for food and nutritional security. *Heredity*, 128(6): 387–390.

Prakash, P., Rajakani, R., and Gupta, V. (2016). Transcriptome-wide identification of *Rauvolfia serpentina* microRNAs and prediction of their potential targets. *Computational Biology and Chemistry*, 61: 62–74.

Raharjo, T. J., Widjaja, I., Roytrakul, S., and Verpoorte, R. (2004). Comparative proteomics of *Cannabis sativa* plant tissues. *Journal of Biomolecular Techniques*, 15(2): 97–106.

Rajakani, R., Narnoliya, L., Sangwan N. S., Sangwan R. S., and Gupta V. (2014). Subtractive transcriptomes of fruit and leaf reveal differential representation of transcripts in *Azadirachta indica. Tree Genetics & Genomes*, 10: 1331–1351.

Rastogi, S., Kalra, A., Gupta, V., Khan, F., Lal, R. K., Tripathi, A. K., *et al.* (2015). Unravelling the genome of Holy basil: an "incomparable" "elixir of life" of traditional Indian medicine. *BMC Genomics*, 16(1): 1–15.

Rastogi, S., Meena, S., Bhattacharya, A., Ghosh, S., Shukla, R. K., Sangwan, N. S., *et al.* (2014). De novo sequencing and comparative analysis of holy and sweet basil transcriptomes. *BMC Genomics*, 15(1): 1–18.

Saito, K. (2013). Phytochemical genomics: a new trend. *Current Opinion in Plant Biology*, 16(3): 373–380.

Shah, M., Alharby, H. F., Hakeem, K. R., Ali, N., Rahman, I. U., Munawar, M., *et al.* (2020). De novo transcriptome analysis of *Lantana camara* L. revealed candidate genes involved in phenylpropanoid biosynthesis pathway. *Scientific Reports*, 10(1): 1–14.

Sun, W., Xu, Z., Song, C., and Chen, S. (2022). Herbgenomics: decipher molecular genetics of medicinal plants. *The Innovation*, 3(6): 1–2.

Upadhyay, A. K., Chacko, A. R., Gandhimathi, A., Ghosh, P., Harini, K., Joseph, A. P., *et al.* (2015). Genome sequencing of herb Tulsi (Ocimumtenuiflorum) unravels key genes behind its strong medicinal properties. *BMC Plant Biology*, 2015 Dec; 15: 1–20.

Wang, X., Hu, H., Wu, Z., Fan, H., Wang, G., Chai, T., and Wang, H. (2021). Tissue-specific transcriptome analyses reveal candidate genes for stilbene, flavonoid and anthraquinone biosynthesis in the medicinal plant *Polygonum cuspidatum. BMC Genomics*, 22(1): 1–17.

Wang, Z., Gerstein, M., and Snyder, M. (2009). RNA-seq: a revolutionary tool for transcriptomics. *Nature Reviews Genetics*, 10: 57–63.

Wu, Y., Guo, J., Zhou, Q., Xin, Y., Wang, G., and Xu, L. A. (2018). De novo transcriptome analysis revealed genes involved in flavonoid biosynthesis, transport and regulation in *Ginkgo biloba. Industrial Crops and Products*, 124: 226-235.

Xiong, L. L., Du, R. L., Xue, L. L., Jiang, Y., Huang, J., Chen, L., *et al.* (2020). Anti-colorectal cancer effects of scutellarin revealed by genomic and proteomic analysis. *Chinese Medicine*, 15(1): 1–15.

Ye, J., Zhang, X., Tan, J., Xu, F., Cheng, S., Chen, Z., *et al.* (2020). Global identification of *Ginkgo biloba* microRNAs and insight into their role in metabolism regulatory network of terpene trilactones by high-throughput sequencing and degradome analysis. *Industrial Crops and Products*, 148: 112289.

Zhou, R., Jiang, F., Niu, L., Song, X., Yu, L., Yang Y., *et al.* (2022). Increase crop resilience to heat stress using omic strategies. *Frontiers in Plant Science*, 13: 1–10.

Zhu, S., Li, J., Li, W., Ashok, M., Jundoria, A., Zhang, Y., *et al.* (2012). Mung Bean (Vigna radiata) extract is protective against lethal sepsis in mice. *Shock,* 2012 Jan 1; 37: 25–6.

6 Application of Biotechnological Methods for the Sustainable Use of Plants

Rakesh Kumar, Vikas Sharma[1], Sudesh Kumar[2]

[1]Department of Life Sciences, Sant Baba Bhag Singh University Jalandhar, Punjab, India

[2]Department of Zoology, Govt. Degree College Kilhotran, Doda, J&K- India

6.1 INTRODUCTION

Plant biodiversity provides a number of natural products for food and medicine (Tomlinson et al. 2015). According to the World Health Organization (WHO), about 80% of people living in developing nations depend on plants. Plants offer many fundamental raw materials and fresh genetic information that are beneficial for breeding initiatives, the development of more resilient plants and crops, and increased crop productivity. New therapeutic compounds have been introduced primarily through the use of medicinal plants. Research on medicinal plants has been conducted all over the world to produce novel herbal remedies. The reduction of biodiversity can be attributed to a number of factors, such as habitat degradation, invasive species introduction, weed infestation, climate change, and human pressure, all of which increase the number of threatened species (Jhariya et al. 2022). Medicinal plants can be of such importance to humans that they can lay the critical groundwork for the preservation of natural ecosystems and ecological services (Astutik et al. 2019). Since the beginning of time, substances derived from plants have been used to treat a wide range of human ailments. As a result, the need to produce secondary metabolites originating from plants is growing (Elshafie et al. 2023). But a number of species of medicinal plants are now threatened because we are not protecting natural resources with the attention that they need (Volenzo and Odiyo 2020). Alternative techniques are therefore required to preserve key plant species for medicine. Because biotechnology has so many benefits in the forestry, pharmaceutical, food, and agricultural sectors, it has become a focal point of interest (Alamgir et al. 2017).

Plant diversity is being characterized by the growing use of modern biotechnology technologies, which is undoubtedly beneficial to plant conservation efforts (Hamilton 2013). However, biotechnological approaches are only as effective as their capacity to be precisely targeted and used in conjunction with other technologies as enabling and complementary ones. Potential "conservation" users may find it difficult to assess the advantages and relevance of emerging techniques and practices within their own specialized field due to the rapid advancement of biotechnology (Adenle et al. 2012). It is imperative to acknowledge that the successful incorporation of biotechnology in conservation programs requires multidisciplinary collaboration (Barrett et al. 2011). Biotechnological techniques like organogenesis, somatic embryogenesis, protoplast culture, micrografting, micropropagation, and cryopreservation have emerged recently and have not only offered alternatives but also sustainable approaches toward the conservation and regeneration of medicinally important plant species.

DOI: 10.1201/9781032697550-6

6.2 BIOTECHNOLOGICAL METHODS

There are various techniques in biotechnology that are useful for the conservation and regeneration of plants:

- Micrografting
- Micropropagation
- Organogenesis
- Somatic embryogenesis
- Protoplast culture
- Cryopreservation

6.2.1 Micrografting

Micrografting is the process in which the mature scion shoot tips are attached to a rootstock. This method is performed both in vivo and in vitro. This technique was created by Jonard in 1986. Micrografting is useful in order to produce disease-free scions, invigorate mature shoot materials, increase the possibility of cloning mature plants true to type, and investigate graft unions (Onay et al. 2003).

Fruit trees have been improved and multiplied by the use of in vitro micrografting, a process that combines the benefits of quick in vitro multiplication with higher production. The percentage of effective micrografts has been increased using specialized methods that include silicon tubes, growth regulators, antioxidants, etiolation therapies, and greater sucrose levels. Commercial applications include the generation of virus-free, disease-free crops, especially those opposing soil-borne infections, multiply hard-to-root plants, study histology, forecast incompatibility between grafting partners, and index diseases (Hussain et al. 2014). As per Singh et al. (2008), the grafting method was utilized to regenerate the Kinnow mandarin in vitro without any virus, while *Argania spinosa* was regenerated by Koufan et al. (2020). Freedom as the rootstock and Flame Seedless scion have been used for quick in vitro grapevine plantlet propagation (El-Zaher et al. 2022).

6.2.2 Micropropagation

A method called "plant micropropagation" is employed in vitro to preserve plant species and even hard-to-germinate economically significant plant seeds. To create plantlets, various plant elements, including nodes, leaves, seeds, anthers, tubers, and ovaries, were employed in vitro. It might assist in producing virus-free plants, which are hard to propagate with traditional techniques. It is the first commercial biotechnology-based method for propagating clonal plants (Rai 2010). For micropropagation to be successful, the explants used are crucial since they determine whether the produced plantlets will eventually develop calluses.

According to Juliani et al. (2011), the plant species with unique physical characters shows direct insertion of regenerated plantlets into their native habitat. It is helpful for the development of many plant species in a constrained amount of time and space. The creation of pathogen-free cloned plants for use in forestry and agriculture is one of its uses. Regenerated plantlets should, from a conservation perspective, have as few somatic changes as possible through the use of micropropagation. *Lavandula pedunculata* was micropropagated without harming natural resources in order to produce essential oils (Zuzarte et al. 2010).

Techniques have been developed for in vitro propagation of endangered medicinal plants (Shahzad and Saeed 2013). Gailīte et al. (2010) and Verma et al. (2012) conducted research using seeds of *Saussurea esthonica*. Plant specimens for which experiments on micropropagation have been conducted include *Prunus africana* (Komakech et al. 2020), *Rungia pectinata* (Shekhawat et al. 2016), *Saussurea costus* (Mehalaine and Chenchouni 2021), *Panax vietnamensis* var. *langbianensis* (Anh et al. 2022), *Decalepis arayalpathra* (Srivastava et al. 2022), and *Campomanesia xanthocarpa* (Machado et al. 2020).

6.2.3 Protoplast Culture

Protoplast culture is a process in which whole plants are developed from the culture of cells without a cell wall. The fusion takes place either on its own or in response to an agent that induces fusion (Sajja et al. 2017). Certain beneficial genes can be passed from one species to another, including those that fight disease; fix nitrogen; develop quickly and produce more products; and are frost-hardy, drought-resistant, or herbicide-resistant. Thus, to develop strains with desirable features, a biotechnological approach such as this has been employed to combine genes from diverse organisms. This technique involves the removal of plant cell walls through the use of mechanical or enzymatic techniques. After that, they undergo purification, the wall undergoes another artificial regeneration, and it is moved to appropriate media where it will eventually mature into a fully grown plant. The process is mostly employed to create genetically modified or hybrid plants. Two protoplasts from two distinct species are typically fused together and cultivated in an appropriate environment during the procedure (Table 6.1).

6.2.4 Somatic Embryogenesis

A group of somatic cells or tissues are combined to generate somatic embryos through a process known as somatic embryogenesis. According to Gambino et al. (2007) and Prado et al. (2010), somatic embryos are created from explants such as stigmas, ovaries, anthers, and whole flowers. It is helpful to apply both internal and exterior regulatory components to control somatic embryogenesis in plants (Mujib et al. 2022).

6.2.4.1 Types of Somatic Embryogenesis (Table 6.2)

6.2.4.1.1 Direct Somatic Embryogenesis

Direct somatic embryogenesis is a process in which embryos are formed directly from isolated cells, without the production of callus tissue. This process is rarely used.

TABLE 6.1
Plants in Which Protoplast Culture Was Studied

Plant	Reference
Centella asiatica	Aziz et al. 2006
Phellodendron amurense	Azad et al. 2006
Nigella damascene L.	Klimek-Chodacka et al. 2020
Physalis angulata	Mastuti and Rosyidah 2018

TABLE 6.2
Plants in Which Somatic Embryogenesis Was Studied

Plant	Reference
Santalum album L.	Faisal et al. 2024
Polygonatum cyrtonema	Yang et al. 2024
Centaurium erythraea	Simonović et al. 2020
Kelussia odorotissima	Ebrahimi et al. 2018

6.2.4.1.2 Indirect Somatic Embryogenesis

Indirect somatic embryogenesis is characterized by the formation of a callus from the explant, and then embryo are produced from the callus tissue (Ramírez-Mosqueda, 2022).

The first step is the induction of proembryonic masses, which is often promoted by first exposing the tissue to growth regulators of the auxin type, such as 2,4-D. Subsequently, the proembryonic cultures are shifted to a medium that has been altered such that the somatic embryos fully differentiate. This medium contains less auxin, either eliminated or at a lower concentration than before. Cultivating somatic embryogenesis in an agitated liquid medium is frequently advantageous. There is great conservation in the patterns of differentiation. Initially, "globular" embryonic structures develop. This is followed by the "torpedo" stage, when elongation of the shoot, and especially the root, occurs and vascular connections between the two are established. This stage follows the "heart" stage, in which the shoot and root meristems can be readily separated at the two poles. After some time, this coordinated growth yields a well-organized somatic embryo or plantlet (Loberant and Altman 2010).

6.2.4.2 Advantages of Somatic Embryogenesis

Somatic embryogenesis is a method of plant multiplication that yields a high number of plants regardless of the season and has various advantages over conventional techniques. Elite clones can only be produced by somatic embryogenesis and cannot be produced by zygotic embryos. The primary benefit of Somatic Embryogenesis is that it can produce several plantlets from a single cell, giving rise to the possibility of screening and assessing them. Maximizing plant output can be achieved with minimum labor costs, also with low labor inputs. The kinds of explants, growth regulator combinations and concentrations, and culture media employed at various stages all affect the regeneration of plantlets. This approach can be used to create haploid plantlets, which may be useful for studying mutations.

6.2.5 Organogenesis

Organogenesis is the process of creating new organs from cultured tissues, which include roots, buds, and shoots. The high ratio of cytokinin-to-auxin leads to the induction of shoot organogenesis, whereas a high ratio of auxin-to-cytokinin induces the development of roots (Pernisová et al. 2009). De novo organogenesis requires an equilibrium number of auxin and cytokinin in the media (Skoog and Miller 1957) (Table 6.3).

Organogenesis is divided into three stages: the first is the response of explants to phytohormones; the second is the re-entry of dormant cells into the cell cycle and the determination of their cell fate

TABLE 6.3
Plants in Which Organogenesis Was Done

Plant	Reference
Thymus species (*Th. vulgaris* and *Th. serpyllum*)	Cherednichenko et al. 2024
Swertia minor	Kshirsagar et al. 2021
Ocimum tenuiflorum L.	Aggarwal et al. 2020
Brucea mollis Wall. ex Kurz	Das et al. 2018
Lilium pumilum DC. Fisch.	Zhang et al. 2016
Salvia miltiorrhiza Bunge	Tsai et al. 2016
Lepidium sativum	Pande et al. 2002
Zingiber officinale	Rout and Das 1997
Hyoscyamus muticus	Basu and Chand 1996

to form particular organ primordia; and the third and final phase is morphogenesis. Organogenesis, then, is the control of cell proliferation, cell division, differentiation of tissues and cell types, and overall organ patterning (Khan et al. 2012).

During cultivation, these two processes may occur simultaneously, although more often than not, either shoots or, less frequently, roots form first. Organogenesis begins with a unique arrangement of a few meristematic cells from the callus or directly within the explants, which subsequently develop into a shoot or root meristem. The creation of a shoot or root meristem depends on the kind of explant used, the makeup of the culture media, and the cultural environment. After meristems for shoots or roots are arranged, they start to grow, producing tiny roots and shoots. Plantlets are ultimately produced in vitro through the establishment of functional vascular connections between the developing shoots and roots during subsequent stages of organization, differentiation, and growth. After going through this series of steps and procedures, fresh plantlets are taken out of the culture vessels, given an ex vitro acclimatization period, and then grown in a traditional greenhouse or in the field (Loberant and Altman 2010).

6.2.5.1 Advantages

An efficient method for inducing exuberant shoots in plants in vitro is called organogenesis. An organ generated by meristem development is less likely to have mutations. An effective and efficient method for inducing high-fidelity clonal replication is direct organogenesis.

6.2.5.2 Disadvantages

The induction of somaclonal variation in plants is a significant topic regarding in vitro organogenesis. A restricted number of plant species can utilize the in vitro organogenesis process, particularly those that are resistant to it. It takes a lot of time, and you need some practical experience.

6.2.6 Cryopreservation

For long-term conservation in liquid nitrogen (–196°C), the cryopreservation approach is helpful. According to Jain et al. (2012), it also entails freezing materials at extremely low temperatures. Embryos, seeds, anthers, meristems, and even protoplasts have all been successfully preserved via cryopreservation (Tripathi and Tripathi 2003). It doesn't affect the preserved materials' viability, genetic composition, or ability to survive (Ogbu 2010). Before extremely quickly freezing the tissues, the cryopreservation method depends on physically or osmotically drying the tissues to eliminate any water that could freeze (Kaviani 2011). Clonally propagated plant species yield refractory seeds that are difficult to preserve using traditional techniques (Benson 2004). It is beneficial for certain plant varieties (Reed 2008). Genetic or physiological factors affect the outcomes of cryopreservation; so, a deeper comprehension of the predisposing behaviors of these factors will aid in the development of cryopreservation methods (Johnston et al. 2009). A study of molecular markers shows that cryopreservation preserves genetic stability (Liu et al. 2008).

6.3 CONCLUSION

Biotechnology is crucial to the conservation and enhancement of plant species, especially those with substantial commercial and medicinal significance. By gathering, processing, and trading medicinal plants, the industry has the potential to increase locals' incomes. The impoverished may be able to find work opportunities through the development of small and microenterprises, particularly those focused on the processing and commercialization of medicinal herbs. There are several methods that are useful in the cultivation and preservation of therapeutic plants. Since many species are misused in the pharmaceutical industry, they also need to be protected.

Numerous types of therapeutic plants are facing extinction due to various reasons, making them rare, susceptible, or endangered. Hence, biotechnological techniques such as micropropagation, cryopreservation, somatic embryogenesis, organogenesis, protoplast culture, and micrografting should be used effectively for the regeneration of medicinal plants in order to promote sustainable growth and development of the industry, people, and planet.

REFERENCES

Adenle, A. A., Sowe, S. K., Parayil, G., & Aginam, O. (2012). Analysis of open source biotechnology in developing countries: An emerging framework for sustainable agriculture. *Technology in Society*, *34*(3), 256–269.

Aggarwal, D., Neeti, N., Reddy, M. S., & Kumar, A. (2020). Shoot organogenesis and assessment of clonal fidelity of regenerated plants of *Ocimum tenuiflorum* L.: queen of herbs. *Vegetos*, *33*, 420–429.

Alamgir, A. N. M., & Alamgir, A. N. M. (2017). Cultivation of herbal drugs, biotechnology, and in vitro production of secondary metabolites, high-value medicinal plants, herbal wealth, and herbal trade. *Therapeutic Use of Medicinal Plants and Their Extracts: Volume 1: Pharmacognosy*, *1*, 379–452.

Anh, T. T. L., Tung, H. T., Khai, H. D., Mai, N. T. N., Luan, V. Q., Cuong, D. M., ... & Nhut, D. T. (2022). Micropropagation of Lang Bian ginseng: an endemic medicinal plant. *Plant Cell, Tissue and Organ Culture (PCTOC)*, *151*(3), 565–578.

Astutik, S., Pretzsch, J., & Ndzifon Kimengsi, J. (2019). Asian medicinal plants' production and utilization potentials: A review. *Sustainability*, *11*(19), 5483.

Azad, M. A. K., Yokota, S., Ishiguri, F., & Yoshizawa, N. (2006). Plant regeneration from mesophyll protoplasts of a medicinal plant, Phellodendron amurense rupr. *In Vitro Cellular & Developmental Biology–Plant*, *42*, 502–507.

Aziz, Z. A., Davey, M. R., Lowe, K. C., & Power, J. B. (2006). Isolation and culture of protoplasts from the medicinal plant *Centella asiatica*. *Revista Brasileira de Plantas Medicinais*, *8*, 105–109.

Barrett, M. A., Bouley, T. A., Stoertz, A. H., & Stoertz, R. W. (2011). Integrating a One Health approach in education to address global health and sustainability challenges. *Frontiers in Ecology and the Environment*, *9*(4), 239–245.

Basu, P., & Chand, S. (1996). Regeneration of plantlets from root-derived callus of Egyptian Henbane. *Cell Chromosome Res*, *19*, 31–34.

Benson, E. E. (2004). Cryoconserving algal and plant diversity: historical perspectives and future challenges. In *Life in the Frozen State* (pp. 325–354). CRC Press.

Cherednichenko, M. Y., Drevova, A. N., Husainova, A. R., Polivanova, O. B., & Khlebnikova, D. A. (2024). Callusogenesis and ōrganogenesis of thymus cultivars in vitro. In *E3S Web of Conferences* (Vol. 494, p. 04011). EDP Sciences.

Das, P., Tanti, B., & Borthakur, S. K. (2018). In vitro callus induction and indirect organogenesis of *Brucea mollis* Wall. ex Kurz–A potential medicinal plant of Northeast India. *South African Journal of Botany*, *119*, 203–211.

Ebrahimi, M., Mokhtari, A., & Amirian, R. (2018). A highly efficient method for somatic embryogenesis of *Kelussia odorotissima* Mozaff., an endangered medicinal plant. *Plant Cell, Tissue and Organ Culture (PCTOC)*, *132*, 99–110.

Elshafie, H. S., Camele, I., & Mohamed, A. A. (2023). A Comprehensive review on the biological, agricultural and pharmaceutical properties of secondary metabolites based-plant origin. *International Journal of Molecular Sciences*, *24*(4), 3266.

El-Zaher, A., Sahar, M. H., Elwahab, M. A., El-Gany, A., El-Sattar, A., & Fathy, H. A. (2022). Differences in morphological, biochemical, and genetic characteristics of grape explants grafted using various micrografting techniques. *Journal of Applied Horticulture*, *25*(1), 42–46.

Faisal, M., Alatar, A. A., Doležal, K., & Shekhawat, M. S. (2024). Strigolactone analogue GR24 mediated somatic embryogenesis from leaf tissues of Santalum album L. *In Vitro Cellular & Developmental Biology–Plant*, *60*(1), 39–49.

Gailīte, A., Kļaviņa, D., & Ievinsh, G. (2010). In vitro propagation of an endangered plant Saussurea esthonica. *Environmental and Experimental Botany*, *8*, 43–48.

Gambino, G., Ruffa, P., Vallania, R., & Gribaudo, I. (2007). Somatic embryogenesis from whole flowers, anthers and ovaries of grapevine (Vitis spp.). *Plant Cell, Tissue and Organ Culture*, *90*, 79–83.

Hamilton, A. (2013). *Plant Conservation: An Ecosystem Approach*. Routledge.

Hussain, G., Wani, M. S., Mir, M. A., Rather, Z. A., & Bhat, K. M. (2014). Micrografting for fruit crop improvement. *African Journal of Biotechnology*, *13*(25), 2474–2483.

Jain, M., Johnson, T. S., & Krishnan, P. (2012). Biotechnological approaches to conserve the wealth of nature: endangered and rare medicinal plant species: A review. *Journal of Natural Remedies*, *12*(2), 93–102.

Jhariya, M. K., Banerjee, A., Raj, A., Meena, R. S., Khan, N., Kumar, S., & Bargali, S. S. (2022). Species invasion and ecological risk. In *Natural Resources Conservation and Advances for Sustainability* (pp. 503–531). Elsevier.

Johnston, J. W., Benson, E. E., & Harding, K. (2009). Cryopreservation induces temporal DNA methylation epigenetic changes and differential transcriptional activity in Ribes germplasm. *Plant Physiology and Biochemistry*, *47*(2), 123–131.

Jonard, R. (1986). Micrografting and its applications to tree improvement. In *Trees I* (pp. 31–48). Berlin, Heidelberg: Springer Berlin Heidelberg.

Juliani, H. R., Koroch, A. R., Zygadlo, J. A., & Trippi, V. S. (2011). Evaluation of micropropagation for the introduction into cultivation and conservation of Lippia junelliana, an endemic aromatic plant from Argentina. *Industrial Crops and Products*, *34*(2), 1353–1357.

Kaviani, B. (2011). Conservation of plant genetic resources by cryopreservation. *Australian Journal of Crop Science*, *5*(6), 778–800.

Khan, S., Al-Qurainy, F., & Nadeem, M. (2012). Biotechnological approaches for conservation and improvement of rare and endangered plants of Saudi Arabia. *Saudi Journal of Biological Sciences*, *19*(1), 1–11.

Klimek-Chodacka, M., Kadluczka, D., Lukasiewicz, A., Malec-Pala, A., Baranski, R., & Grzebelus, E. (2020). Effective callus induction and plant regeneration in callus and protoplast cultures of *Nigella damascena* L. *Plant Cell, Tissue and Organ Culture (PCTOC)*, *143*(3), 693–707.

Komakech, R., Kim, Y. G., Kim, W. J., Omujal, F., Yang, S., Moon, B. C., ... & Kang, Y. (2020). A micropropagation protocol for the endangered medicinal tree *Prunus africana* (Hook f.) Kalkman: genetic fidelity and physiological parameter assessment. *Frontiers in Plant Science*, *11*, 548003.

Koufan, M., Mazri, M. A., Essatte, A., Moussafir, S., Belkoura, I., El Rhaffari, L., & Toufik, I. (2020). A novel regeneration system through micrografting for *Argania spinosa* (L.) Skeels, and confirmation of successful rootstock-scion union by histological analysis. *Plant Cell, Tissue and Organ Culture (PCTOC)*, *142*, 369–378.

Kshirsagar, P. R., Mohite, A., Suryawanshi, S., Chavan, J. J., Gaikwad, N. B., & Bapat, V. A. (2021). Plant regeneration through direct and indirect organogenesis, phyto-molecular profiles, antioxidant properties and Swertiamarin production in elicitated cell suspension cultures of Swertia minor (Griseb.) Knobl. *Plant Cell, Tissue and Organ Culture (PCTOC)*, *144*, 383–396.

Liu, Y. G., Liu, L. X., Wang, L., & Gao, A. Y. (2008). Determination of genetic stability in surviving apple shoots following cryopreservation by vitrification. *CryoLetters*, *29*(1), 7–14.

Loberant, B., & Altman, A. (2010). Micropropagation of plants. *Encyclopedia of Industrial Biotechnology: Bioprocess, Bioseparation, and Cell Technology. Wiley, New York*, 3499–3515.

Machado, J. S., Degenhardt, J., Maia, F. R., & Quoirin, M. (2020). Micropropagation of *Campomanesia xanthocarpa* O. Berg (Myrtaceae), a medicinal tree from the Brazilian Atlantic Forest. *Trees*, *34*, 791–799.

Mastuti, R., & Rosyidah, M. (2018, October). In vitro enzymatic isolation of protoplasts from tissues of the medicinal plant *Physalis angulata* L. In *AIP Conference Proceedings* (Vol. 2019, No. 1). AIP Publishing.

Mehalaine, S., & Chenchouni, H. (2021). New insights for the production of medicinal plant materials: Ex vitro and in vitro propagation of valuable Lamiaceae species from northern Africa. *Current Plant Biology*, *27*, 100216.

Mujib, A., Bansal, Y., Malik, M. Q., Syeed, R., Mamgain, J., & Ejaz, B. (2022). Internal and external regulatory elements controlling somatic embryogenesis in Catharanthus: A model medicinal plant. In *Somatic Embryogenesis: Methods and Protocols* (pp. 11–27). New York, NY: Springer US.

Ogbu, J. U., Essien, B. A., Essien, J. B., & Anaele, M. U. (2010). Conservation and management of genetic resources of horticultural crops in Nigeria: Issues and biotechnological strategies. *Journal of Horticulture and Forestry*, *2*(9), 214–222.

Onay, A., Pirinc, V., Adiyaman, F., Işikalan, Ç., Tilkat, E., & Başaran, D. (2003). In Vivo and in Vitro Micrografting of Pistachio, *Pistacia vera* L. cv." Siirt". *Turkish Journal of Biology*, *27*(2), 95–100.

Pande, D., Malik, S., Bora, M., & Srivastava, P. S. (2002). A rapid protocol for in vitro micropropagation of *Lepidium sativum* Linn. and enhancement in the yield of Lepidine. *In Vitro Cellular & Developmental Biology-Plant*, *38*, 451–455.

Pernisová, M., Klíma, P., Horák, J., Válková, M., Malbeck, J., Souček, P., ... & Hejátko, J. (2009). Cytokinins modulate auxin-induced organogenesis in plants via regulation of the auxin efflux. *Proceedings of the National Academy of Sciences*, *106*(9), 3609–3614.

Prado, M. J., Grueiro, M. P., González, M. V., Testillano, P. S., Domínguez, C., López, M., & Rey, M. (2010). Efficient plant regeneration through somatic embryogenesis from anthers and ovaries of six autochthonous grapevine cultivars from Galicia (Spain). *Scientia Horticulturae, 125*(3), 342–352.

Rai, M. K. (2010). Biotechnological strategies for conservation of rare and endangered medicinal plants. *Biodiversitas, 11*(3), 157–166.

Ramírez-Mosqueda, M. A. (2022). Overview of somatic embryogenesis. *Somatic Embryogenesis: Methods and Protocols, 2527*, 1–8.

Reed, B. M. (Ed.). (2008). *Plant Cryopreservation: A Practical Guide* (p. 513). New York: Springer.

Rout, G. R., & Das, P. (1997). In vitro organogenesis in ginger (*Zingiber officinale* Rosc.). *Journal of Herbs, Spices & Medicinal Plants, 4*(4), 41–51.

Sajja, U., Ranjan, T., & Prasad, B. D. (2017). Protoplast Isolation and Fusion. In *Plant Biotechnology, Volume 1* (pp. 161–183). Apple Academic Press.

Shahzad, A., & Saeed, T. (2013). In vitro conservation protocols for some rare medicinal plant species. *Recent Trends in Biotechnology and Therapeutic Applications of Medicinal Plants*, 263–291.

Shekhawat, M. S., Manokari, M., & Ravindran, C. P. (2016). Micropropagation, micromorphological studies, and in vitro flowering in *Rungia pectinata* L. *Scientifica, 2016*(1), 5813851.

Simonović, A. D., M. Trifunović-Momčilov, M., Filipović, B. K., Marković, M. P., Bogdanović, M. D., & Subotić, A. R. (2020). Somatic embryogenesis in *Centaurium erythraea* Rafn—Current status and perspectives: A review. *Plants, 10*(1), 70.

Singh, B., Sharma, S., Rani, G., Hallan, V., Zaidi, A. A., Virk, G. S., & Nagpal, A. (2008). In vitro micrografting for production of Indian citrus ringspot virus (ICRSV)-free plants of kinnow mandarin (*Citrus nobilis* Lour× *C. deliciosa* Tenora). *Plant Biotechnology Reports, 2*, 137–143.

Skoog, F. and Miller, C. O. (1957). Chemical regulation of growth and organ formation in plant tissue cultured in vitro. *Symposia of the Society for Experimental Biology,* 11, 118–131).

Srivastava, S., Nalla, V. K., Singh, V. P., & Prasad, R. (2022). Optimization of in vitro micropropagation and root establishment through combinatorial approaches for enhanced production of secondary metabolites in the endangered species *Decalepis arayalpathra* KMA 05 clones. *Brazilian Journal of Botany, 45*(3), 869–881.

Tomlinson, T. R., & Akerele, O. (Eds.). (2015). *Medicinal Plants: Their Role in Health and Biodiversity.* University of Pennsylvania Press.

Tripathi, L., & Tripathi, J. N. (2003). Role of biotechnology in medicinal plants. *Tropical Journal of Pharmaceutical Research, 2*(2), 243–253.

Tsai, K. L., Chen, E. G., & Chen, J. T. (2016). Thidiazuron-induced efficient propagation of *Salvia miltiorrhiza* through in vitro organogenesis and medicinal constituents of regenerated plants. *Acta Physiologiae Plantarum, 38*, 1–9.

Verma, P., Mathur, A. K., Jain, S. P., & Mathur, A. (2012). In vitro conservation of twenty-three overexploited medicinal plants belonging to the Indian sub continent. *The Scientific World Journal, 2012*(1), 929650.

Volenzo, T., & Odiyo, J. (2020). Integrating endemic medicinal plants into the global value chains: The ecological degradation challenges and opportunities. *Heliyon, 6*(9): e04970.

Yang, G., Huang, L. J., Jiang, D., Huang, J., Cui, C., & Li, N. (2024). Development of indirect somatic embryogenesis and plant regeneration system with immature embryos of the cultivated traditional Chinese medicinal herb *Polygonatum cyrtonema. Industrial Crops and Products, 214*, 118557.

Zhang, J., Gai, M., Li, X., Li, T., & Sun, H. (2016). Somatic embryogenesis and direct as well as indirect organogenesis in *Lilium pumilum* DC. Fisch., an endangered ornamental and medicinal plant. *Bioscience, Biotechnology, and Biochemistry, 80*(10), 1898–1906.

Zuzarte, M. R., Dinis, A. M., Cavaleiro, C., Salgueiro, L. R., & Canhoto, J. M. (2010). Trichomes, essential oils and in vitro propagation of *Lavandula pedunculata* (Lamiaceae). *Industrial Crops and Products, 32*(3), 580–587.

7 Orchestrating Orchid Research through Metabolomic and Proteomic Approaches

Neha Thakur[1], Arshpreet Kaur, Jaspreet K. Sembi
[1]Department of Botany, Panjab University, Chandigarh

7.1 INTRODUCTION

Orchids represent one of the largest flowering families in the plant kingdom, exhibiting intricate floral architecture and remarkable diversity of form and habit, with around 30,000 species distributed across various ecosystems and habitats worldwide (Atwood, 1986; Govaerts et al., 2023). About 70% of orchid species are found to be epiphytes, whereas 25% are terrestrial, while the remaining 5% depend on different supports for their growth (Atwood, 1986; Dressler, 1981). The unique shapes and color patterns of orchid flowers have popularized them as ornamental plants of immense floricultural significance (Chugh et al., 2009). Some of the commercially important orchid genera are *Aranda*, *Aranthera*, *Ascocenda*, *Bulbophyllum*, *Cattleya*, *Cymbidium*, *Dendrobium*, *Epidendrum*, *Habenaria*, *Miltoniopsis*, *Mokara*, *Oncidium*, *Paphiopedilum*, *Phalaenopsis*, *Renanstylis*, *Renanthera*, *Rhynchostylis*, *Spathoglottis*, *Tolumnia*, *Vanda* and *Zygopetalum* (Yuan et al., 2021). Besides being prized for their aesthetic value, orchids hold medicinal importance as well. Some of the medicinally important orchid genera include *Anoectochilus*, *Bletilla*, *Bulbophyllum*, *Calanthe*, *Coelogyne*, *Cymbidium*, *Cypripedium*, *Dendrobium*, *Eria*, *Eulophia*, *Gastrodia*, *Habenaria*, *Liparis*, *Ludisia*, *Paphiopedilum*, *Spiranthes*, *Thunia* and *Vanda* (Choudhary et al., 2023; Ghai et al., 2022). The presence of a wide range of phytochemicals, such as alkaloids, phenolic compounds, flavonoids, triterpenoids, stilbenoids, bibenzyls, phenanthrenes and anthocyanins, is responsible for various pharmacological properties observed in orchids (Pant, 2013; Kaur et al., 2023). Thus, the immense importance of orchids has led to their unscrupulous collection and declining populations, resulting in their inclusion in both Appendix I and II of the Convention on International Trade in Endangered Species of Wild Fauna and Flora (CITES) (Hinsley et al., 2018).

With the advent of modern scientific techniques, it has become easy to propagate and conserve plants and utilize them for commercial purposes. Omics technology provides molecular insights into the various phytochemical and enzymatic production pathways, which help to monitor the therapeutic potential of plants. Functional investigations of the metabolome and proteome are the main aspects of metabolomic and proteomic studies (Mumtaz et al., 2017). As orchids have a high potential as a medicinal resource, such studies become imperative for discovering active principles and the underlying pathways. Metabolite profiling of the bioactive compounds has been reported using metabolomic studies. Such analyses help in the qualitative and quantitative assessment of metabolites. Proteomic studies in orchids provide insights into the developmental aspects based on the analysis of differentially expressed proteins (DEPs). These reports are discussed later in the chapter.

DOI: 10.1201/9781032697550-7

7.2 ORCHIDS AS AN IMPORTANT FLORICULTURAL RESOURCE

Orchids are known for their exquisite floral features, which makes them commercially important plants of immense ornamental value. Orchid genera such as *Cymbidium*, *Dendrobium*, *Oncidium*, *Paphiopedilum*, *Phalaenopsis* and *Vanda* are extensively used in floriculture (Janakiram and Baskaran, 2018). Owing to their weak reproductive barriers, a large number of natural hybrids exist in orchids. The occurrence of the first natural hybrid was reported to be a cross between *Phalaenopsis aphrodite* and *P. rosea*, resulting in the hybrid *P. intermedia* (cf. Li et al., 2021). The first artificial hybrid, *Calanthe dominyi*, was produced by crossbreeding *C. masuca* and *C. furcata* in the year 1856 (cf. Jangyukala and Hemanta, 2021). Hybridization in orchids involves intraspecific, intrageneric and intergeneric crossbreeding. Intergeneric crosses include two, three, four and up to five genera for producing bigeneric, trigeneric, tetrageneric and pentageneric hybrids (Li et al., 2021). For example, *Aeridovanda* (*Aerides* × *Vanda*), *Brassocattleya* (*Brassavola* × *Cattleya*) and *Vandanopsis* (*Phalaenopsis* × *Vanda*) are bigeneric hybrids (Jangyukala and Hemanta, 2021). *Epilaeliocattleya* is a trigeneric hybrid obtained by crossing *Epidendrum* × *Laelia* × *Cattleya*. Similarly, *Kirchara* is a tetrageneric hybrid produced by crossing *Cattleya* × *Epidendrum* × *Laelia* × *Sophronitis* (Li et al., 2021). Some of the commercially important hybrid genera are *Aranda*, *Ascocenda* and *Mokara* (Janakiram and Baskaran, 2018).

Globally, orchids are marketed as cut flowers and potted plants with increasing demands for exotic varieties (Hinsley et al., 2018). Based on the global import value from 2007 to 2018, Japan is the world's largest importer of cut orchids, followed by the United States, Italy, China, Vietnam, the United Kingdom, France, Germany, Singapore and Australia. On the other hand, the global export value of the Netherlands is the highest, followed by Thailand, Taiwan, Singapore, New Zealand, Vietnam, China, Malaysia, South Korea and Belarus. In the same study, it was reported that India ranked among the top ten countries that imported orchids (*Dendrobium* and *Paphiopedilum*) from Taiwan and Thailand (Yuan et al., 2021). Besides this, there also exist reports of undocumented trade of wild species of ornamental orchids (Phelps and Webb, 2015). Reports also suggest that most of the species of *Bulbophyllum*, *Cleisostoma*, *Eria* and *Pholidota* with small and non-attractive floral characteristics are also prevalent in commercial trade (Phelps and Webb, 2015), as these orchids are extensively employed in breeding programs and as progenitors of elite hybrids. Since orchids are an indispensable part of the floriculture industry, many conventional and modern propagation methods have been developed for the production and conservation of commercially valuable orchids (Zhang et al., 2022).

7.3 USE OF ORCHIDS IN TRADITIONAL MEDICINE SYSTEMS

Owing to the presence of a plethora of bioactive compounds, orchids are known for their anti-inflammatory, antimicrobial, antioxidant, anticancer, antidiabetic, antihepatotoxic, antimutagenic, antipyretic, antiallergic, antiaging, anthelmintic, diuretic, immunomodulatory, neuroprotective and wound healing properties (Gutiérrez, 2010; Ghai et al., 2022). The earliest record of Greeks using orchids for curative and restorative purposes dates back to 2000 years ago, thus popularizing orchids throughout Europe with people using orchids for their therapeutic potential during the 18th–20th centuries (Duggal, 1971). The Chinese were the first to systematically describe the medicinal uses of orchids like *Dendrobium* and *Bletilla striata*, which have been listed in the Chinese *materia medica* (Reinikka, 1995; Bulpitt, 2005). Traditional Chinese medicine uses *Dendrobium nobile*, *Bletilla striata*, *Gastrodia elata* and several species of *Anoectochilus* and *Bulbophyllum kwangtungense* (Pant, 2013). The Indian system of traditional medicine, Ayurveda, has reported the medicinal uses of *Dendrobium macraei*, *Orchis latifolia* and *Eulophia campestris* (Singh, 2001). Ashtavarga, a group of eight herbs, is reported to use four orchid species—*Habenaria edgeworthii*,

Habenaria intermedia, *Malaxis acuminata* and *Malaxis muscifera*—as major ingredients in the preparation of Chavyanprasa (Singh and Duggal, 2009; Pant, 2013). Some of the other orchid species used in Indian indigenous medicine system include *Dendrobium fimbriatum*, *D. nobile*, *D. moschatum*, *Eria musicucola*, *Eulophia compestris*, *Satyrium nepalense*, *Liparis odorata*, *Vanda cristata*, *V. coerula*, *V. tessallata*, *V. spathulata*, *Cymbidium gigantieum*, *C. aloifolium*, *C. williomsoni* and *Zeuxine strateumatica* (Tiwari et al., 2022).

7.4 ORCHIDS AS AN IMPORTANT FOOD AND FLAVORING AGENT

A number of orchids have been used as an important food source. Salep is primarily a beverage made from tubers of wild orchids and is used as an aphrodisiac and a restorative (Tamer et al., 2006). The widely used orchid genera for salep production include *Anacamptis*, *Barlia*, *Dactylorhiza*, *Himantoglossum*, *Neotinea*, *Ophyris*, *Orchis* and *Serapias*. In Turkey, salep is consumed as a hot drink prepared in milk and is used in the production of Kahramanmaras ice cream. It also finds use in traditional medicine as a cure for stomach ulcers and respiratory tract congestion (Kasparek and Grimm, 1999). Indian salep is chiefly prepared from tubers of *Eulophia*, *Dactylorhiza* and *Satyrium* species (Duggal, 1971). In Africa, chikanda cake is a popular snack prepared using tubers of terrestrial orchids such as *Brachycorythis*, *Disa*, *Eulophia*, *Habenaria*, *Roeperocharis* and *Satyrium* (Veldman et al., 2018). On the Isle of Reunion and Mauritius, a drink is prepared from the fragrant leaves of *Angraecum fragrans*, called "Faham tea," which leaves a long-lasting taste (Schiff, 2018). These leaves are also used to flavor custards, ice creams and other homemade delicacies (Duggal, 1971). Vanillin, extracted from processed seed capsules of *Vanilla planifolia*, is used as a flavoring agent worldwide and is a popular ingredient in ice creams, confectioneries, beverages, soaps, perfumes and pharmaceuticals (Correll, 1953).

7.5 THREATS TO ORCHID DIVERSITY: MANIFESTATION AND PROGNOSIS

The immense economic importance of orchids has led to their overexploitation at the global level. Illegal, uncontrolled and unsustainable collection for commercial purposes is a major factor contributing to the decline in the orchid population. Countries like China, India and Turkey have a high demand for medicinal orchids, which leads to the overharvesting of these orchids (Teoh, 2019). Additionally, orchid habitat degradation as a consequence of anthropogenic activities, such as the clearing of forests for housing, agriculture and commercial purposes, has led to endangering the natural populations of orchids (Teoh, 2016). During the Victorian era in England, orchid collecting, known as orchidelirium, began as a hobby of affluent people (Hansen, 2016). Illegal trading to meet high demands has posed a risk of extinction of rare orchids (Tiwari et al., 2022). *Paphiopedilum* is one such example that faces survival challenges due to intensive collection. *Paphiopedilum glaucophyllum*, an endemic orchid of Java, Indonesia, has lost much of its population due to overcollection for sale (Whitten et al., 1996). Similarly, *P. canhii*, which is endemic to Vietnam, lost about 99.5% of its population due to excessive commercial collection and is categorized as critically endangered, facing a high risk of extinction as per International Union for Conservation of Nature (IUCN) criteria (Averyanov et al., 2014). Another similar instance is that of *Phragmipedium kovachii*, which was declared critically endangered in 2001 by the IUCN due to its limited range and horticultural collection activities (Cribb, 2005). In India, the high demand for *Habenaria edgeworthii*, *Habenaria intermedia*, *Malaxis acuminata*, *Malaxis muscifera* (the ashtavarga herbs) and *Eulophia spectabilis* (an aphrodisiac) has led to the scarcity of these orchids (Teoh, 2019). Mining operations in the Niyamgiri hill ranges of Orissa have put the whole ecosystem, including the 20 orchid species found in the area, under threat (Dash et al., 2008). Besides these, the collection of orchids for botanical studies also leads to the depletion of native orchids in the wild (Lokesha and Vasudeva, 1992). Since orchids possess specialized pollination mechanisms and symbiotic associations with mycorrhizal fungi for seed germination, the nonavailability of specific pollinators and fungal species in the habitat increases the risk of orchid flora depletion (Phillips et al., 2020).

Orchids are one of the most threatened species in the plant kingdom. Most of them are critically endangered and hence, all the orchids have been listed in Appendix I and II of CITES (Pant et al., 2013; Popova et al., 2016). Appendix I lists all species that are threatened with extinction. *Aerangis ellipsis*, *Dendrobium cruentum*, *Laelia jongheana*, *Laelia lobata*, *Peristeria elata*, *Renanthera imschootiana* and two genera (*Paphiopedilum* and *Phragmipedium*) have been listed in Appendix I of CITES (Hinsley et al., 2018). The rest of the species of orchids are included in Appendix II (Hinsley et al., 2018). An automated assessment of 13,910 orchid species, based on a novel deep learning method, suggested that 4342 of the assessed species (31.2%) were possibly threatened with extinction. Detailed evaluation classified 68 species (0.5%) as vulnerable, 2567 species (18.5%) as endangered and 718 species (5.2%) as critically endangered (Zizka et al., 2021). A recent quantitative assessment of 25,434 orchid species suggested that 278 species need to be prioritized for immediate conservation (Vitt et al., 2023). Only 37 of the evaluated species have an IUCN Red List assessment and most of them are identified as threatened with extinction (Vitt et al., 2023). Therefore, in situ and ex situ conservation strategies have been employed to protect these valuable plants. In situ conservation practices involve protecting and preserving orchids in their natural environment, whereas ex situ conservation focuses on the conservation of orchids outside their natural habitats.

7.6 OMICS TECHNOLOGY AS AN AID TO ORCHID RESEARCH

7.6.1 Metabolomics

While tissue culture technology and bioreactor technology are effective strategies for the extraction and mass production of secondary metabolites, metabolomics aids in the quantification and structural elucidation of these important phytochemicals. Besides identifying key genes responsible for metabolite gene expression using genomic surveys, the analysis of metabolic output is necessary for understanding the regulatory network. Metabolomics is an emerging analytical tool for metabolite profiling under omics technology. While transcriptomics and proteomics are concerned with the flow of genetic information, metabolomics is associated with metabolites (Fukusaki and Kobayashi, 2005). Metabolomic studies help in establishing a relationship between metabolic output and the metabolic pathway, along with its various regulatory factors controlled by the genotype (Hong et al., 2016; Abdelhafez et al., 2020). An integrative approach combining metabolomics with transcriptomic studies can be applied to elucidate the regulation of metabolic networks in plants. Such analyses have been used to study metabolite concentrations and corresponding changes in metabolic flux in plants (Toubiana et al., 2013). Metabolomics employs various techniques for the qualitative and quantitative assessment of metabolites. Fractionation and isolation of metabolites are usually performed using chromatographic techniques. The conventional approach employs column chromatography (CC) and thin-layer chromatography (TLC), whereas modern techniques include vacuum liquid chromatography (VLC) and high-performance liquid chromatography (HPLC). The next step in metabolomic studies is structure elucidation of the isolated compounds. Commonly used methods to determine molecular structures, functional groups and molecular masses include spectroscopic techniques such as ultraviolet spectroscopy (UV), infrared spectroscopy (IR), mass spectrometry (MS) and nuclear magnetic resonance (NMR) spectroscopy (Sarker et al., 2006). Further, molecular mass and molecular formula can be determined by various MS techniques such as electron impact mass spectrometry (EIMS), electrospray ionization mass spectrometry (ESI-MS) and fast atom bombardment mass spectrometry (FABMS). Additionally, hyphenated techniques, coupling chromatography and spectroscopy techniques are also commonly used for metabolite profiling. These include LC-HRMS, HPLC-DAD-MS-SPE-UV-NMR, HPLC-DAD-ESI-QTOF-MS/MS and UPLC-Q-TOF-MS/MS (Sarker and Nahar, 2005).

In orchids, the detection of bioactive compounds such as polyphenols, flavonoids, anthocyanins, stilbenoids, terpenoids and phenanthrenes using metabolomic techniques has been reported by various researchers (Table 7.1, Figure 7.1). VLC and CC techniques have been used to isolate polyphenols and phenolic compounds from *Aerides multiflora* (Thant et al., 2021), *Dendrobium*

TABLE 7.1
Metabolites Isolated from Orchids through Metabolomic Approaches

Compound	Plant Species	Reference
Polyphenols	*Aerides multiflora*	Thant et al., 2021
	Dendrobium brymerianum	Klongkumnuankarn et al., 2015
	Dendrobium catenatum	Zhang et al., 2019
	Dendrobium formosum	Inthongkaew et al., 2017
	Pleione formosana	Shiao et al., 2009
	Vanilla fragrans	Sun et al., 2001
	Vanilla planifolia	Palama et al., 2010
Phenanthrenes or its derivatives	*Aerides rosea*	Cakova et al., 2015
	Bulbophyllum reptans	Majumder et al., 1999a
	Bletilla formosana	Lin et al., 2005
	Bulbophyllum odoratissimum	Xu et al., 2009
	Calanthe arisanensis	Lee et al., 2009
	Coelogyne flaccida	Majumder and Maiti, 1991
	Coelogyne cristata	Majumder et al., 2001
	Cremastra appendiculata	Xue et al., 2006
	Cymbidium pendulum	Majumder and Sen, 1991
	Dendrobium chrysotoxum	Jie et al., 2023
	Dendrobium draconis	Sritularak et al., 2011
	Dendrobium nobile	Kim et al., 2015
	Eulophia nuda	Tuchinda et al., 1988
	Pleione bulbocodioides	Bai et al., 1996; Shao et al., 2019
	Pleione yunnanensis	Dong et al., 2010
	Thunia alba	Yan et al., 2016
Triterpenoids	*Agrostophyllum brevipes*	Majumder et al., 2003
	Agrostophyllum callosum	Majumder et al., 2003
	Agrostophyllum callosum	Majumder et al., 1999b
	Pleione formosana	Shiao et al., 2009
Stilbenoids	*Agrostophyllum callosum*	Majumder et al., 1996
	Arundina graminifolia	Liu et al., 2004
	Bletilla striata	Bai et al., 1993
	Bletilla ochracea	Yang et al., 2012
	Pleione bulbocodioides	Bai et al., 1998
	Thunia alba	Yan et al., 2016
Bibenzyls	*Bletilla striata*	Jiang et al., 2019
	Dendrobium catenatum	Zhu et al., 2021
	Dendrobium chrysotoxum	Jie et al., 2023
	Dendrobium falconeri	Sritularak and Likhitwitayawuid, 2009
	Dendrobium nobile	Zhang et al., 2007
	Pleione bulbocodioides	Bai et al., 1997
Phenylpropanoids	*Bulbophyllum retusiusculum*	Fang et al., 2018
Phenolic acids	*Dendrobium chrysotoxum*	Jie et al., 2023
Flavonoids	*Dendrobium officinale*	Zhan et al., 2020
Anthocyanins	*Dracula chimaera*	Fossen and Øvstedal, 2003
	Dracula cordobae	Fossen and Øvstedal, 2003
Glycosides	*Anoectochilus formosanus*	Wu et al., 2007
	Calanthe discolor	Yoshikawa et al., 1998
	Calanthe liukiuensis	Yoshikawa et al., 1998
	Dendrobium nobile	Zhao et al., 2001

(*Continued*)

TABLE 7.1 (CONTINUED)
Metabolites Isolated from Orchids through Metabolomic Approaches

Compound	Plant Species	Reference
Type II arabinogalactan	*Anoectochilus formosanus*	Lai et al., 2015
Polysaccharides	*Anoectochilus roxburghii*	Zeng et al., 2016
Alkaloids	*Calanthe discolor*	Yoshikawa et al., 1998
	Calanthe liukiuensis	Yoshikawa et al., 1998
	Cremastra appendiculata	Ikeda et al., 2005
	Dendrobium crepidatum	Xu et al., 2020; Hu et al., 2020; Ding et al., 2021
	Dendrobium nobile	Wang et al., 2016
Steroid	*Coelogyne uniflora*	Majumder and Pal, 1990
	Dendrobium ochreatum	Banerjee and Karati, 2022

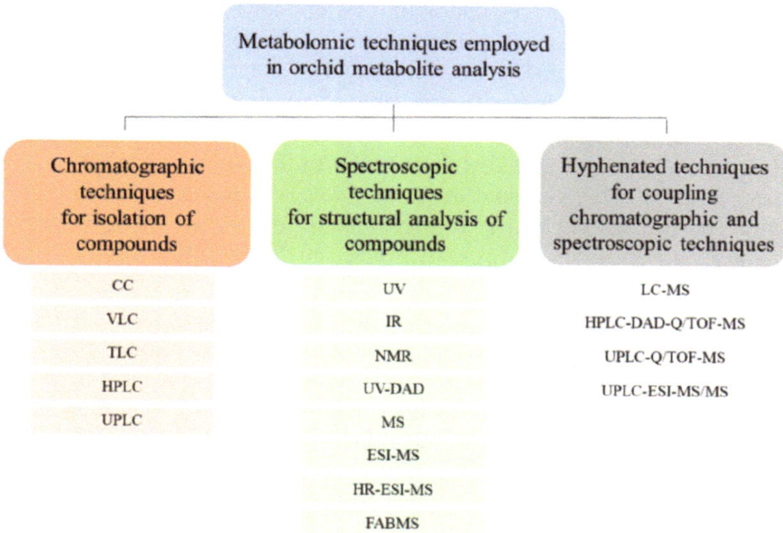

FIGURE 7.1 Flowchart depicting the metabolomic techniques used in orchid research.

brymerianum (Klongkumnuankarn et al., 2015), *Dendrobium formosum* (Inthongkaew et al., 2017) and *Vanda bensonii* (Jimoh et al., 2022). Similarly, CC and TLC have been reported for isolation and purity testing of compounds such as triterpenoids in *Agrostophyllum brevipes* (Majumder et al., 2003), *Agrostophyllum callosum* (Majumder et al., 1999a; Majumder et al., 2003), phenanthrenes and their derivatives in *Bletilla formosana* (Lin et al., 2005), *Bulbophyllum reptans* (Majumder et al., 1999b), *Calanthe arisanensis* (Lee et al., 2009), *Coelogyne cristata* (Majumder et al., 2001), *Cremastra appendiculata* (Xue et al., 2006), *Dendrobium nobile* (Kim et al., 2015), *Eulophia nuda* (Tuchinda et al., 1988), *Pleione bulbocodioides* (Shao et al., 2019), *Thunia alba* (Yan et al., 2016), alkaloids in *Dendrobium crepidatum* (Xu et al., 2020; Hu et al., 2020; Ding et al., 2021), phenolic compounds in *Cymbidium* sp. (Axiotis et al., 2021), *Vanilla fragrans* (Sun et al., 2001), stilbenoids in *Agrostophyllum callosum* (Majumder et al., 1996), *Arundina graminifolia* (Liu et al., 2004), *Bletilla striata* (Bai et al., 1993), *Bletilla ochracea* (Yang et al., 2012), *Thunia alba* (Yan et al., 2016), butenolide in *Goodyera schlechtendaliana* (Dai et al., 2021), steroid in *Dendrobium ochreatum* (Banerjee and Karati, 2022) and bibenzyl derivatives in *Dendrobium nobile* (Zhang et al., 2007). Certain antifungal compounds have also been isolated using these techniques in *Cypripedium macranthos* var. *rebunense* (Shimura et al., 2007). The isolation of compounds by CC in conjunction with preparative or semi-preparative HPLC for molecular fractionation of compounds has been conducted successfully in various orchids like *Aerides rosea* (Cakova et al., 2015), *Anoectochilus formosanus* (Wu et al., 2007), *Anoectochilus roxburghii* (Zeng et al., 2016), *Bletilla striata* (Jiang et al., 2019), *Bletilla ochracea* (Yang et al., 2012), *Bulbophyllum retusiusculum* (Fang et al., 2018), *Calanthe discolor* and *C. liukiuensis* (Yoshikawa et al., 1998), *Calanthe arisanensis* (Lee et al., 2009), *Cremastra appendiculata* (Ikeda et al., 2005; Xue et al., 2006), *Cymbidium* sp. (Axiotis et al., 2021), *Cypripedium macranthos* var. *rebunense* (Shimura et al., 2007), *Dendrobium catenatum* (Zhang et al., 2019; Zhu et al., 2021), *Dendrobium crepidatum* (Xu et al., 2020; Hu et al., 2020; Ding et al., 2021), *Dendrobium nobile* (Zhang et al., 2007; Kim et al., 2015), *Dracula chimaera* and *D. cordobae* (Fossen and Øvstedal, 2003), *Goodyera schlechtendaliana* (Dai et al., 2021), *Pleione formosana* (Shiao et al., 2009), *Pleione yunnanensis* (Dong et al., 2010; Dong et al., 2013) and *Thunia alba* (Yan et al., 2016).

A combination of chromatographic and spectroscopic techniques is useful for the complete characterization of metabolites. After the separation of compounds, the next step involves the determination of molecular structure for identification and characterization of the isolated compound. Spectroscopic techniques such as UV, IR, NMR and MS have been used to predict functional groups, molecular weights and chemical structures of triterpenoids in *Agrostophyllum callosum* (Majumder et al., 1999a), stilbenoids in *Bletilla striata* (Bai et al., 1993), phenanthrenes in *Eulophia nuda*, *Cymbidium pendulum* and *Bulbophyllum reptans* (Tuchinda et al., 1988; Majumder and Sen, 1991; Majumder et al., 1999a), steroidal esters in *Coelogyne uniflora* (Majumder and Pal, 1990) and phenanthrene derivatives in *Coelogyne flaccida* and *Coelogyne cristata* (Majumder and Maiti, 1991; Majumder et al., 2001). In addition to NMR and MS spectral analysis, X-ray crystallography was reported to determine the structure of a steroid in *Dendrobium ochreatum* (Banerjee and Karati, 2022). In *Vanilla planifolia*, identification of amino acids, carbohydrates, organic acids and phenolic compounds was determined by NMR and UV-DAD (diode array detector) spectral analyses (Palama et al., 2010). For quick and high-resolution analysis, ultra-high-performance liquid chromatography (UHPLC/UPLC) is used for the detection and identification of compounds. For instance, in *Dendrobium catenatum*, isolation and structure prediction of C-glycosyltransferase were done by UPLC-MS/MS analysis (Ren et al., 2020). In another study, quantitative detection of one coumarin, two alkaloids and two bibenzyls was performed by the UPLC-PDA (photo diode array) method, and a chromatographic fingerprint of 15 species of *Dendrobium* was developed (Xu et al., 2010).

For accurate quantitative measurements of the molecular mass of biomolecules, ESI-MS proves to be an efficient and reliable analytical tool (Ho et al., 2003). Some instances of ESI-MS spectral analysis in orchids include the detection of phenolic compounds in *Dendrobium brymerianum* (Klongkumnuankarn et al., 2015), phenanthrenes in *D. nobile* (Kim et al., 2015), flavonoids in *D. officinale* (Zhan et al., 2020) and polyphenols and triterpenes in *Pleione formosana* (Shiao et al., 2009). Furthermore, high-resolution electrospray ionization mass spectrometry (HR-ESI-MS) for precise estimation of molecular mass and structure elucidation has been employed in *Aerides multiflora* (Thant et al., 2021), *Bletilla striata* (Jiang et al., 2019), *Bletilla ochracea* (Yang et al., 2012), *Bulbophyllum retusiusculum* (Fang et al., 2018), *Bulbophyllum odoratissimum* (Xu et al., 2009), *Calanthe arisanensis* (Lee et al., 2009), *Dendrobium catenatum* (Zhu et al., 2021), *Dendrobium crepidatum* (Xu et al., 2020; Hu et al., 2020; Ding et al., 2021), *Dendrobium draconis* (Sritularak et al., 2011), *Dendrobium falconeri* (Sritularak and Likhitwitayawuid, 2009), *Dendrobium formosum* (Inthongkaew et al., 2017), *Goodyera schlechtendaliana* (Dai et al., 2021), *Pleione bulbocodioides* (Shao et al., 2019), *Pleione yunnanensis* (Dong et al., 2010; Dong et al., 2013) and *Vanda bensonii*

(Jimoh et al., 2022). Another technique for producing effective structural information is the use of FABMS, as reported by Liu et al. (2004) for analyzing functional groups and the structure of a stilbenoid in *Arundina graminifolia*. Similarly, this technique has been reported to identify glycosides and alkaloids in *Calanthe discolor* and *Calanthe liukiuensis* (Yoshikawa et al., 1998), phenanthrene derivatives in *Cremastra appendiculata* (Xue et al., 2006) and sesquiterpene glycosides in *Dendrobium nobile* (Zhao et al., 2001).

Additionally, hyphenated techniques for analyzing metabolomic constituents have also been reported in orchid phytochemical research. Extracts from *Dendrobium catenatum* were subjected to HPLC-DAD-ESI-QTOF-MS/MS analysis, facilitating identification and characterization of phenols and flavonoids (Zhang et al., 2019). Likewise, in *Aerides rosea*, detection and characterization of phenanthrenes and minor compounds were carried out by HPLC-DAD-MS-SPE (solid-phase extraction)-UV-NMR spectral analysis (Cakova et al., 2015). UPLC coupled with triple-quadrupole mass spectrometry (UPLC-Q/TOF-MS) for higher resolution and fast analysis has been used for comparative analysis of differential metabolites among different species of *Dendrobium* (Zhang et al., 2023). Likewise, identification of bibenzyls, phenanthrenes and phenolic acids in *Dendrobium chrysotoxum* was carried out by UPLC-Q-TOF-MS/MS analysis (Jie et al., 2023). UPLC coupled with ESI/QTOF-MS facilitated the identification and characterization of sesquiterpene alkaloids in *Dendrobium nobile* (Wang et al., 2016). Likewise, in *Dendrobium officinale*, detection of flavonoids by UPLC-ESI-MS/MS was reported by Zhan et al. (2020). Thus, quantitative and qualitative measurements of phytochemicals using reliable and accurate metabolomic tools and techniques have opened wide opportunities for identifying and characterizing bioactive compounds among orchid species.

7.6.2 Proteomics

Proteomics involves the comprehensive study of proteins in biological systems at a specific time and condition, ranging from their identification and quantification to the investigation of their post-translational modifications, interactions, localization and functions (van Wijk, 2001). Proteomics helps in the complete understanding of a complex biological process at a molecular level. It provides insights into the differential expression profiles for both plant development processes and stress response mechanisms and elucidates the interactions within various signaling pathways in a biological system (Chen and Harmon, 2006; Baginsky, 2009). In plants, different proteomic techniques like 2-DE (two-dimensional gel electrophoresis), MALDI-TOF (matrix-assisted laser desorption/ionization-time of flight), isobaric tagging and nanoLC-MS (mass spectrometry) have been employed to identify and characterize the proteins present in a biological sample. In a gel-based analysis, first the proteins are separated using gel electrophoresis followed by quantification and identification using mass spectrometry. On the other hand, a gel-free analysis involves the initial degradation of the protein mixtures using proteases followed by chromatographic separation and mass spectrometric identification (Champagne and Boutry, 2013; Tan et al., 2017).

In orchids, proteomic studies using diverse methodologies have been done to evaluate various aspects of growth and development in various species (Table 7.2, Figure 7.2). In *Bipinnula fimbriata*, the adaptation mechanism and role of mycorrhiza in a heavy metal-polluted ecosystem were studied by comparing the proteome profiles of mycorrhizal and non-mycorrhizal root segments (Herrera et al., 2018). Likewise, a comparison of the proteome profiles of mycorrhizal and non-mycorrhizal root segments elucidated the role of *Mycena dendrobii* (a mycorrhizal fungus) in promoting the growth and survival of tissue-cultured derived seedlings of *Dendrobium officinale* (Xu et al., 2015). Similarly, the stress responses have been studied through proteomics, such as identifying the proteins involved in adaptation mechanisms toward biotic stress in *Phalaenopsis amabilis* (Lai et al., 2013), drought stress in *Cymbidium sinense* and *Cymbidium tracyanum* (Li et al., 2018) and salt stress in *Dendrobium huoshanense* (Song et al., 2022). Further, stress-related antioxidant enzymes have been identified in *Bulbophyllum morphologlorum* through comparative proteomics approach (Huehne et al., 2020).

TABLE 7.2
Comparative Analysis of Differentially Expressed Protein Studies in Orchids

Tissues under Study	Plant Species	Proteomic Techniques*	Reference
Mycorrhizal and non-mycorrhizal root segments	*Bipinnula fimbriata*	2-DE ESI-Q-TOF MS	Herrera et al., 2018
	Dendrobium officinale	2-D DIGE MALDI-TOF-MS	Xu et al., 2015
	Oeceoclades maculata	2D-DIGE LC-MS/MS ITRAQ MALDI-TOF/TOF MS	Valadares et al., 2020
Floral parts	*Cymbidium ensifolium*	2-DE MALDI-TOF/TOF MS	Li et al., 2014
	Cymbidium goeringii	Reverse-phase HPLC LC-MS/MS	Chen et al., 2021
	Dendrobium chrysanthum	2-DE MALDI-TOF/TOF MS	Wang et al., 2017
	Ophrys exaltata, *O. garganica*, *O. sphegodes*	SDS-PAGE LC-MS/MS	Sedeek et al., 2013
	Phalaenopsis amabilis	ITRAQ LC-ESI MS/MS MRM	Chen et al., 2020
	Phalaenopsis 'Red Dragon'	2-DE MALDI-TOF/TOF-MS	Chen et al., 2018
Protocorms/ protocorm-like bodies	*Cattleya tigrina*	2-DE MALDI-TOF/MS	De Conti et al., 2020
	Dendrobium officinale	ITRAQ MS/MS	Chen et al., 2017
	Gastrodia elata	SDS-PAGE ITRAQ HPLC MS/MS	Zeng et al., 2018
	Oncidium sphacelatum	2-DE LC-MS/MS iTRAQ	Valadares et al., 2014
		2D-PAGE MALDI-TOF-MS	López-Chávez et al., 2016
	Vanilla planifolia	2-DE MALDI TOF-TOF MS	Palama et al., 2010
Callus	*Vanilla planifolia*	2-DE MALDI TOF-TOF MS	Tan et al., 2013
Leaves	*Bletilla striata*	SDS-PAGE LC-MS/MS	Lu et al., 2020
	Cymbidium sinense, *C. tracyanum*	2-DE MALDI-TOF/TOF MS	Li et al., 2018
	Phalaenopsis amabilis	2-DE MALDI-TOF-MS	Lai et al., 2013
	Phalaenopsis hybrida cv. Redsky	2-DE LC-MS/MS	Chen et al., 2012
Seedlings	*Dendrobium huoshanense*	LC/MS	Song et al., 2022
	Dendrobium officinale	Reverse-phase HPLC LC-MS/MS	Feng et al., 2017

* *2-DE:* two-dimensional gel electrophoresis; 2D-DIGE: two-dimensional difference gel electrophoresis; 2D-PAGE: two-dimensional polyacrylamide gel electrophoresis; ESI-Q-TOF MS: electrospray ionization quadrupole-time of flight mass spectrometry; HPLC: high-performance liquid chromatography; iTRAQ: isobaric tags for relative and absolute quantitation; MS/MS: tandem mass spectrometry; MALDI-TOF/TOF MS: matrix-assisted laser desorption/ionization time of flight/time of flight mass spectrometry; MRM: mass spectrometry multireaction monitoring; LC-MS/MS: liquid chromatography with tandem mass spectrometry; nanoLC-LTQ Orbitrap MS: nanoliquid chromatography linear ion trap triple quadrupole Orbitrap mass spectrometry; SDS-PAGE: sodium dodecyl-sulfate polyacrylamide gel electrophoresis.

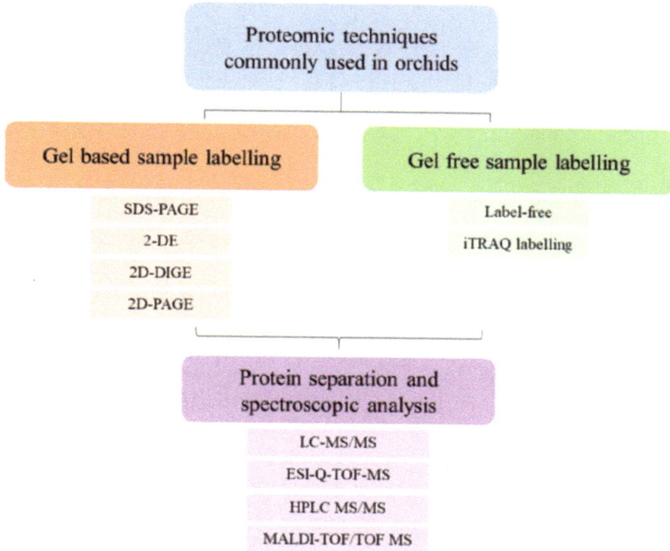

FIGURE 7.2 Flowchart depicting the commonly used proteomic techniques in orchids.

To understand the mechanism of symbiotic seed germination in orchids, the change in the expression of the key proteins regulating this phenomenon has been studied in *Oncidium sphacelatum* (Valadares et al., 2014), *Dendrobium officinale* (Chen et al., 2017) and *Gastrodia elata* (Zeng et al., 2018). Identification of endophytic bacteria as growth promoters through proteomics analysis was done in *Epipactis albensis*, *E. helleborine* and *E. purpurata* (Jakubska-Busse et al., 2021). The PLBs (protocorm-like bodies) development has been assessed by evaluating the differentially expressed proteins related to energy and carbohydrate metabolism, cell proliferation, protein processing, secondary metabolism and phytohormones in *Cattleya tigrina* (De Conti et al., 2020). Proteins related to floral organ development have been identified in *Phalaenopsis amabilis* (Chen et al., 2020) and *Cymbidium goeringii* by proteomics approaches (Chen et al., 2021). Further, the molecular basis of pollinator attraction has also been explored in sexually deceptive orchids like *Ophrys exaltata*, *Ophrys sphegodes* and *Ophrys garganica* (Sedeek et al., 2013). Thus, proteomics methodologies offer valuable insights into numerous processes occurring within complex biological events, making them a powerful tool for tracking the physiological status of plants under distinct developmental conditions.

7.7 CONCLUSIONS

Orchids are recognized as a valuable resource of horticultural, pharmaceutical, and ecological significance on a global scale. Many advanced tools and techniques of metabolomics and proteomics have been adopted to discover and analyze the underlying molecular interactions of various biosynthetic pathways in orchids. An integrated omics approach can help in elucidating the complex network and mechanism of biological processes. Furthermore, such scientific explorations can play a role in improving genotypic traits and resulting phenotypes for enhancing the orchid resource value.

REFERENCES

Abdelhafez, O. H., Othman, E. M., Fahim, J. R., Desoukey, S. Y., Pimentel-Elardo, S. M., Nodwell, J. R., ... & Abdelmohsen, U. R. (2020). Metabolomics analysis and biological investigation of three Malvaceae plants. *Phytochemical Analysis*, *31*(2), 204–214.

Atwood, J. T. (1986). The size of the Orchidaceae and the systematic distribution of epiphytic orchids. *Selbyana*, 171–186.

Averyanov, L. V., Pham, V. T., Ke, L. P., Tien, H. N., Xuan, C. C., Tien, V. N., & Quan, H. N. (2014). Field survey of *Paphiopedilum canhii*: from discovery to extinction. *Slipper Orchids*, *24*(4), 16–26.

Axiotis, E., Angelis, A., Antoniadi, L., Petrakis, E. A., & Skaltsounis, L. A. (2021). Phytochemical analysis and dermo-cosmetic evaluation of Cymbidium sp.(Orchidaceae) cultivation by-products. *Antioxidants*, *11*(1), 101.

Baginsky, S. (2009). Plant proteomics: concepts, applications, and novel strategies for data interpretation. *Mass Spectrometry Reviews*, *28*(1), 93–120.

Bai, L., Kato, T., Inoue, K., Yamaki, M., & Takagi, S. (1993). Stilbenoids from Bletilla striata. *Phytochemistry*, *33*(6), 1481–1483.

Bai, L., Yamaki, M., Yamagata, Y., & Takagi, S. (1996). Shanciol, a dihydrophenanthropyran from Pleione bulbocodioides. *Phytochemistry*, *41*(2), 625–628.

Bai, L., Masukawa, N., Yamaki, M., & Takagi, S. (1997). Two bibenzyl glucosides from Pleione bulbocodioides. *Phytochemistry*, *44*(8), 1565–1567.

Bai, L., Masukawa, N., Yamaki, M., & Takagi, S. (1998). Four stilbenoids from Pleione bulbocodioides. *Phytochemistry*, *48*(2), 327–331.

Banerjee, J., & Karati, D. (2022). In-silico pharmacology against cancer of a novel phytosterol Dendrosterone isolated from plant *Dendrobium ochreatum*. *Pharmacological Research-Modern Chinese Medicine*, *4*, 100149.

Bulpitt, C. J. (2005). The uses and misuses of orchids in medicine. *QJM*, *98*(9), 625–631.

Cakova, V., Urbain, A., Antheaume, C., Rimlinger, N., Wehrung, P., Bonté, F., & Lobstein, A. (2015). Identification of phenanthrene derivatives in *Aerides rosea* (Orchidaceae) using the combined systems HPLC–ESI–HRMS/MS and HPLC–DAD–MS–SPE–UV–NMR. *Phytochemical Analysis*, *26*(1), 34–39.

Champagne, A., & Boutry, M. (2013). Proteomics of nonmodel plant species. *Proteomics*, *13*(3–4), 663–673.

Chen, S., & Harmon, A. C. (2006). Advances in plant proteomics. *Proteomics*, *6*(20), 5504–5516.

Chen, G., Chen, D., Wang, T., Xu, C., & Li, L. (2012). Analysis of the proteins related to browning in leaf culture of Phalaenopsis. *Scientia Horticulturae*, *141*, 17–22.

Chen, J., Liu, S. S., Kohler, A., Yan, B., Luo, H. M., Chen, X. M., & Guo, S. X. (2017). iTRAQ and RNA-Seq analyses provide new insights into regulation mechanism of symbiotic germination of *Dendrobium officinale* seeds (Orchidaceae). *Journal of Proteome Research*, *16*(6), 2174–2187.

Chen, C., Zeng, L., & Ye, Q. (2018). Proteomic and biochemical changes during senescence of Phalaenopsis 'Red Dragon'petals. *International Journal of Molecular Sciences*, *19*(5), 1317.

Chen, C., Zeng, L., Zhao, H., & Ye, Q. (2020). Proteomic analysis of the early development of the *Phalaenopsis amabilis* flower bud under low temperature induction using the iTRAQ/MRM approach. *Molecules*, *25*(5), 1244.

Chen, Y., Xu, Z., Shen, Q., & Sun, C. (2021). Floral organ-specific proteome profiling of the floral ornamental orchid (*Cymbidium goeringii*) reveals candidate proteins related to floral organ development. *Botanical Studies*, *62*, 1–12.

Choudhary, D., Mashkey, V. K., Goutam, E., Shrivastava, M., Rawat, M., Kumari, A., & Tripathi, V. (2023). Medicinal orchids: Traditional uses and recent advances. *Annals of Phytomedicine*, *12*(1), 1–9.

Chugh, S., Guha, S., & Rao, I. U. (2009). Micropropagation of orchids: a review on the potential of different explants. *Scientia Horticulturae*, *122*(4), 507–520.

Correll, D. S. (1953). Vanilla-its botany, history, cultivation and economic import. *Economic Botany*, *7*(4), 291–358.

Cribb, P. (2005). 511. Phragmipedium Kovachii: Orchidaceae. *Curtis's Botanical Magazine*, *22*(1), 8–11.

Dai, L. Y., Yin, Q. M., Qiu, J. K., Zhang, Z. Y., Li, G., Huang, M. N., & Liu, L. (2021). Goodyschle A, a new butenolide with significant BchE inhibitory activity from Goodyera schlechtendaliana. *Natural Product Research*, *35*(23), 4916–4921.

Dash, P. K., Sahoo, S., & Bal, S. (2008). Ethnobotanical studies on orchids of Niyamgiri hill ranges, Orissa, India. *Ethnobotanical Leaflets*, *2008*(1), 9.

De Conti, D., Corredor-Prado, J. P., Bueno, J. C. F., Cangahuala-Inocente, G. C., Guerra, M. P., & Pescador, R. (2020). Differentially expressed proteins in the development of protocorm-like bodies of *Cattleya tigrina* A. Richard. *Acta Physiologiae Plantarum*, *42*, 1–14.

Ding, X. Q., Zou, Y. Q., Liu, J., Wang, X. C., Hu, Y., Liu, X., & Zhang, C. F. (2021). Dendrocrepidamine, a novel octahydroindolizine alkaloid from the roots of *Dendrobium crepidatum*. *Journal of Asian Natural Products Research*, *23*(11), 1085–1092.

Dong, H., Wang, C., Li, Y., Guo, S., & Yang, J. (2010). Complete assignments of 1H and 13C NMR data of three new dihydrophenanthrofurans from *Pleione yunnanensis. Magnetic Resonance in Chemistry, 48*(3), 256–260.

Dong, H. L., Liang, H. Q., Wang, C. L., Guo, S. X., & Yang, J. S. (2013). Shancigusins E-I, five new glucosides from the tubers of *Pleione yunnanensis. Magnetic Resonance in Chemistry, 51*(6), 371–377.

Dressler, R. L. (1981). *The Orchids: Natural History and Classification.* Harvard University Press. https://www.academia.edu/download/84561214/the-orchids-natural-history-and-classification.pdf.

Duggal, S. C. (1971). Orchids in human affairs (a review). *Quarterly Journal of Crude Drug Research, 11*(2), 1727–1734.

Fang, Y. S., Yang, M. H., Cai, L., Wang, J. P., Yin, T. P., Yu, J., & Ding, Z. T. (2018). New phenylpropanoids from *Bulbophyllum retusiusculum. Archives of Pharmacal Research, 41*, 1074–1081.

Feng, S., Jiao, K., Guo, H., Jiang, M., Hao, J., Wang, H., & Shen, C. (2017). Succinyl-proteome profiling of *Dendrobium officinale*, an important traditional Chinese orchid herb, revealed involvement of succinylation in the glycolysis pathway. *BMC Genomics, 18*, 1–12.

Fossen, T., & Øvstedal, D. O. (2003). Anthocyanins from flowers of the orchids *Dracula chimaera* and *D. cordobae. Phytochemistry, 63*(7), 783–787.

Fukusaki, E., & Kobayashi, A. (2005). Plant metabolomics: potential for practical operation. *Journal of Bioscience and Bioengineering, 100*(4), 347–354.

Ghai, D., Kaur, A., Kahlon, P. S., Pawar, S. V., & Sembi, J. K. (2022). A walk through the maze of secondary metabolism in orchids: a transcriptomic approach. *Frontiers in Plant Science, 13*, 837563.

Govaerts, R., Bernet, P., Kratochvil, K., et al. (2023). World checklist of Orchidaceae. Kew: Facilitated by the Royal Botanic Gardens http://wcsp.science.kew.org/ (13 December 2023).

Gutiérrez, R. M. P. (2010). Orchids: A review of uses in traditional medicine, its phytochemistry and pharmacology. *Journal of Medicinal Plants Research, 4*(8), 592–638.

Hansen, E. (2016). *Orchid Fever: A Horticultural Tale of Love, Lust, and Lunacy.* Vintage.

Herrera, H., Valadares, R., Oliveira, G., Fuentes, A., Almonacid, L., do Nascimento, S. V., ... & Arriagada, C. (2018). Adaptation and tolerance mechanisms developed by mycorrhizal Bipinnula fimbriata plantlets (Orchidaceae) in a heavy metal-polluted ecosystem. *Mycorrhiza, 28*(7), 651–663.

Hinsley, A., De Boer, H. J., Fay, M. F., Gale, S. W., Gardiner, L. M., Gunasekara, R. S., ... & Phelps, J. (2018). A review of the trade in orchids and its implications for conservation. *Botanical Journal of the Linnean Society, 186*(4), 435–455.

Ho, C. S., Lam, C. W. K., Chan, M. H., Cheung, R. C. K., Law, L. K., Lit, L. C. W., ... & Tai, H. (2003). Electrospray ionisation mass spectrometry: principles and clinical applications. *The Clinical Biochemist Reviews, 24*(1), 3.

Hong, J., Yang, L., Zhang, D., & Shi, J. (2016). Plant metabolomics: an indispensable system biology tool for plant science. *International Journal of Molecular Sciences, 17*(6), 767.

Hu, Y., Yang, H., Ding, X., Liu, J., Wang, X., Hu, L., ... & Zhang, C. (2020). Anti-inflammatory octahydroindolizine alkaloid enantiomers from Dendrobium crepidatum. *Bioorganic Chemistry, 100*, 103809.

Huehne, P. S., Bhinija, K., Srisomsap, C., Chokchaichamnankit, D., Weeraphan, C., Svasti, J., & Mongkolsuk, S. (2020). Detection of superoxide dismutase (Cu–Zn) isoenzymes in leaves and pseudobulbs of *Bulbophyllum morphologlorum* Kraenzl orchid by comparative proteomic analysis. *Biochemistry and Biophysics Reports, 22*, 100762.

Ikeda, Y., Nonaka, H., Furumai, T., & Igarashi, Y. (2005). Cremastrine, a pyrrolizidine alkaloid from *Cremastra appendiculata. Journal of Natural Products, 68*(4), 572–573.

Inthongkaew, P., Chatsumpun, N., Supasuteekul, C., Kitisripanya, T., Putalun, W., Likhitwitayawuid, K., & Sritularak, B. (2017). α-Glucosidase and pancreatic lipase inhibitory activities and glucose uptake stimulatory effect of phenolic compounds from *Dendrobium formosum. Revista Brasileira De Farmacognosia, 27*, 480–487.

Jakubska-Busse, A., Kędziora, A., Cieniuch, G., Korzeniowska-Kowal, A., & Bugla-Płoskońska, G. (2021). Proteomics-based identification of orchid-associated bacteria colonizing the *Epipactis albensis, E. helleborine* and *E. purpurata* (Orchidaceae, Neottieae). *Saudi Journal of Biological Sciences, 28*(7), 4029–4038.

Janakiram, T., & Baskaran, V. (2018). Commercialisation and conservation aspects of orchids. *The Journal of The Orchid Society of India, 32*, 55–61.

Jangyukala, M., & Hemanta, L. (2021). Review on advances in production of hybrids in orchids. *Research & Reviews: Journal of Crop Science and Technology, 10*(3), 15–20. *Review on Advances in Production of Hybrids in Orchids Jangyukala and Hemanta STM Journals, 10,* 2.

Jiang, S., Wan, K., Lou, H. Y., Yi, P., Zhang, N., Zhou, M., ... & Pan, W. D. (2019). Antibacterial bibenzyl derivatives from the tubers of *Bletilla striata*. *Phytochemistry*, *162*, 216–223.

Jie, X., Feng, Y., Jiahao, F., Ganggui, L., Jiani, Y., Zhongyu, X., ... & Zongsuo, L. (2023). Comprehensive chemical profiling of two Dendrobium species and identification of anti-hepatoma active constituents from *Dendrobium chrysotoxum* by network pharmacology. *BMC Complementary Medicine and Therapies*, *23*(1), 217.

Jimoh, T. O., Nuamnaichati, N., Sungthong, R., Chansriniyom, C., Chanvorachote, P., Likhitwitayawuid, K., ... & Sritularak, B. (2022). Phytochemicals from *Vanda bensonii* and their bioactivities to inhibit growth and metastasis of non-small cell Lung cancer cells. *Molecules*, *27*(22), 7902.

Kasparek, M., & Grimm, U. (1999). European trade in Turkish salep with special reference to Germany. *Economic Botany*, *53*, 396–406.

Kaur, A., Verma, J., Yadav, V.G., Pawar, S.V., Sembi, J.K. (2023). Exploring the potential of in vitro cultures as an aid to the production of secondary metabolites in medicinal orchids. In: Tiwari, P., Chen JT. (eds) *Advances in Orchid Biology, Biotechnology and Omics*. Springer, Singapore.

Kim, J. H., Oh, S. Y., Han, S. B., Uddin, G. M., Kim, C. Y., & Lee, J. K. (2015). Anti-inflammatory effects of Dendrobium nobile derived phenanthrenes in LPS-stimulated murine macrophages. *Archives of Pharmacal Research*, *38*, 1117C1126.

Klongkumnuankarn, P., Busaranon, K., Chanvorachote, P., Sritularak, B., Jongbunprasert, V., & Likhitwitayawuid, K. (2015). Cytotoxic and antimigratory activities of phenolic compounds from Dendrobium brymerianum. *Evidence-Based Complementary and Alternative Medicine*, 1–9, *2015*.

Lai, T., Deng, Y., Zhang, P., Chen, Z., Hu, F., Zhang, Q., ... & Shi, N. (2013). Proteomics-based analysis of Phalaenopsis amabilis in response toward Cymbidium mosaic virus and/or Odontoglossum ringspot virus infection. *American Journal of Plant Sciences*, *4*(09), 1853.

Lai, C. Y., Yang, L. C., & Lin, W. C. (2015). Type II arabinogalactan from *Anoectochilus formosanus* induced dendritic cell maturation through TLR2 and TLR4. *Phytomedicine*, *22*(14), 1207–1214.

Lee, C. L., Chang, F. R., Yen, M. H., Yu, D., Liu, Y. N., Bastow, K. F., ... & Lee, K. H. (2009). Cytotoxic phenanthrenequinones and 9, 10-dihydrophenanthrenes from *Calanthe arisanensis*. *Journal of Natural Products*, *72*(2), 210–213.

Li, X., Xu, W., Chowdhury, M. R., & Jin, F. (2014). Comparative proteomic analysis of labellum and inner lateral petals in *Cymbidium ensifolium* flowers. *International Journal of Molecular Sciences*, *15*(11), 19877–19897.

Li, J. W., Chen, X. D., Hu, X. Y., Ma, L., & Zhang, S. B. (2018). Comparative physiological and proteomic analyses reveal different adaptive strategies by *Cymbidium sinense* and *C. tracyanum* to drought. *Planta*, *247*, 69–97.

Li, C., Dong, N., Zhao, Y., Wu, S., Liu, Z., & Zhai, J. (2021). A review for the breeding of orchids: current achievements and prospects. *Horticultural Plant Journal*, *7*(5), 380–392.

Lin, Y. L., Chen, W. P., & Macabalang, A. D. (2005). Dihydrophenanthrenes from *Bletilla formosana*. *Chemical and Pharmaceutical Bulletin*, *53*(9), 1111–1113.

Liu, M. F., Han, Y., Xing, D. M., Shi, Y., Xu, L. Z., Du, L. J., & Ding, Y. (2004). A new stilbenoid from *Arundina graminifolia*. *Journal of Asian Natural Products Research*, *6*(3), 229–232.

Lokesha, R., & Vasudeva, R. (1992). Commercial exploitation—a threat to Indian orchids?. *Current Science*, *63*, 740–743.

López-Chávez, M. Y., Guillén-Navarro, K., Bertolini, V., Encarnación, S., Hernández-Ortiz, M., Sánchez-Moreno, I., & Damon, A. (2016). Proteomic and morphometric study of the in vitro interaction between *Oncidium sphacelatum* Lindl.(Orchidaceae) and Thanatephorus sp. RG26 (Ceratobasidiaceae). *Mycorrhiza*, *26*, 353–365.

Lu, B. W., An, F. X., Cao, L. J., Yang, Y. J., Liu, P. M., Wang, X., ... & Liu, J. (2020). Proteomic profiling uncovered the cytosolic superoxide dismutase BsSOD1 associated with plant defence in the herbal orchid Bletilla striata. *Functional Plant Biology*, *47*(10), 937–944.

Majumder, P. L., & Pal, S. (1990). A steroidal ester from Coelogyne uniflora. *Phytochemistry*, *29*(8), 2717–2720.

Majumder, P. L., & Maiti, D. C. (1991). Isoflaccidinin and isooxoflaccidin, stilbenoids from *Coelogyne flaccida*. *Phytochemistry*, *30*(3), 971–974.

Majumder, P. L., & Sen, R. C. (1991). Pendulin, a polyoxygenated phenanthrene derivative from the orchid Cymbidium pendulum. *Phytochemistry*, *30*(7), 2432–2434.

Majumder, P. L., Banerjee, S., & Sen, S. (1996). Three stilbenoids from the orchid *Agrostophyllum callosum*. *Phytochemistry*, *42*(3), 847–852.

Majumder, P. L., Pal, S., & Majumder, S. (1999a). Dimeric phenanthrenes from the orchid Bulbophyllum reptans. *Phytochemistry*, *50*(5), 891–897.

Majumder, P. L., Sen, S., & Banerjee, S. (1999b). Agrostophyllol and isoagrostophyllol, two novel diastereomeric 9, 10-dihydrophenanthropyran derivatives from the orchid Agrostophyllum callosum. *Tetrahedron*, *55*(21), 6691–6702.

Majumder, P. L., Sen, S., & Majumder, S. (2001). Phenanthrene derivatives from the orchid *Coelogyne cristata. Phytochemistry*, *58*(4), 581–586.

Majumder, P. L., Majumder, S., & Sen, S. (2003). Triterpenoids from the orchids *Agrostophyllum brevipes* and *Agrostophyllum callosum. Phytochemistry*, *62*(4), 591–596.

Mumtaz, M. W., Hamid, A. A., Akhtar, M. T., Anwar, F., Rashid, U., & AL-Zuaidy, M. H. (2017). An overview of recent developments in metabolomics and proteomics–phytotherapic research perspectives. *Frontiers in Life Science*, *10*(1), 1–37.

Palama, T. L., Menard, P., Fock, I., Choi, Y. H., Bourdon, E., Govinden-Soulange, J., ... & Kodja, H. (2010). Shoot differentiation from protocorm callus cultures of Vanilla planifolia (Orchidaceae): proteomic and metabolic responses at early stage. *BMC Plant Biology*, *10*, 1–18.

Pant, B. (2013). Medicinal orchids and their uses: tissue culture a potential alternative for conservation. *African Journal of Plant Science*, *7*(10), 448–467.

Phelps, J., & Webb, E. L. (2015). "Invisible" wildlife trades: Southeast Asia's undocumented illegal trade in wild ornamental plants. *Biological Conservation*, *186*, 296–305.

Phillips, R. D., Reiter, N., & Peakall, R. (2020). Orchid conservation: from theory to practice. *Annals of Botany*, *126*(3), 345–362.

Popova, E., Kim, H. H., Saxena, P. K., Engelmann, F., & Pritchard, H. W. (2016). Frozen beauty: the cryobiotechnology of orchid diversity. *Biotechnology Advances*, *34*(4), 380–403.

Ren, Z., Ji, X., Jiao, Z., Luo, Y., Zhang, G. Q., Tao, S., ... & Wei, G. (2020). Functional analysis of a novel C-glycosyltransferase in the orchid *Dendrobium catenatum. Horticulture Research*, *7*: 1–18.

Reinikka, Merle A. (1995). *A History of the Orchid* . Timber Press.

Sarker, S. D., Latif, Z., & Gray, A. (2006). An introduction to natural products isolation. *Methods in Molecular Biology*, *864*, 1–25.

Sarker, S. D., & Nahar, L. (2005). Hyphenated techniques. *Natural Products Isolation*, *20*, 233–267.

Schiff, J. L. (2018). *Rare and Exotic Orchids: Their Nature and Cultural Significance* (pp. vii+–115). Springer International Publishing AG, India.

Sedeek, K. E., Qi, W., Schauer, M. A., Gupta, A. K., Poveda, L., Xu, S., ... & Schlüter, P. M. (2013). Transcriptome and proteome data reveal candidate genes for pollinator attraction in sexually deceptive orchids. *PloS One*, *8*(5), e64621.

Shao, S. Y., Wang, C., Han, S. W., Sun, M. H., & Li, S. (2019). Phenanthrenequinone enantiomers with cytotoxic activities from the tubers of Pleione bulbocodioides. *Organic & Biomolecular Chemistry*, *17*(3), 567–572.

Shiao, Y. J., Chen, W. P., & Lin, Y. L. (2009). New polyphenols and triterpene from the pseudobulbs of Pleione formosana. *Journal of the Chinese Chemical Society*, *56*(4), 828–833.

Shimura, H., Matsuura, M., Takada, N., & Koda, Y. (2007). An antifungal compound involved in symbiotic germination of *Cypripedium macranthos* var. rebunense (Orchidaceae). *Phytochemistry*, *68*(10), 1442–1447.

Singh, D. K. (2001). Orchid diversity in India: an overview. *Orchids: Science and Commerce*, Pathak P, RN Sehgal, N Shekhar, M Sharma and A Sood (Eds), 35–65.

Singh, A., & Duggal, S. (2009). Medicinal orchids-an overview. *Ethnobotanical Leaflets*, *2009*(3), 3.

Song, C., Zhang, Y., & Chen, R. (2022). Label-free quantitative proteomics unravel the impacts of salt stress on *Dendrobium huoshanense. Frontiers in Plant Science*, *13*, 874579.

Sritularak, B., & Likhitwitayawuid, K. (2009). New bisbibenzyls from *Dendrobium falconeri. Helvetica Chimica Acta*, *92*(4), 740–744.

Sritularak, B., Anuwat, M., & Likhitwitayawuid, K. (2011). A new phenanthrenequinone from *Dendrobium draconis. Journal of Asian Natural Products Research*, *13*(03), 251–255.

Sun, R., Sacalis, J. N., Chin, C. K., & Still, C. C. (2001). Bioactive aromatic compounds from leaves and stems of *Vanilla fragrans. Journal of Agricultural and Food Chemistry*, *49*(11), 5161–5164.

Tamer, C.E., Karaman, B., & Utku Copur, O. (2006). A traditional Turkish beverage: salep. *Food Reviews International*, *22*(1), 43–50.

Tan, B. C., Chin, C. F., Liddell, S., & Alderson, P. (2013). Proteomic analysis of callus development in *Vanilla planifolia* Andrews. *Plant Molecular Biology Reporter*, *31*, 1220–1229.

Tan, B. C., Lim, Y. S., & Lau, S. E. (2017). Proteomics in commercial crops: a overview. *Journal of Proteomics*, *169*, 176–188.

Teoh, E.S. (2016). Sources of medicinal orchids and conservation. In: Teoh, E.S. (ed.) *Medicinal Orchids of Asia,* 691–727. Springer, Cham.

Teoh, E.S. (2019). The challenge: orchid conservation. In: Teoh, E.S. (ed.) *Orchids as Aphrodisiac, Medicine or Food,* 363–376. Springer, Cham.

Thant, M. T., Sritularak, B., Chatsumpun, N., Mekboonsonglarp, W., Punpreuk, Y., & Likhitwitayawuid, K. (2021). Three novel biphenanthrene derivatives and a new phenylpropanoid ester from Aerides multiflora and their α-glucosidase inhibitory activity. *Plants, 10*(2), 385.

Tiwari, P., Sharma, A., Bose, S. K., & Gautam, A. (2022). *Biotechnological interventions in Orchids: Recent updates.* Translational success, and Commercial outcomes.

Toubiana, D., Fernie, A. R., Nikoloski, Z., & Fait, A. (2013). Network analysis: tackling complex data to study plant metabolism. *Trends in Biotechnology, 31*(1), 29–36.

Tuchinda, P., Udchachon, J., Khumtaveeporn, K., Taylor, W. C., Engelhardt, L. M., & White, A. H. (1988). Phenanthrenes of *Eulophia nuda. Phytochemistry, 27*(10), 3267–3271.

Valadares, R. B. S., Perotto, S., Santos, E. C., & Lambais, M. R. (2014). Proteome changes in *Oncidium sphacelatum* (Orchidaceae) at different trophic stages of symbiotic germination. *Mycorrhiza, 24,* 349–360.

Valadares, R. B., Perotto, S., Lucheta, A. R., Santos, E. C., Oliveira, R. M., & Lambais, M. R. (2020). Proteomic and transcriptomic analyses indicate metabolic changes and reduced defense responses in mycorrhizal roots of *Oeceoclades maculata* (Orchidaceae) collected in nature. *Journal of Fungi, 6*(3), 148.

van Wijk, K. J. (2001). Challenges and prospects of plant proteomics. *Plant Physiology, 126*(2), 501–508.

Veldman, S., Kim, S. J., Van Andel, T. R., Bello Font, M., Bone, R. E., Bytebier, B., ... & De Boer, H. J. (2018). Trade in Zambian edible orchids—DNA barcoding reveals the use of unexpected orchid taxa for chikanda. *Genes, 9*(12), 595.

Vitt, P., Taylor, A., Rakosy, D., Kreft, H., Meyer, A., Weigelt, P., & Knight, T. M. (2023). Global conservation prioritization for the Orchidaceae. *Scientific Reports, 13*(1), 6718.

Wang, Y. H., Avula, B., Abe, N., Wei, F., Wang, M., Ma, S. C., ... & Khan, I. A. (2016). Tandem mass spectrometry for structural identification of sesquiterpene alkaloids from the stems of Dendrobium nobile using LC-QToF. *Planta Medica, 82*(07), 662–670.

Wang, W., Yu, H., Li, T., Li, L., Zhang, G., Liu, Z., ... & Zhang, Y. (2017). Comparative proteomics analyses of pollination response in endangered orchid species Dendrobium chrysanthum. *International Journal of Molecular Sciences, 18*(12), 2496.

Whitten, T., Soeriaatmadja, R. E., & Afiff, S. A. (1996). *Ecology of Java & Bali* (Vol. 2). Oxford University Press: UK.

Wu, J. B., Lin, W. L., Hsieh, C. C., Ho, H. Y., Tsay, H. S., & Lin, W. C. (2007). The hepatoprotective activity of kinsenoside from *Anoectochilus formosanus. Phytotherapy Research: An International Journal Devoted to Pharmacological and Toxicological Evaluation of Natural Product Derivatives, 21*(1), 58–61.

Xu, J., Yu, H., Qing, C., Zhang, Y., Liu, Y., & Chen, Y. (2009). Two new biphenanthrenes with cytotoxic activity from *Bulbophyllum odoratissimum. Fitoterapia, 80*(7), 381–384.

Xu, J., Zhao, W. M., Qian, Z. M., Guan, J., & Li, S. P. (2010). Fast determination of five components of coumarin, alkaloids and bibenzyls in Dendrobium spp. using pressurized liquid extraction and ultra-performance liquid chromatography. *Journal of Separation Science, 33*(11), 1580–1586.

Xu, X. B., Ma, X. Y., Lei, H. H., Song, H. M., Ying, Q. C., Xu, M. J., ... & Wang, H. Z. (2015). Proteomic analysis reveals the mechanisms of *Mycena dendrobii* promoting transplantation survival and growth of tissue culture seedlings of Dendrobium officinale. *Journal of Applied Microbiology, 118*(6), 1444–1455.

Xu, X., Chen, X., Yang, R., Li, Z., Zhou, H., Bai, Y., ... & Ding, G. (2020). Crepidtumines A and B, two novel indolizidine alkaloids from *Dendrobium crepidatum. Scientific Reports, 10*(1), 9564.

Xue, Z., Li, S., Wang, S., Wang, Y., Yang, Y., Shi, J., & He, L. (2006). Mono-, Bi-, and Triphenanthrenes from the Tubers of *Cremastra appendiculata. Journal of Natural Products, 69*(6), 907–913.

Yan, H. G., Zhao, H. R., Hu, J., Lu, A. M., Fu, X. M., Jia, B., & Yang, M. H. (2016). Determination of phenanthrenes and stilbenoid in the ethyl acetate extract of Thunia alba (Lindl) by HPLC-DAD. *Analytical Methods, 8*(24), 4867–4871.

Yang, X., Tang, C., Zhao, P., Shu, G., & Mei, Z. (2012). Antimicrobial constituents from the tubers of Bletilla ochracea. *Planta Medica, 78*(06), 606–610.

Yoshikawa, M., Murakami, T., Kishi, A., Sakurama, T., Matsuda, H., Nomura, M., ... & Kubo, M. (1998). Novel indole S, O-bisdesmoside, calanthoside, the precursor glycoside of tryptanthrin, indirubin, and isatin, with increasing skin blood flow promoting effects, from two Calanthe species (Orchidaceae). *Chemical and Pharmaceutical Bulletin, 46*(5), 886–888.

Yuan, S. C. *et al.* (2021). The global orchid market. In: Chen, FC., Chin, SW. (eds) *The Orchid Genome.* Compendium of Plant Genomes. Springer, Cham.

Zeng, B., Su, M., Chen, Q., Chang, Q., Wang, W., & Li, H. (2016). Antioxidant and hepatoprotective activities of polysaccharides from *Anoectochilus roxburghii. Carbohydrate Polymers, 153*, 391–398.

Zeng, X., Li, Y., Ling, H., Chen, J., & Guo, S. (2018). Revealing proteins associated with symbiotic germination of *Gastrodia elata* by proteomic analysis. *Botanical Studies, 59*, 1–7.

Zhan, X., Qi, J., Zhou, B., & Mao, B. (2020). Metabolomic and transcriptomic analyses reveal the regulation of pigmentation in the purple variety of *Dendrobium officinale. Scientific Reports, 10*(1), 17700.

Zhang, X., Xu, J. K., Wang, J., Wang, N. L., Kurihara, H., Kitanaka, S., & Yao, X. S. (2007). Bioactive bibenzyl derivatives and fluorenones from *Dendrobium nobile. Journal of Natural Products, 70*(1), 24–28.

Zhang, X., Zhang, S., Gao, B., Qian, Z., Liu, J., Wu, S., & Si, J. (2019). Identification and quantitative analysis of phenolic glycosides with antioxidant activity in methanolic extract of Dendrobium catenatum flowers and selection of quality control herb-markers. *Food Research International, 123*, 732–745.

Zhang, D., Zhao, X. W., Li, Y. Y., Ke, S. J., Yin, W. L., Lan, S., & Liu, Z. J. (2022). Advances and prospects of orchid research and industrialization. *Horticulture Research, 9*, uhac220.

Zhang, T., Yang, X., Wang, F., Liu, P., Xie, M., Lu, C., ... & Fan, B. (2023). Comparison of the metabolomics of different Dendrobium species by UPLC-QTOF-MS. *International Journal of Molecular Sciences, 24*(24), 17148.

Zhao, W., Ye, Q., Tan, X., Jiang, H., Li, X., Chen, K., & Kinghorn, A. D. (2001). Three new sesquiterpene glycosides from *Dendrobium nobile* with immunomodulatory activity. *Journal of Natural Products, 64*(9), 1196–1200.

Zhu, L. J., Wang, M. Q., Qin, Y., Wang, M. N., Zhang, G. Q., Niu, L. T., ... & Yao, X. S. (2021). Two new dibenzyl derivatives from the stems of *Dendrobium catenatum. Journal of Asian Natural Products Research, 23*(10), 955–960.

Zizka, A., Silvestro, D., Vitt, P., & Knight, T. M. (2021). Automated conservation assessment of the orchid family with deep learning. *Conservation Biology, 35*(3), 897–908.

8 Synthetic Biology and Its Application in Biodiversity Conservation

Radha[1], *Niharika Sharma*[1], *Neeraj Kumari*[1], *Suraj Prakash*[1], *Ashok Pundir*[2], *Sunil Puri*[1]

[1]School of Biological and Environmental Sciences, Shoolini University of Biotechnology and Management Sciences, Solan, India

[2]Department of Civil Engineering, Shoolini University, Solan Himachal Pradesh, India

8.1 INTRODUCTION

Synthetic biology is a rapidly advancing field, drawing inspiration from the explosive growth and decentralized nature of developments in computing (Carlson, 2011; Church and Regis, 2014). This interdisciplinary domain blends elements of engineering and biology, and its definitions are expansive and open-ended, though not all are explicitly geared toward practical applications. In essence, synthetic biology can be described as an evolution and expansion of modern biotechnology, bringing together science, technology, and engineering to enhance and expedite the comprehension, design, redesign, manufacturing, and/or modification of genetic materials, living organisms, and biological systems (Benner and Sismour, 2005). Recent technological strides have opened up new possibilities for leveraging synthetic biology in ecological conservation efforts, providing a complementary and reinforcing dimension to existing conservation strategies. These emerging technologies in synthetic biology hold promise for curbing biodiversity loss by mitigating the impact of human-induced hazards, such as those stemming from habitat destruction. Biodiversity comprises each of these numerous species of life. Biodiversity is a fundamental element of ecosystem functioning, and it is critical for both naturally occurring and artificially created ecosystems. Furthermore, it is indispensable for the free ecosystem services that nature offers to human civilization (Sobti et al., 2023). The encroachment of commercial activities into sea and land habitats limits wildlife access, posing a threat that synthetic biology may help address (Sharma et al., 2022). Various applications within synthetic biology are strategically crafted to contribute to the attainment of Sustainable Development Goals (SDGs). For instance, innovations targeting "invasive species" align with the objectives of conserving terrestrial and marine biodiversity (SDGs 14 and 15). Meanwhile, applications tackling disease vectors like "mosquitos" not only advance goals related to human health but also contribute to poverty alleviation (SDGs 1 and 3). It's crucial to note, however, that some hazards associated with synthetic biology may pose challenges in achieving these aims, introducing a nuanced perspective to the overall impact (Sharma et al., 2022). The pace of Earth's biodiversity loss is accelerating, with species disappearing a thousand times faster than the baseline rate (Díaz et al., 2019). In tandem, biotechnology, encompassing synthetic biology, has emerged as a powerful set of techniques empowering humans to rapidly manipulate DNA, influencing the forms and functions of cells and organisms. This transformative field has already made waves in business, industry, and medicine, with the global synthetic biology market reaching a value of $5.3 billion in

 DOI: 10.1201/9781032697550-8

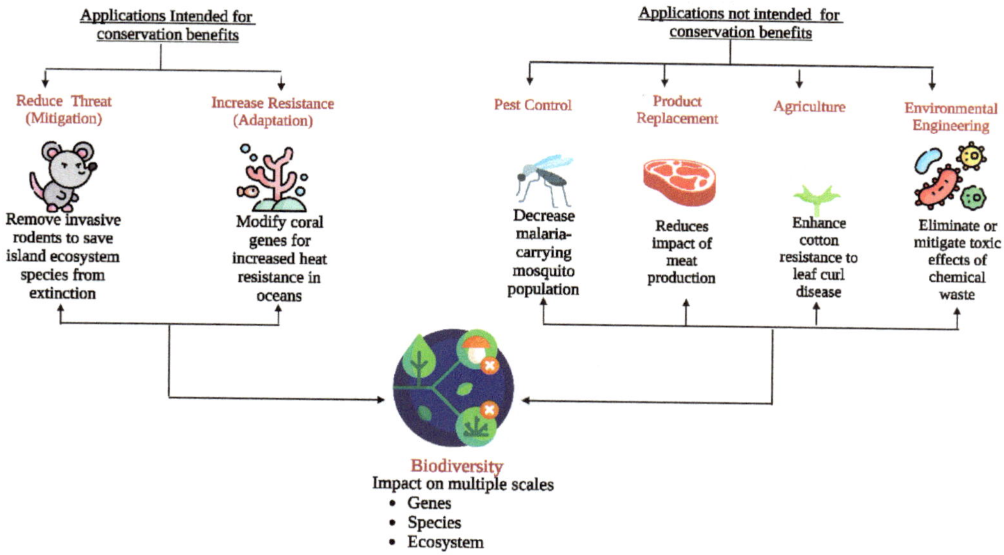

FIGURE 8.1 Impact of synthetic biology on conservation strategies.

2019 and projected to soar to $4 trillion annually within the next two decades (Chui et al., 2023). The dynamic landscape of synthetic biology holds the potential to reshape biodiversity conservation in both overt and subtle ways, offering both positive and negative impacts (Figure 8.1). From gene and genome editing using CRISPR-Cas9 to gene synthesis and drives, synthetic biology tools can be instrumental in addressing conservation challenges like invasive species, wildlife trade, and disease. However, these interventions come with the caveat of potential adverse effects (McPhee et al., 2014; Maloney et al., 2018; Novak et al., 2018; Godwin et al., 2019). Simultaneously, efforts are underway to revolutionize production methods and raw materials for consumer products such as omega-3 oils and vanillin. These changes may, in turn, alter land use patterns and nature-based supply chains, presenting both risks and opportunities for biodiversity (Bomgardner, 2016; Sprague et al., 2017; Redford et al., 2019). It's essential to recognize that like all new technologies, synthetic biology and its industrial applications have far-reaching implications for biodiversity conservation. The driving forces of markets and human consumption play a pivotal role in shaping the biosphere, and synthetic biology is no exception to this rule.

8.2 SYNTHETIC BIOLOGY TECHNIQUES

8.2.1 DNA Synthesis and Assembly

In the realm of synthetic biology, the last four decades have witnessed significant advancements in *de novo* chemical DNA synthesis and enzymatic gene assembly technologies. These innovations have greatly expanded the ability to create DNA assemblies, with the capacity growing from sequences of under 100 base pairs to the remarkable achievement of assembling DNA constructs exceeding 1 megabase pair (Mb) in length (Gibson et al., 2010a, 2010b). Synthetic DNA, with its specific nucleotide sequences and functions, finds applications across a broad spectrum, spanning metabolic engineering, gene circuit design, and genome synthesis (Ro et al., 2006; Danino et al., 2010; Steen et al., 2010). However, the applications of synthetic DNA within the field of synthetic biology are diverse and impactful, encompassing the construction of intricate genetic circuits, the engineering of complex metabolic pathways, and even the synthesis of entire artificial genomes

(Gibson et al., 2008, 2010a). Notably, scientists at the J. Craig Venter Institute achieved a groundbreaking feat by successfully performing the *de novo* synthesis and assembly of a one-million-base-pair bacterial genome. This synthetic genome, when introduced inside a bacterial cell host devoid of its own genetic material, was able to successfully generate an artificial living cell of its own genetic material, resulting in the creation of an artificial living cell, marking a significant leap forward in the field of synthetic biology (Gibson et al., 2010a). Nevertheless, it is essential to acknowledge that while DNA synthesis capabilities have seen significant advancements, the true challenge lies in comprehending the intricate biological processes underlying genome functionality. Even minor errors in genome construction can lead to the loss of essential functions, underscoring the need for a deeper understanding of these complex systems.

8.2.2 Metabolic Engineering

Metabolic engineering stands out as a valuable domain within synthetic biology, where researchers can program cells for practical applications, such as the cost-effective synthesis of chemicals or drugs. This is achieved by engineering parallel metabolic pathways that interact with the cell's natural metabolic machinery (Jiang et al., 2008; Ma et al., 2009; Steen et al., 2010). An illustrative example of this capability lies in the construction of a biosynthetic pathway for artemisinin in yeast, resulting in the microbial production of this antimalarial compound at a fraction of the cost compared to traditional methods involving extraction from the rare *Artemisia annua* plant (Ro et al., 2006). When considering whole genome synthesis, viral genomes emerge as particularly amenable targets. Their relatively small size and potential applications in vaccine development make them an attractive focus. While some endeavors seek to optimize codon usage for higher gene expression in heterologous hosts, vaccine developers employ a contrasting approach by deliberately de-optimizing codon usage to produce live attenuated forms of viruses. The SAVE (synthetic attenuated virus engineering) process, developed by Coleman and colleagues, perturbs intrinsic viral genome bias through large-scale computer-aided genome redesign and has shown promise in creating effective vaccine candidates, such as SAVE-attenuated influenza viruses (Coleman et al., 2008; Mueller et al., 2010). Until now, few milestones have been achieved, including the complete synthesis of an entire yeast chromosome (Annaluru et al., 2014), the synthesis of an entire bacterial genome (Gibson et al., 2008), and the subsequent creation of a minimal bacterial genome (Hutchison et al., 2016). These attempts serve as powerful demonstrations of the potential of synthetic DNA and the proficiency of current gene synthesis techniques in facilitating large-scale synthetic biology initiatives. Despite these remarkable advancements, existing DNA fabrication technologies still face challenges such as low throughput and a risk of errors. They are not yet mature enough for the convenient and cost-effective engineering of large genomes. Nevertheless, interdisciplinary efforts persist in exploring new chemistries and strategies for DNA synthesis and gene assembly, with the goal of achieving the widespread and economical construction of complex libraries and genomes. These ongoing endeavors hold the promise of making significant impacts on both research and society, paving the way for novel applications in synthetic biology and beyond.

8.2.3 Genetic Engineering and Modification

Genetic engineering stands at the forefront of synthetic biology, where recent breakthroughs have given rise to engineered gene drive systems. This interdisciplinary marvel, merging biotechnology, engineering, computation, and chemistry, empowers scientists to craft new DNA sequences, shaping cells and organisms with unprecedented functionalities including the precise manipulation of DNA, a process fundamental to obtaining specific functions and advancing applications in medicine, agriculture, and environmental sustainability (NASEM, 2017). In the realm of biodiversity conservation, genetic engineering emerges as a linchpin, offering innovative solutions to myriad challenges faced by ecosystems and endangered species. Noteworthy contributions include bolstering disease resistance among vulnerable species, with a focus on combating devastating pathogens like chytrid fungus in amphibians (Woodruff and Gall, 1992; Sandler, 2020; WHO, 2020; Leachman, 2021; McCormick, 2023). Genetic modification also holds promise in managing invasive species, notorious for disrupting native ecosystems. Synthetic biology enables the development of organisms with altered traits, hindering their reproductive capabilities and competitive edge (Teem et al., 2020; Bernos et al., 2022). Additionally, genetic engineering contributes to habitat restoration efforts by enhancing the growth and resilience of key plant species, aiding in the recovery of degraded ecosystems critical for biodiversity (Pyšek and Richardson, 2010; Reynolds et al., 2012; Carroll et al., 2014; Dumroese et al., 2015). Endangered species, grappling with issues such as low genetic diversity and susceptibility to diseases, benefit from genetic engineering interventions. Techniques like in vitro fertilization, cloning, and gene editing address reproductive challenges, enhancing genetic variability within small populations (Mastromonaco and Songsasen, 2020; Iqbal et al., 2021; Erwin, 2022a; Kosch et al., 2022). Synthetic biology emerges as a critical tool amid climate change, enabling the development of organisms better suited to evolving environmental conditions. For instance, modifying the genes of corals to enhance heat tolerance offers a potential lifeline to vulnerable coral reefs facing rising sea temperatures (Weston and Hood, 2004; Santoro et al., 2021; Hobbs et al., 2022). In pest control, biocontrol agents designed through synthetic biology offer targeted solutions without harming non-target species (Murty and Banerjee, 2011; Sharma, 2015). Additionally, genetic engineering plays a crucial role in conservation monitoring efforts, utilizing DNA barcoding and genetic markers for accurate species identification and population monitoring (Kress, 2017; Breed et al., 2019; Phillips et al., 2019). However, the ethical and ecological implications of these technologies must be carefully considered, ensuring responsible and sustainable deployment in the ongoing mission to preserve Earth's diverse biological heritage (Gamborg and Sandøe, 2010; Bruce and Bruce, 2014; Babale and Atoi, 2021). Scientists are currently exploring the potential of engineered gene drive systems to spread genetic changes through wild populations. Despite the rapid development of this technology, deployment remains years away. Genetic engineering tools may insert coral or microbial genes encoding antioxidant enzymes or introduce synthetic microbes for enhanced thermal tolerance. These endeavors signify a frontier in genetic engineering's potential contributions to biodiversity conservation, emphasizing its dynamic and evolving role in the face of contemporary environmental challenges (Levin et al., 2017; van Oppen et al., 2017; Baker et al., 2018; Redford et al., 2019).

8.2.4 Genome Editing

Genome editing, a pivotal area within the realm of technological advancements, is often closely associated with the field of synthetic biology. At the forefront of these developments stands CRISPR, an acronym denoting "clustered regularly interspaced short palindromic repeats" (Jiang and Doudna, 2017). CRISPR technology has ushered in a new era of possibilities in genetic engineering, manifesting in diverse applications such as organisms featuring engineered gene drives (Legros et al., 2013), the de novo domestication of species (Zsögön et al., 2018), and the innovative concept of multiplex editing (Scientific Committees, 2014; Sánchez-León et al., 2018; Borrelli et al., 2018; Feng et al., 2018; Guo et al., 2018; Li A. et al., 2018a, 2018b; Zsögön et al., 2018). However, characterizing genome editing as a subset of synthetic biology can be a challenging task, primarily due to the diverse outcomes it can yield. For instance, in the context of crop breeding, genome editing encompasses a spectrum of changes, ranging from alterations in DNA sequences akin to those arising from spontaneous or induced mutations to the precise insertion of genes, reminiscent of transgenic crop development (Custers et al., 2019). This spectrum of outcomes underscores the broader applicability of genome editing techniques in comparison to conventional biotech and non-biotech methods for generating genetic diversity. Notably, scientific committees advising the European Commission have highlighted the potential of multiplexed genome editing. This approach offers a means for modifying entire genomes with remarkable accuracy and precision, surpassing the capabilities of conventional techniques. The salient feature of concern, as identified by these committees, pertains more to the ease of using this technology and the rapid pace at which new organisms can be developed. This speed and convenience in organism creation may pose challenges in terms of regulatory oversight and ensuring a robust risk assessment framework (Scientific Committees, 2015). The connection of genome editing and synthetic biology is marked by significant advancements and innovative applications, such as CRISPR technology and its varied uses. Nevertheless, the diverse outcomes of genome editing, particularly in areas like crop breeding, raise questions about its classification as a purely synthetic biological endeavor. Moreover, the focus shifts toward regulatory considerations, as the ease and speed of organism development present a more prominent challenge than the technology or the characteristics of the modified organisms themselves.

8.3 APPLICATIONS IN BIODIVERSITY CONSERVATION

Synthetic biology explores new areas, providing an interesting opportunity for possibly bringing back species that have been extinct for a long time. Special attention is given to the significant impacts that reintroducing these plants may have on how ecosystems work and the plans for conserving them (Valdez et al., 2019). For instance, the woolly mammoth or the passenger pigeon act as informative references in this context (Novak, 2018). These projects not only display the used genetic engineering methods ranging from copying to changing the genome but also emphasize the delicate balance between challenges and achievements that are naturally part of such endeavors (Jorgensen, 2022). Beyond the controlled environments of the laboratory, this exploration requires a careful examination of the detailed ecological interactions that may happen between brought-back plant organisms and existing plants and animals (Odenbaugh, 2023). As human activities increasingly harm ecosystems, the urgent need to restore damaged habitats places synthetic biology at the forefront of conservation strategies. Human impact on ecosystems often begins with the destruction of species regeneration niches, disrupting seed production and dispersal (Hoffmann, 2022). This leads to the survival of non-recruiting adult plants, termed "living dead." This phenomenon, known as "extinction debt," results in the eventual extinction of populations in degraded habitats, even without further deterioration. This applies to various species when habitat quality, area, and connectivity thresholds are crossed due to alteration or fragmentation (Volis, 2019). Synthetic biology offers a versatile set of tools, ranging from microorganisms that enhance soil health to plants that aid in habitat recovery (Ke et al., 2021). Ongoing research projects and successful case studies shine a light on the way forward in restoring ecosystems affected by human activities. In the context of biodiversity conservation, the current standing of "synthetic biology" within Convention on Biological Diversity (CBD) discussions line up with the broader categorization of "biotechnology" as defined in Article 2. According to the Cartagena Protocol outlined in Article 3, many organisms that have been developed or are presently under development "through techniques of synthetic biology" are classified as living modified organisms (LMOs). However, there is some ambiguity for certain organisms, such as "transiently modified organisms" and those created using specific applications of genome editing, as highlighted in the 2019 report of the Ad Hoc Technical Expert Group (AHTEG) on Synthetic Biology (Keiper & Atanassova, 2020). Moreover, regulatory systems typically require prior authorization for genetic engineering or genetically modified organism's release. Lower-risk activities may involve notification or reporting obligations for lenient oversight. Synthetic biology applications undergo phased regulation from laboratory stages to full deployment, including confined field trials (Redford et al., 2019) (Figure 8.2).

In 2020, despite economic challenges such as a recession and high unemployment, the biotech industry, especially synthetic biology companies, saw a remarkable increase in investment. These companies received a staggering $7.8 billion in funding, nearly 2.5 times more than in 2019 and almost double the amount in 2018 (Figure 8.3).

Notably, National Resilience, Inc. led with an $800 million funding round. As we navigate a subdued economy and the ongoing global pandemic in 2021, the surge in biotech investment raises

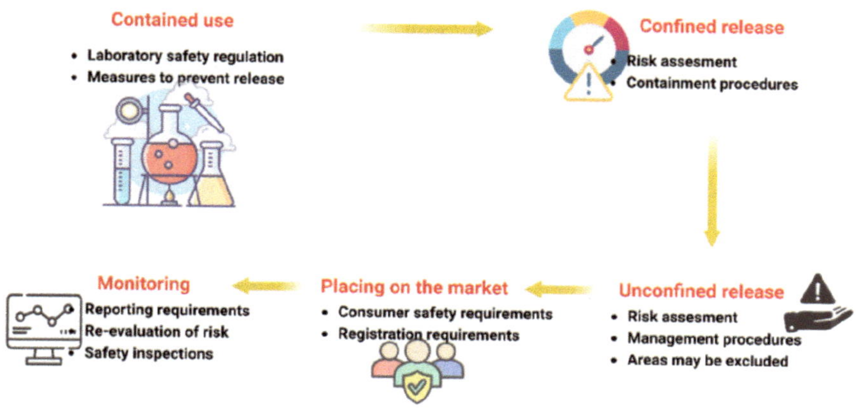

FIGURE 8.2 Key phases in the risk regulation process for synthetic biology. (Modified from Redford et al., 2019.)

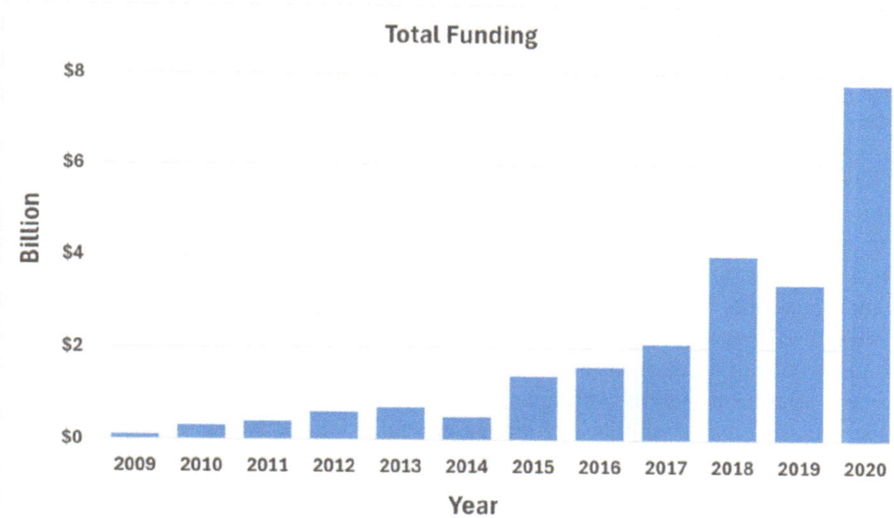

FIGURE 8.3 Aggregate funding for synthetic biology from 2009 to 2020, including private investment, public funding, and non-dilutive government grants. (Data from Synbiobeta, 2021.)

questions about the driving factors behind this growth (Synbiobeta, 2021). Among the ongoing sixth mass extinction, a crisis threatening biodiversity and posing substantial risks to ecosystems and humanity, the importance of genetic diversity is often overlooked in conservation efforts. Despite its recognized significance at various biological levels, genomics remains underutilized in biodiversity assessments (Theissinger et al., 2023). This gap can be attributed to factors such as the prioritization of limited funds for addressing human-induced threats, limited collaboration between genomic researchers and conservation practitioners, and a lack of explicit references to genetic diversity in key international regulations. Recent indications, however, suggest a potential change in this landscape, as seen in updates from the Convention for Biological Diversity (Convention on Biological

Diversity, 2022). Integrating population genomic data into conservation efforts provides a robust tool to assess genetic diversity within and between species, map diversity across landscapes, and understand the impact of neutral evolutionary processes. This information also illuminates genetic changes due to natural selection and informs the interplay between adaptation and neutral processes in shaping biodiversity (Hoffmann et al., 2015). As genomics becomes routine in conservation, it reveals phenomena like incomplete allele sorting and adaptive divergence, offering valuable insights into climate change adaptation and hybridization effects (Lovell et al., 2021). This knowledge, when applied to synthetic biology in biodiversity conservation, enhances the precision of interventions. By leveraging population genomics, synthetic biology approaches can be tailored to specific genetic diversity patterns, contributing to more effective strategies for biodiversity resilience in a changing climate. Table 8.1 presents an overview of major conservation challenges and potential synthetic biology solutions.

8.4 CASE STUDIES

In a case study, the application of synthetic biology emerges as a pivotal and transformative force in the conservation efforts of the black-footed ferret (*Mustela nigripes*). Faced with enduring threats such as genetic drift, inbreeding depression, and susceptibility to sylvatic plague, traditional conservation methodologies have proven insufficient to ensure the sustained recovery of this endangered species (Antolin et al., 2002; Abbott et al., 2012; Wisely et al., 2015). In response, scientists at the San Diego Frozen Zoo have leveraged cryopreserved cell lines from the last remaining wild ferrets in Wyoming, employing whole genome sequencing to uncover significant genetic variation within these lines. The subsequent endeavor involves the development of laboratory methods for cloning these frozen cell lines, with the aim of introducing newfound genetic diversity to counteract the effects of genetic drift and inbreeding (Revive and Restore, San Diego Zoo Global and Intrexon, 2016). Additionally, the integration of precise genome editing techniques offers a revolutionary approach to confer inheritable immunity to the persistent threat of sylvatic plague (Wisely et al., 2015). The issuance of an endangered species recovery permit in July 2018 underscores the regulatory acknowledgment of the potential of synthetic biology in addressing the multifaceted challenges faced by the black-footed ferret (Novak et al., 2018). While uncertainties and potential incompatibility issues remain, the synthetic biology interventions proposed in this case study signify a paradigm shift, demonstrating the indispensability of cutting-edge genetic technologies in enhancing the resilience and long-term viability of endangered species conservation. Furthermore, the prospect of utilizing precise genome editing techniques to confer inheritable immunity to sylvatic plague presents a groundbreaking avenue for disease management. Synthetic biology's ability to address these critical challenges showcases its instrumental role in enhancing the efficacy of conservation efforts, providing a nuanced and strategic approach to preserving biodiversity and safeguarding endangered species like the black-footed ferret (Wisely et al., 2015; Novak et al., 2018). In another study, synthetic biology plays a pivotal and transformative role in the restoration efforts of case of American chestnut (*Castanea dentata*) which is a critically endangered species. The invasive blight fungus, which nearly drove the American chestnut to extinction, posed significant challenges for traditional breeding methods (Anagnostakis, 1987; Jacobs et al., 2013). Synthetic biology, however, offers a cutting-edge solution by introducing a single gene from wheat, encoding the enzyme oxalate oxidase, into a new line of American chestnut trees. This innovation provides enhanced tolerance to blight infections without directly harming the fungus, representing a groundbreaking avenue for disease management (Chen et al., 2010; Zhang et al., 2013). However, conventional breeding approaches, constrained by the slow and unpredictable nature of the process (Goldspiel et al., 2019), are supplemented by synthetic biology's precision and efficiency. The introduction of a specific gene allows for the breakdown of the blight fungus's toxin, offering a nuanced and targeted solution to the threat of extinction (Chen et al., 2010). Furthermore, synthetic biology turns transgenic

TABLE 8.1
Key Conservation Challenges and Potential Synthetic Biology Solutions

Conservation Issues	Biodiversity Issues	Solution through Synthetic Biology	Reference
Habitat conversion	Soil productivity diminishes due to the use of pesticides and herbicides, as well as mining activities like gold or strip mining	Regenerate soil microbiomes synthetically to facilitate habitat restoration, and develop plants engineered to necessitate fewer pesticides and herbicides during production	(Sutherland et al., 2016)
	The process of extracting and utilizing fossil fuels	Explore alternative solutions to alleviate pressure on fossil fuel resources and mitigate associated damages like habitat loss and pollution	(Sutherland et al., 2014)
		Develop and modify microorganisms capable of consuming hydrocarbons to facilitate the cleanup of oil spills	
Environmental pollution	Presence of microplastics in oceans and soil	Develop or alter microorganisms with the capability to consume or break down microplastics	(Rands et al., 2010)
	Release of carbon dioxide (CO_2) or other greenhouse gases into the atmosphere	Biofuel produced from synthetic algae	(Redford et al., 2014)
	Substituting items crafted from petroleum and synthetic rubber	Genetically modify plants to produce equivalent products	(Sutherland et al., 2014)
	Bleaching of coral reefs	Modify the genome of coral reefs to enhance resilience by incorporating genetic pathways from coral species that can withstand higher temperatures and increased acidity	(Jin et al., 2016)
Overexploitation	Ivory from rhino horns and squalene from deep-sea sharks	Create a synthetic substitute material	(Piaggio et al., 2017)
Invasive species	Brown tree snake (*Boiga irregularis*) in Guam	Employ Y chromosome alterations and gene drives to disrupt reproduction in the brown tree snake species in Guam	(Piaggio et al., 2017; Levine et al., 2021)
Loss of biodiversity	Revive and restore extinct species	"Revival of extinct species" (e.g., woolly mammoth) involves modifying the genome of an existing species (e.g., elephant) by incorporating genetic code from the extinct species	(Novak, 2018)
		Process aims to create a proxy species that can potentially fulfill the same ecological role as the extinct species	
	Limits of agriculture in providing nutrition and habitat (forests) for an expanding human population	Novel food sources or methods of food production that eliminate the need for pesticides and extensive arable land	(Muhie, 2022; Zenda et al., 2023)
Pathogens	Fungal pathogens affecting North American bats as white-nose syndrome and amphibians and snakes as chytrid fungus	Genetically engineer resistance to fungal diseases in affected species	(Taylor and Gemmell, 2016)
	Occurrence of plague in black-footed ferrets	Apply CRISPR-Cas9 technology to excise the susceptible portion of the genome in black-footed ferrets and replace it with genetic code conferring disease resistance	(Novak et al., 2018)

Source: Adapted from Piaggio et al. (2017).

chestnut plantings into refugia, reducing the likelihood of blight evolution over time (NRC, 1986). Moreover, synthetic biology provides a pathway to overcome limitations associated with traditional breeding, such as inconsistent blight resistance in offspring and the extended time required for multiple generations of breeding. The application of synthetic biology plays a role in enhancing the resilience and long-term viability of the American chestnut, demonstrating that advanced genetic technologies are indispensable for effective conservation efforts. Moreover, synthetic biology has also played a significant role in case studies such as the adaptation of corals to climate change/acidification, creating proxies of extinct species (Redford et al., 2019). However, by addressing concerns related to genetic diversity, environmental safety, and public acceptance, synthetic biology emerges as a powerful tool for restoring not only the American chestnut but also serving as a model for the conservation of other threatened species and ecosystems.

8.5 FUTURE PROSPECTS

The field of synthetic biology holds profound implications for biodiversity conservation, aligning seamlessly with the overarching theme of "synthetic biology and its application in biodiversity conservation." A critical aspect of this domain lies in the prospect of de-extinction, where extinct species can be resurrected through genetic engineering (Novak, 2018). This groundbreaking approach not only offers the potential restoration of lost biodiversity but also provides an avenue for reintegrating pivotal species into ecosystems, thereby contributing significantly to the restoration of ecological equilibrium (Thulin and Röcklinsberg, 2020). Beyond de-extinction, synthetic biology, with its genetic engineering capabilities, extends its influence to enhance the resilience of organisms against environmental stressors. This includes the modification of traits such as disease resistance and adaptability to shifting climatic conditions, presenting proactive measures for the safeguarding of vulnerable species (Erwin, 2022b; Ghildiyal et al., 2023). Precision conservation strategies, a hallmark of synthetic biology, involve meticulous tailoring of interventions to the specific genetic characteristics of target species or populations. This precision optimizes the efficacy of conservation initiatives, aligning them intricately with the unique genetic profiles of each species (Becker et al., 2022; Hirschi et al., 2022). In the context of ecosystem restoration, synthetic biology emerges as a potent tool, enabling the engineering of organisms–whether plants, microbes, or animals—to actively participate in the rehabilitation of damaged habitats (Sun et al., 2022). This multifaceted approach supports the recovery of native species and fosters overall ecosystem health (Yaashikaa et al., 2022). Furthermore, the application of synthetic biology extends to climate-resilient organisms, characterized by genetically induced adaptations designed to empower species to withstand the challenges imposed by climate change. This proactive stance toward climate adaptation assumes paramount importance in light of the evolving environmental conditions (Tan et al., 2022; Zenda et al., 2023). However, as synthetic biology takes precedence in conservation, ethical considerations, international collaborations, and robust regulatory frameworks become imperative. The responsible and sustainable deployment of synthetic biology technologies hinges upon shared research efforts, adherence to ethical guidelines, and a concerted global commitment to environmental stewardship (Bohua et al., 2023; Macfarlane et al., 2022). As this scientific landscape unfolds, synthetic biology not only promises a paradigm shift in biodiversity preservation but necessitates a conscientious and collaborative approach to ensure its harmonious integration into conservation practices.

8.6 CONCLUSION

In conclusion, the chapter underscores the pivotal role of synthetic biology in biodiversity conservation. From the potential resurrection of extinct species to the strategic reintroduction of key organisms into ecosystems, synthetic biology offers groundbreaking avenues for restoring ecological equilibrium. Its genetic engineering capabilities extend to fortifying organismal resilience, addressing traits like disease resistance and adaptability to dynamic climates, providing proactive measures for safeguarding vulnerable species. The precision of synthetic biology in tailoring conservation strategies to specific genetic characteristics optimizes intervention efficacy, intricately aligning efforts with the unique genetic profiles of each species. This precision extends to ecosystem restoration, where engineered organisms actively contribute to habitat rehabilitation, aiding in the recovery of native species and bolstering overall ecosystem health. Moreover, synthetic biology's application to climate-resilient organisms, equipped with genetically induced adaptations, becomes paramount in navigating the challenges of climate change. As synthetic biology takes center stage in conservation, ethical considerations, international collaboration, and robust regulatory frameworks become imperative for its responsible and sustainable deployment. In essence, synthetic biology not only promises a transformative shift in biodiversity preservation but demands a conscientious and collaborative approach for its harmonious integration into conservation practices.

REFERENCES

Abbott RC, Osorio JE, Bunck CM, Rocke TE. Sylvatic plague vaccine: a new tool for conservation of threatened and endangered species? *EcoHealth*. 2012 Sep; 9:243–50.

Anagnostakis SL. Chestnut blight: the classical problem of an introduced pathogen. *Mycologia*. 1987 Jan 1;79(1):23–37.

Annaluru N, Muller H, Mitchell LA, Ramalingam S, Stracquadanio G, Richardson SM, Dymond JS, Kuang Z, Scheifele LZ, Cooper EM, Cai Y. Total synthesis of a functional designer eukaryotic chromosome. *Science*. 2014 Apr 4;344(6179):55–8.

Antolin MF, Gober P, Luce B, Biggins DE, Van Pelt WE, Seery DB, Lockhart M, Ball M. The influence of sylvatic plague on North American wildlife at the landscape level, with special emphasis on black-footed ferret and prairie dog conservation. 2002.

Babale YK, Atoi EN. Ethical concerns and risk perceptions associated with the application of genetic engineering. *KIU Journal of Social Sciences*. 2021 Oct 11;7(3):231–8.

Baker DM, Freeman CJ, Wong JC, Fogel ML, Knowlton N. Climate change promotes parasitism in a coral symbiosis. *The ISME Journal*. 2018 Mar;12(3):921–30.

Becker J, Fakhiri J, Grimm D. Fantastic AAV gene therapy vectors and how to find them - random diversification, rational design and machine learning. *Pathogens* 2022;11. https://doi.org/10.3390/pathogens11070756.

Benner, SA, Sismour, AM. Synthetic biology. *Nature Reviews Genetics*. 2005;6(7): 533–43.

Bernos TA, Jeffries KM, Mandrak NE. Aquatic invasive species specialists' perceptions on the importance of genetic tools and concepts to inform management. *Biological Invasions*. 2022 Jun;24(6):1863–79.

Bohua L, Yuexin W, Yakun O, Kunlan Z, Huan L, Ruipeng L. Ethical framework on risk governance of synthetic biology. *The Journal of Biosafety and Biosecurity*. 2023;5:45–56. https://doi.org/https://doi.org/10.1016/j.jobb.2023.03.002.

Bomgardner MM. The problem with vanilla. *Chemical & Engineering News*. 2016 Sep 12;94(36):38–42.

Borrelli VM, Brambilla V, Rogowsky P, Marocco A, Lanubile A. The enhancement of plant disease resistance using CRISPR/Cas9 technology. *Frontiers in Plant Science*. 2018 Aug 24;9:1245.

Breed MF, Harrison PA, Blyth C, Byrne M, Gaget V, Gellie NJ, Groom SV, Hodgson R, Mills JG, Prowse TA, Steane DA. The potential of genomics for restoring ecosystems and biodiversity. *Nature Reviews Genetics*. 2019 Oct;20(10):615–28.

Bruce D, Bruce A, editors. *Engineering Genesis: Ethics of Genetic Engineering in Non-Human Species*. Routledge; 2014 Jan 27.

Carlson RH. *Biology is Technology: The Promise, Peril, and New Business of Engineering Life*. Harvard University Press; 2011 Apr 15.

Carroll SP, Jørgensen PS, Kinnison MT, Bergstrom CT, Denison RF, Gluckman P, Smith TB, Strauss SY, Tabashnik BE. Applying evolutionary biology to address global challenges. *Science*. 2014 Oct 17;346(6207):1245993.

Chen C, Sun Q, Narayanan B, Nuss DL, Herzberg O. Structure of oxalacetate acetylhydrolase, a virulence factor of the chestnut blight fungus. *Journal of Biological Chemistry*. 2010 Aug 20;285(34):26685–96.

Chui M, Evers M, Manyika J, Zheng A, Nisbet T. The bio revolution: innovations transforming economies, societies, and our lives. In *Augmented Education in the Global Age*, 2023 Mar 23 (pp. 48–74). Routledge.

Church GM, Regis E. *Regenesis: how synthetic Biology will Reinvent Nature and Ourselves.* Basic Books; 2014 Apr 8.

Coleman JR, Papamichail D, Skiena S, Futcher B, Wimmer E, Mueller S. Virus attenuation by genome-scale changes in codon pair bias. *Science.* 2008 Jun 27;320(5884):1784–7.

Convention on Biological Diversity. COP15: Nations Adopt Four Goals, 23 Targets for 2030 In Landmark UN Biodiversity Agreement | Convention on Biological Diversity 2022. https://www.cbd.int/article/cop15-cbd-press-release-final-19dec2022 (accessed November 28, 2023).

Custers R, Casacuberta JM, Eriksson D, Sági L, Schiemann J. Genetic alterations that do or do not occur naturally; consequences for genome edited organisms in the context of regulatory oversight. *Frontiers in Bioengineering and Biotechnology.* 2019 Jan 16;6:213.

Danino T, Mondragón-Palomino O, Tsimring L, Hasty J. A synchronized quorum of genetic clocks. *Nature.* 2010 Jan 21;463(7279):326–30.

Díaz S, Settele J, Brondízio ES, Ngo HT, Agard J, Arneth A, Balvanera P, Brauman KA, Butchart SH, Chan KM, Garibaldi LA. Pervasive human-driven decline of life on Earth points to the need for transformative change. *Science.* 2019 Dec 13;366(6471):eaax3100.

Dumroese RK, Williams MI, Stanturf JA, Clair JB. Considerations for restoring temperate forests of tomorrow: forest restoration, assisted migration, and bioengineering. *New Forests.* 2015 Nov;46:947–64.

Erwin A. Changing Genes for a Changing Climate: Genetic Intervention as a Tool for Biodiversity Conservation in the Era of Climate Change. *SSRN Electronic Journal.* 2022a: 1–34. https://doi.org/10.2139/SSRN.4188993.

Erwin JA. Building better species: assisted evolution, genetic engineering, and the endangered species Act. *Cornell Law Review.* 2022b; 108:1117.

Feng X, Zhao D, Zhang X, Ding X, Bi C. CRISPR/Cas9 assisted multiplex genome editing technique in Escherichia coli. *Biotechnology Journal.* 2018 Sep;13(9):1700604.

Gamborg C, Sandøe P. Ethical considerations regarding genetically modified trees. *IUFRO Forests and Genetically Modified Trees.* Food and Agriculture Organization of the United Nations, Rome. 2010:163–76.

Ghildiyal K, Nayak SS, Rajawat D, Sharma A, Chhotaray S, Bhushan B, et al. Genomic insights into the conservation of wild and domestic animal diversity: A review. *Gene.* 2023;886:147719. https://doi.org/https://doi.org/10.1016/j.gene.2023.147719.

Gibson DG, Benders GA, Andrews-Pfannkoch C, Denisova EA, Baden-Tillson H, Zaveri J, Stockwell TB, Brownley A, Thomas DW, Algire MA, Merryman C. Complete chemical synthesis, assembly, and cloning of a *Mycoplasma genitalium* genome. *Science.* 2008 Feb 29;319(5867):1215–20.

Gibson DG, Glass JI, Lartigue C, Noskov VN, Chuang RY, Algire MA, Benders GA, Montague MG, Ma L, Moodie MM, Merryman C. Creation of a bacterial cell controlled by a chemically synthesized genome. *Science.* 2010a Jul 2;329(5987):52–6.

Gibson DG, Smith HO, Hutchison III CA, Venter JC, Merryman C. Chemical synthesis of the mouse mitochondrial genome. *Nature Methods.* 2010b Nov;7(11):901–3.

Godwin J, Serr M, Barnhill-Dilling SK, Blondel DV, Brown PR, Campbell K, Delborne J, Lloyd AL, Oh KP, Prowse TA, Saah R. Rodent gene drives for conservation: opportunities and data needs. *Proceedings of the Royal Society B.* 2019 Nov 6;286(1914):20191606.

Goldspiel HB, Newhouse AE, Powell WA, Gibbs JP. Effects of transgenic American chestnut leaf litter on growth and survival of wood frog larvae. *Restoration Ecology.* 2019 Mar;27(2):371–8.

Guo X, Chavez A, Tung A, Chan Y, Kaas C, Yin Y, Cecchi R, Garnier SL, Kelsic ED, Schubert M, DiCarlo JE. High-throughput creation and functional profiling of DNA sequence variant libraries using CRISPR–Cas9 in yeast. *Nature Biotechnology.* 2018 Jul;36(6):540–6.

Hirschi S, Ward TR, Meier WP, Müller DJ, Fotiadis D. Synthetic biology: Bottom-up assembly of molecular systems. *Chemical Reviews.* 2022;122:16294–328. https://doi.org/10.1021/acs.chemrev.2c00339.

Hobbs JP, Richards ZT, Popovic I, Lei C, Staeudle TM, Montanari SR, DiBattista JD. Hybridisation and the evolution of coral reef biodiversity. *Coral Reefs.* 2022 Jun;41(3):535–49.

Hoffmann A, Griffin P, Dillon S, Catullo R, Rane R, Byrne M, et al. A framework for incorporating evolutionary genomics into biodiversity conservation and management. *Climate Change Responses.* 2015;2:1. https://doi.org/10.1186/s40665-014-0009-x.

Hoffmann S. Challenges and opportunities of area-based conservation in reaching biodiversity and sustainability goals. *Biodiversity and Conservation.* 2022;31:325–52. https://doi.org/10.1007/s10531-021-02340-2.

Hutchison III CA, Chuang RY, Noskov VN, Assad-Garcia N, Deerinck TJ, Ellisman MH, Gill J, Kannan K, Karas BJ, Ma L, Pelletier JF. Design and synthesis of a minimal bacterial genome. *Science.* 2016 Mar 25;351(6280):aad6253.

Iqbal A, Ping J, Ali S, Zhen G, Kang JZ, Yi PZ, Huixian L, Zhihui Z. Conservation of endangered species through somatic cell nuclear transfer (SCNT). *Conservation Genetics Resources*. 2021 Sep; 13:349–57.

Jacobs DF, Dalgleish HJ, Nelson CD. A conceptual framework for restoration of threatened plants: The effective model of American chestnut (*Castanea dentata*) reintroduction. *New Phytologist*. 2013 Jan;197(2):378–93.

Jiang F, Doudna JA. CRISPR–Cas9 structures and mechanisms. *Annual Review of Biophysics*. 2017 May 22;46:505–29.

Jiang L, Althoff EA, Clemente FR, Doyle L, Rothlisberger D, Zanghellini A, Gallaher JL, Betker JL, Tanaka F, Barbas III CF, Hilvert D. De novo computational design of retro-aldol enzymes. *Science*. 2008 Mar 7;319(5868):1387–91.

Jin YK, Lundgren P, Lutz A, Raina J-B, Howells EJ, Paley AS, Willis BL, van Oppen MJH. Genetic markers for antioxidant capacity in a reef-building coral. *Science Advances*. 20162(5):e1500842. https://doi.org/10.1126/sciadv.1500842

Jorgensen D. Extinction and the end of futures*. *History Theory*. 2022;61:209–18. https://doi.org/https://doi.org/10.1111/hith.12258.

Ke J, Wang B, Yoshikuni Y. Microbiome engineering: Synthetic biology of plant-associated microbiomes in sustainable agriculture. *Trends in Biotechnology*. 2021;39:244–61. https://doi.org/https://doi.org/10.1016/j.tibtech.2020.07.008.

Keiper F, Atanassova A. Regulation of synthetic biology: Developments under the convention on biological diversity and its protocols. *Frontiers in Bioengineering and Biotechnology*. 2020;8:310. https://doi.org/10.3389/fbioe.2020.00310.

Kosch TA, Waddle AW, Cooper CA, Zenger KR, Garrick DJ, Berger L, Skerratt LF. Genetic approaches for increasing fitness in endangered species. *Trends in Ecology & Evolution*. 2022 Apr 1; 37(4): 332–345.

Kress WJ. Plant DNA barcodes: Applications today and in the future. *Journal of Systematics and Evolution*. 2017 Jul;55(4):291–307.

Leachman S. The Current: Survival Lessons. University of California, Santa Barbara. September 23, 2021. Available online at https://news.ucsb.edu/2021/020405/survival-lessons Accessed on 16, Nov. 2023.

Legros M, Xu C, Morrison A, Scott TW, Lloyd AL, Gould F. Modeling the dynamics of a non-limited and a self-limited gene drive system in structured Aedes aegypti populations. *PLoS One*. 2013 Dec 10;8(12):e83354.

Levin RA, Voolstra CR, Agrawal S, Steinberg PD, Suggett DJ, Van Oppen MJ. Engineering strategies to decode and enhance the genomes of coral symbionts. *Frontiers in Microbiology*. 2017 Jun 30;8:1220.

Levine BA, Yackel Adams AA, Douglas MR, Douglas ME, Nafus MG. Female persistence during toxicant treatment predicts survival probability of offspring in invasive brown treesnakes (Boiga irregularis). *Global Ecology and Conservation*. 2021;31:e01827. https://doi.org/https://doi.org/10.1016/j.gecco.2021.e01827

Li A, Jia S, Yobi A, Ge Z, Sato SJ, Zhang C, Angelovici R, Clemente TE, Holding DR. Editing of an alpha-kafirin gene family increases digestibility and protein quality in sorghum. *Plant Physiology*. 2018a Aug 1;177(4):1425–38.

Li L, Wei K, Zheng G, Liu X, Chen S, Jiang W, Lu Y. CRISPR-Cpf1-assisted multiplex genome editing and transcriptional repression in Streptomyces. *Applied and Environmental Microbiology*. 2018b Sep 15;84(18):e00827–18.

Lovell JT, MacQueen AH, Mamidi S, Bonnette J, Jenkins J, Napier JD, et al. Genomic mechanisms of climate adaptation in polyploid bioenergy switchgrass. *Nature*. 2021;590:438–44. https://doi.org/10.1038/s41586-020-03127-1.

Ma SM, Li JW, Choi JW, Zhou H, Lee KM, Moorthie VA, Xie X, Kealey JT, Da Silva NA, Vederas JC, Tang Y. Complete reconstitution of a highly reducing iterative polyketide synthase. *Science*. 2009 Oct 23;326(5952):589–92.

Macfarlane NBW, Adams J, Bennett EL, Brooks TM, Delborne JA, Eggermont H, et al. Direct and indirect impacts of synthetic biology on biodiversity conservation. *IScience*. 2022;25:105423. https://doi.org/https://doi.org/10.1016/j.isci.2022.105423.

Maloney T, Phelan R, Simmons N. Saving the horseshoe crab: A synthetic alternative to horseshoe crab blood for endotoxin detection. *PLoS Biology*. 2018 Oct 12;16(10):e2006607.

Mastromonaco GF, Songsasen N. Reproductive technologies for the conservation of wildlife and endangered species. In *Reproductive Technologies in Animals,* 2020 Jan 1 (pp. 99–117). Academic Press: Netherlands. doi: https://doi.org/10.1016/B978-0-12-817107-3.00007-2.

McCormick, G. Vaccine against deadly chytrid fungus primes frog microbiome for future exposure. June 12, 2023. Available online at https://www.psu.edu/news/eberly-college-science/story/vaccine-against-deadly-chytrid-fungus-primes-frog-microbiome-future/ Accessed on 16, Nov. 2023.

Mcphee D, Pin A, Kizer L, Perelman L. Squalane from sugarcane. *Cosmetics & Toiletries Magazine.* 2014;129(6):1–6.

Mueller S, Coleman JR, Papamichail D, Ward CB, Nimnual A, Futcher B, Skiena S, Wimmer E. Live attenuated influenza virus vaccines by computer-aided rational design. *Nature Biotechnology.* 2010 Jul;28(7):723–6.

Muhie SH. Novel approaches and practices to sustainable agriculture. *The Journal of Agriculture and Food Research.* 2022;10:100446. https://doi.org/https://doi.org/10.1016/j.jafr.2022.100446.

Murty US, Banerjee AK. Bioinformatics with solutions in pest management science: An insight into the evolving technologies. *Pests and Pathogens: Management Strategies.* 2011;521:536–90.

National Academies of Sciences, Engineering, and Medicine (NASEM). Preparing for future products of biotechnology. National Academies Press; 2017 Jun 28.

National Research Council (NRC). *Pesticide resistance: strategies and tactics for management.* National Academies Press. 1986.

Novak BJ, Maloney T, Phelan R. Advancing a new toolkit for conservation: From science to policy. *Cris of Journal.* 2018;1:11–5. https://doi.org/10.1089/crispr.2017.0019.

Novak BJ. De-extinction. *Genes.* 2018;9(11):1–33. https://doi.org/10.3390/genes9110548

Odenbaugh J. Philosophy and ethics of de-extinction. *Cambridge Prism Extinction.* 2023;1:e7. https://doi.org/10.1017/ext.2023.4.

Phillips JD, Gillis DJ, Hanner RH. Incomplete estimates of genetic diversity within species: Implications for DNA barcoding. *Ecology and Evolution.* 2019 Mar;9(5):2996–3010.

Piaggio AJ, Segelbacher G, Seddon PJ, Alphey L, Bennett EL, Carlson RH, Friedman RM, Kanavy D, Phelan R, Redford KH, Rosales M, Slobodian L, Wheeler K. Is It time for synthetic biodiversity conservation? *Trends in Ecology & Evolution.* 2017;32(2):97–107. https://doi.org/10.1016/j.tree.2016.10.016

Pyšek P, Richardson DM. Invasive species, environmental change and management, and health. *Annual Review of Environment and Resources.* 2010 Nov 21;35:25–55.

Rands MRW, Adams WM, Bennun L, Butchart SHM, Clements A, Coomes D, Entwistle A, Hodge I, Kapos V, Scharlemann JPW, Sutherland WJ, Vira B. Biodiversity conservation: Challenges beyond 2010. *Science.* 2010;329(5997):1298–1303. https://doi.org/10.1126/science.1189138

Redford K, Brooks T, Macfarlane N, Adams J, Alphey L, Bennet E, Delborne J, Eggermont H, Esvelt K, KinGirl A, Kokotovich A. Genetic frontiers for conservation: An assessment of synthetic biology and biodiversity conservation. *IUCN.* 2019.

Redford KH, Adams W, Carlson R, Mace GM, Ceccarelli B. Synthetic biology and the conservation of biodiversity. *Oryx.* 2014;48:330–6. https://doi.org/DOI: 10.1017/S0030605314000040.

Revive & Restore, San Diego Zoo Global and Intrexon. 'A genetic rescue proposal for the black-footed ferret: addressing genetic diversity erosion and enhancing disease resistance in the endangered black-footed ferret with a new form of genetic rescue–employing interspecies somatic cell nuclear transfer an', 2016, 20.

Reynolds LK, McGlathery KJ, Waycott M. Genetic diversity enhances restoration success by augmenting ecosystem services. *PLoS One.* 2012 Jun 25;7(6):e38397.

Ro DK, Paradise EM, Ouellet M, Fisher KJ, Newman KL, Ndungu JM, Ho KA, Eachus RA, Ham TS, Kirby J, Chang MC. Production of the antimalarial drug precursor artemisinic acid in engineered yeast. *Nature.* 2006 Apr 13;440(7086):940–3.

Sánchez-León S, Gil-Humanes J, Ozuna CV, Giménez MJ, Sousa C, Voytas DF, Barro F. Low-gluten, non-transgenic wheat engineered with CRISPR/Cas9. *Plant Biotechnology Journal.* 2018 Apr;16(4):902–10.

Sandler R. The ethics of genetic engineering and gene drives in conservation. *Conservation Biology.* 2020 Apr;34(2):378–85.

Santoro EP, Borges RM, Espinoza JL, Freire M, Messias CS, Villela HD, Pereira LM, Vilela CL, Rosado JG, Cardoso PM, Rosado PM. Coral microbiome manipulation elicits metabolic and genetic restructuring to mitigate heat stress and evade mortality. *Science Advances.* 2021 Aug 13;7(33):eabg3088.

Scientific Committees. Scientific Committee on Emerging, and Newly Identified Health Risks, Scientific Committee on Consumer Safety, Scientific Committee on Health and Environmental Risks (Scientific Committees). Opinion on Synthetic Biology I: Definition, September 2014. Brussels: Scientific Committees. 2014.

Scientific Committees. Scientific Committee on Emerging, and Newly Identified Health Risks. Scientific Committee on Consumer Safety, Scientific Committee on Health and Environmental Risks (Scientific Committees) Final Opinion on Synthetic Biology III: Risks to the Environment and Biodiversity Related to Synthetic Biology and Research Priorities in the Field of Synthetic Biology, December 2015. Brussels: Scientific Committees

Sharma HC. Application of Modern Tools of Biotechnology for Pest Management-Prospects and Limitations. *Biological and Molecular Approaches in Pest Management.* 2015;1:1–11.

Sharma P, Verma N, Dhar PK. Synthetic biology and biodiversity. *Asian Biotechnology & Development Review.* 2022 Mar 1;24(1):71–96.

Sobti RC, Thakur M, Kaur T, Mishra S. *Biodiversity: Threats and conservation strategies.* In *Biodiversity.* CRC Press. 2023.

Sprague M, Betancor MB, Tocher DR. Microbial and genetically engineered oils as replacements for fish oil in aquaculture feeds. *Biotechnology Letters.* 2017;39:1599–1609.

Steen EJ, Kang Y, Bokinsky G, Hu Z, Schirmer A, McClure A, Del Cardayre SB, Keasling JD. Microbial production of fatty-acid-derived fuels and chemicals from plant biomass. *Nature.* 2010 Jan 28;463(7280):559–62.

Sun T, Song J, Wang M, Zhao C, Zhang W. Challenges and recent progress in the governance of biosecurity risks in the era of synthetic biology. *The Journal of Biosafety and Biosecurity.* 2022;4:59–67. https://doi.org/https://doi.org/10.1016/j.jobb.2022.02.002.

Sutherland WJ, Aveling R, Brooks TM, Clout M, Dicks L V, Fellman L, et al. A horizon scan of global conservation issues for 2014. *Trends in Ecology & Evolution.* 2014;29:15–22. https://doi.org/https://doi.org/10.1016/j.tree.2013.11.004.

Sutherland WJ, Broad S, Caine J, Clout M, Dicks L V, Doran H, et al. A Horizon Scan of Global Conservation Issues for 2016. *Trends in Ecology & Evolution.* 2016;31:44–53. https://doi.org/https://doi.org/10.1016/j.tree.2015.11.007.

Synbiobeta. Synthetic Biology Investment Reached a New Record of Nearly $8 Billion in 2020 — What Does This Mean For 2021? - SynBioBeta 2021. https://www.synbiobeta.com/read/synthetic-biology-investment-reached-a-new-record-of-nearly-8-billion-in-2020-what-does-this-mean-for-2021 (accessed November 28, 2023).

Tan C, Kalhoro MT, Faqir Y, Ma J, Osei MD, Khaliq G. Climate-resilient microbial biotechnology: A perspective on sustainable agriculture. *Sustainability.* 2022;14:1–29. https://doi.org/10.3390/su14095574.

Taylor HR, Gemmell NJ. Emerging technologies to conserve biodiversity: Further opportunities via genomics. Response to Pimm et al. *Trends in Ecology & Evolution.* 2016;31:171–2. https://doi.org/10.1016/j.tree.2016.01.002.

Teem JL, Alphey L, Descamps S, Edgington MP, Edwards O, Gemmell N, Harvey-Samuel T, Melnick RL, Oh KP, Piaggio AJ, Saah JR. Genetic biocontrol for invasive species. *Frontiers in Bioengineering and Biotechnology.* 2020 May 25;8:452.

Theissinger K, Fernandes C, Formenti G, Bista I, Berg PR, Bleidorn C, et al. How genomics can help biodiversity conservation. *Trends in Genetics.* 2023;39:545–59. https://doi.org/10.1016/j.tig.2023.01.005.

Thulin C-G, Röcklinsberg H. Ethical considerations for wildlife reintroductions and rewilding. *Frontiers in Veterinary Science.* 2020;7:1–10. https://doi.org/10.3389/fvets.2020.00163.

Valdez RX, Kuzma J, Cummings CL, Nils Peterson M. Anticipating risks, governance needs, and public perceptions of de-extinction. *Journal of Responsible Innovation.* 2019;6:211–31. https://doi.org/10.1080/23299460.2019.1591145.

Van Oppen MJ, Gates RD, Blackall LL, Cantin N, Chakravarti LJ, Chan WY, Cormick C, Crean A, Damjanovic K, Epstein H, Harrison PL. Shifting paradigms in restoration of the world's coral reefs. *Global Change Biology.* 2017 Sep;23(9):3437–48.

Volis S. Conservation-oriented restoration – a two for one method to restore both threatened species and their habitats. *Plant Divers.* 2019;41:50–8. https://doi.org/https://doi.org/10.1016/j.pld.2019.01.002.

Weston AD, Hood L. Systems biology, proteomics, and the future of health care: toward predictive, preventative, and personalized medicine. *Journal of Proteome Research.* 2004 Apr 12;3(2):179–96.

Wisely SM, Ryder OA, Santymire RM, Engelhardt JF, Novak BJ. A road map for 21st century genetic restoration: gene pool enrichment of the black-footed ferret. *Journal of Heredity.* 2015 Sep;106(5):581–92.

Woodruff DS, Gall GA. Genetics and conservation. *Agriculture, Ecosystems & Environment.* 1992 Oct 1;42(1–2):53–73.

World Health Organization (WHO). Evaluation of genetically modified mosquitoes for the control of vector-borne diseases: position statement, October 2020.

Yaashikaa PR, Devi MK, Kumar PS. Engineering microbes for enhancing the degradation of environmental pollutants: A detailed review on synthetic biology. *Environmental Research.* 2022;214:113868. https://doi.org/https://doi.org/10.1016/j.envres.2022.113868.

Zenda T, Wang N, Yan X, Dong A, Yang Q, Zhong Y, et al. Opportunities and avenues for achieving crop climate resilience. *Environmental and Experimental Botany.* 2023;213:105414. https://doi.org/https://doi.org/10.1016/j.envexpbot.2023.105414.

Zhang B, Oakes AD, Newhouse AE, Baier KM, Maynard CA, Powell WA. A threshold level of oxalate oxidase transgene expression reduces Cryphonectria parasitica-induced necrosis in a transgenic American chestnut (Castanea dentata) leaf bioassay. *Transgenic Research*. 2013 Oct;22:973–82.

Zsögön A, Čermák T, Naves ER, Notini MM, Edel KH, Weinl S, Freschi L, Voytas DF, Kudla J, Peres LE. De novo domestication of wild tomato using genome editing. *Nature Biotechnology*. 2018 Dec;36(12):1211–6.

9 Microorganisms for Habitat Restoration

Shivani Guleria[1], Kriti Singh[1], Ranbir Chander Sobti[2], Swati Yadav[3], Suman Mishra[3], Kamal Jaiswal[3]

[1]Shoolini University of Biotechnology and Management Sciences, Solan - Himachal Pradesh, India

[2]Department of Biotechnology, Panjab University, Chandigarh India

[3]Department of Zoology, School of Life Sciences, Babasaheb Bhimrao Ambedkar University, Lucknow, Uttar Pradesh, India

9.1 INTRODUCTION

Ecosystem restoration is increasingly acknowledged as a prominent remedy for the biodiversity issue. Nevertheless, there is a disparity between the field of restoration research and its practical application. More precisely, we need a conceptual framework that will enhance our comprehension of the recuperation of ecosystems and enable us to improve the design of restoration efforts (Gawecka and Bascompte, 2023). The primary goal of most restoration initiatives is to preserve biodiversity. This can be as simple as reintroducing a single species of unusual plant or animal or as complex as trying to reintroduce a large variety of species (Jordan *et al.*, 1988, 1990; Falk *et al.*, 1996; Volis *et al.*, 2019). The process of manipulating ecological processes and animals to establish self-organizing, sustainable, native ecosystems as essential components of the landscape is known as restoration (Jones, 2017). Moreover, ecological restoration aims to address the environmental degradation caused by growing human populations (Dietz *et al.*, 2003; Jones, 2017).

Microorganisms have existed since the beginning of Earth, for about 3.5 billion years, and their diversity and abundance serve as the foundation for sustaining healthy ecosystems and providing life support for the biosphere (Knoll, 2015; Cavicchioli *et al.*, 2019). As per the studies, microorganisms are essential for nutritional cycles including nitrogen and carbon, as well as in plant and animal health, food security, and even agricultural yields (Harindintwali *et al.*, 2021). Furthermore, seasonal variations and the occurrence of extreme events may also be exacerbated by climate change, which may have an impact on the variety of microbes (Khursheed, 2016; Banerjee *et al.*, 2020; Ibáñez *et al.*, 2023). Indeed, microorganisms somehow can survive in the harshest conditions on Earth, including the deep sea's highest pressure, hypersaline and acidic environments, and even in warm and extremely cold habitats (Boetius *et al.*, 2015; Hiraoka *et al.*, 2020; Manea *et al.*, 2019). Considering how fast ecosystems and biodiversity are disappearing, ecological restoration efforts seem to be the most promising way to restore them (Corinaldesi *et al.*, 2023). According to Gann *et al.* (2019), this process can help to recover degraded, damaged, and destroyed ecosystems to restore ecosystem services and lost biodiversity such as nutrient cycling, habitat and genetic maintenance, primary and secondary production, pollution control, carbon storage, and prevention of erosion.

Since very few natural ecosystem-dwelling microorganism species are cultivated, they are not even an acceptable choice for restoration (Gann *et al.*, 2019). A significant exception is symbionts (mycorrhizae or N-fixing bacteria), which are frequently employed in reforestation and agriculture but less frequently in restoration (Allen *et al.*, 2001). It has been found that a range of benign, non-pathogenic, and effective microorganisms live with each other to form effective microorganisms

DOI: 10.1201/9781032697550-9

(EM Trading, 2000). It consists mostly of a mixture of groups that are common in all ecosystems, both anaerobic and aerobic. Natural microorganisms that can renew and purify the environment are used in this effective microorganism (EM) technology (Zakaria *et al.*, 2010). The roughly 80 species of microorganisms found in EM are classified into five groups: actinomycetes, yeasts, fermenting fungi, photosynthesizing bacteria, and lactic acid bacteria (Balogun *et al.*, 2016). The primary species that are typically involved are the actinomycetes *Streptomyces griseus* and *Ablus;* the lactic acid bacteria including *Lactobacillus plantarum*, *L. casei*, and *Streptoccus lactis*; the photosynthetic bacteria such as *Rhodopseudomonas palustris* and *Rhodobacter spaeroides;* the yeasts *Saccharomyces cerevisiae* and *Candida utilis*; and the fermenting fungi *Aspergillus oryzae*, *Penicillium* sp., and *Mucor hiemalis* (Naik *et al.*, 2019).

9.2 MICROORGANISMS IN AQUATIC HABITAT RESTORATION

Nowadays, the goods and services provided by riverine systems are becoming critical to humanity. With the increase in population and demands, our dependence also grows (McIntyre *et al.*, 2016). These systems continue to deteriorate globally as our activities grow. Even though we are making efforts to restore these systems, many efforts have fallen short. Since microbial communities are mostly responsible for the breakdown of organic matter, removal or transformation of nutrients, removal of pollutants, and the production of biomass in river ecosystems, a possible approach to improve the efficacy of restorations would be to specifically design systems to support microbial communities (Hilderbrand *et al.*, 2023). Many techniques and strategies for the ecological restoration of various aquatic habitats and ecosystems, such as macroalgae forests, reef-forming corals, and seagrass beds, have been accounted for. As microbial communities are mostly responsible for many activities, as discussed earlier, they are very beneficial and help to improve the efficacy of restorations (Bentley, 2021). Numerous ecologically and commercially significant ecosystem services are offered by marine ecosystems, such as nutrient cycling, coastal protection, habitat provision, and fisheries enhancement (Buonocore *et al.*, 2021). The microorganisms living in these ecosystems are intrinsically linked to the health and services provided, such as disease, pollution remediation, and drug discovery (Dash *et al.*, 2013; Singh *et al.*, 2019). The organisms that form the foundation of these ecosystems, including sponges, seagrass beds, mangroves, salt marshes, and corals, are primarily responsible for aquatic ecosystem services (Trevathan-Tackett *et al.*, 2019).

The range of roles that microbiota play in aquatic ecosystems is far more extensive than what is depicted in Figure 9.1. Numerous ecosystem processes are both directly and indirectly impacted by microbiota. Some of the effects include animal and plant health (such as triggering immune responses, outcompeting opportunistic pathogens, and generating antibiotic compounds), resistance to drought stress (for example, systemic tolerance can be biochemically induced by plant growth-promoting microorganisms), nutrient cycling (such as regulating belowground carbon by stabilizing it in the soil mineral matrix or by decomposing organic matter), hormone production in animals and plants (including microbe-derived auxin as a signaling molecule in the growth and development of plants), regulating climatic conditions (through consuming as well as producing CO_2, NO_2, and CH_4), and inter- and intrakingdom communication (via biochemical lures and quorum sensing, respectively) (Gao *et al.*, 2010; Garcia *et al.*, 2018; Ma *et al.*, 2020; Ethridge *et al.*, 2021; Eichmann *et al.*, 2021; Wu *et al.*, 2022).

The effectiveness of restoration and the overall resilience of the ecosystem to the stresses that caused the restoration could be enhanced by concentrating on the diversity and function of microbes (Singh and Gupta, 2018). Recent developments in next-generation sequencing have made it possible to evaluate the ecological conditions of streams using microbial omics techniques such as metagenomics and metatranscriptomics, in a manner similar to that of fish and macrovertebrates (Hilderbrand *et al.*, 2023). It has been proved that microbial populations are basic components of the aquatic environment as they contribute to dynamics, biological processes, and the food web (Bowman *et al.*, 2016). These microbial communities are mainly comprised of viruses, microbial

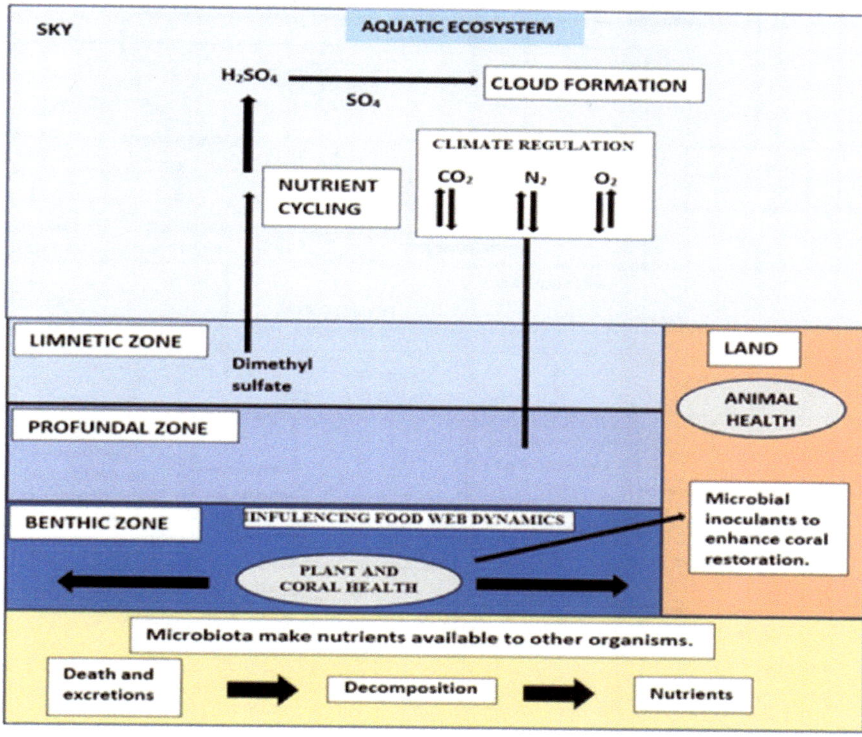

FIGURE 9.1 The importance of microbial communities to aquatic ecosystems (Robinson *et al.*, 2023).

eukaryotes, bacteria, and archaea, representing almost all unfathomable levels of the aquatic life cycle (Nissimov, 2017). Moreover, next-generation tools, like ecological metatranscriptomics, could improve our comprehension of how particular restoration strategies contribute to ecological services and help evaluate the efficacy of restorations (Robinson *et al.*, 2023). Scientists have nowadays widely adopted metagenomic analysis, which can be specified as the analysis and production of shotgun genomic data from microbial assemblages and are now regularly used in studies of a variety of aquatic habitats (Grossart *et al.*, 2020). Gilbert and Dupont (2011) stated that the term "metagenomics" encompasses single gene marker studies. Some of the recent impactful metagenomics studies from the freshwater and marine ecosystems contain overarching challenges and concepts in which metagenomics is involved. These mainly include (a) the development and validation of metabolic and biogeochemical models of aquatic microbial systems using metagenomics data, (b) to facilitate more mechanistic hypothesis testing and deeper interpretation of meta-omics datasets, appropriate model organisms and systems are required, and (c) the significance of including metagenomic techniques in experimental laboratory and field research (Grossart *et al.*, 2020). The concept of freshwater and marine ecosystems with respect to metagenomics is shown in Figure 9.2.

Figure 9.2 is mainly a summary of the meta-omics toolkit and the information that can be obtained from microbial genotype to phenotype using the meta-omics technique. However, microbial ecology can transition from a descriptive to a more predictive field by integrating biogeochemical and ecological information of aquatic microbiomes with tools and techniques from microbiology.

Nowadays, the constant findings of phylogenetic and metabolic diversity in microbial systems are a classic contribution of metagenomics. Discoveries such as the marine archaea capacity for ammonia oxidation, sulfur oxidation in abundant oxygen minimum zone bacteria, and, moreover,

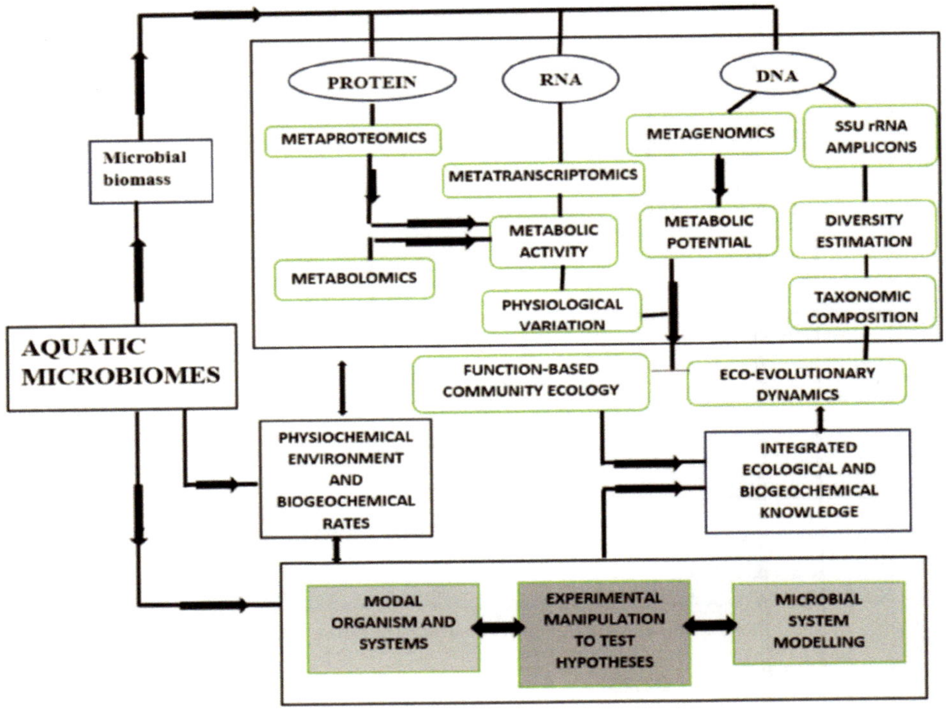

FIGURE 9.2 A conceptual structure for biogeochemistry and studies of aquatic microbial ecology that incorporates a variety of meta-omics approaches (Grossart *et al.*, 2020).

rhodopsin-based phototrophy in the surface ocean are noteworthy (Hallam *et al.*, 2006; Béja *et al.*, 2000; Walsh *et al.*, 2009). Based on MAG (metagenome-assembled genome) and SAG (single-cell amplified genome) analysis, freshwater bacteria may have a role in extracellular electron transfer in freshwater (He *et al.*, 2019). Moreover, SAR202 has now been linked by analysis of MAGs and SAGs to the degradation of land organic matter in the Arctic Ocean, as well as the oxidation of resistant organic matter and sulfur in the deep sea (Landry *et al.*, 2017; Mehrshad *et al.*, 2018; Colatriano *et al.*, 2018). Furthermore, metagenomics and metatranscriptomics are promising techniques to investigate the ecology, physiology, and evolution of microbial eukaryotes through gene content, while also revealing metabolic features of archaea and bacteria (Caron *et al.*, 2017).

In addition, a great way to understand the basic concepts of community dynamics and assembly is by studying aquatic microbial communities (Gralka *et al.*, 2020). The vast phylogenetic diversity of microorganisms and our limited knowledge of numerous taxon-specific traits significantly challenge the interpretation of community membership patterns and functional implications in microbial community ecology (Krause *et al.*, 2014; Hall *et al.*, 2018). It has been acknowledged that processes such as diversification, dispersal, drift, and selection are all are crucial for the formation of microbial and bacterial communities. Moreover, using metagenomic profiling, their capacity to influence the functional and taxonomic composition of various microbial groups, systems, and scales can be examined (Nemergut *et al.*, 2013; Grossart *et al.*, 2020). It has been stated that the supply of bioavailable nitrogen in exchange for carbon compounds produced by methanogenic assemblages and metagenomic evidence supports a beneficial interaction between nitrogen-fixing bacteria and methanogenic archaea (Fernandez *et al.*, 2020). Additionally, researchers inspected the bacterial interaction associated with microcystis, while others studied the functional gene profiling and SSU rRNA to investigate the biotic interactions and microbial community assembly that are different in shallow lakes with macrophyte and phytoplankton-dominated regimes (Cook *et al.*, 2020; Wang *et al.*, 2020). Different biotic interactions shape the functional and taxonomic composition of two systems in different ways, as the researchers have described. This is mainly because they influence structure–function relationships and control biogeological processes, evolution, ecology, and the emergence of microbial interactions.

9.3 MICROORGANISMS IN RESTORATION OF LAKE AQUATIC–TERRESTRIAL ECOTONES

The degradation of lake water habitats, particularly the rise in the eutrophication of freshwater lakes, is a widespread phenomenon that occurs anywhere there is water (Bhagowati and Ahamad, 2019). The process of eutrophication is responsible for several unfavorable outcomes, including the abundance of algae blooms, the decrease in aquatic vegetation, the death of fish, and the final deterioration and collapse of ecosystems (Amorim and Moura, 2021). One of the primary reasons for eutrophication is the deterioration of aquatic–terrestrial ecotones (Xie *et al.*, 2018). The transitional zones among terrestrial ecosystems and aquatic ecosystems are referred to as aquatic–terrestrial ecotones (Hennings *et al.*, 2021). They are essential to the biological, geochemical, and hydrological cycles that connect terrestrial and aquatic ecosystems. These cycles include runoff management, biodiversity preservation, microclimate regulation, and the movement and transformation of nutrients, such as nitrogen and phosphorus (Zhu *et al.*, 2020; Deere et al., 2022). Lake aquatic–terrestrial ecotones serve as both a recipient and provider of nutrients, and they also serve as a genetic reservoir and crucial habitat for lake fauna (Mander *et al.*, 2017; Arif *et al.*, 2023). Lake aquatic–terrestrial ecotones possess the distinctive characteristic of experiencing seasonal fluctuations in water levels, which give rise to anaerobic conditions that are vital for the survival of plants and animals in these areas (Penton *et al.*, 2013). Freshwater lake ecosystems provide significant value as both human resources and vital environmental systems (Cheng *et al.*, 2023; Huang *et al.*, 2020; Yang *et al.*, 2016). Hence, it is very significant to investigate the impact of lake aquatic–terrestrial ecotones on eutrophication via ecological restoration studies. To successfully regulate or minimize pollutants and accomplish water purification, a lake ecotone requires a vegetation buffer zone with a minimum width requirement. The primary methods of action consist of the accumulation of pollutants in water runoff, absorption by plants, adsorption by soil, and the breakdown of pollutants via microorganisms (Liang *et al.*, 2019). In lake ecotones, microbial degradation is essential for the removal of phosphorus and nitrogen. Microorganisms play a crucial role in ecosystems by decomposing organic matter, which helps to eliminate pollutants and recycle nutrients. Microorganisms play a significant role in nitrogen dynamics in high-nutrient habitats, since plants are unable to directly absorb and use nitrogen (Dai *et al.*, 2023).

The microbial community in lake ecotones is diverse and includes metazoans, actinomycetes, bacteria, mycobiota, and protozoa. Of all the organisms, bacteria are the most widespread and intricate group. The bacterial population density in soil may reach several hundred million or even billions per gram. Microorganisms often adhere to solid surfaces, including the surfaces of plants (Zhou *et al.*, 2020). The plants found in lake ecotones can eliminate nitrogen and phosphorus by directly absorbing them, as well as indirectly via the stimulation of microbial activity by changing the redox potential in the wetland. Kickuth's root zone theory suggests that plants in lake ecotones can transfer oxygen through photosynthesis to the root zone via sparsely distributed tissue. This process results in the formation of an aerobic zone near the root zone, while areas farther away from the root zone become deprived of oxygen (anoxic), and areas even farther away become devoid of oxygen (anaerobic) (Winter and Kickuth, 1989). Thus, a variety of microorganisms, including facultative anaerobic bacteria, aerobic bacteria, and anaerobic bacteria, may be found in this zone. A lake ecotone harbors separate anaerobic and aerobic bacterial populations. Furthermore, the indigenous microbial community is influenced by an array of factors including the concentration of oxygen in the water, the release of chemical compounds by plants, and the characteristics of the soil. Pollutants operate as both carbon and energy supplies for the proliferation and reproduction of microbes, or generate intermediate metabolites that offer vital nutrients, thereby promoting the release of active enzymes for the breakdown of pollutants (Wen *et al.*, 2011). Microorganisms play a crucial role in the removal of nitrogen and phosphorus pollutants. They directly contribute to the breakdown of organic matter via mineralization and ammonia oxidation. This process is essential for mitigating nitrogen pollution and promoting ecological services in vegetated buffer environments.

Water bodies that are eutrophic are constantly bombarded by nitrogen, phosphorus, and organic contaminants transported by microorganisms. It is crucial for ecological rejuvenation that microorganisms that eliminate nitrogen and phosphorus in lake ecotones are as efficient and functional as possible (Dai *et al.*, 2023). To increase the ability of microorganisms in the ecotones of eutrophic water bodies, strategies may involve introducing novel bacteria, utilizing the varied physiological traits of different microorganisms, methodically identifying the most suitable bacteria, combining several microbial strains into a composite microbial treatment, and employing appropriate bacterium encapsulation technology (to increase their ecological efficiency). It is important to note that practical uses of incorporating microbial communities to improve microbial habitats in eutrophic water bodies are subject to some constraints. Some of them include the fact that microbes are susceptible to being washed away by moving water, that their activity durations are relatively short, and that it may be difficult to stabilize treated water since other organisms are more likely to prey on them in stagnant water (Dai *et al.*, 2023). Furthermore, the absence of transporters for adsorption, development, and reproduction makes it challenging to preserve and transmit these organisms across the environment (Sun *et al.*, 2020). The induction and development of technology treatment for microbial immobilization have, fortunately, been able to successfully solve these issues. Microbial immobilization methods have potential for the restoration of eutrophic lakes and rivers. Immobile microorganisms effectively restrict the growth of algae, therefore aiding in water filtration. Encapsulation is a well-researched technique for immobilization (Dong *et al.*, 2014).

9.4 MICROORGANISMS IN TERRESTRIAL HABITAT RESTORATION

Microorganisms are widespread in the environment, where they perform several critical tasks, some of which are shown in Figure 9.3. Microbial communities perform a crucial role in shaping land and are responsible for many ecosystem services associated with terrestrial ecosystems (Rajendhran and Gunasekaran, 2008; Torsvik and Øvreas, 2002). These services include plant production, protection of drinking water, and carbon sequestration. They are directly influenced by the activities and functional characteristics of microorganisms (Schulz *et al.*, 2013).

Soil is the skin of Earth and provides essential ecosystem services for existence. It filters and stores water, provides nutrients to plants and heterotrophs, serves as a growing medium for plants and heterotrophs, provides surroundings to a wide variety of organisms, and contains many antibiotics we use (Coban *et al.*, 2022).

Microbial communities are integral components of many ecosystems and have a crucial impact on their operation. Therefore, prior to exploring the contribution of microorganisms in the recovery of impaired ecosystems, it is crucial to comprehend their varied functions in the operation of ecosystems. Soil microorganisms play a crucial role in the process of biogeochemical cycling (Smith *et al.*, 2015) and significantly contribute to the productivity of terrestrial ecosystems (Kotas *et al.*, 2018). The soil microbiota mostly consists of microorganisms such as fungi and bacteria, which discharge nutrients such as phosphorus, nitrogen, and carbon into the soil, thereby preserving soil quality and fertility. Plant microbiota in the rhizosphere, phyllosphere, and endosphere play many important roles, including promoting plant growth, protecting against pathogens, and enhancing host resistance (Singh Rawat *et al.*, 2023). The study of plant–microbe interactions and the comprehension of how microbial communities are formed and evolve provide insights into the

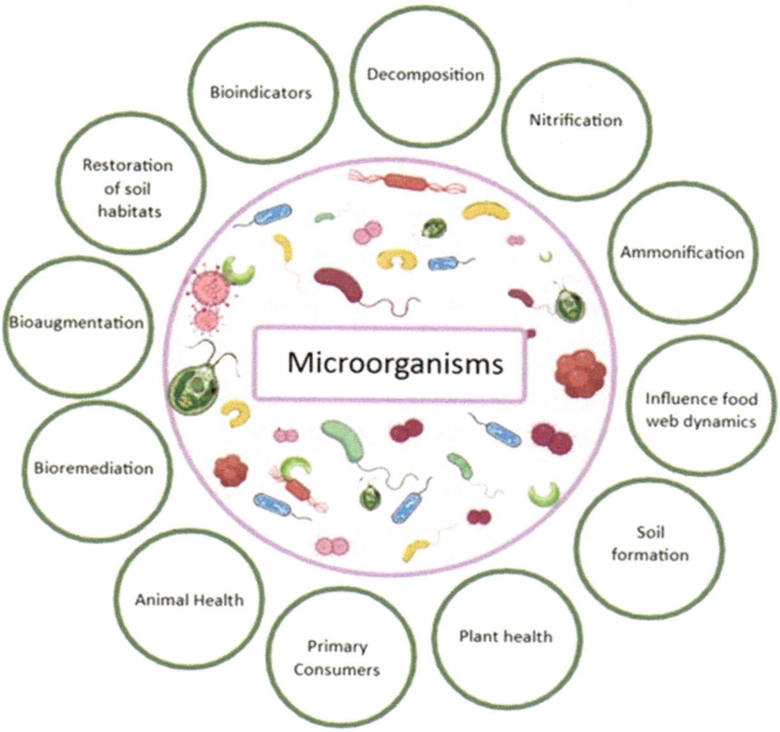

FIGURE 9.3 Functional roles of microorganisms in the terrestrial ecosystem.

advantages that bacteria provide to plants (Vishwakarma *et al.*, 2020). Disruption of ecosystems adversely affects microbial functions, which need to be restored to recover the disturbed/damaged ecosystems.

As Sala *et al.* (2000) have previously anticipated, the primary factor causing the loss of terrestrial biodiversity (Prince *et al.*, 2018), including ecosystem services, regeneration capacity, and soil biodiversity, is habitat degradation brought on by changes in land use for lodging and agricultural services. The urgency of soil conservation and regeneration is paramount in ensuring the preservation of biodiversity and promoting human well-being (Montgomery, 2007; De Deyn & Kooistra, 2021). For the sake of human health and biodiversity, soil regeneration and conservation are critical. There are considerable differences between natural and agricultural habitats, yet both are necessary for sustainable growth. To achieve multifunctionality and prevent imposing stress, proper landscape management is necessary for the coexistence of many habitat types. Because natural ecosystems support coevolved and connected above- and belowground biodiversity as well as elemental cycles, they continue to be crucial for the conservation of biodiversity. These systems are also necessary for the absorption of carbon, the purification and storage of water, and the preservation of soil. The effectiveness of conservation greatly depends on the extent of natural habitat fragmentation (De Deyn & Kooistra, 2021). Small fragments experience ecosystem deterioration resulting in a lower percentage of biodiversity compared to larger habitat fragments, as well as a diminished capacity for carbon absorption (Chase *et al.*, 2020; Ordway & Anser, 2020). Agricultural systems are purposefully structured to provide sustenance, animal feed, and natural fibers, hence necessitating physical disruption and the extraction of nutrients, water resources, and energy. Nevertheless, the degree of disruption, and therefore the quality of soil habitat, is heavily contingent upon the specific kind of land management (De Deyn & Kooistra, 2021).

9.5 APPLYING MICROBIOMICS TO RESTORE A TERRESTRIAL ECOSYSTEM

Microbiomics is a scientific discipline that involves the comprehensive analysis and measurement of microbial communities, referred to as microbiomes, which reside inside various ecosystems and often play a major role in the operation of these ecosystems. Microbiomics encompasses the use of DNA- and RNA-based methodologies along with bioinformatics and biochemistry. The main objective of this analysis is to achieve a comprehensive interpretation of the compositional, temporal, and geographical dynamics, as well as the functional properties of microbial populations. In addition, microbiomic methods were used to examine the functionalities and biochemical characteristics of microbial populations. The field of microbiomics is revolutionizing our understanding of ecosystems and their potential for restoration. The microbiota performs several functional roles in the management of ecosystems, including nutrient cycling, temperature management, and the well-being of humans, plants, and animals (Bardgett *et al.*, 2008; Jiang *et al.*, 2020; Robinson *et al.*, 2023; Renz & Skevaki, 2021). Developing a comprehensive knowledge of these activities and processes would significantly augment our capacity to effectively rehabilitate ecosystems that have undergone degradation (Robinson *et al.*, 2023).

The process of restoration has three key steps, namely, planning, execution, and monitoring. Microbiomics provides significant insights along this continuum. For example, the achievement of effective ecosystem restoration often necessitates the reintroduction of functioning vegetation communities, such as forests, meadows, woods, and grasslands (Elliott *et al.*, 2022). The use of microbiomics might provide valuable insights into the mechanisms by which the microbiota impacts several aspects of vegetative functioning, such as plant immunology, dispersion, growth, and stress tolerance (Fischhoff *et al.*, 2020; Egamberdieva *et al.*, 2017).

9.5.1 Restoration Planning

The process of ecosystem restoration necessitates the implementation of well-informed judgments on interventions and the careful selection of targets in order to effectively make progress toward predetermined objectives (Suding *et al.*, 2016). When formulating plans, it is advisable to consider the incorporation of microbiomics and complementary non-omic methodologies, if feasible.

Incorporating microbiomics into the initial evaluation during the preparatory phase will contribute to establishing diversity, microbial functionality, and complex characteristics of the ecological foundation (for example, via soil metagenomics). The potential consequences of this phenomenon may subsequently impact the course of recovery when interactions between bacteria and plants or animals become established (Heneghan *et al.*, 2008). To effectively incorporate soil microbiota into restoration efforts, it is crucial to prioritize the identification of specific microbial compositions present in reference sites. This can be achieved through the utilization of amplicon sequencing approaches, which require comprehensive sample coverage. Additionally, capturing the functionality of soil microbiota should be emphasized, which can be accomplished through the application of metagenomic or metatranscriptomic approaches (Peddle *et al.*, 2023). The assessment of soil microbial diversity and functioning, often referred to as microbiomics, is essential for restoration planning. This process involves selecting the appropriate methods to test and monitor treatments, hence having significant consequences for restoration strategies.

9.5.2 Restoration Implementation

One potential strategy for ecosystem restoration involves implementing a microbiome-centric intervention, such as the reintroduction of soil bacteria that may be deficient in damaged ecosystems. The promotion of the colonization of suitable soil microbiota has the potential to alter the composition of plant communities, enhance plant growth and establishment, and restore essential activities of the soil ecosystem, including nutrient cycling and decomposition (Delgado-Baquerizo *et al.*, 2021; Liddicoat *et al.*, 2019).

Plant microbiota experiments that employ microbiomics have the potential to contribute to restoration endeavors by offering valuable insights into the preservation of the core microbiota, which refers to the unique microbial communities, of plant communities. Additionally, these experiments can shed light on the crucial microbial metabolites that play a significant role in promoting the growth of plants, bolstering immunity, as well as enhancing stress tolerance. One potential approach to enhance the success of restoration efforts at a location is to pre-inoculate organisms with microbiota before planting. As an example, the use of aquatic microbiota has the potential to be employed in the inoculation of aquatic species such as seagrass, coral, and macroalgae prior to transplanting. Additionally, it may be utilized to introduce native beneficial microbiota to a restoration site, hence facilitating the proliferation of these microorganisms before the actual planting process takes place. Mangrove restoration initiatives may potentially derive advantages from the use of bioaugmentation techniques, including the introduction of petroleum-degrading microorganisms. This approach aims to minimize the possible adverse effects caused by active oil spills, which pose a prevalent risk to ecosystems (Martin *et al.*, 2020). The use of petroleum-degrading bacteria in conjunction with biosurfactant-producing bacteria has been employed for the purpose of bioaugmentation and subsequent evaluation of its impact on the decomposition of petrochemicals in saltwater. The researchers observed a disintegration rate of 72% over a period of 45 days, which was much greater than the degradation rate of 38% seen in the treatment group that did not undergo bioaugmentation (Shi *et al.*, 2020).

9.5.3 Restoration Monitoring

The assessment of ecosystem recovery often involves the comparison of restoration sites with reference sites, using metrics such as community structure and abundance of species. Evaluations of microbial functional restoration necessitate surpassing taxonomic data and instead employing whole genome sequencing along with other metaomic methodologies (such as metagenomics and metatranscriptomics) that can furnish comprehensive functional insights alongside compositional metrics (for example, alpha and beta diversity). It should be emphasized that there is often a weak correlation between metatranscriptomic and metagenomic data obtained from the same environment. The selection between metatranscriptomics and metagenomics for restoration monitoring will mostly rely on the objectives of the monitoring process. Metatranscriptomics reveals the immediate functioning of the microbial community, while metagenomics reveals the genetic capabilities of the microbial population. The selection of the method will mostly depend on allocated finances and the proficiency of the research team.

Amplicon sequencing is becoming more often used for ecosystem restoration monitoring due to its ability to provide biodiversity data quickly and accurately across a wide variety of taxa. Amplicon sequencing is a cost-effective method, but it does have significant drawbacks. For example, taxonomy is often limited to the level of genus, and it is not possible to make reliable conclusions regarding the functions and activity levels of microorganisms, particularly in situations where there is a lack of reference material (Robinson *et al.*, 2023). Furthermore, it is not possible to accurately determine the actual biological abundance of taxa. In contrast, metaomic techniques such as metatranscriptomics and metagenomics provide a more accurate evaluation of the taxonomic makeup and functional capabilities. They provide a significant degree of assurance in identifying the communities present throughout the initial and post-intervention phases, as well as the functional functions these organisms may have. Nevertheless, metaomic methodologies sometimes need a considerable amount of proficiency to handle the intricate bioinformatics and analysis involved (Van der Heyde *et al.*, 2022).

Soil microbiota in terrestrial systems have been the focus of biodiversity monitoring efforts because of their significant contributions to the restoration of ecosystem functioning (Coban *et al.*,

2022). Moreover, bacteria play essential roles as facilitators in the process of the recovery of vegetation (Harris, 2009).

9.6 FUNCTION OF MICROORGANISMS IN BIOREMEDIATION

Bioremediation is the process by which biological organisms transform and repurpose waste materials into a usable form for other species. Currently, the global community is grappling with many forms of environmental contamination. Microorganisms are crucial for a fundamental alternative approach to surmount difficulties. Microorganisms can thrive in all locations on the biosphere due to their remarkable metabolic activity, allowing them to persist in a wide variety of environmental circumstances (Tang *et al.*, 2007). Microorganisms possess diverse feeding capabilities, making them valuable for the bioremediation of environmental contaminants (Abatenh *et al.*, 2017).

Aerobic bacteria such as *Pseudomonas*, *Alcaligens*, *Sphingomonas*, *Rhodococcus*, and *Mycobacterium* are frequently recognized for their ability to degrade hydrocarbons (both alkanes and polyaromatic) and pesticides (Pande *et al.*, 2020) (Table 9.1). When it comes to the process, aerobic bacteria are used more often than anaerobic bacteria (Nielsen *et al.*, 2021). It is known that anaerobic bacteria break down trichloroethylene (TCE), chloroform, dechlorinate, and polychlorinated biphenyls (PCBs) in sediments (such as lake and riverbeds) (Kahlon, 2016). *Phanerochaete chrysosporium* is one of the ligninolytic fungi that has been found to be helpful in the bioremediation of several recalcitrant toxic environmental pollutants, including TNT, DDT, plastics, and several high molecular weight polynuclear aromatics (Pande *et al.*, 2020).

TABLE 9.1
Microorganisms Involved in Toxic Pollution Bioremediation

Microorganisms	Pollutant	References
Acinetobacter species	Atrazine	Singh *et al.* (2004a)
Enterobacter spp., *Bacillus* spp., *Providencia* spp., and *Pseudomonas* spp.		El-Bestawy *et al.* (2014)
Bacterium strain B-14	Chlorpyrifos	Singh *et al.* (2004b)
Bacillus spp.	Halogenated hydrocarbons, phenoxyacetates	Cybulski *et al.* (2003)
Hydrogenophaga fava ENV735	Methyl *tert*-butyl ether	Streger *et al.* (2002)
β-Proteobacterium strain PM1		Smith *et al.* (2005)
Pseudomonas stutzeri KC	Carbon tetrachloride	Dybas *et al.* (2002)
Pleurotus ostreatus	Orange 3,4-(4-nitrophenylazo)aniline	Zhao *et al.* (2006)
Pseudomonas putida MHF 7109	Benzene, toluene, and *o*-xylene (BTX)	Singh & Fulekar (2010)
Pseudomonas spp.	Benzene, anthracene, hydrocarbons PCBs	Cybulski *et al.* (2003)
Pseudomonas puteda, *P. aeruginosa* and *Acinetobacter faecalis*	Glyphosate	Olawale *et al.* (2011)
Ralstonia eutropha (pJP4)	2,4-Dichlorophenoxyacetic acid	Daane and Haggblom (1999)
Ralstonia eutropha JMP134		Roane *et al.* (2001)
Rhodococcus spp.	Hydrocarbons aromatics	Dean-Ross *et al.* (2002)
Streptomyces spp.	Organochlorine (lindane)	Fuentes *et al.* (2010)
Staohylococcus aureua	Organophosphorous pesticide-malathion	Akilandeswari and Sona (2013)

Source: Pande *et al.* (2020).

9.7 CONCLUSION

The microbiota could swiftly adjust and adapt to changes in the climate and anthropogenic stresses, surpassing the adaptability of their associated multicellular organisms. This heightened adaptability has the potential to enhance the overall resilience of the holobiont under unfamiliar environmental situations (Martin *et al.*, 2019). Microbiome-based techniques provide significant promise in facilitating the restoration of damaged ecosystems, thereby enhancing their resilience and aligning with the United Nations' sustainability objectives and habitat restoration goals (Corinaldesi *et al.*, 2023). The incorporation of bacteria into ecosystem restoration endeavors has the potential to enhance the overall effectiveness of restoration solutions. Microbiomics offers novel perspectives that are not accessible via conventional methodologies. The field of microbiomics presents novel prospects for boosting restoration efforts through improved evaluation of ecosystem complexity and functions, as well as a deeper understanding of crucial contributions of microbiota throughout the restoration process (Robinson *et al.*, 2023). The successful implementation of this endeavor necessitates a novel comprehension at the genomic level of distinct microbial taxa, as well as the interactions between microorganisms and their hosts, and the interconnections among different microorganisms. Additionally, it demands an awareness of the functional capabilities that may be operative at the level of the complete microbial community (Robinson *et al.*, 2023).

REFERENCES

Abatenh E, Gizaw B, Tsegaye Z, Wassie M. The role of microorganisms in bioremediation-A review. *Open Journal of Environmental Biology*. 2017 Nov 10;2(1):038–46.

Akilandeswari K, Sona V. Efficiency of Staphylococcus aureus in the degradation of an organo phosphorous pesticide, malathion. *Journal of Pharmaceutical and Scientific Innovation (JPSI)*. 2013;2(6):13–21.

Allen EB, Brown JS, Allen MF. Restoration of animal, plant and microbial diversity. *Encyclopedia of Biodiversity*. 2001 Jan 1;5:185–202.

Amorim CA, do Nascimento Moura A. Ecological impacts of freshwater algal blooms on water quality, plankton biodiversity, structure, and ecosystem functioning. *Science of the Total Environment*. 2021 Mar 1;758:143605.

Arif M, Jiajia L, Tahir M, Jie Z, Li C. Environmental literacy scenarios lead to land degradation and changes in riparian zones: Implications for policy in China. *Land Degradation & Development*. 2023 Jan 15;34(1):156–72.

Bhagowati B, Ahamad KU. A review on lakeeutrophication dynamics and recent developments in lake modeling. *Ecohydrology & Hydrobiology*. 2019 Jan 1;19(1):155–66.

Balogun RB, Ogbu JU, Umeokechukwu EC, Kalejaiye-Matti RB. Effective micro-organisms (EM) as sustainable components in organic farming: principles, applications and validity. *Organic Farming for Sustainable Agriculture*. 2016;9:259–91.

Banerjee A, Cornejo J, Bandopadhyay R. Emergent climate change impact throughout the world: call for "Microbiome Conservation" before it's too late. *Biodiversity and Conservation*. 2020 Jan;29:345–8.

Bardgett RD, Freeman C, Ostle NJ. Microbial contributions to climate change through carbon cycle feedbacks. *The ISME journal*. 2008 Aug;2(8):805–14.

Béja O, Aravind L, Koonin EV, Suzuki MT, Hadd A, Nguyen LP, Jovanovich SB, Gates CM, Feldman RA, Spudich JL, Spudich EN. Bacterial rhodopsin: evidence for a new type of phototrophy in the sea. *Science*. 2000 Sep 15;289(5486):1902–6.

Bentley, S. 2021 *Quantifying soil, microbial, and plant community changes following wetland restoration on private land* (Doctoral dissertation, Open Access Te Herenga Waka-Victoria University of Wellington).

Boetius A, Anesio AM, Deming JW, Mikucki JA, Rapp JZ. Microbial ecology of the cryosphere: sea ice and glacial habitats. *Nature Reviews Microbiology*. 2015 Nov;13(11):677–90.

Bowman JS, Vick-Ma,jors TJ, Morgan-Kiss R, Takacs-Vesbach C, Ducklow HW, Priscu JC. Microbial community dynamics in two polar extremes: The lakes of the McMurdo Dry Valleys and the West Antarctic Peninsula marine ecosystem. *BioScience*. 2016 Oct 1;66(10):829–47.

Buonocore E, Grande U, Franzese PP, Russo GF. Trends and evolution in the concept of marine ecosystem services: An overview. *Water*. 2021 Jul 29;13(15):2060.

Caron DA, Alexander H, Allen AE, Archibald JM, Armbrust EV, Bachy C, Bell CJ, Bharti A, Dyhrman ST, Guida SM, Heidelberg KB. Probing the evolution, ecology and physiology of marine protists using transcriptomics. *Nature Reviews Microbiology*. 2017 Jan;15(1):6–20.

Cavicchioli R, Ripple WJ, Timmis KN, Azam F, Bakken LR, Baylis M, Behrenfeld MJ, Boetius A, Boyd PW, Classen AT, Crowther TW. Scientists' warning to humanity: microorganisms and climate change. *Nature Reviews Microbiology*. 2019 Sep;17(9):569–86.

Chase JM, Blowes SA, Knight TM, Gerstner K, May F. Ecosystem decay exacerbates biodiversity loss with habitat loss. *Nature*. 2020 Aug 13;584(7820):238–43.

Cheng L, Gao X, Wang G, Ding Z, Xue B, Zhang C, Liu J, Jiang Q. Intensified sensitivity and adaptability of zooplankton Bosminidae in subtropical shallow freshwater lakes with increasing trophic level. *Frontiers in Ecology and Evolution*. 2023 Mar 2;11:1121632.

Coban O, De Deyn GB, van der Ploeg M. Soil microbiota as game-changers in restoration of degraded lands. Science. 2022 Mar 4;375(6584):abe0725.

Colatriano D, Tran PQ, Guéguen C, Williams WJ, Lovejoy C, Walsh DA. Genomic evidence for the degradation of terrestrial organic matter by pelagic Arctic Ocean Chloroflexi bacteria. *Communications Biology*. 2018 Jul 5;1(1):90.

Cook KV, Li C, Cai H, Krumholz LR, Hambright KD, Paerl HW, Steffen MM, Wilson AE, Burford MA, Grossart HP, Hamilton DP. The global Microcystis interactome. *Limnology and Oceanography*. 2020 Jan;65:S194–207.

Corinaldesi C, Bianchelli S, Candela M, Dell'Anno A, Gambi C, Rastelli E, Varrella S, Danovaro R. Microbiome-assisted restoration of degraded marine habitats: a new nature-based solution?. *Frontiers in Marine Science*. 2023 Sep 20;10:1227560.

Cybulski Z, Dziurla E, Kaczorek E, Olszanowski A. The influence of emulsifiers on hydrocarbon biodegradation by Pseudomonadacea and Bacillacea strains. *Spill Science & Technology Bulletin*. 2003 Jan 1;8(5–6):503–7.

Daane LL, Haggblom MM. Earthworm egg capsules as vectors for the environmental introduction of biodegradative bacteria. *Applied and Environmental Microbiology*. 1999 Jun 1;65(6):2376–81.

Dai T, Liu R, Zhou X, Zhang J, Song M, Zou P, Bi X, Li S. Role of Lake Aquatic–Terrestrial ecotones in the ecological restoration of eutrophic water bodies. *Toxics*. 2023 Jun 26;11(7):560.

Dash HR, Mangwani N, Chakraborty J, Kumari S, Das S. Marine bacteria: potential candidates for enhanced bioremediation. *Applied Microbiology and Biotechnology*. 2013 Jan;97(2):561–71.

De Deyn GB, Kooistra L. The role of soils in habitat creation, maintenance and restoration. *Philosophical Transactions of the Royal Society B*. 2021 Sep 27;376(1834):20200170. https://doi.org/10.1098/rstb.2020.0170

Dean-Ross D, Moody J, Cerniglia CE. Utilization of mixtures of polycyclic aromatic hydrocarbons by bacteria isolated from contaminated sediment. *FEMS Microbiology Ecology*. 2002 Jul 1;41(1):1–7.

Deere NJ, Bicknell JE, Mitchell SL, Afendy A, Baking EL, Bernard H, Chung AY, Ewers RM, Heroin H, Joseph N, Lewis OT. Riparian buffers can help mitigate biodiversity declines in oil palm agriculture. *Frontiers in Ecology and the Environment*. 2022 Oct;20(8):459–66.

Delgado-Baquerizo M, Eldridge DJ, Liu YR, Sokoya B, Wang JT, Hu HW, He JZ, Bastida F, Moreno JL, Bamigboye AR, Blanco-Pastor JL. Global homogenization of the structure and function in the soil microbiome of urban greenspaces. *Science Advances*. 2021 Jul 9;7(28):eabg5809. https://doi.org/10.1126/sciadv.abg5809

Dietz T, Ostrom E, Stern PC. The struggle to govern the commons. *Science*. 2003 Dec 12;302(5652):1907–12.

Dong YW, Miao JZ, Chen AH. Research on immobilized ammonia-oxidizing bacteria by sodium alginate. *Applied Mechanics and Materials*. 2014 Oct 22;618:283–7.

Dybas MJ, Hyndman DW, Heine R, Tiedje J, Linning K, Wiggert D, Voice T, Zhao X, Dybas L, Criddle CS. Development, operation, and long-term performance of a full-scale biocurtain utilizing bioaugmentation. *Environmental Science & Technology*. 2002 Aug 15;36(16):3635–44.

Egamberdieva D, Wirth SJ, Alqarawi AA, Abd_Allah EF, Hashem A. Phytohormones and beneficial microbes: essential components for plants to balance stress and fitness. *Frontiers in Microbiology*. 2017 Oct 31;8:2104.

Eichmann R, Richards L, Schäfer P. Hormones as go-betweens in plant microbiome assembly. *The Plant Journal*. 2021 Jan;105(2):518–41.

El-Bestawy E, Sabir J, Mansy AH, Zabermawi N. Comparison among the efficiency of different bioremediation technologies of Atrazine-contaminated soils. *Journal of Bioremediation and Biodegradation*. 2014 Jul 1;5(5):237.

Elliott BB, Olds AD, Henderson CJ, Rummell AJ, Gilby BL. Optimising restoration and rehabilitation using environmental and spatial drivers of plant assemblages. *Landscape and Urban Planning.* 2022 Oct 1;226:104484.

Ethridge AD, Bazzi MH, Lukacs NW, Huffnagle GB. Interkingdom communication and regulation of mucosal immunity by the microbiome. *The Journal of Infectious Diseases.* 2021 Jun 15;223(Supplement_3):S236–40.

Falk DA, Millar CI, Olwell M, Noss RF. Restoring diversity: strategies for reintroduction of endangered plants. (No Title). 1996.

Fernandez L, Bertilsson S, Peura S. Non-cyanobacterial diazotrophs dominate nitrogen-fixing communities in permafrost thaw ponds. Limnology and Oceanography. 2020 Jan;65:S180–93.

Fischhoff IR, Huang T, Hamilton SK, Han BA, LaDeau SL, Ostfeld RS, Rosi EJ, Solomon CT. Parasite and pathogen effects on ecosystem processes: a quantitative review. *Ecosphere.* 2020;11(5):e03057. https://doi.org/10.1002/ecs2.3057.

Fuentes MS, Benimeli CS, Cuozzo SA, Saez JM, Amoroso MJ. Microorganisms capable of degrading organochlorine pesticides. *Current Research, Technology and Education Topics in Applied Microbiology and Microbial Biotechnology.* 2010;2:1255–64.

Gann GD, McDonald T, Walder B, Aronson J, Nelson CR, Jonson J, Hallett JG, Eisenberg C, Guariguata MR, Liu J, Hua F. International principles and standards for the practice of ecological restoration. *Restoration Ecology.* 2019;27(S1):S1–46.

Gao FK, Dai CC, Liu XZ. Mechanisms of fungal endophytes in plant protection against pathogens. *African Journal of Microbiology Research.* 2010 Jul 4;4(13):1346–51.

Garcia J, Kao-Kniffin J. Microbial group dynamics in plant rhizospheres and their implications on nutrient cycling. *Frontiers in microbiology.* 2018 Jul 11;9:1516.

Gawecka KA, Bascompte J. Habitat restoration and the recovery of metacommunities. *Journal of Applied Ecology.* 2023 Jan 1;60(8):1622–36.

Gilbert JA, Dupont CL. Microbial metagenomics: beyond the genome. *Annual Review of Marine Science.* 2011 Jan 15;3:347–71.

Gralka M, Szabo R, Stocker R, Cordero OX. Trophic interactions and the drivers of microbial community assembly. *Current Biology.* 2020 Oct 5;30(19):R1176–88.

Grossart HP, Massana R, McMahon KD, Walsh DA. Linking metagenomics to aquatic microbial ecology and biogeochemical cycles. *Limnology and Oceanography.* 2020 Jan;65:S2–0.

Hall EK, Bernhardt ES, Bier RL, Bradford MA, Boot CM, Cotner JB, Del Giorgio PA, Evans SE, Graham EB, Jones SE, Lennon JT. Understanding how microbiomes influence the systems they inhabit. Nature microbiology. 2018 Sep;3(9):977–82.

Hallam SJ, Mincer TJ, Schleper C, Preston CM, Roberts K, Richardson PM, DeLong EF. Pathways of carbon assimilation and ammonia oxidation suggested by environmental genomic analyses of marine Crenarchaeota. *PLoS Biology.* 2006 Apr;4(4):e95.

Harindintwali JD, Zhou J, Muhoza B, Wang F, Herzberger A, Yu X. Integrated eco-strategies towards sustainable carbon and nitrogen cycling in agriculture. *Journal of Environmental Management.* 2021 Sep 1;293:112856.

Harris J. Soil microbial communities and restoration ecology: facilitators or followers?. *Science.* 2009 Jul 31;325(5940):573–4. 10.1126/science.1172975

He S, Lau MP, Linz AM, Roden EE, McMahon KD. Extracellular electron transfer may be an overlooked contribution to pelagic respiration in humic-rich freshwater lakes. *Msphere.* 2019 Feb 27;4(1):10–128.

Heneghan L, Miller SP, Baer S, Callaham Jr MA, Montgomery J, Pavao-Zuckerman M, Rhoades CC, Richardson S. Integrating soil ecological knowledge into restoration management. *Restoration Ecology.* 2008 Dec;16(4):608–17.

Hennings N, Becker JN, Guillaume T, Damris M, Dippold MA, Kuzyakov Y. Riparian wetland properties counter the effect of land-use change on soil carbon stocks after rainforest conversion to plantations. *Catena.* 2021 Jan 1;196:104941.

Hilderbrand RH, Bambakidis T, Crump BC. The roles of microbes in stream restorations. *Microbial Ecology.* 2023 Jan 25:1–9.

Hiraoka S, Hirai M, Matsui Y, Makabe A, Minegishi H, Tsuda M, Juliarni, Rastelli E, Danovaro R, Corinaldesi C, Kitahashi T. Microbial community and geochemical analyses of trans-trench sediments for understanding the roles of hadal environments. *The ISME Journal.* 2020 Mar;14(3):740–56.

Huang X, Xu X, Guan B, Liu S, Xie H, Li Q, Li K. Transformation of aquatic plant diversity in an environmentally sensitive area, the Lake Taihu drainage basin. *Frontiers in Plant Science.* 2020 Nov 12;11:513788.

Ibáñez A, Garrido-Chamorro S, Barreiro C. Microorganisms and climate change: a not so invisible Effect. *Microbiology Research.* 2023 Jul 25;14(3):918–47.

Jones TA. Ecosystem restoration: recent advances in theory and practice. *The Rangeland Journal.* 2017 Aug 30;39(6):417–30.

Jordan WR, Gilpin ME, Aber JD, editors. *Restoration ecology: a synthetic approach to ecological research.* Cambridge University Press; 1990 Aug 16.

Jordan WR, Peters RL, Allen EB. Ecological restoration as a strategy for conserving biological diversity. *Environmental Management.* 1988 Jan;12:55–72.

Kahlon RS. Biodegradation and bioremediation of organic chemical pollutants by Pseudomonas. *Pseudomonas: Molecular and Applied Biology.* 2016 Jun 2:343–417.

Khursheed S. Soil biodiversity and climate change. *Advances in Plants & Agriculture Research.* 2016;3(5):00113.

Knoll AH. Paleobiological perspectives on early microbial evolution. *Cold Spring Harbor Perspectives in Biology.* 2015 Jul 1;7(7):a018093.

Kotas P, Šantrůčková H, Elster J, Kaštovská E. Soil microbial biomass, activity and community composition along altitudinal gradients in the High Arctic (Billefjorden, Svalbard). *Biogeosciences.* 2018 Mar 29;15(6):1879–94.

Krause S, Le Roux X, Niklaus PA, Van Bodegom PM, Lennon JT, Bertilsson S, Grossart HP, Philippot L, Bodelier PL. Trait-based approaches for understanding microbial biodiversity and ecosystem functioning. *Frontiers in Microbiology.* 2014 May 27;5:251.

Landry Z, Swan BK, Herndl GJ, Stepanauskas R, Giovannoni SJ. SAR202 genomes from the dark ocean predict pathways for the oxidation of recalcitrant dissolved organic matter. *MBio.* 2017 May 3;8(2):10–128.

Liang S, Yuan X, Lu X, Xu G, Qin V, Jiang Y. Soilphysico-chemical properties and distribution characteristics in an aquatic-terrestrial ecotone of the Lijiang River, Southwest China. *Acta Ecologica Sinica.* 2019;39(8):2752–61.

Liddicoat C, Weinstein P, Bissett A, Gellie NJ, Mills JG, Waycott M, Breed MF. Can bacterial indicators of a grassy woodland restoration inform ecosystem assessment and microbiota-mediated human health?. *Environment International.* 2019 Aug 1;129:105–17. https://doi.org/10.1016/j.envint.2019.05.011

Ma Y, Dias MC, Freitas H. Drought and salinity stress responses and microbe-induced tolerance in plants. *Frontiers in Plant Science.* 2020 Nov 13;11:591911.

Mander Ü, Tournebize J, Tonderski K, Verhoeven JT, Mitsch WJ. Planning and establishment principles for constructed wetlands and riparian buffer zones in agricultural catchments. *Ecological Engineering.* 2017 Jun 1;103:296–300.

Manea E, Dell'anno A, Rastelli E, Tangherlini M, Nunoura T, Nomaki H, et al. Viral infections boost prokaryotic biomass production and organic C cycling in hadal trench sediments. *Frontiers in Microbiology.* 2019;10:1952. doi: 10.3389/fmicb.2019.01952

Martin BC, Bougoure J, Ryan MH, Bennett WW, Colmer TD, Joyce NK, Olsen YS, Kendrick GA. Oxygen loss from seagrass roots coincides with colonization of sulphide-oxidizing cable bacteria and reduces sulfide stress. *The ISME Journal.* 2019 Mar;13(3):707–19.

Martin BC, Middleton JA, Fraser MW, Marshall IP, Scholz VV, Hausl B, Schmidt H. Cutting out the middle clam: lucinid endosymbiotic bacteria are also associated with seagrass roots worldwide. *The ISME Journal.* 2020 Nov;14(11):2901–5. https://doi.org/10.1038/s41396-020-00771-3

McIntyre PB, Reidy Liermann CA, Revenga C. Linking freshwater fishery management to global food security and biodiversity conservation. *Proceedings of the National Academy of Sciences.* 2016 Nov 8;113(45):12880–5.

Mehrshad M, Rodriguez-Valera F, Amoozegar MA, López-García P, Ghai R. The enigmatic SAR202 cluster up close: shedding light on a globally distributed dark ocean lineage involved in sulfur cycling. The ISME journal. 2018 Mar;12(3):655–68.

Montgomery DR. Soil erosion and agricultural sustainability. *Proceedings of the National Academy of Sciences.* 2007 Aug 14;104(33):13268–72.

Naik K, Mishra S, Srichandan H, Singh PK, Sarangi PK. Plant growth promoting microbes: Potential link to sustainable agriculture and environment. Biocatalysis and Agricultural Biotechnology. 2019 Sep 1;21:101326.

Nemergut DR, Schmidt SK, Fukami T, O'Neill SP, Bilinski TM, Stanish LF, Knelman JE, Darcy JL, Lynch RC, Wickey P, Ferrenberg S. Patterns and processes of microbial community assembly. *Microbiology and Molecular Biology Reviews.* 2013 Sep;77(3):342–56.

Nielsen DA, Fierer N, Geoghegan JL, Gillings MR, Gumerov V, Madin JS, Moore L, Paulsen IT, Reddy TB, Tetu SG, Westoby M. Aerobic bacteria and archaea tend to have larger and more versatile genomes. *Oikos*. 2021 Apr;130(4):501–11.

Nissimov J. A communal catalog reveals Earth's multiscale microbial diversity. *Nature*. 2017 Nov 1;551:457-63.

Olawale AK, Akintobi OA. Biodegradation of glyphosate pesticide by bacteria isolated from agricultural soil. *Report and Opinion*. 2011;3(1):124–8.

Ordway EM, Asner GP. Carbon declines along tropical forest edges correspond to heterogeneous effects on canopy structure and function. *Proceedings of the National Academy of Sciences*. 2020 Apr 7;117(14):7863–70. https://doi.org/10.1073/pnas.1914420117

Pande V, Pandey SC, Sati D, Pande V, Samant M. Bioremediation: an emerging effective approach towards environment restoration. *Environmental Sustainability*. 2020 Mar;3:91–103.

Peddle SD, Bissett A, Borrett RJ, Bullock P, Gardner MG, Liddicoat C, Tibbett M, Breed MF, Krauss SL. Soil DNA chronosequence analysis shows bacterial community re-assembly following post-mining forest rehabilitation. *Restoration Ecology*. 2023 Mar;31(3):e13706. https://doi.org/10.1111/rec.13706

Penton CR, Deenik JL, Popp BN, Bruland GL, Engstrom P, Louis DS, Tiedje J. Importance of sub-surface rhizosphere-mediated coupled nitrification–denitrification in a flooded agroecosystem in Hawaii. *Soil Biology and Biochemistry*. 2013 Feb 1;57:362–73.

Prince S, Von Maltitz G, Zhang F, Byrne K, Driscoll C, Eshel G, Kust G, Martínez-Garza C, Metzger JP, Midgley G, Moreno-Mateos D. Status and trends of land degradation and restoration and associated changes in biodiversity and ecosystem functions. *Intergovernmental Science-Policy Platform on Biodiversity and Ecosystem Services (IPBES)*. 2018.

Rajendhran J, Gunasekaran P. Strategies for accessing soil metagenome for desired applications. *Biotechnology Advances*. 2008 Nov 1;26(6):576-90. https://doi.org/10.1016/j.biotechadv.2008.08.002

Renz H, Skevaki C. Early life microbial exposures and allergy risks: opportunities for prevention. *Nature Reviews Immunology*. 2021 Mar;21(3):177–91.

Roane TM, Josephson KL, Pepper IL. Dual-bioaugmentation strategy to enhance remediation of contaminated soil. *Applied and Environmental Microbiology*. 2001 Jul 1;67(7):3208–15.

Robinson JM, Hodgson R, Krauss SL, Liddicoat C, Malik AA, Martin BC, Mohr JJ, Moreno-Mateos D, Muñoz-Rojas M, Peddle SD, Breed MF. Opportunities and challenges for microbiomics in ecosystem restoration. *Trends in Ecology & Evolution*. 2023 Aug 28;38(12):1189–202. https://doi.org/10.1016/j.tree.2023.07.009

Sala OE, Stuart Chapin FI, Armesto JJ, Berlow E, Bloomfield J, Dirzo R, Huber-Sanwald E, Huenneke LF, Jackson RB, Kinzig A, Leemans R. Global biodiversity scenarios for the year 2100. *Science*. 2000 Mar 10;287(5459):1770–4.

Schulz S, Brankatschk R, Dümig A, Kögel-Knabner I, Schloter M, Zeyer J. The role of microorganisms at different stages of ecosystem development for soil formation. *Biogeosciences*. 2013 Jun 18;10(6):3983–96. https://doi.org/10.5194/bg-10-3983-2013

Shi K, Zhang Q, Xue J, Chen X, Chen Y, Qiao Y, Yang Y, Sun J. Study on the degradation performance and bacterial community of bioaugmentation in petroleum-pollution seawater. *Journal of Environmental Chemical Engineering*. 2020 Aug 1;8(4):103900. https://doi.org/10.1016/j.jece.2020.103900

Singh AK, Sisodia A, Sisodia V, Padhi M. Role of microbes in restoration ecology and ecosystem services. In *New and future developments in microbial biotechnology and bioengineering*, 2019 Jan 1 (pp. 57–68). Elsevier.

Singh BK, Walker A, Morgan JA, Wright DJ. Biodegradation of chlorpyrifos by Enterobacter strain B-14 and its use in bioremediation of contaminated soils. *Applied and Environmental Microbiology*. 2004b Aug;70(8):4855–63.

Singh D, Fulekar MH. Biodegradation of petroleum hydrocarbons by *Pseudomonas putida* strain MHF 7109. *Clean–Soil, Air, Water*. 2010 Aug;38(8):781–6.

Singh JS, Gupta VK. Soil microbial biomass: A key soil driver in management of ecosystem functioning. *Science of the Total Environment*. 2018 Sep 1;634:497–500.

Singh P, Suri CR, Cameotra SS. Isolation of a member of Acinetobacter species involved in atrazine degradation. *Biochemical and Biophysical Research Communications*. 2004a May 7;317(3):697–702.

Singh Rawat V, Kaur J, Bhagwat S, Arora Pandit M, Dogra Rawat C. Deploying microbes as drivers and indicators in ecologicalrestoration. *Restoration Ecology*. 2023 Jan;31(1):e13688.

Smith AE, Hristova K, Wood I, Mackay DM, Lory E, Lorenzana D, Scow KM. Comparison of biostimulation versus bioaugmentation with bacterial strain PM1 for treatment of groundwater contaminated with methyl tertiary butyl ether (MTBE). *Environmental Health Perspectives*. 2005 Mar;113(3):317-22.

Smith P, Cotrufo MF, Rumpel C, Paustian K, Kuikman PJ, Elliott JA, McDowell R, Griffiths RI, Asakawa S, Bustamante M, House JI. Biogeochemical cycles and biodiversity as key drivers of ecosystem services provided by soils. *Soil.* 2015 Nov 19;1(2):665–85.

Streger SH, Vainberg S, Dong H, Hatzinger PB. Enhancing transport of Hydrogenophaga flava ENV735 for bioaugmentation of aquifers contaminated with methyl tert-butyl ether. *Applied and Environmental Microbiology.* 2002 Nov;68(11):5571–9.

Suding K, Spotswood E, Chapple D, Beller E, Gross K. Ecological dynamics and ecological restoration. *Foundations of Restoration Ecology.* 2016:27–56.

Sun X, Liu Y, Wang F, Tan Xia S, Yang L, Fang WA. Application of immobilized microorganism technology for the biotreatment of Eutrophic water. *Journal of Ecology and Rural Environment.* 2020;36(4):433–41.

Tang CY, Fu QS, Criddle CS, Leckie JO. Effect of flux (transmembrane pressure) and membrane properties on fouling and rejection of reverse osmosis and nanofiltration membranes treating perfluorooctane sulfonate containing wastewater. *Environmental Science & Technology.* 2007 Mar 15;41(6):2008–14.

Torsvik V, Øvreås L. Microbial diversity and function in soil: from genes to ecosystems. *Current Opinion in Microbiology.* 2002 Jun 1;5(3):240–5. https://doi.org/10.1016/S1369-5274(02)00324-7

Trading EM. Effective Microorganisms (EM) from Sustainable Community Development. Effective Microorganisms@ emtrading. com. 2000.

Trevathan-Tackett SM, Sherman CD, Huggett MJ, Campbell AH, Laverock B, Hurtado-McCormick V, Seymour JR, Firl A, Messer LF, Ainsworth TD, Negandhi KL. A horizon scan of priorities for coastal marine microbiome research. *Nature Ecology & Evolution.* 2019 Nov;3(11):1509–20.

Van der Heyde M, Bunce M, Nevill P. Key factors to consider in the use of environmental DNA metabarcoding to monitor terrestrial ecological restoration. *Science of the Total Environment.* 2022 Nov 20;848:157617.

Vishwakarma K, Kumar N, Shandilya C, Mohapatra S, Bhayana S, Varma A. Revisiting plant–microbe interactions and microbial consortia application for enhancing sustainable agriculture: a review. *Frontiers in Microbiology.* 2020 Dec 21;11:560406.

Volis S. Conservation-oriented restoration–a two for one method to restore both threatened species and their habitats. *Plant Diversity.* 2019 Apr 1;41(2):50–8.

Walsh DA, Zaikova E, Howes CG, Song YC, Wright JJ, Tringe SG, Tortell PD, Hallam SJ. Metagenome of a versatile chemolithoautotroph from expanding oceanic dead zones. *Science.* 2009 Oct 23;326(5952):578–82.

Wang Y, Cao X, Zeng J, Li H, Zhao D, Wu QL. Distinct shifts in bacterioplankton community composition and functional gene structure between macrophyte-and phytoplankton-dominated regimes in a large shallow lake. *Limnology and Oceanography.* 2020 Jan;65:S208–19.

Winter M, Kickuth R. Elimination of sulfur compounds from wastewater by the root zone process—I. Performance of a large-scale purification plant at a textile finishing industry. *Water Research.* 1989 May 1;23(5):535–46.

Wen J, Gao D, Zhang B, Liang H. Co-metabolic degradation of pyrene by indigenous white-rot fungus *Pseudotrametes gibbosa* from northeast China. *International Biodeterioration & Biodegradation.* 2011 Jul 1;65(4):600–4.

Wu L, Zhang Y, Guo X, Ning D, Zhou X, Feng J, Yuan MM, Liu S, Guo J, Gao Z, Ma J. Reduction of microbial diversity in grassland soil is driven by long-term climate warming. *Nature Microbiology.* 2022 Jul;7(7):1054–62.

Xie A, Chen H, You S. Advance of nitrogen removal in constructed wetlands. In *IOP Conference Series: Materials Science and Engineering* 2018 (Vol. 301, No. 1, p. 012120). IOP Publishing.

Yang Y, Zhang J, Zhao Q, Zhou Q, Li N, Wang Y, Xie S, Liu Y. Sediment ammonia-oxidizing microorganisms in two plateau freshwater lakes at different trophic states. *Microbialogy of Ecology.* 2016 Feb;71:257–65.

Zakaria Z, Gairola S, Shariff NM. Effective microorganisms (EM) technology for water quality restoration and potential for sustainable water resources and management. In the proceedings of International Congress on Environmental Modelling and Software. 2010, Fifth Biennial Meeting, Ottawa, Canada.

Zhao X, Hardin IR, Hwang HM. Biodegradation of a model azo disperse dye by the white rot fungus Pleurotus ostreatus. *International Biodeterioration & Biodegradation.* 2006 Jan 1;57(1):1–6.

Zhou Q, Liu Y, Li T, Zhao H, Alessi DS, Liu W, Konhauser KO. Cadmium adsorption to clay-microbe aggregates: implications for marine heavy metals cycling. *Geochimica et Cosmochimica Acta.* 2020 Dec 1;290:124–36.

Zhu Z, Chen Z, Li L, Shao Y. Response of dominant plant species to periodic flooding in the riparian zone of the Three Gorges Reservoir (TGR), China. *Science of The Total Environment.* 2020 Dec 10;747:141101.

10 Introduction to Bioresources and Sustainable Utilization

Naresh Thakur[1], *Priyanka Kumari*[2]*, *Swati Yadav*[3], *Suman Mishra*[3], *Kamal Jaiswal*[3]

[1]Shoolini University of Biotechnology and Management Sciences, Solan – Himachal Pradesh, India
[2]Faculty of Basic Sciences, Shoolini University of Biotechnology and Management Sciences, Solan – Himachal Pradesh, India
[3]Department of Zoology, School of Life Sciences, Babasaheb Bhimrao Ambedkar University, Lucknow, Uttar Pradesh, India
*Corresponding Author: Dr. Priyanka Kumari

10.1 INTRODUCTION

Energy is a fundamental component of society and, by improving living standards and quality of life, it is essential to the socioeconomic development of that society. The pattern and quality of energy use in any region can be used to gauge its level of economic growth. The comparable residue production rate for poor nations is 69 exajoules annually. However, as the majority of the residues have other uses, not all of them are used to produce energy. Nonetheless, these wastes can provide up to 90% of the energy used in households in villages. As the economy expands, there is a corresponding shift in the consumption pattern, which is contingent upon the energy's source and availability, conversion loss, and efficiency of use. Humanity has experimented with a variety of energy sources throughout its evolution, such as coal, oil, wood, petroleum, and nuclear power. The irresponsible resource extraction and uncontrolled growth have resulted in significant environmental and ecological issues (Ramachandra TV *et al.*, 2004). Since bioresources are among the most accessible renewable energy sources for lowering our reliance on fossil fuels, they are receiving more attention.

Large amounts of bioresources are produced in municipal, forestry, marine, and agricultural sectors of the economy. Industries such as oil palm mills process and manufacture the feedstock of these resources. All bioproducts derived from agricultural plants can be used as platform chemicals, specialty goods, or energy sources (Rashid MH, 2015). The process of transferring carbon mass into different states of use is continuously occurring, as illustrated by the sustainable carbon life cycles in Figure 10.1. Depending on the needs, the energy obtained from the sun is absorbed by the carbon cycle and transformed into materials, energy, and food.

10.2 BIORESOURCES (TERRESTRIAL AND MARINE)

10.2.1 Classification and Potential of Terrestrial Resources

10.2.1.1 Jatropha

The project commenced in 2007, and three phases have been completed. Phase 4 was launched in 2012 and was carried out in the Indian state of Tamil Nadu before the most recent objectives. The main aim of the project is to decrease CO_2 emissions by utilizing more biodiesel. Additionally, the

DOI: 10.1201/9781032697550-10

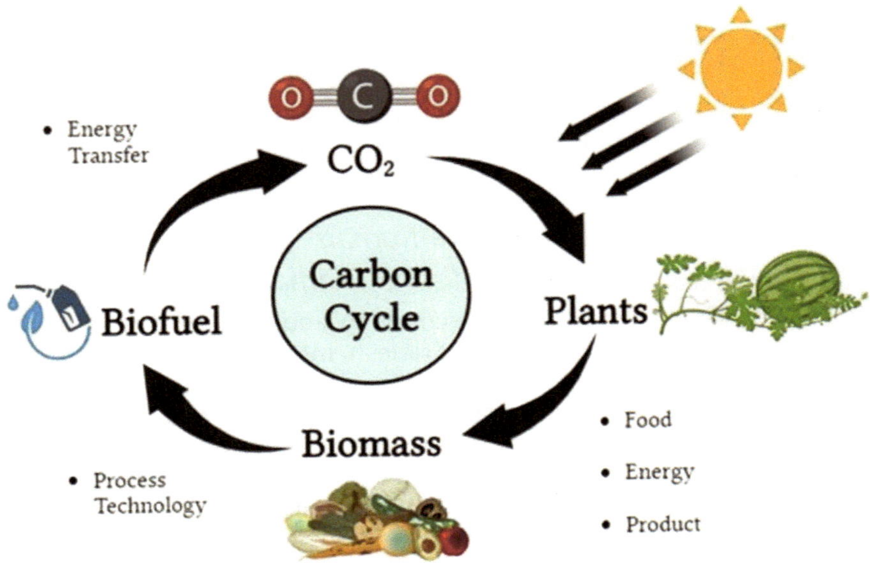

FIGURE 10.1 Sustainable carbon life cycle for food, energy, and material production.

campaign sought to educate youth in developing countries about the damaging consequences of greenhouse gas emissions on the environment (Ramachandra TV *et al.*, 2013).

10.2.1.2 Agricultural By-Products

An improved knowledge of the biomass collected from agricultural by-products has made it possible to use ethanol, methane, and hydrogen as secondary energy sources. In addition to this, microbiological action can also produce energy from the biomass of agricultural by-products such as leftovers, making them sustainable alternative energy sources. There are many different types of lignocellulosic substrates in nature, and according to their abundance, they can be divided into one major category known as agricultural products. About 60–70% of the cellulose and hemicellulose is present mostly in agricultural biomass, which is primarily made up of lignocellulose.

The biomass contains between 10% and 25% of its dry weight in lignin. Numerous research studies made it evident which components of biomass came from different sources—lignin, hemicellulose, and cellulose (Klinke HB *et al.*, 2004). These compounds' role in the production of bioenergy from biomass has not yet received much attention.

10.2.1.3 Energy Plants

The committed crops that are regularly grown are the energy plants. A single planting of certain crops, including grasses, may yield many harvests depending on the crop selection. The use of various biomass resources, however, is mostly determined by the product that needs to be produced and the availability of those resources locally (Millati R *et al.*, 2002; Lin Y & Tanaka S, 2006; Najafi G *et al.*, 2009; Balat M *et al.*, 2008).

10.2.2 Classification and Potential of Marine Resources

The earliest class of vegetation to emerge from the pre-Cambrian period is marine algae. They are important primary producers in the maritime food web and account for 90% of all marine plant species (Dhargalkar VK & Neelam P, 2005).

TABLE 10.1
Chemical Constitution of Various Forms of Plants Biomass

Component	Hardwood (%)	Softwood (%)	Herbaceous Plants (%)
Cellulose	40–50	41–53	24–50
Hemicellulose	25–35	25–30	12–38
Lignin	20–25	25–35	6–29
Pectin	1–0	1–2	Trace
Starch	Trace	Trace	Trace

Sources: Klinke HB et al. (2004) and Dwivedi P et al. (2009).

10.2.2.1 Microalgae (Seaweed Biomass)

Since the seaweeds that come from the maritime environment have a considerable cellulose content and a low lignin content, they are being exploited as valuable biomass for the production of renewable energy. The low lignin content indicates that biomass has the potential to advance toward the production of biofuels and develop into a third-generation process with considerable advantages. Anaerobic digestion has been tried with algal biomass to produce methane. Seaweeds have been shown to have the following benefits for biofuel research:

1. Their growth rate and biomass production were much higher than those of angiosperms.
2. Compared to terrestrial plants, the cultivation requires comparatively less land.
3. Growth can occur without the need for fertilizers, pesticides, or fresh water. On "seaweed forest" land, we can harvest 50 tonnes of biomass per acre. Food and fuel do not conflict (Bruton T. *et al.*, 2009).
4. Land clearing and deforestation are not necessary for cultivation. The culture area is easily transportable because maintaining the temperature is not too difficult due to the watering of the cultivation media. However, severe climate catastrophes, such as fires, floods, and droughts, can completely devastate seaweed farming (Bruton T *et al.*, 2009).

10.2.2.2 Microalgae

Microalgae can contain 8–31% (dry weight) of lipids. According to multiple research studies, microalgae is a more effective and promising source for the extraction of oil than biofuel crops. According to Ramachandra TV *et al.* (2013), microalgae have almost 9000 Lt/ha of energy output production in one year.

10.2.2.3 Cyanobacteria

Cyanobacteria are found in both fresh and marine phytoplankton and play a major role in photosynthesis, which releases oxygen into the atmosphere. Because the microorganism grows considerably more quickly than plants and, like macroalgae, doesn't require land, it might be exploited as a resource for pyrolysis to obtain bio-oil (Xiong W *et al.*, 2015; Sarsekeyeva F *et al.*, 2015).

10.3 BIORESOURCE INVENTORYING

The quality, quantity, change, productivity, and condition of an area's bioresources can all be described using a bioresource inventory. These inventories might be used for assessments at the national or regional level.

10.3.1 Forest Inventory

Forest inventories, which vary in scope and purpose, are most frequently employed for assessing forest biomass. In order to gather data on the forest's various uses—such as recreation, grazing, wildlife, and water conservation—inventory designs are being made. Their purpose is to measure biomass from forests instead of or in addition to traditional volume. A few hundred to several thousand hectares, or one or more management units, make up the forest inventory area. With regard to stem frequency, merchantable wood volume is the main characteristic of interest (Ramachandra TV *et al.*, 2004).

10.3.2 Agro-Residue Inventory

To guarantee that assessments of crop output and residue are based on completely ripe crops, field sampling is done no later than one week before harvest (Ramachandra TV *et al.*, 2004). The residue-yield ratio generated along with this crop yield is also calculated by the area sampling frame methodology.

10.3.3 Plantation Inventory

An energy plantation would be managed more like a farming enterprise than a traditional forestry one. A plantation's spatial size, mean annual increment, annual productivity, and cycling time are all evaluated as part of the inventory process. Selected fuel wood species that can be gathered quickly could be grown to meet the rising population's energy needs. The species have been carefully selected to yield enormous amounts of wood, a high tolerance or adaptation, resilience to pests and drought, rapid growth, and a good survival rate. Most species with dual uses are favored (Ramachandra TV *et al.*, 2004).

10.4 BIOMASS

Biomass is the organic matter that comes from living entities such as plants and aquatic and terrestrial animals. Bioenergy is released when this organic carbon reacts with oxygen. Bioenergy is produced from various feedstocks, which include wood from sustainable forests, forest residues, agricultural by-products, and wastes from homes, businesses, and industries. Photosynthesis is the main process that builds up biomass (Johansson TB *et al.*, 1993). The following equation can be used to illustrate the procedure:

$$6CO_2 + 6H_2O \rightarrow \text{Sunlight} + C_6H_{12}O_6 + 6O_2$$

With a conservation efficiency of about 1%, global photosynthesis is estimated to produce 220 billion dry tons of biomass annually (Johansson TB *et al.*, 1993). These carbon reserves are converted by conversion technologies into more flexible forms like steam, heat, or other solid, liquid, or gaseous fuels to increase their efficiency of use. These fuels are commonly referred to as biofuels due to their biological origin. The utilization of biomass as a renewable energy source is crucial for nations with finite sources of fossil fuels. These fuels are also frequently referred to as biofuels because of their biological origin (Ramachandra TV *et al.*, 2004). The organic leftovers from the creation of food, fiber, and forests are known as biomass residues.

10.4.1 Woody Biomass

Urban trees, farm trees, bush trees, forest and agro-industry, and plantations are examples of woody biomass. Branches, bark, wood, and leaves make up the aboveground biomass of wood. Woody biomass is typically a highly valued commodity with a variety of applications, including cooking fuel, pencil and matchstick manufacturing, pulp and paper industries, and lumber.

10.4.2 Nonwoody Biomass

Nonwoody biomass includes household wastes (food, trash, sewage), crop leftovers (straw, leaves, and plant stems), and processing residue (nutshells and husks). They are utilized as fuel for cooking or as fodder, and they are harvested at the village level.

10.4.3 Animal Waste

The trash that results from raising animals is called animal waste. Dried animal dung contains the same amount of energy as wood, making it a potentially significant biomass resource. It is only approximately 10% efficient when burned for heat. Forty percent of the 150 Mt of dry cow dung is used annually as a fuel worldwide comes from India (UN Publication, 1980). However, dung can only be easily recovered from livestock that is kept in confinement (Ramachandra TV *et al.*, 2004).

10.4.4 Composition of Biomass

From an energy perspective, the chemical makeup of biomass is crucial since it affects the different processes of conversion. Alpha cellulose is typically the main structural component of a variety of biomass forms. The percentage contribution from fat and protein content is lower than that from the carbohydrate component. Despite having the largest energy content, fat makes up a smaller percentage of the composition of most biomass categories.

The moisture and ash contents of nonwoody biomass have an impact on its energy value. Since wood may be utilized without any special preparation or alteration other than cutting it into small pieces, it is most frequently used for fuel (Tillman DA *et al.*, 1981). The physical attributes of fuel, such as its specific gravity, void volume, and moisture content, influence its qualities. Depending on how much the wood has been dried, the moisture content varies. Fuel wood's energy value is low and varies from 10.9 to 21.3 GJ per tonne; oven-dried wood with a moisture percentage of 0% has an average of 16.9 GJ. Wood's combustion efficiency is determined by its chemical composition and structure since combustion involves a sequence of chemical processes. In most species, extractives are also found, albeit in trace amounts. Wood is classified as softwood or hardwood based on its composition. Softwoods typically include about 24–37% hemicellulose, 25–30% lignin, and 40–45% cellulose at their core. Roughly 40–50% cellulose and 22–40% hemicellulose is found in hardwoods (Egneus H & Ellegard A, 1984).

TABLE 10.2

Carbon Percentage and Energy Contribution of Some Biomass Components

Components	Carbon (%)	MJ/Kg
Monosaccharides	40	15.6
Disaccharides	42	16.7
Polysaccharides	44	17.5
Lignin	63	25.1
Crude protein	53	24
Fat	75	39.8
Carbohydrate	41–44	16.7–17.7
Crude fiber	47–50	18.8–19.8

Source: Bisio A and Boots S (1997).

TABLE 10.3
Composition of Intact Wood (Oven Dry Weight Percentage)

Components	Hardwood	Softwood
Lignin	21–22%	27– 28%
Sugars	50–52%	50%
Glucose	1.5–2.3%	50%
Mannose	10%	4.5%
Galactose	20–21%	6–7%
Xylose	0.3%	2–3%
Arabinose		

Source: Egneus H and Ellegard A (1984).

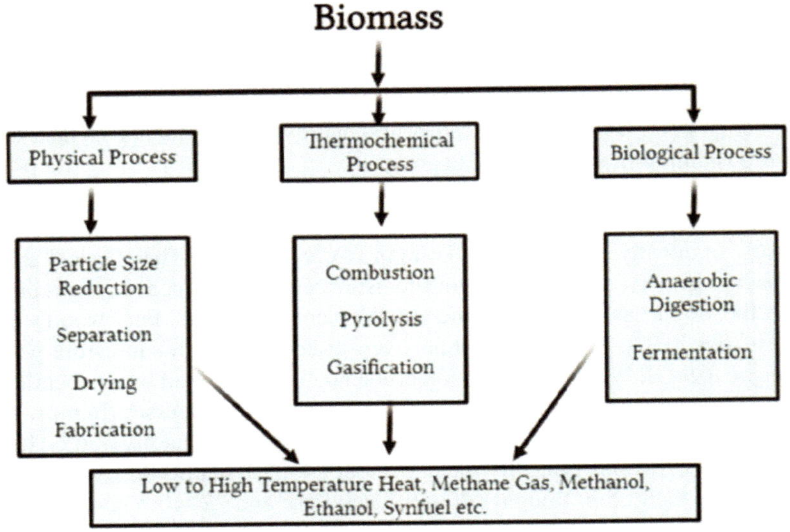

FIGURE 10.2 Overview of biomass conversion.

10.4.5 Biomass Conservation Technologies

Fresh biomass has various disadvantages compared to fossil fuels, such as low bulk density, variable physical form, and high moisture content that lowers combustion efficiency. The process of converting biomass enhances the material's fuel-related properties. Reducing the water content and enhancing the material's handling properties are the main goals of the conversion operations. The energy so obtained can be used for household needs, and small-scale enterprises such as papermaking, paddy drying, coffee/tea processing, sericulture, jaggery making, and agriculture. A summary of the routes for converting biomass is shown in Figure 10.2 (TV Ramachandra *et al.*, 2004).

10.5 WASTES

10.5.1 Organic Solid Waste

The biggest issue that modern society is facing today is managing the final disposal of solid waste (Barles, 2014). The current linear manufacturing method does not employ a comprehensive approach to waste and resource management that addresses all aspects of the production chain, including product design, extraction of raw materials, consumption, production, and ultimate disposal. Typically, governments take charge of all the waste management systems, and it is their responsibility to take on the problem of offering the populace an efficient and successful system. According to Yukalang *et al.* (2017), they frequently deal with issues that extend outside the purview of municipal government, such as a lack of funding and organizational capacity, as well as complicated and multifaceted subject matter.

10.5.2 Urban Waste Management

Municipal solid waste management must be operated and maintained, and a first-rate urban infrastructure is needed (Alm, 2015). Having a proper management system is a big challenge because of collection, transportation, and disposal (Sanjeevi& Shahabudeen, 2015) as well as keeping track of the effects on the economy, society, and environment (Deus *et al.*, 2017). Municipalities can execute waste management with people engagement when it comes to the social dimension. Through this method, organic waste can help local business initiatives and provide green energy. According to Velis *et al.* (2009), the collection of selective solid waste in urban regions is a methodical process that involves separating and classifying elements that are already found in garbage-generating sources to promote their reuse and ensure their quality. Since the start of the 20th century, this practice has been mostly well known and used in Europe and the US (Velis *et al.*, 2009). Garbage collectiors are typically employed by the informal sector in poor nations to recycle solid waste (Gutberlet, 2012).

10.5.3 Agriculture Waste

Material derived from plant growth or crop production is known as agricultural waste. However, because it has such a high potential for energy conversion, biomass made from agricultural waste is being employed to generate electricity. Since biomass is widely available worldwide, we must find alternative energy sources to meet our energy consumption requirements. One viable way to turn agricultural waste into energy is through efficient utilization. The ability to produce electricity from biomass can open up new opportunities for farmers to diversify their farming operations. Some crops might be grown on marginal land or even in ecologically degraded areas, which would benefit the environment (Harshwardhan K & Upadhyay K, 2017). The world's energy reserves are finite and are running out as a result of the rising global population and a lack of advancement in technology (Holechek, 2022). Worldwide, biomass resources are abundant (Yaning Z *et al.*, 2012). The use of biomass, particularly agricultural residue, is very low. The leftover biomass was disposed of in an environmentally unfavorable manner by burning it in the field, which resulted in the production of a black cloud that was saturated with carbon dioxide (Elfeki M & Tkadlec T, 2015; Blin J *et al.*, 2007). The advantages of biomass renewal and abundance make it a desirable source of renewable energy. Because it uses plants to produce biofuel and energy, biomass energy, or "bioenergy," is seen as a promising technology (Yaning Z *et al.*, 2012). Maize is regarded as a staple food in most regions of the world; its yearly global production is 864.96 million metric tonnes. Each year, these cultivations yield over 1730 million metric tonnes of residues (leaves, stalks, and straw) (FAO, 2002). About 2.88 million kilograms of the dry leaves were used in the numerous studies conducted to manufacture biofuels from biomass using a wide range of technologies, including thermochemical conversion. Only wood biomass is needed for the quick pyrolysis process to produce bio-oil; no other agricultural waste is needed (Baker EG & Elliott DC, 1988; Yanik J *et al.*, 2007; Bridgwater AV & Peacocke GVC 2000).

10.6 OVERVIEW OF BIOFUEL

Currently, fossil fuels account for more than 80% of primary energy consumption, with the transport sector accounting for 58% of this share (Escobar JC *et al.*, 2009). Fossil fuels release toxic gasses into the atmosphere, and the negative effects of these gases include biodiversity loss, sea level rise, glacier retreat, and climate change. The fast-paced modern world is primarily responsible for the fluctuating fuel demand since it is both industrialized and motorized (Agrawal AK, 2007). Biofuels are among the numerous alternative energy sources that are currently accessible. Because biofuels can effectively replace nonrenewable fuels, researchers are always trying to make biofuel from sustainable biomass (Weldemichael Y & Assefa G, 2016). Following are the benefits of biofuels over petroleum fuels:

1. They are more environmentally friendly.
2. They can be readily harvested from biomass.
3. Their combustion is based on the carbon dioxide cycle.
4. They are very sustainable.

Because biofuel has advantages for the environment, this would undoubtedly lead to a significant increase in the agricultural sector's output and related by-products (Kim S & Dale BE, 2005; Demirbas A, 2008). Biofuel can be classified as a solid, liquid, and gaseous fuel with biomass serving as the primary source of production. According to reports, biomass ranks as the fourth largest global energy source (Acma H & Yaman S, 2010). Additionally, biomass can be thought of as an affordable, natural way to store energy that can be used whenever needed (Castro JFMD, 2007; Arthe R *et al.*, 2008). India now has about 500 million metric tonnes of biomass available annually, comprising both forest and agricultural biomass, with an estimated 18,000 MW of energy potential (MNRE, 2016). Researchers are figuring out how to efficiently extract the potential energy that can be saved from biomass. Tonnes of biomass produced by industry, forestry, agriculture, marine, and urban solid waste are decomposing uncontrollably and releasing hazardous gases into the atmosphere, which is causing environmental issues (Acma H & Yaman S, 2010; Taherzadeh MJ & Karimi K, 2007). When these renewable resources are used efficiently, there are several immediate advantages, such as lower greenhouse gas emissions from burning green fuels, less pollution in the environment, and enhanced rural economies (Ragauskas JA *et al.*, 2006; Karakashev D *et al.*, 2007).

The gasoline produced by photosynthesis has emerged as yet another promising area for biofuels. Plants and algae efficiently absorb CO_2 from the atmosphere and store it as biomass, which may

TABLE 10.4
Biomass Generated from Main Agriculture Crop in India

Crop	Biomass Residue Million Tons (2009–2015)	Biomass Average Million Tons (2009–2015)
Rise husk	303	50.5
Wheat straw	729.4	119.9
Maize cobs	39.3	6.51667
Pearl millet straw	74.8	14.96
Sugarcane bagasse	675.3	112.55
Coconut shell	12.5	2.5
Coconut fiber	15.1	3.02
Coir pith	22.7	4.54
Groundnut shell	13.8	2.3
Cotton stalks	171.6	34.32
Jute sticks	13.2	2.642

Source: Agricultural Research Data Book (2016).

TABLE 10.5
Chemical Constituents (% DW) of Agriculture By-Products

Lignocellulosic Biomass	Cellulose	Hemicellulose	Lignin	Extractives	Reference
Oil palm shell	20.8	22.7	50.7	4.8	(Koba Y & Ishizaki, 1990)
Pineapple waste	19.4	22.4	4.7	—	(Ban-Koffi L & Han YW,1990)
Forest (logging) residues	46.3	28.3	26.5	4.4	(Tsoumis G, 1991; Barneto AG *et al.*, 2009)
Coir pith	26–30	3–7	30–40	—	(NCLN, 1993; Rowell RM & Stout HP, 1998)
Sugarcane bagasse	26–50	25–28	23–25	—	(Pandey A, 2000; Geo GL *et al.*, 2009)
Empty fruit palm bunches	54.4	28	17.6	2.8	(Sun RC & Tomkinson J, 2001)
Oil palm fronds	62.3	24.2	14.8	1.8	(Wan Rosli WD *et al.*, 2004)
Coconut husk	44.2	12.1	32.8	6.4	(Hpsa Khalil *et al.*, 2006)
Soybean hulls	29–51	10–20	1–4	—	(Corredor DY *et al.*, 2006; Lee JW *et al.*, 2007; Mielenz JR, 2001)
Corn stover	31–41	20–34	16–23	—	
Wheat straw	32–49	23–39	5–19	—	
Oil palm fibers	20.8	38.8	28.5	6.3	(Law KN *et al.*, 2007)
Oil palm trunks	41.2	34.4	17.1	2.8	(Kelly-Yong TL *et al.*, 2007)
Silk cotton	80–85	10–15	3.5–5	—	(Korenberg C, 2007)
Rice husk	35.1	20.9	17.6	—	(Nordin R *et al.*, 2007)
Banana residues	31.9	18	22.4	8	(Bilba K, *et al.*, 2007)
Paddy straw	35	21–25	6–13	—	(Geo GL *et al.*, 2009)
Reed	39.5	29.8	24	—	(Wati L *et al.*, 2007)
Rapessed	27.6	20.2	18.3	—	

Source: Goh CS *et al.* (2010).

be transformed back into any other form of energy. Without producing biomass, the photosynthetic microorganism cyanobacteria may directly convert CO_2 into the fuel molecule ethylene. A research team at NREL created genetically modified cyanobacteria that produce ethylene at a high rate (Xiong W *et al.*, 2015).

10.7 GLOBAL VIEW

It has been noted on a global scale that estimations of biofuels for future energy systems are relatively high. To replace oil-based fuels with biomethanol and bioethanol, biomass can help move the economy from one based on hydrocarbons to one based on carbohydrates.

10.8 BIOFUEL ECONOMY

The cost of biofuels has not increased in comparison to traditional fuels in the modern world. Given that biofuels have inherent benefits for the environment, their significant contribution to the transportation industry is expected to result in substantial expansion shortly. Both developed and developing nations view biofuels as relevant technology for several reasons. Two primary strategies for promoting the supply of biomass/feedstock for the production of biofuel are biofuel obligation and tax reduction (Demirbas A, 2008).

The production of energy has been a major factor in the development of agriculture in recent years. Agriculture's ability to provide energy is crucial for our economic growth. Nowadays, it is highly costly to produce methanol in agriculture from natural gas and to synthesize ethanol from ethylene. From an economic perspective, it makes sense to produce bioethanol concurrently with sugar juice biomethanol in regions where hydroelectricity is more affordable. We suggest that large, environmentally friendly biomass be used as the best alternative to meet our energy needs (Demirbas A, 2008).

10.9 TECHNOLOGIES FOR THE CONSERVATION OF BIORESOURCES (TERRESTRIAL AND MARINE) AND BIOFUELS

Depending on the first drying stage, there are two ways to obtain energy from seaweed and agricultural waste (Milledge JJ *et al.*, 2014).

10.9.1 Pretreatment

Pretreatment methods for lignocellulosic materials have included physical, physicochemical, chemical, and biological procedures. The processing of lignocellulosic materials is also not simple due to their complexity. The optimum pretreatment technique and parameters are highly dependent on the lignocellulose type.

10.9.2 Enzyme Conversion Technology

The cellulose molecules present in the cell wall of biomass-producing plants are firmly embedded in the amorphous molecules known as lignin and hemicellulose. The complex material known as lignocellulose, which makes up about 90% of the total dry weight of the plant cell, is created when these polymers form strong non-covalent bonds with one another. It has been suggested that enzymatic hydrolysis is a viable method for converting biomass (Wati L *et al.*, 2007).

10.9.3 Microbes in Conversion of Biomass

For many years, researchers have explored enzyme production for biomass processing, focusing on cultivating enzymes on specific substrates like cellulose and hemicellulose. Studies by Ishaque and Kluepfel (1980) and Kluepfel et al. (1986) demonstrated that cultivating enzymes on these substrates enhances their activity and adaptability to break down complex plant materials, thus optimizing biomass conversion processes. Therefore, in order to manufacture enzymes at a high rate, genetically engineered bacteria have become the main focus of cutting-edge biotechnological research. The study of other bacteria's ability to produce cellulose may also provide insight into how to use it to hydrolyze polysaccharides. Numerous microorganisms, such as fungi, yeast, and bacteria, ferment carbohydrates to produce energy. Because of how quickly they ferment, bacteria have caught the interest of researchers. Bacteria generally require less fermenting time than yeast. Every microorganism has its limitations, such as low final-product yields (Lynd LR, 1996).

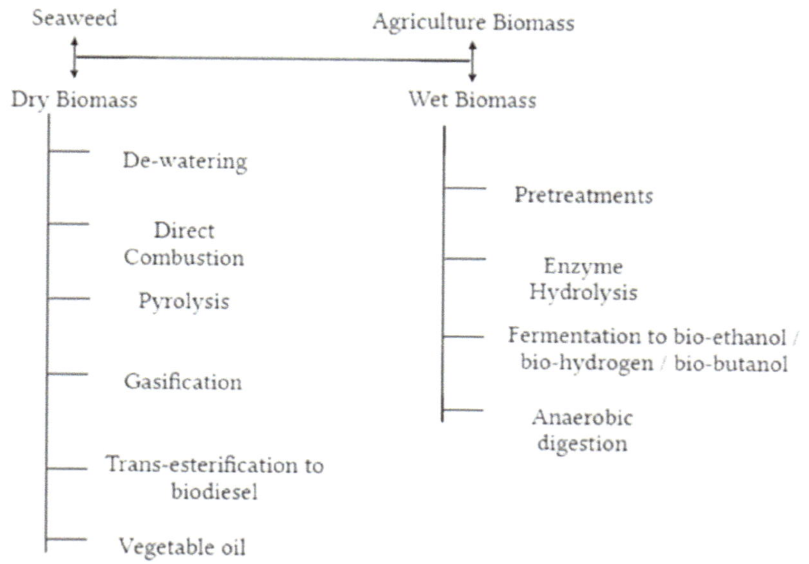

FIGURE 10.3 Process of energy extraction from bioresources.

10.10 PROCESS OF ENERGY EXTRACTION FROM BIORESOURCES

The process of producing biofuels is divided into three distinct categories, which are discussed next.

10.10.1 First Generation

The standard technique of producing ethanol and biodiesel is a part of the first-generation manufacturing process. The transesterification process helps to extract the oil from plants and then convert it into fuel that can be used directly by the engines. Redesigned engines may run on straight vegetable oils as fuel. Glycerine and fatty acids are produced as a by-product of the transesterification process, which employs enzyme catalysts or acids, alkaline, ethanol, or methanol (Elshahed MS, 2010; Hassan & Kalam, 2013).

10.10.2 Second Generation

Second-generation biofuel production methods include the hydrolysis of cellulose and subsequent fermentation of sugar. When using the gasification process to produce syngas, or synthesis gas, biological materials can be very helpful. Several catalytic methods can help transform it into liquid biofuel. The anaerobic digestion process creates natural gas along with methane. Crop by-products or agricultural waste are broken down during the process (Elshahed MS, 2010).

10.10.3 Third Generation

Third-generation processes are used in making biofuel from algae. Algae can easily create oil, which can then be refined further to make diesel and certain petrol constituents. The carbon metabolic process may be directed toward ethanol as the final product by genome and metabolic engineering techniques. Both open raceway ponds and photobioreactors can also be used to produce algal biomass. Third-generation methods have the drawback of producing non-stable biofuel (Bruton T *et al.*, 2009).

10.11 CHALLENGES OF BIOFUEL PRODUCTION

10.11.1 Economic Pretreatment

The following conditions should be satisfied by an efficient and affordable pretreatment:

1. Preservation of cellulose and hemicelluloses
2. Prevention of potential inhibitors of fermenting microorganisms and hydrolytic enzymes
3. Reduction of energy consumption
4. Reduction of feedstock size reduction costs
5. Reduction of material costs for pretreatment reactor construction
6. Production of fewer residues
7. Use of inexpensive chemicals with little or no chemical consumption

10.11.2 Cost of Enzymes

To lower the price of producing biofuel, an operative hydrolytic enzyme manufacturing unit must be set up. Large-scale fuel generation is not financially feasible for fuel demand commercialization due to the high cost of commercial enzymes (Barles, 2014).

10.11.3 Efficiency of Fermentation

According to reports, new screening technologies and genetic engineering will introduce bacteria and yeast that can ferment both glucose and xylose. Technology advancements help in less hydrolysate detoxification and more ethanol production in less time (Sonderegger M & Sauer U, 2003).

10.12 SUSTAINABLE FEEDSTOCK

10.12.1 Terrestrial Resource

The largest energy storage includes a variety of land and marine resources together referred to as "biomass," which has drawn the majority of study interest due to its special advantages. Although they provide about 14% of the world's main energy, coal, petroleum and natural gas are not very sustainable (Hamelinck CN, *et al.*, 2005). According to Saxena RC *et al.* (2009), the expected yearly primary production of terrestrial biomass was 220 billion oven-dry tonnes (odt). Approximately 200×10^9 tonnes of plant biomass are produced annually, of which 90% are lignocelluloses.

10.12.2 Marine Resources

India is a tropical country in South Asia with coordinates 08.04–37.06 N and 68.07–97.25 E. India is also considered a mega-diverse country. More than 70% of the global biodiversity is found in these nations (Oliver R *et al.*, 2009). Its 7500 miles of shoreline feature an extremely rare marine biota. When it comes to the quantity of particular and intraspecific seaweed species that have been documented, India leads South Africa and Australia among nations bordering the Indian Ocean (Leary SJ *et al.*, 2013; Sahoo D *et al.*, 2001). About 844 species of marine algae, mostly found in deep water and intertidal areas along the Indian coast, were previously documented. These species are classified into 217 genera (Mantri VA, 2006). The number of marine algae species and genera has recently expanded to 271 and 1153, respectively, and they are classified into four groups: Rhodophyceae, Cyanophyceae, Phaeophyceae, and Chlorophyceae (Oza RM & Zaidi SH., 2000). Seaweeds thrive quite well in India, particularly along the Tamil Nadu and Andaman and Nicobar coasts.

10.12.3 Sustainable Production

While renewable resources are evenly spread around the world, nuclear and fossil fuel resources are running out quickly and have limited availability. A lot of uncertainty surrounds the sustainability of the current energy systems due to concerns about the economy, equality, and environment. Numerous sustainability programs and biofuels schemes are being developed to safeguard the land, communities, and biodiversity. Organizations from the public, commercial, and non-governmental sectors have participated in these projects. The US government offers funding to hasten the market's research and development for biofuels. India's "National Biofuel Policy" aims to replace diesel with biodiesel in order to meet the country's 20% diesel demand. Additionally, they have suggested using bioethanol in place of 10–20% of petrol (Oza & Zaidi, 2000; Kim SK, 2011).

10.13 SUSTAINABILITY, POLICY, AND COMMERCIAL ASPECTS OF BIOFUEL

10.13.1 Sustainable Energy

Biofuel has become a significant source of sustainable fuel and is thought to help with the development of new energy sources, air quality improvements, and greenhouse gas emission reductions. The demand for fuel production has skyrocketed since people continually require it for transportation, heating, and other needs. Although alternative sources of sustainable oil and gas production are being developed, the world's supply of these resources has reached saturation. So, the alternative to fossil fuel is being considered to be biofuel (Kim SK, 2011; Ghosh SK, 2016).

10.13.2 Biofuel Policy and Analysis

Energy plays an important role in a nation's economic growth. The goal of India's national biofuels policy, which was formulated in December 2009, is to use biomass and other renewable resources to produce bioenergy as a motor fuel by 2017 with a 20% blending ratio. From the standpoint of biofuels, people living in rural areas of developing nations can benefit from new opportunities presented by oil imports. Overall, the primary goals of the biofuels strategy are employment development, environmental preservation, and increased corporate sector efficiency (Demirbas A, 2007; Ghosh SK, 2016). The Ministry of New and Renewable Energy (MNRE), New Delhi, India, has suggested a biofuel policy and its regulation to achieve an economically viable biofuel output.

10.13.3 Commercial Aspects of Biofuel

Projects aimed at commercializing biofuels would create more jobs directly and indirectly. These initiatives provide methods for hiring locals of all skill levels for a variety of departments. Unlike wind energy and solar energy initiatives, biomass-based biofuel programs must boost the local economy.

10.14 DISCUSSION AND FUTURE SCOPE

Today we realize that using biofuel to meet all our energy needs is a more environmentally friendly option. We obtain environmental benefits from engine combustion, such as reduced carbon dioxide emissions relative to the quantity isolated from the atmosphere. As a result, a closed carbon cycle is created. Despite the advantages that come with the use of biofuel, it's important to remember that their manufacturing and use have a negative influence on the environment. Some of these effects include the degradation of soil, food productivity, destruction of forests, and excessive water use. In many parts of India, water scarcity already limits the amount of food that can be produced. Water pollution and scarcity are caused by monoculture farming, which uses more water to grow biofuel crops like sugarcane (MNRE, 2015–16; Weldemichael Y & Assefa G, 2016). Massive volumes of wood can be accumulated as solid waste and utilized as a feedstock to produce adsorbents for the treatment of wastewater, thereby combating water pollution (Devasena *et al.*, 2024; Cutillas-Barreiro *et al.*, 2016). The adoption of alternative fuel choices is a top priority in emerging countries such as India in order to meet future oil demands. Seaweeds can be used for this because they are most abundant around the Indian coast.

10.15 CONCLUSION

The economics of a variety of interconnected factors will determine how biofuel is produced globally in the future. Around the world, a large number of governments are pushing for the usage of biofuels. The adoption of alternative fuel choices is a top priority in emerging countries such as India in order to meet the demands of oil. It goes without saying that each nation must make use of the biomass that is already available and grow crops that produce oil and are suited to their particular environment. To meet the socioeconomic energy needs of nations, large-scale seaweed cultivation and meticulous extraction process research are necessary for the manufacture of biofuels. The biofuel industry has a lot of promise and has gained international acceptance. Therefore, sustainable biofuel production will be beneficial for improving the economy globally, reducing climate change, and improving the environmental condition. Thus, the initial discovery and advances in the field of alternative fuel assist both the industry and the environment.

REFERENCES

Acma H, Yaman S. Interaction between biomass and different rank coals during co-pyrolysis. *Renew Energy* 2010; 35:288–92.

Agrawal AK. Biofuels (alcohols and biodiesel) applications as fuels for internal combustion engines. *Prog Energy Combust Sci* 2007; 33:233–71.

Agricultural Research Data Book. Pusa, New Delhi, India: ICAR (Indian Agricultural Statistics Research Institute; 2016), ⟨http://iasri.res.in/agridata/16data/HOME_16.HTML/;2016⟩, [accessed 16.11.2016].

Alm, J. Financing urban infrastructure: knowns, unknowns, and a way forward. *J Econ Surv* 2015; 29:230e262. https://doi.org/10.1111/joes.12045.

Arthe R, Rajesh R, Rajesh EM, Rajendran S, Jeyachandran S. Production of bioethanol from cellulosic cotton waste through microbial extracellular enzymatic hydrolysis and fermentation. *EJEAF Che* 2008; 2008(7):2984–92.

Baker EG, Elliott DC, Catalytic Hydrotreating of Biomass-derived Oils, In: *Pyrolysis Oils from Biomass.* American Chemical Society, 1988; 228–240.

Balat M, Balata H, Cahide OZ. Progress in bioethanol processing. *Prog Energy Combust Sci* 2008; 34:551–73.

Ban-Koffi L, Han YW. Alcohol production from pineapple waste. *World J Microbiol Biotechnol* 1990; 1990(6):281–4.

Barles S. History of waste management and the social and cultural representations of waste. *Basic Environ His* 2014 Oct 16; 4: 199–226. https://doi.org/10.1063/1.2234308.

Barneto AG, Carmona JA, Alfonso JEM, Alcaide LJ. Use of autocatalytic kinetics to obtain composition lignocellulosic materials. *Bioresour Technol* 2009;100:3963–73.

Bilba K, Arsene MA, Ouensanga A. Study of banana and coconut fibers: botanical composition, thermal degradation and textural observations. *Bioresour Technol* 2007; 98:58–68.

Bisio A, Boots S. *Encyclopedia of Energy and the Environment.* New York: John Wiley & Sons, Inc; 1997.

Blin J, Volle G, Girard P, Bridgwater T, Meier D, *et al.* Biodegradability of biomass pyrolysis oils: comparison to conventional petroleum fuels and alternative fuels in current use. *Fuel* 2007; 86:2679–2686.

Bridgwater AV, Peacocke GVC, Fast pyrolysis processes for biomass. *Renew Sustain Energy Rev* 2000; 1:1–73.

Bruton T, Lyons H, Lerat Y, Stanley M, Bo Rasmussen M. A Review of the Potential of Marine Algae as a Source of Biofuel in Ireland. Sustainable Energy Ireland report; p.1–88. ⟨http://www.seai.ie/Publications/RenewablesPublications/Bioenergy/Algaereport.pdf/⟩; [accessed 30.08.16]; 2009.

Castro JFMD. Biofuels – An overview, Final report, Prepared for: DGIS/DMW/IB 2007. ⟨https://np-et.pbworks.com/f/BZOS+(2007)+Biofuels+in+Africa_overview.pdf⟩; [accessed 30.08.16].

Corredor DY, Bean SR, Schober T, Wang D. Effect of Decorticating Sorghum on ethanol production and composition of DDGS1. *Cereal Chem* 2006; 83:17–21.

Cutillas-Barreiro L, Paradelo R, Igrexas-Soto A, Núñez-Delgado A, Fernández-Sanjurjo MJ, Álvarez-Rodriguez E, ... Arias-Estévez M. Valorization of biosorbent obtained from a forestry waste: competitive adsorption, desorption and transport of Cd, Cu, Ni, Pb and Zn. *EcotoxicolEnviron Safety* 2016; 131:118–26.

Demirbas A. *Biodiesel: a realistic fuel alternative for diesel engines*. Springer-Verlag London Limited. 2007.

Demirbas A. Biofuels sources, biofuel policy, biofuel economy and global biofuel projections. *Energy Convers Manag* 2008; 49:2106–16.

Deus, R.M., Battistelle, R.A.G., Silva, G.H.R., Scenario evaluation for the management of household solid waste in small Brazilian municipalities. *Clean Technol Environ* 2017; 19:205e214. https://doi.org/10.1007/s10098-016- 1205-0.

Devasena SS, Padmavathy P, Rani V, Ganesan P, Pereira JJ. A review on various bio-resources as a tool to reduce water pollution via eco-friendly adsorption technology. *Asian JBiotechnol Bioresource Technol* 2024; 10(1):34-46.

Dhargalkar VK, Neelam P. Seaweed: promising plant of the millennium. *SciCultur* 2005; 71:60–6.

Egneus H, Ellegard A. *Bioenergy-84, Bioenergy state of the art*. London: Elsevier Applied Science Publication; 1984.

Elfeki M, Tkadlec T, Treatment of Municipal Organic Solid Waste in Egypt. *J Mater Environ Sci* 6: 2015; 756-764.

Elshahed MS. Microbiological aspects of biofuel production: current status and future directions. *J Adv Res* 2010; 1(2):103–11.

Escobar JC, Lora ES, Venturini OJ, Yanez EE, Castillo EF, Almazan O. Biofuels: environment, technology, and food security. *Renew Sustain Energy Rev* 2009; 13:1275–87.

FAO, Food and Agriculture Organization of the United Nations. Date of Palm Cultivation, 2002

Geo GL, Hsu DC, Chen WH, Chen WH, Hwang WS. Characterization of enzymatic saccharification for acid-pretreated lignocellulosic materials with different lignin composition. *Enzy Microb Technol* 2009;45(2):80–7.

Ghosh SK. Biomass & bio-waste supply chain sustainability for bio-energy and bio-fuel production. *Procedia Environ Sci* 2016; 31:31–9.

Goh CS, Tan KT, Lee KT, Bhatia S. Bio-ethanol from lignocellulose: status, perspectives and challenges in Malaysia. *Bioresour Technol* 2010; 101:4834–41.

Gutberlet J. Informal and cooperative recycling as a poverty eradication strategy. *Geogr Compass* 2012; 6:19–34. https://doi.org/10.1111/j.1749-8198.2011.00468.x.

Hamelinck CN, Van Hooijdonk G, Faaij APC. Ethanol from lignocellulosic biomass: techno-economic performance in short, middle and long term. *Biomass Bioenergy* 2005; 28:384–410

Harshwardhan K, Upadhyay K. Effective utilization of agricultural waste: review. *J Fundam Renew Energy Appl* 2017; 7:237.

Hassan MH, Kalam MA. An overview of biofuel as a renewable energy source: development and challenges. *Procedia Eng* 2013; 56:39–53.

Holechek JL, Geli HM, Sawalhah MN, Valdez R. A global assessment: can renewable energy replace fossil fuels by 2050?. *Sustainability* 2022 Apr 16; 14(8):4792.

Hpsa Khalil, Alwani MS, Akm Omar. Chemical composition, anatomy, lignin distribution, and cell wall structure of Malaysian plant waste fibers. *Bioresources* 2006; 1:220–32.

Ishaque M, Kluepfel D. Cellulase complex of a mesophilic Streptomyces strain. *Can J Microbiol* 1980; 26:183–9.

Johansson TB, Kelly H, Reddy AKN, Williams RH (Eds.). Renewable energy sources for fuels and electricity. Earthscan. 1993.

Karakashev D, Belinda A, Angelidaki I. Anaerobic biotechnological approaches for production of liquid energy carriers from biomass. *Biotechnol Lett* 2007; 29:1005–12.

Kelly-Yong TL, Lee KT, Mohamed AR, Bhatia S. Potential of hydrogen from oil palm biomass as a source of renewable energy worldwide. *Energy Policy* 2007; 35:5692–701.

Kim S, Dale B. Life cycle inventory information of the United States electricity system (11/17 pp). *The International Journal of Life Cycle Assessment*. 2005 Jul;10:294–304.

Kim, SK. Eco-Biochemical Studies of Common Seaweeds in the Lower Gangetic Delta. In: Ghosh R, Banerjee K, and Mitra A. *Handbook of marine macroalgae: biotechnology and applied phycology*. Se-Kwon Kim. 2011. p. 45–57.

Klinke HB, Thomsen AB, Ahring BK. Inhibition of ethanol-producing yeast and bacteria by degradation products produced during pre-treatment of biomass. *Appl Microbiol Biotechnol* 2004; 66:10–26.

Kluepfel D, Shareck F, Mondou F, Morosoli R. Characterization of cellulose and xylanase activities of *Streptomyces lividans. Appl Microbiol Biotechnol* 1986; 24:230–4.

Koba Y, Ishizaki A. Chemical composition of palm fiber and its feasibility as cellulosic raw material for sugar production. *Agric Biol Chem* 1990; 54:1183–7.

Korenberg C. The effect of ultraviolet-filtered light on the mechanical strength of fabrics. *Tech Res Bull* 2007; 1:23–7.

Law KN, Wanrosli WD, Ghazali A. Morphological and chemical nature of fiber strands of Oil Palm Empty-Fruit-Bunch (OPEFB). *Bioresources* 2007; 2:351–62.

Leary SJ, Hice LA, Feldheim KA, Frisk MG, McElroy AE, Fast MD, *et al.* Severe inbreeding and small effective number of breeders in a formerly abundant marine fish. *PLoS One* 2013; 8(6):e66126. http://dx.doi.org/10.1371/journal.pone.0066126.

Lee JW, Gwakl KS, Park JY, Choi DH, Kwon M, Choil IG. Biological pretreatment of softwood Pinus densiflora by three white rot fungi. *J Microbiol* 2007; 45:485–91.

Lin Y, Tanaka S. Ethanol fermentation from biomass resources: current state and prospects. *Appl Microbiol Biotechnol* 2006; 69:627–42.

Lynd LR. Overview and evaluation of fuel ethanol production from cellulosic biomass: technology, economics, the environment, and policy. *Ann Rev Energy Environ* 1996; 21:403–65.

Mantri VA. Seaweed floristic studies along tsunami-affected Indian coasts: a litmus test scenario after 26th December 2004. *J Earth Syst Sci* 2006; 115:371–8.

Mielenz JR. Ethanol production from biomass: technology and commercialization status. *Curr Opin Microbiol* 2001; 4:324–9.

Millati R, Niklasson C, Taherzade MJ. Effect of pH, time and temperature of overliming on detoxification of dilute-acid hydrolyzates for fermentation by *S. cerevisiae. Process Biochem* 2002;38:515–22.

Milledge JJ, Smith B, Dyer PW, Harvey P. Macroalgae-derived biofuel: a review of methods of energy extraction from seaweed biomass. *Energies* 2014; 7:7194–222.

MNRE, Annual Report 2015-16, Ministry of New and Renewable Energy, Ministry of Government of India, New Delhi, India. http://mnre.gov.in/file-manager/ annual-report/2015-2016/EN/Chapter%205/chapter_5.htm/; 2016; [accessed 16.11.2016]

MNRE, Biomass and Cogeneration Program, Ministry of New and Renewable Energy, Ministry of Government of India, New Delhi, India. ⟨http://mnre.gov.in/related-links/grid-connected/biomass-powercogen/⟩; 2016; [accessed 16.11.16].

Najafi G, Ghobadian B, Tavakoli T, Yusaf T. Potential of bioethanol production from agricultural wastes in Iran. *Renew Sustain Energy Rev* 2009; 13:1418–27.

NCLN Charyulu, Coir RavirajN. pith as a source of power alcohol and active carbon. Workshop on utilization of coir pith. Pollachi, p. 6–10; 1993.

Nordin R, Said CMS, Ismail H. Properties of rice husk powder/natural rubber composite. *Solid State Sci Technol* 2007; 15:83–91.

Oliver R, David I, King A. Quo vadis biofuels. *Energy Environ Sci* 2009; 2:343.

Oza RM, Zaidi SH. *A revised checklist of Indian Marine Algae*. Bhavanagar, India: Central Salt and Marine Chemicals Research Institute; 2000. p. 296.

Pandey A, Soccola CR, Mitchel D. New developments in solid state fermentation: bioprocess and products. *Process Biochem* 2000; 35:1153–69.

Ragauskas JA, Williams CK, Davison BH, Britovsek G, Cairney J, Charles A, *et al.* The path forward for biofuels and biomaterials. *Science* 2006; 311:484–9.

Ramachandra TV, Kamakshi G, Shruthi BV. Bioresource status in Karnataka. *Renewable Sustainable Energy Rev* 2004 Feb 1;8(1):1-47.

Ramachandra TV, Madhab MD, Shilpi S, Joshi NV. Algal biofuel from urban wastewater in India: scope and challenges. *Renew Sustain Energy Rev* 2013; 21:767–77.

Rashid MH. *Electric renewable energy systems.* Cambridge (MA): Academic Press; 2015; Nov 25.

Rowell RM, Stout HP. Jute and Kenaf [Chapter 7]. In: Lewin M, Pearce EM, editors. 2nd ed. *Handbook of Fiber Chemistry.* New York: Marcel Dekker Inc; 1998. p. 465-504.

Sahoo D, Sahu N, Sahoo D. *Seaweeds of Indian coast.* New Delhi: A. P. H. Publishing Corporation; 2001. p. 283.

Sanjeevi, V., Shahabudeen, P., Development of performance indicators for municipal solid waste management (PIMS): a review. *Waste Manag Res* 2015; 33:1052e1065. https://doi.org/10.1177/0734242X15607428.

Sarsekeyeva F, Zayadan BK, Usserbenov A, Bedbenov VS, Sinetova MA, Los DA. Cyanofuels: biofuels from Cyanobacteria. Reality and perspectives. *Photosynth Res* 2015;125(1–2), 329–40

Saxena RC, Adhikari DK, Goyal HB. Biomass-based energy fuel through biochemical routes: a review. *Renew Sustain Energy Rev* 2009; 13:167–78.

Sonderegger M, Sauer U. Evolutionary engineering of Saccharomyces cerevisiae for anaerobic growth on xylose. *Appl Environ Microbiol* 2003; 69:1990–8.

Sun RC, Tomkinson J. Fractional separation and physico-chemical analysis of lignins from the black liquor of oil palm trunk fiber pulping. *Sep Purif Technol* 2001; 24:529–39.

Taherzadeh MJ, Karimi K. Acid-based hydrolysis processes for ethanol from lignocellulosic materials: a review. *BioResources* 2007; 2:472–99.

Tsoumis G. *Science and technology of wood: structure: properties and utilization.* New York: Van Nostrand Reinhold; 1991.

UN Publication, 1980. Guidebook on biogas development.

Velis, C.A., Wilson, D.C., Cheeseman, C.R., 19th century London dust-yards: a case study in closed-loop resource efficiency. *Waste Manag* 2009; 29:1282e1290. https://doi.org/10.1016/j.wasman.2008.10.018.

Wan Rosli WD, Law KN, Zainuddin Z, Asro R. Effect of pulping variables on the characteristics of oil-palm frond fiber. *Bioresour Technol* 2004; 93:233–40.

Wati L, Kumari S, Kundu BS. Paddy straw as substrate for ethanol production. *Ind J Microbiol* 2007; 47:26–9.

Weldemichael Y, Assefa G. Assessing the energy production and GHG (greenhouse gas) emissions mitigation potential of biomass resources for Alberta. *J Clean Prod* 2016; 112:4257–64.

Xiong W, Morgan JA, Ungerer J, Wang B, Maness PC, Yu J. The plasticity of cyanobacterial metabolism supports direct CO2 conversion to ethylene. *Nat Plants* 2015; 1:15053. http://dx.doi.org/10.1038/NPLANTS.2015.53

Yanik J, Kornmayer C, Saglam M, Yuksel M. Fast pyrolysis of agriculture waste; characterization of pyrolysis products. *Fuel Process Technol* 2007; 88:942–947.

Yaning Z, Ghaly AE, Bingxi Li, Availability and physical properties of residues from major agricultural crops for energy conversion through thermochemical processes. *Am J Agri Biol Sci* 2012; 7:312–321.

Yukalang, N., Clarke, B., Ross, K., Barriers to effective municipal solid waste management in a rapidly urbanizing area in Thailand. *Int J Environ Res* 2017; 14:9–14. https://doi.org/10.3390/ijerph14091013.

11 Sustainable Use of Forest Resources in India under Changing Environment

*Pardeep Kumar[1], Isha[1], Abhishek Kumar[1], Meenu Patil[1], Sabir Hussain[1], Sheenu Sharma[1], Anand Narain Singh[1]**

[1]Soil Ecosystem and Restoration Ecology Lab, Department of Botany, Panjab University, Chandigarh India
*Corresponding author: Anand Narain Singh, Soil Ecosystem and Restoration Ecology Lab, Department of Botany, Panjab University, Chandigarh India

11.1 INTRODUCTION

Forest ecosystems are one of the critical components of the world's biodiversity as they cover approximately 31% of the global land area (FAO, 2020). Forests are home to millions of species found on Earth and render valuable services to society, including food, water, medicine, fibre, timber, aesthetic and spiritual values (Pan et al., 2013). In addition, they are a storehouse of one-third of terrestrial gross primary productivity (Beer et al., 2010) and therefore act as a significant terrestrial carbon sink (Anderson-Teixeira et al., 2016). At the same time, forest ecosystems are facing multiple threats in the form of natural calamities, deforestation, degradation, land use change, invasive species and pollution. No doubt the economic development and modernisation of society have lifted billions of people out of poverty but at the cost of altering nature (FAO, 2020). Hosonuma et al. (2012) highlighted that commercial agriculture is the most prevalent deforestation driver, accounting for 40% of deforestation, followed by local/subsistence agriculture (33%), mining (7%), infrastructure (10%) and urban expansion (10%) of the total deforestation.

India, due to its varied topography, land use types and environmental conditions, supports a wide variety of natural resources, and forests are one of the key elements of India's natural wealth. They support millions of people, including tribal and ethnic communities, for their livelihood and offer a variety of ecosystem services. Owing to increasing population pressure, fuel and fodder demand, industrialisation, and subsequent energy demand, India's forests are under unprecedented pressure to meet the rising needs. Forests act as the backbone for the economy of any country as all sections of society are dependent on them directly or indirectly. Therefore, there is a global call to conserve these natural resources. Conservation and sustainable development of all types of forests worldwide have now emerged as priority items on the international policy agenda, particularly in the context of the United Nations Conference on Environment and Development (UNCED) held in Brazil in June 1992 (Hoogeveen and Verkooijen, 2010).

India is currently in a critical position where it has to align its forestry objectives with economic, environmental, and social interests and obligations, clearly identifying the synergies and trade-offs (Aggarwal et al., 2009). The Indian government has started initiatives such as intended nationally

DOI: 10.1201/9781032697550-11

determined contributions (INDCs), focused on enhancing the country's green cover and creating an additional carbon sink of 2.5–3 billion tonnes by the year 2030. According to the International Tropical Timber Organization (ITTO), sustainable forest management (SFM) is the process of managing forest to achieve one or more clearly specified objectives of management with regard to the production of a continuous flow of desired forest products and services without undue reduction of its inherent values and future productivity and without undue undesirable effects on the physical and social environment. (https://www.atibt.org/en/members/8/itto-international-tropical-timber-organization)

India made a voluntary pledge in 2010 to reduce the emission intensity of its gross domestic product (GDP) by 20-25% from 2005 levels by 2020 (excluding emissions from agriculture). Later in 2015, India submitted its nationally determined contributions (NDCs) under the Paris Agreement, wherein India voluntarily pushed up its target of reducing the emission intensity of its GDP by 33–35% from 2005 levels by 2030. As a result of India's multiple mitigation actions, the emission intensity has already reduced by 21% between 2005 and 2014 (MOEFCC, 2018). In this chapter, we aim to discuss the sustainable use of forest resources in India with the following objectives: (1) What is the existing state of forests in India and what are the prevailing threats? (2) How can forest resources be used in a sustainable manner? (3) What are the initiatives taken by the government of India for the sustainable utilisation of forest resources?

11.2 FOREST COVER IN INDIA

The total forest cover of India as per the Indian State of Forest Report (FSI, 2021) is 80.9 million hectares, making up about 24.62% of its entire geographic area. The country's overall forest and tree cover has increased by 2261 sq km since 2019. Approximately 1540 square kilometres have been added to the forest cover, and 721 square kilometres have been added to the tree cover. The most recent evaluation divides the country's forest cover into three categories: very dense forest (VDF), moderately dense forest (MDF) and open forest (OF), depending on canopy density. In terms of distribution, tropical dry forests constitute the highest percentage (34.80%) of geographical area followed by tropical moist deciduous (33.19%), semi-evergreen (7.72%) and wet evergreen (7.54%) forests (Figure 11.1) (Reddy et al., 2015).

11.3 FOREST RESOURCES AND ECOSYSTEM SERVICES

Ecosystem services are the benefits that ecosystems provide to human well-being, both directly and indirectly (Figure 11.2). These benefits primarily relate to the provisioning, regulating, cultural and sustaining services that are necessary to maintain other services (Tian et al., 2015). First, any economic activity that is based on the tangible resources that an ecosystem provides, such as food, raw materials and drinking water, is referred to as provisioning services. Second, regulating services are the unintended advantages that an ecosystem offers, such as air cleaning, soil erosion avoidance and climate regulation. Third, cultural services are intangible benefits that come from encounters and enhance one's identity and culture. Last, supporting services are the remaining operations that enable ecosystems to deliver all of the aforementioned advantages (Martín-López and Montes, 2010). These varied services are connected in a variety of ways, from cooperative to tolerant, antagonistic and mutually exclusive. Since the idea of ecosystem services is multidisciplinary by nature, it naturally combines the social and natural sciences. Thorough evaluations necessitate cooperation between several disciplines (McDonough et al., 2017).

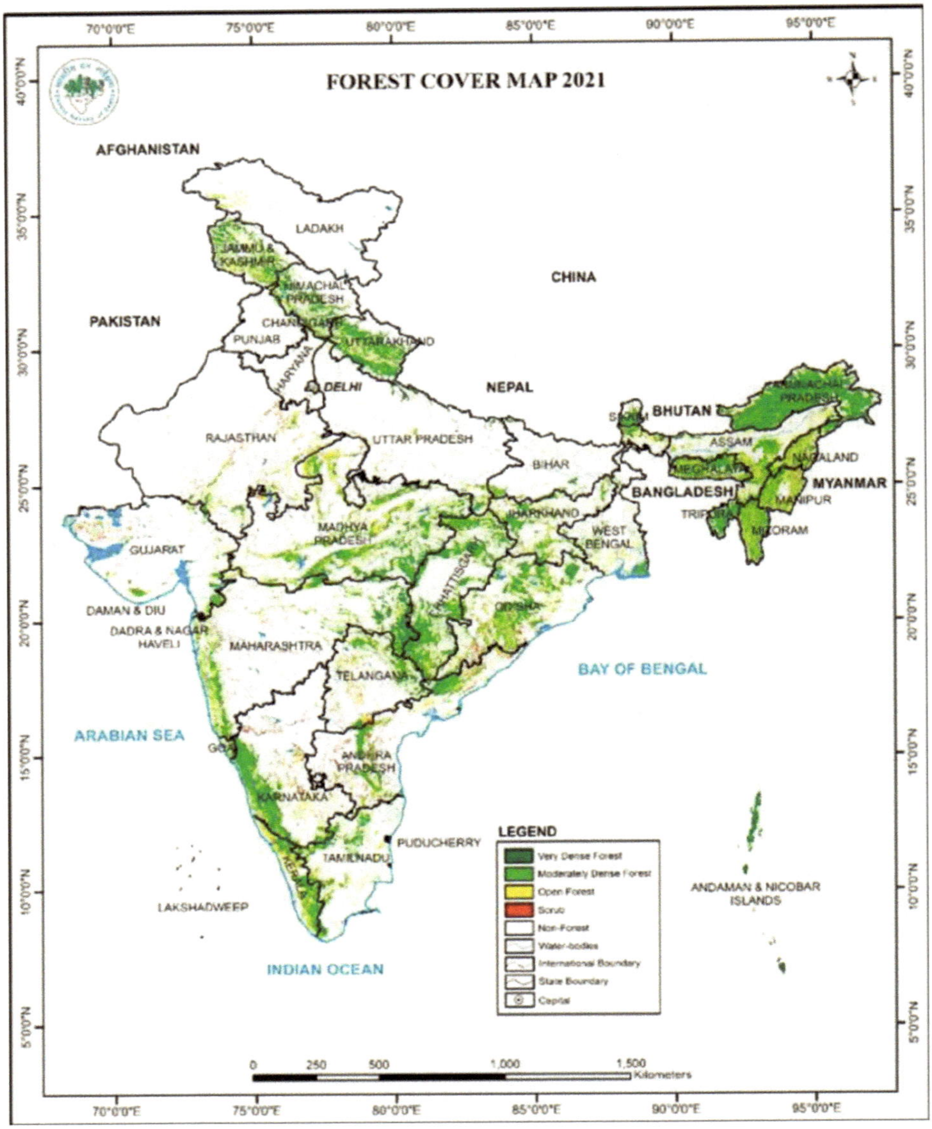

FIGURE 11.1 Forest cover map of India (FSI, 2021).

11.4 IMPACT OF CHANGING ENVIRONMENTS ON FOREST RESOURCES

In the larger context of global change and sustainable development, forests are essential for resolving issues related to climate change. There are three ways that forests and climate change are related: (1) forests release greenhouse gases (GHGs); (2) they provide chances for mitigation to stabilise GHG concentrations; and (3) climate change influences forests (Chaturvedi et al., 2008). The fragile forest ecosystems of India are predicted to face increased pressure from climate change. According to a recent study using the Regional Climate Model (RCM) and the BIOME model (BIOME 4), by 2085, over 50% of India's vegetation will not be as well adapted to its current location, increasing its susceptibility to both increased biotic stresses and unfavourable weather (Ravindranath et al., 2007). For example, the study projects that greater rainfall and mild temperature will

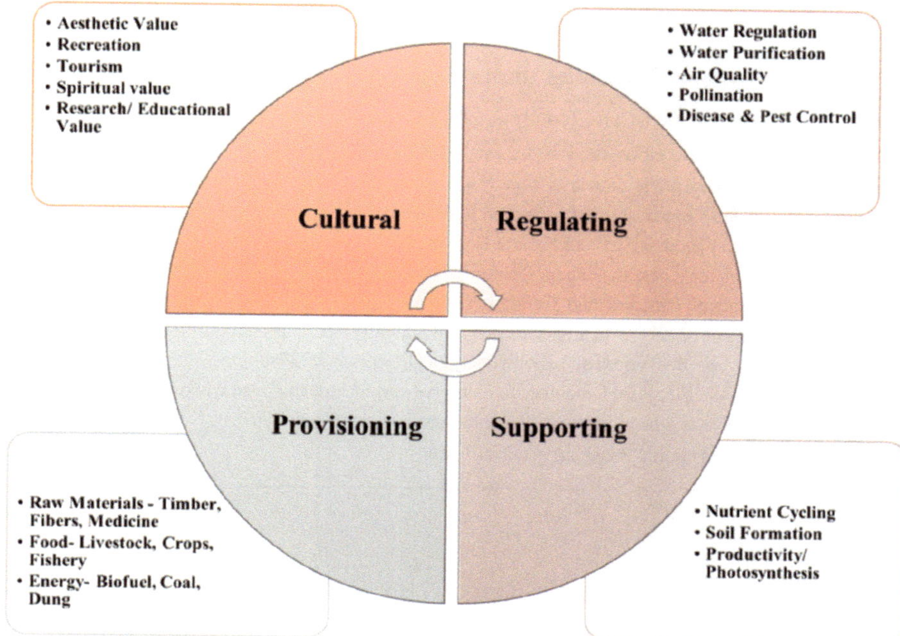

FIGURE 11.2 Ecosystem services provided by forests to mankind.

increases the area covered by tropical evergreen forests, at the expense of tropical deciduous and semi-deciduous forests. Tropical savanna and deciduous woods will replace a sizable portion of the tropical xerophytic shrubland, which will experience a significant area reduction owing to increased rainfall. With the great majority of cold and temperate forests giving way to warm mixed forests, the study also highlights the greater susceptibility of cold and temperate forests relative to tropical forests.

11.5 SUSTAINABLE MANAGEMENT AND USE OF FOREST RESOURCES

A report of the World Commission on Environment and Development (WCED) is credited with popularising the term "environmentally sustainable economic development," also referred to as "sustainable development." Sustainable development is described as "economic development that meets present needs without compromising the ability of future generations to meet their own needs" in this report (WCED, 1987). The globe is confronted with enormously complex environmental, social and economic concerns. Approximately 800 million people suffer from hunger (Von Grember et al., 2016) and 1.2 billion people are extremely impoverished (Suresh and Johnson, 2015). The effects of climate change on human welfare are widespread and, in certain places, such as island nations flooded by rising sea levels and populations uprooted by severe droughts or altered precipitation patterns, they even pose a threat to fundamental human life. In many areas, including sections of sub-Saharan Africa, inequality is growing and weakening social cohesion, resulting in civil unrest and aggravating governance issues.

In this context, the world's countries in 2015 agreed to adopt the 2030 Agenda for Sustainable Development, which provides a shared blueprint for peace and prosperity for people and the planet, now and into the future. The 2030 agenda encompasses 17 Sustainable Development Goals (SDGs), which are an urgent call for action by all countries—developed and developing—in a global partnership. The aim of these goals is to alleviate poverty, improve health and education, reduce inequality, and spur economic growth. Among these, goal 15 (commonly referred to as SDG 15) aims to

TABLE 11.1
An Overview of Different Objectives under SDG15

Target 15.1	Conserve and Restore Terrestrial and Freshwater Ecosystems
Target 15.2	End Deforestation and Restore Degraded Forests
Target 15.3	End Desertification and Restore Degraded Land
Target 15.4	Ensure Conservation of Mountain Ecosystems
Target 15.5	Protect Biodiversity and Natural Habitats
Target 15.6	Promote Access to Genetic Resources and Fair Sharing of the Benefits
Target 15.7	Eliminate Poaching and Trafficking of Protected Species
Target 15.8	Prevent Invasive Alien Species on Land and in Water Ecosystems
Target 15.9	Integrate Ecosystem and Biodiversity in Governmental Planning
Target 15.A	Increase Financial Resources to Conserve and Sustainably Use Ecosystems and Biodiversity
Target 15.B	Finance and Incentivise Sustainable Forest Management
Target 15.C	Combat Global Poaching and Trafficking

promote, protect and sustainably manage forests. It also calls for halting and reversing land degradation, fighting desertification and stopping biodiversity loss (https://sdgs.un.org/goals) (Table 11.1).

11.5.1 Key Initiatives by Government of India

Forests in India are legally protected by constitutional provisions, as Article 48A and 51-A(g) guarantee the protection and improvement of the environment and safeguarding of forests and wildlife and protection of the natural environmental as a fundamental duty of every citizen. Furthermore, India has an effective legal and policy framework guiding and directing the sustainable management of forests (Murthy and Kumar, 2019). Brief details about key policies, acts and rules are given in Tables 11.2–11.4.

TABLE 11.2
Key policies for Sustainable Management of Forests in India

Policy	Aim
National Forest Policy, 1952	To support rural economy and promote forest-based industries, 33% of forest-based industries. 33% of country's land area should be under forest cover.
Joint Forest Management, 1990	Afforestation programme under social forestry
National Environment Policy, 2006	Environment management seeks to extend the coverage, fill in gaps that still exist in environment management and mainstream environmental concerns in all development activities.
Draft Wetland Policy, 2016	Effective conservation and management of wetland sites
National Agroforestry Policy, 2014	Policy to integrate the practice of planting trees and crops, and on agricultural land, to achieve sustainability in agriculture, optimise productivity and mitigate climate change impact.
Green Highways (Plantation, Transplantation, Beautification and Maintenance) Policy, 2015	Policy to promote greening of highway corridors with the participation of communities, farmers and private sectors, and to generate employement.

Sources: Adapted from various sources (Murthy and Kumar, 2019; Dandabathula et al., 2021).

TABLE 11.3
Various Acts/Rules for Protection of Forests

Act	Aim
Indian Forest Act, 1927	Consolidation of forests, regulation of transit, forest offences, penalty, powers of forest officers etc.
Wildlife (Protection) Act, 1972	Protection of Wildlife, National Board for Wild Life, State Boards for Wild Life, Standing Committees, Protected Areas, Possession of wildlife articles, hunting, trade of wildlife, Schedules I to VI, Penal provisions, etc.
Forest (Conservation) Act, 1980	Strictly restricts and regulates the de-reservation of forests or the use of forest land for non-forest purposes without the prior approval of the central government
Environment (Protection) Act, 1986	Protection and improvement of the environment and for matters connected therewith
Biological Diversity Act, 2002	Conservation of biological diversity, sustainable use of its components and fair equitable sharing of benefits arising out of the use of biological resources
Forest Rights Act, 2006	Recognition of forest rights to scheduled tribes and other traditional forest dwellers

Source: Adapted from various sources (Murthy and Kumar, 2019; Dandabathula et al., 2021).

TABLE 11.4
Forestry Schemes Associated with Targets of SDGs 15

Scheme	Objective
National Afforestation Programme (NAP)	To restore ecologically damaged forests and develop forest resources with people's participation, with an emphasis on improving the livelihoods of forest-fringe communities, particularly the poor
Green India Mission (GIM)	Aims at protecting, restoring and enhancing India's diminishing forest cover and responding to climate change by a combination of adaptation and mitigation measures
Intensification of Forest Management Scheme (IFMS)	Forest fire control and management, survey, demarcation and preparation of working plans, and improvement of infrastructure
Integrated Development of Wildlife Habitat (IDWH)	Provides financial and technical assistance to state and union territory (UT) governments for the conservation of wildlife and its habitats
Project Tiger	Conservation of tigers
Project Elephant	To manage free-ranging populations of Asian elephants
Compensatory Afforestation Fund Management and Planning Authority (CAMPA)	Promote afforestation activities and reduce the impact of diverting forest land for non-forest purposes

Source: Adapted from various sources (Murthy and Kumar, 2019; Dandabathula et al., 2021).

11.6 CONCLUSION

The contribution of forests to society is immense and invaluable. They act as a life support system for billions of species on Earth as they render important ecosystem services such as food, water, medicine and climate regulation. At the same time, forests are under immense pressure due to anthropogenic activities, climate change and invasion. In this regard, countries have set targets to reduce existing pressure on forests and to ensure their sustainable management. The Indian government is also taking key measures for the sustainable management of forest resources and has formulated several legal provisions and policies. India's commitment to achieving SDGs by the year 2030 is strong. However, despite a strong legal framework and policies, there is a gap in their implementation and enforcement. This could be due to a lack of manpower or the communication gap between local communities and authorities.

REFERENCES

Aggarwal, A., Paul, V., & Das, S. (2009). Forest resources: Degradation, livelihoods, and climate change. *Looking Back to Change Track*, *219*, 91–108.

Anderson-Teixeira, K. J., Wang, M. M. H., McGarvey, J. C. & LeBauer, D. S. (2016). Carbon dynamics of mature and regrowth tropical forests derived from a pantropical database (T rop F or C-db). *Global Change Biology*, *22*, 1690–1709. https://doi.org/10.1111/gcb.13226.

Beer, C., Reichstein, M., Tomelleri, E., Ciais, P., Jung, M., Carvalhais, N., Rödenbeck, C., et al. (2010). Terrestrial gross carbon dioxide uptake: global distribution and covariation with climate. *Science*, *329*, 834–838. https://doi.org/10.1126/ science.1184984.

Chaturvedi, R. K., Tiwari, R., & Ravindranath, N. H. (2008). Climate change and forests in India. *International Forestry Review*, *10*(2), 256–268.

Dandabathula, G., Chintala, S. R., Ghosh, S., Balakrishnan, P., & Jha, C. S. (2021). Exploring the nexus between Indian forestry and the Sustainable Development Goals. *Regional Sustainability*, *2*(4), 308–323.

FAO and UNEP. 2020. *The State of the World's Forests 2020. Forests, biodiversity and people*. Rome. https://doi.org/10.4060/ca8642en.

Forest Survey of India (2021). India State of Forest Report (ISFR). FSI, Dehradun. https://fsi.nic.in/forest-report-2021-details.

Hoogeveen H, & Verkooijen P. (2010). Transforming sustainable development diplomacy: lessons learned from global forest governance. Wageningen University and Research.

Hosonuma, N., Herold, M., De Sy, V., De Fries, R. S., Brockhaus, M., Verchot, L., ... & Romijn, E. (2012). An assessment of deforestation and forest degradation drivers in developing countries. *Environmental Research Letters*, *7*(4), 044009.

Maini, J. S. (1992). Sustainable development of forests. *Unasylva*, *43*(169), 3–8.

Martín-López, B., & Montes, C. (2010). Funciones y servicios de los ecosistemas: una herramienta para la gestión de los espacios naturales. *Guía científica de Urdaibai*, *1*, 13–32.

McDonough, K., Hutchinson, S., Moore, T., & Hutchinson, J. S. (2017). Analysis of publication trends in ecosystem services research. *Ecosystem Services*, *25*, 82–88.

MoEFCC. (2018). India: Second Biennial Update Report to the United Nations Framework Convention on Climate Change. Ministry of Environment, Forest and Climate Change, Government of India.

Murthy, I. K., & Kumar, P. (2019). Forests policies and programmes in India: Implications for climate change adaptation. *Open Journal of Forestry*, *9*(3), 226–240.

Pan, Y., Birdsey, R. A., Phillips, O. L. & Jackson, R. B. (2013). The structure, distribution, and biomass of the world's forests. *Annual Review of Ecology, Evolution, and Systematics*, *44*, 593–622. https://doi.org/10.1146/annurev-ecolsys-110512-135914.

Ravindranath, N.H., Joshi, N.V., Sukumar, R. & Saxena, A. (2006). Impact of climate change on forests in India. *Current science, 90(3),* 354–361.

Reddy, C. S., Jha, C. S., Diwakar, P. G., & Dadhwal, V. K. (2015). Nationwide classification of forest types of India using remote sensing and GIS. *Environmental Monitoring and Assessment*, *187*, 1–30.

von Stechow, C., Minx, J. C., Riahi, K., Jewell, J., McCollum, D. L., Callaghan, M. W., ... & Baiocchi, G. (2016). 2° C and SDGs: united they stand, divided they fall?. *Environmental Research Letters*, *11*(3), 034022.

Suresh, P., & Johnson, L. R. (2015). The post-2015 agenda: from millennium development goals (MDGs) to sustainable development goals (SDGs). *International Journal of Current Research and Review*, *7*(15), 62.

Tian, N., Poudyal, N. C., Hodges, D. G., Young, T. M., & Hoyt, K. P. (2015). Understanding the factors influencing nonindustrial private forest landowner interest in supplying ecosystem services in Cumberland Plateau, Tennessee. *Forests*, *6*(11), 3985–4000.

World Commission on Environment and Development. (1987). *Our common future*. Oxford: Oxford University Press.

12 Conservation and Management of Bioresources in India

*Sabir Hussain[1], Sheenu Sharma[1], Abhishek Kumar[1], Meenu Patil[1], Pardeep Kumar[1], Isha[1], Anand Narain Singh[1]**

[1]Soil Ecosystem and Restoration Ecology Lab, Department of Botany, Panjab University, Chandigarh India

*Corresponding author: Anand Narain Singh, Soil Ecosystem and Restoration Ecology Lab, Department of Botany, Panjab University, Chandigarh India

12.1 INTRODUCTION

Bioresources are natural, non-fossil materials derived from living organisms, which humans use for food, essential products, and energy purposes (Gaurav et al. 2017). They encompass a diverse range of materials like plants, animals, microorganisms, and waste from different industries. Secondary bioresources include leftovers or by-products from green spaces or biological organisms, while tertiary bioresources consist of parts of raw materials separated during processing, which are naturally occurring biological-based materials utilized by humans for important functions in the soil (Akanmu et al. 2023). India has a rich and diverse array of bioresources due to its vast geographical expanse, varied geographic and climatic conditions, and diverse ecosystems (Dar et al. 2022; Kulyal et al. 2022; Madhusudan & Vanak 2023). These bioresources include a wide range of flora, fauna, and microbial species, which play crucial roles in sustaining livelihoods, supporting ecosystems, and contributing to the economy and cultural heritage of the country (Ojija & Mayowela 2022; Raj et al. 2020; Dasgupta et al. 2022). The flora of India is particularly rich, with thousands of plant species ranging from towering trees in dense forests to delicate wildflowers in alpine meadows (Tewari et al. 2017; Hussain et al. 2023). India is home to diverse forest types, including tropical rainforests, deciduous forests, temperate forests, and alpine forests (Awasthi et al. 2003; Shankar & Tripathi 2017; Tarakeswara Naidu et al. 2018; Kumar et al. 2020). These plants provide essential ecosystem services such as oxygen production, soil stabilization, and water regulation, while also serving as sources of food, medicine, and raw materials for various industries. These forests harbor a wide range of plant species, many of which hold medicinal, nutritional, or economic significance (Ray et al. 2010). Further, in terms of medicinal plant assets, India is one of the world's richest countries with a long history of traditional medicine systems like Ayurveda, Siddha, and Unani medicinal systems (Mukherjee 2001). To sustain a comfortable living, many regions in India act as agricultural hubs, cultivating and growing various agricultural plant species across various agroclimatic zones. India is also a major producer of agricultural crops, ranging from staple grains like rice and wheat to cash crops like sugarcane, cotton, and tea. The region has diverse agro-climatic zones that support the cultivation of a wide variety of crops, ensuring food security and livelihoods for millions of people. Horticultural crops such as fruits, vegetables, and spices are also cultivated extensively, adding to the nutritional and economic diversity of the agricultural sector. The faunal diversity of

DOI: 10.1201/9781032697550-12

India is equally impressive, with numerous species of mammals, birds, reptiles, amphibians, and insects inhabiting its forests, grasslands, mountains, and wetlands (Chettri et al. 2010; Acharya et al. 2011; Dorji et al. 2019; Kubendran et al. 2019). The region is well known for providing refugia to some of the iconic and endangered faunal diversity, including Bengal tigers, Asian elephants, Indian rhinoceroses, Asiatic lions, and Indian leopards. The faunal diversity has deep roots in the Indian culture, and some of them are even included as heritage animals (Bradshaw, 2007). Moreover, the region is bestowed with a large number of avifaunal diversity, both resident and migratory, taking refuge in the wetlands, forests, grasslands, and coastal areas, which provide crucial nesting and feeding grounds for birds.

Similarly, India's bioresources extend beyond land-based ecosystems and incorporate aquatic resources that are found along its extensive coastline, rivers, lakes, and wetlands. The marine ecosystem has many diverse marine organisms, including fish, crustaceans, mollusks, corals, and marine mammals (Venkataraman & Raghunathan 2015). Such resources play vital roles in supporting coastal communities and contribute to the nation's economy through activities such as fisheries, aquaculture, and tourism (Chen et al. 2020). Inland water bodies are equally significant, providing water resources that are used for irrigation, drinking, and industrial use, while also serving as habitats for a wide array of aquatic species (Datta et al. 2022).

Bioenergy, defined as the conversion of organic materials such as plants, crops, agricultural residues, forestry products, algae, and organic waste into usable energy forms, is a renewable energy source. It comprises various processes and technologies that produce heat, electricity, and transportation fuels. Biofuels, biogas, and biomass are among the common sources of bioenergy, playing vital roles in supporting livelihoods across the globe. Biofuels derived mainly from biomass come in solid, liquid, and gaseous forms. They're categorized into three generations: first, second, and third, based on the complexity of the biomass they're derived from (Ho et al. 2014; Gaurav et al. 2017). First-generation fuels like biodiesel and vegetable oils are made from crop plants. Second-generation fuels, which require fertile lands for cultivation, include bioethanol and biohydrogen that are generated from agricultural by-products and energy plants. Furthermore, marine resources such as seaweeds and cyanobacteria are promising sources for third-generation biofuels like biogas, bioethanol, and biobutanol because they yield significant biomass in short periods (Sims et al. 2010; Sudhakar et al. 2018; Leya 2022).

Microbial resources, despite often being overlooked, are also essential components of India's bioresources (Bhattacharyya et al. 2011). Soil microbes, including bacteria, fungi, actinomycetes, and algae, perform essential functions such as nutrient cycling, maintaining soil fertility, and promoting plant health (Sahu et al. 2017). They contribute significantly to agricultural productivity and the resilience of ecosystems. Industrial microbes possess significant biotechnological potential and are used in various sectors including pharmaceuticals, enzyme production, biofuel production, and bioremediation (Saeed et al. 2022). Some of the prominent roles of bioresource in the context of human livelihood and their welfare are mentioned in Figure 12.1. Their utilization could definitely contribute to the growth of India's biotechnology industry.

However, the alarming sign is the numerous challenges that India's bioresources have faced in the last decade, indicating the necessary and essentiality of effective management and conservation measures. Several significant threats, including habitat loss, fragmentation, and degradation stemming from activities like deforestation, urbanization, and agricultural expansion, pose threats to biodiversity and ecosystem health. Overexploitation of natural resources, along with issues like poaching, illegal wildlife trade, pollution, invasive species, and climate change, further exacerbates these challenges, resulting in the loss of species, ecosystem services, and genetic diversity (Kumar et al. 2022).

12.1.2 Types of Bioresources

Bioresources can be categorized using various criteria based on their origin, characteristics, and applications.

FIGURE 12.1 The significance of bioresources for human livelihood.

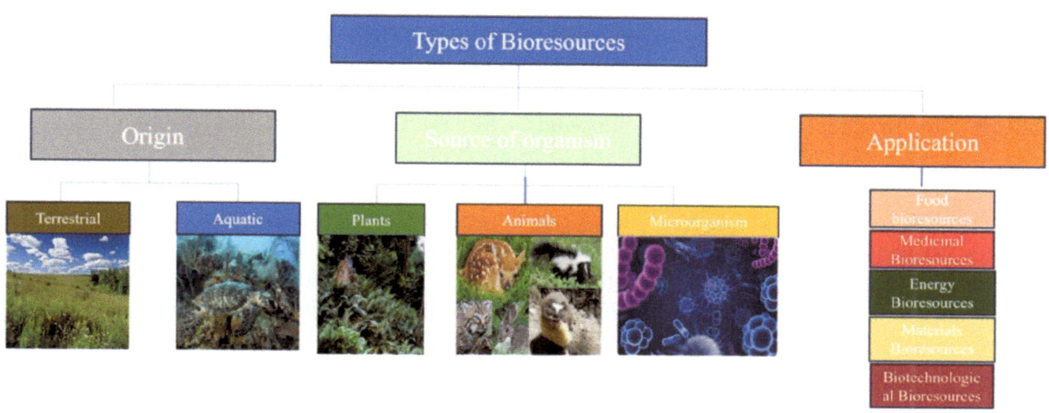

FIGURE 12.2 Different kinds of bioresources based on some prominent criteria.

12.1.2.1 Based on Origin

On the basis of origin, the bioresources can be distinguished into terrestrial and aquatic bioresources. The terrestrial bioresources are derived from land organisms, such as plants, animals, and microorganisms. They are obtained from terrestrial ecosystems such as forests, grasslands, and agricultural lands. The terrestrial bioresources encompass a wide range of products and materials that are employed in agriculture, forestry, and other human activities. They include crops for food and fiber, timber for construction and manufacturing, feed and forage for cattle, as well as microbial agents for soil fertility and bioremediation (Lakerveld et al. 2015). Terrestrial bioresources offer critical ecosystem services essential for human survival, well-being, and livelihood. These services include air and water purification, soil fertility, pollination, nutrient cycling, moderation of extreme events, and climate regulation (Mori et al. 2017; Hussain et al. 2023).

Aquatic bioresources refer to the varied resources that are sourced from aquatic environments that include both marine and freshwater organisms, such as fish, algae, and aquatic plants. These bioresources are vital components of aquatic ecosystems and play crucial roles in marine and freshwater ecosystem functioning. Aquatic bioresources provide valuable food sources, fodder, aesthetic and cultural significance, as well as industrial materials like seaweed for pharmaceuticals and bioplastics. Additionally, they contribute to ecosystem services such as nutrient cycling and water purification (Grizzetti et al. 2016).

12.1.2.2 Source of Organism

Bioresources can also be categorized based on their production sources. They are obtained from plants, animals, and microorganisms. Bioresources are generated from various plant species, including food crops, medicinal herbs, non-forest timber products, forage, and fuelwood (Vaz et al. 2017). Plants are a fundamental source of bioresources, providing a diverse range of materials essential for human life and industry (Gaurav et al. 2017). Food crops supply nourishment and sustenance, while medicinal herbs offer therapeutic compounds for healthcare. Timber from trees serves as a crucial building material for construction and manufacturing. Additionally, plants contribute to energy production through biofuels and biomass, highlighting their multiple roles in supporting human needs and economic activities (van Kleunen et al. 2020). Similarly, bioresources are also obtained from animals, including livestock, aquatic organisms, and insects. Animals play a significant role in providing bioresources essential for human nutrition, healthcare, and economic livelihoods (Herrero et al. 2013; Nimachow et al. 2008). Livestock contribute meat, dairy, and eggs for human consumption, while aquatic organisms provide fish, shellfish, and algae as valuable food sources. Moreover, these animals provide materials such as leather, wool, and silk for textiles and manufacturing (Napolitano et al. 2020). Insects are increasingly recognized for their potential as a sustainable protein source and contribute to pollination and ecosystem services (Liu et al. 2019).

Bioresources are also derived from bacteria, fungi, algae, and other microorganisms. Microbes are primary producers of bioactive compounds with significant applications in industry, agriculture, and medicine. The enzymes obtained from microorganisms are utilized in various industrial processes, including food production, textile manufacturing, and wastewater treatment (Patel et al. 2023). These microbes also produce antibiotics for battling infectious diseases and bioactive compounds for pharmaceuticals and drug industries (Cobongela 2022). Additionally, microorganisms contribute to agricultural productivity through nitrogen fixation, soil fertility enhancement, and biological pest control (Bhardwaj et al. 2014; Kebede 2021), highlighting their immense role in supporting human well-being and ecosystem sustainability.

12.1.2.3 Based on Application

Food bioresources for human or animal consumption include food crops, fish, meat, and dairy products. Such bioresources provide essential nutrients, energy, and sustenance for human health and well-being. Food crops such as grains, fruits, and vegetables form the basis of human diets, providing carbohydrates, proteins, vitamins, and minerals (Yahia et al. 2019). In a similar manner, animal-derived products like meat, eggs, and dairy also contribute high-quality proteins and fats to human populations worldwide. Additionally, fish and seafood are rich sources of omega-3 fatty acids and micronutrients (Badoni et al. 2021). Food bioresources play a crucial role in addressing global food security challenges and ensuring adequate nutrition for populations worldwide (Pachapur et al. 2020).

The medicinal bioresources contain bioactive compounds used for pharmaceuticals, herbal remedies, and healthcare products. Medicinal bioresources cover a wide range of natural substances that are directly or indirectly obtained from plants, animals, marine organisms, and microbial metabolites. These bioactive compounds exhibit therapeutic properties and are utilized in traditional and modern medicine for treating various diseases and health conditions (Wang et al. 2021). These plants produce phytochemicals such as alkaloids, flavonoids, and terpenoids with pharmacological effects (Velu et al. 2018). In a similar context, marine organisms offer novel compounds with potential medicinal applications, including anti-inflammatory, antimicrobial, and anticancer agents (Singh et al. 2020), that are widely used against various diseases. Microbial metabolites, such as antibiotics and enzymes, are essential for treating infections and metabolic disorders (Singh et al. 2017). Research advancements in medicinal bioresources have made significant contributions to drug discovery, development, and healthcare innovation (Suleria et al. 2016).

Energy bioresources refer to the utilization of bioresources for energy production, including biofuels (e.g., biodiesel, bioethanol), biomass, and biogas. Such bioresources play a vital role in diversifying energy sources, reducing reliance on fossil fuels, and mitigating greenhouse gas emissions (Ekpeni et al. 2014; Sri Shalini et al. 2021). Biofuels derived from biomass offer renewable alternatives for transportation fuels and heating. Biomass, including agricultural residues, forestry waste, and energy crops, is used for generating electricity and heat through combustion or conversion technologies. The biogas produced from organic waste through anaerobic digestion provides renewable energy for electricity generation and heating to households (Ardebili, 2020). Similarly, the material bioresources that are used for manufacturing materials such as fibers (e.g., cotton, hemp), bioplastics, and biomaterials, offer sustainable alternatives to conventional materials derived from fossil fuels. Plant-based fibers like cotton, hemp, and jute are utilized in textiles, paper, and construction materials. The bioplastics derived from renewable biomass sources reduce plastic pollution and carbon footprint compared to petrochemical-based plastics (Muthusamy & Pramasivam 2019). Additionally, those including biodegradable polymers and natural composites are applicable in medical devices, packaging, and consumer products and contribute to circular economy principles by promoting resource efficiency, waste reduction, and eco-friendly product design (Rai et al. 2021; Devadas et al. 2021).

The last and final bioresources based on their application are the biotechnological bioresources. These are used as raw materials or tools in biotechnological processes, including enzymes, DNA, and microbial cultures. Biotechnological bioresources empower a wide range of applications in industry, agriculture, and healthcare. Enzymes derived from microorganisms catalyze biochemical reactions in industrial processes such as food processing, textile manufacturing, and biofuel production. DNA molecules serve as templates for genetic engineering, enabling the modification of organisms for desired traits and applications. Microbial cultures produce metabolites and biomolecules with commercial value, including antibiotics, vitamins, and biofuels (Nurfarahin et al. 2018). Biotechnological bioresources drive innovation in biopharmaceuticals, bio-based materials, and sustainable bioprocessing technologies, contributing to economic growth and societal well-being (Hlangwani et al. 2023). These biotechnological resources have a significant impact on people's livelihoods across the globe.

12.2 THREAT TO BIORESOURCES

Threats to bioresources display a diverse challenge that endangers the sustainability, diversity, and resilience of biological systems. These threats arise from human activities, environmental changes, and natural processes, impacting terrestrial, aquatic, and microbial ecosystems worldwide. The assessment and understanding of these challenges and threats are essential for conserving bioresources and maintaining ecosystem integrity. Habitat loss and degradation represent one of the most

FIGURE 12.3 Major threats to global bioresources.

significant threats to bioresources globally. Human activities such as deforestation, urbanization, and agricultural expansion have led to the destruction and fragmentation of natural habitats (Scanes 2018). Such activities reduce biodiversity heterogeneity, disrupt ecological processes, and threaten the survival of existing plant and animal species (Berenguer et al. 2021). Furthermore, overexploitation of bioresources through unsustainable harvesting practices poses a significant threat to biodiversity and ecosystem health (Shafi et al. 2021). The unsustainable fishing, hunting, logging, and collection of wild plants and animals have contributed to population declines, species extinctions, and ecosystem degradation (Young et al. 2016). For instance, the practice of overfishing depletes fish populations, disrupts marine food webs, and undermines the ecological balance of aquatic ecosystems.

Climate change is another primary concern, posing a grave threat to bioresources by altering temperature and precipitation patterns on temporal and spatial scales and ecosystem dynamics. The rising temperatures and changing weather patterns affect the distribution and abundance of species, leading to shifts in biodiversity and ecological communities (Pecl et al. 2017). Climate-related phenomena such as extreme weather events, sea-level rise, and ocean acidification enhance the vulnerability of bioresources, particularly in sensitive habitats like coral reefs, polar regions, and mountain ecosystems. This is further exacerbated by the increasing pollution from various sources, including industrial activities, agricultural runoff, and urbanization, which poses more serious threats to bioresources and ecosystem health. The contamination of air, water, and soil with pollutants such as heavy metals, pesticides, plastics, and chemical waste adversely affects plant and animal life (Sankhla et al. 2016). It also leads to eutrophication of water bodies, algal blooms, and loss of biodiversity in aquatic ecosystems (Amorim & do Nascimento Moura 2021). Addressing pollution requires implementing pollution prevention measures, adopting cleaner production technologies, and promoting sustainable waste management practices to reduce environmental contamination and safeguard bioresources.

Similarly, invasive alien species pose a significant threat to native biodiversity and ecosystem functioning. The introduced species (intentionally or unintentionally) outcompete the native flora and fauna, disrupt ecological processes, and cause serious ecological and economic risks (Strayer 2010). Invasive plants, animals, and microorganisms can spread rapidly and may cause habitat degradation, loss of native species, and changes in ecosystem structure and function. One of the serious threats to biodiversity or bioresources is land use change, which includes the conversion of natural habitats to agriculture, urban areas, and infrastructure expansion. It leads to habitat fragmentation and loss of connectivity among ecosystems, disrupting ecological processes such as migration, dispersal, and gene flow. The fragmented landscapes are more vulnerable to invasive species, habitat degradation, and biodiversity loss (Fischer & Lindenmayer 2007), thus worsening the bioresources generation. Habitat fragmentation occurs when natural habitats are divided into smaller, isolated patches due to human activities such as urbanization, agriculture, and infrastructure development. Furthermore, the illegal wildlife trade also presents a serious threat to bioresources, particularly for endangered species and biodiversity hotspots. Illegal activities such as poaching and trafficking of wildlife products, including ivory, rhino horn, exotic pets, and medicinal plants, drive species decline, disrupt ecosystems, and support crime networks (Chavan et al. 2023). Disease outbreaks and the spread of pathogens also represent significant threats to bioresources, particularly in wildlife populations and agricultural ecosystems. New and emerging infectious diseases, caused by viruses, bacteria, fungi, and parasites, can have devastating impacts on biodiversity, ecosystem dynamics, and food security. All these threats, if neglected, may cause serious impacts on human livelihoods in the near future.

12.3 SUSTAINABLE DEVELOPMENT AND CONSERVATION STRATEGIES OF BIORESOURCES

As discussed earlier in the previous section concerning the challenges faced by bioresources, it becomes vital to devise and adopt strategies regarding their conservation and preservation. This involves proper ways to safeguard these resources from depletion and degradation while ensuring their sustainable use for future generations. Therefore, understanding the threats posed to bioresources, such as habitat destruction, overexploitation, pollution, and climate change, makes conservation and preservation efforts crucial to maintaining ecological balance and safeguarding the benefits provided by these resources. The conservation of bioresources involves actively managing and protecting natural habitats, species, and ecosystems to prevent further decline or extinction. Some of the important conservation strategies are mentioned next.

The creation and management of protected areas are among the most successful and efficient conservation measures. Protected areas, such as national parks, wildlife sanctuaries, and biosphere reserves, are essential for maintaining habitats, biodiversity, and ecosystem resilience. According to Geldmann et al. (2013), protected areas are essential for maintaining habitats and increasing species diversity. Because these areas serve as both habitats for endangered species and hubs for scientific study and teaching, they are crucial for the conservation of bioresources. As such, the execution of efficient management plans, appropriate legal legislation, and community engagement are crucial. Typically, the goal of biodiversity conservation is to preserve biodiversity in its current state or return it to its previous state. It is possible to safeguard areas so that species have a safe haven from overharvesting, habitat loss, and degradation (Ranius et al. 2023). Similarly, targeted conservation programs mainly focused on endangered species would be essential for preventing further extinctions and restoring populations to a sustainable level. The identification and protection of important areas to improve connectivity in support of biodiversity conservation, climate change adaptation, and several other such aspects has been a key aspect for decades in the Convention on Biological Diversity (Mahajan et al. 2023). The sustainable management of resources plays a vital role in maintaining the long-term viability of bioresources while meeting human needs and demands. These practices involve adopting ecosystem-based management strategies (Haugen et al. 2024), harvest limits (Clark et al. 2023), sustainable harvesting techniques, and minimal habitat destruction. In a similar context, the engagement of local communities and respecting indigenous peoples' rights are very crucial to effective conservation. The community-based conservation approaches empower local communities to participate in decision-making, natural resource management, and conservation activities (Ghafran et al. 2024). It also involves the cooperation between indigenous peoples, local communities, governments, and conservation organizations. Moreover, the awareness programs for those communities regarding the conservation of ecosystems and their significance would play a pivotal role in conservation. This includes conducting awareness programs to inform the public, policymakers, and stakeholders about ecosystem degradation and their impacts on human livelihoods.

Strict laws and regulations and employing strong policies are also essential for effective conservation. National legislation and international agreements play a critical role in protecting bioresources and promoting sustainable development (de Lira et al. 2024). Initiatives such as the Convention on Biological Diversity (CBD), the Ramsar Convention on Wetlands, and the Convention on International Trade in Endangered Species of Wild Fauna and Flora (CITES) enable cooperation among countries and thus promote ecosystem conservation efforts. Furthermore, scientific research and monitoring programs are essential for understanding the basics of ecosystem processes, functioning, and ultimately the provision of multiple ecosystem services. Research efforts focused on biodiversity hotspots, threatened species, ecosystem dynamics, and climate change impacts may provide valuable insights into conservation priorities and strategies.

12.4 MANAGEMENT OF NATURAL RESOURCES IN INDIA

India has been at the forefront of ecological conservation efforts and has made many significant strides in conserving its ecological assets. India has enacted several laws and policies to govern the management of bioresources, including the Wildlife Protection Act of 1972, the Forest Conservation Act of 1980, the Biological Diversity Act of 2002, and the National Biodiversity Strategy and Action Plan (NBSAP). These legal frameworks are meant to conserve biological diversity and promote sustainable development to ensure the impartial sharing of benefits obtained from bioresources. Further, India has launched several projects such as Project Tiger, Project Elephant, and projects focusing on renewable energy, water, and forest conservation, all of which were meant to preserve and conserve bioresources.

India has established a network of protected areas, including national parks, wildlife sanctuaries, biosphere reserves, and conservation reserves, to safeguard biodiversity and ecosystems. The protected areas network covers diverse habitats and provides refuge for endangered species, including tigers, elephants, rhinoceroses, and endemic flora and fauna. India has a network of 998 protected areas covering about 5.28% of its total geographic area. These protected areas in India have been categorized according to the Wildlife Protection Act of 1972. It categorizes protected areas into four categories: National Parks (106), Wildlife Sanctuaries (567), Conservation Reserves (105), and Community Reserves (220) (Yadav et al. 2023). Fifty-three tiger reserves have been established due to the vulnerability of tigers and the rapid depletion of their numbers. Further, India promotes sustainable forest management practices to ensure the long-term health and productivity of forest ecosystems while meeting the socio-economic needs of local communities. Sustainable forest management initiatives include afforestation, reforestation, agroforestry, and the adoption of participatory approaches in forest management planning and implementation.

India recognizes the importance of wetlands and coastal zones in supporting biodiversity, providing ecosystem services, and sustaining livelihoods. The National Wetland Conservation Program (NWCP) and the Integrated Coastal Zone Management (ICZM) program aim to conserve wetlands and coastal ecosystems, mitigate habitat degradation, and promote sustainable development in these ecologically sensitive areas. India collaborates with international organizations, such as the Convention on Biological Diversity (CBD), the United Nations Environment Program (UNEP), and the International Union for Conservation of Nature (IUCN), to address global biodiversity conservation challenges, share best practices, and promote international cooperation in managing bioresources. Research institutions, universities, and governmental agencies in India conduct scientific research and monitoring programs to assess biodiversity status, identify conservation priorities, and evaluate the effectiveness of conservation interventions. Monitoring programs track changes in species populations, habitat conditions, and ecosystem health to inform management decisions and adaptive management approaches.

12.5 CONCLUSION

The bioresources of India are very important for sustaining livelihoods, maintaining ecosystem functioning, and supporting the nation's economy and cultural heritage. With numerous plant and animal species, including medicinal herbs and agricultural crops, India boasts many biodiversity hotspots. Furthermore, the faunal diversity includes species like Bengal tigers and Asian elephants, along with myriad bird species, indicating the country's ecological significance. These bioresources play crucial roles in supporting coastal communities, contributing to the nation's economy through fisheries, aquaculture, and tourism, and providing vital ecosystem services. Moreover, the bioenergy derived from organic materials provides renewable alternatives for energy production, while microbial resources are essential for soil fertility, agricultural productivity, and industrial applications. Additionally, India's bioresources serve as sources of food, medicine, raw materials, and energy, highlighting their multifaceted importance. However, despite their significance, India's bioresources face numerous challenges, including habitat loss, overexploitation, pollution, invasive species, and climate change. These threats endanger biodiversity, ecosystem health, and the well-being of communities dependent on these resources. Hence, effective conservation and management strategies are imperative to safeguard India's bioresources for future generations. To address these challenges, India has implemented various conservation measures, including the establishment of protected areas, sustainable forest management practices, wetland conservation programs, and international collaborations. Additionally, strict laws, community engagement, scientific research, and monitoring efforts are crucial components of India's conservation efforts. The conservation and sustainable use of India's bioresources are paramount for maintaining ecological balance, ensuring biodiversity conservation, and promoting human well-being. By adopting holistic approaches and integrating traditional knowledge with modern science, India can effectively manage its bioresources while fostering sustainable development and resilience in the face of global environmental challenges.

REFERENCES

Acharya, B. K., Sanders, N. J., Vijayan, L., & Chettri, B. (2011). Elevational gradients in bird diversity in the Eastern Himalaya: an evaluation of distribution patterns and their underlying mechanisms. *PLoS One*, *6*(12), e29097.

Akanmu, A. O., Olowe, O. M., Phiri, A. T., Nirere, D., Odebode, A. J., Karemera Umuhoza, N. J., ... & Babalola, O. O. (2023). Bioresources in organic farming: implications for sustainable agricultural. *Horticulture*, *9*(6), 659.

Amorim, C. A., & do Nascimento Moura, A. (2021). Ecological impacts of freshwater algal blooms on water quality, plankton biodiversity, structure, and ecosystem functioning. *Science of the Total Environment*, *758*, 143605.

Ardebili, S. M. S. (2020). Green electricity generation potential from biogas produced by anaerobic digestion of farm animal waste and agriculture residues in Iran. *Renewable Energy*, *154*, 29–37.

Awasthi, A., Uniyal, S. K., Rawat, G. S., & Rajvanshi, A. (2003). Forest resource availability and its use by the migratory villages of Uttarkashi, Garhwal Himalaya (India). *Forest Ecology and Management*, *174*(1–3), 13–24.

Badoni, P., Nazir, I., Aier, M., Maity, P. B., Samanta, S., & Das, A. (2021). Significant role of fish nutrients with special emphasis to essential fatty acid in human nutrition. *International Journal of Current Microbiology and Applied Sciences, 10*, 2034–2046.

Berenguer, E., Armenteras, D., Alencar, A., Almeida, C., Aragão, L., Barlow, J., ... & García-Villacorta, R. (2021). Drivers and ecological impacts of deforestation and Forest degradation, 19–1. New York: United Nations Sustainable Development Solutions Network.

Bhardwaj, D., Ansari, M. W., Sahoo, R. K., & Tuteja, N. (2014). Biofertilizers function as key player in sustainable agriculture by improving soil fertility, plant tolerance and crop productivity. *Microbial Cell Factories, 13*, 1–10.

Bhattacharyya, P. N., Baruah, C., Jha, D. K., & Sharma, D. K. (2011). Potential application of microbial resources in North-East India: future prospects and challenges. *NeBIO, 2*(3), 12–18.

Bradshaw, C. J., Field, I. C., Bowman, D. M., Haynes, C., & Brook, B. W. (2007). Current and future threats from non-indigenous animal species in northern Australia: a spotlight on World Heritage Area Kakadu National Park. *Wildlife Research, 34*(6), 419–436.

Chavan, A. S., Muley, E. D., & Naphade, S. R. (2023). Animal trafficking and poaching: major threats to the biodiversity. *Journal of Entomology and Zoology Studies*, 11(5), 45–9.

Chen, S., De Bruyne, C., & Bollempalli, M. (2020). Blue economy: community case studies addressing the poverty–environment nexus in ocean and coastal management. *Sustainability, 12*(11), 4654.

Chettri, B., Bhupathy, S., & Acharya, B. K. (2010). Distribution pattern of reptiles along an eastern Himalayan elevation gradient, India. *Acta Oecologica, 36*(1), 16-22.

Clark, M., Hamad, H. M., Andrews, J., Hillis, V., & Mulder, M. B. (2023). Quantifying local perceptions of environmental change and links to community-based conservation practices. *bioRxiv*, 2023-02.

Cobongela, S.Z.Z. (2022). Applications of Microbes in Antibiotics. In: Inamuddin, Ahamed, M.I., Prasad, R. (eds) *Application of Microbes in Environmental and Microbial Biotechnology. Environmental and Microbial Biotechnology*. Springer, Singapore. https://doi.org/10.1007/978-981-16-2225-0_25

Dar, S. A., Dar, S. A., & Nabi, M. (2022). Conservation of Biodiversity in India: Current status and future strategies. In *Towards Sustainable Natural Resources: Monitoring and Managing Ecosystem Biodiversity* (pp. 195–214). Cham: Springer International Publishing.

Dasgupta, S., Badola, R., Ali, S. Z., & Tariyal, P. (2022). Bioresource, energy and forest sustainability: a case study from Indian Himalayan region. *Journal of Cleaner Production, 337*, 130497.

Datta, S., Sinha, D., Chaudhary, V., Kar, S., & Singh, A. (2022). Water pollution of wetlands: a global threat to inland, wetland, and aquatic phytodiversity. In *Handbook of Research on Monitoring and Evaluating the Ecological Health of Wetlands* (pp. 27–50). IGI Global.

de Lira, G. S., de Noronha Sertori, F., Vargas, J. V. C., Mariano, A. B., & Severo, I. A. (2024). Alginate Extraction from Natural Resources Based on Legal Requirements: An Incentive for Sustainable Development. DOI: 10.5772/intechopen.114217.

Devadas, V. V., Khoo, K. S., Chia, W. Y., Chew, K. W., Munawaroh, H. S. H., Lam, M. K., ... & Show, P. L. (2021). Algae biopolymer towards sustainable circular economy. *Bioresource Technology, 325*, 124702.

Dorji, S., Rajaratnam, R., & Vernes, K. (2019). Mammal richness and diversity in a Himalayan hotspot: the role of protected areas in conserving Bhutan's mammals. *Biodiversity and Conservation, 28*(12), 3277–3297.

Ekpeni, L. E., Benyounis, K. Y., Nkem-Ekpeni, F., Stokes, J., & Olabi, A. G. (2014). Energy diversity through renewable energy source (RES)–a case study of biomass. *Energy Procedia, 61*, 1740–1747.

Fischer, J., & Lindenmayer, D. B. (2007). Landscape modification and habitat fragmentation: a synthesis. *Global Ecology and Biogeography, 16*(3), 265–280.

Gaurav, N., Sivasankari, S., Kiran, G. S., Ninawe, A., & Selvin, J. (2017). Utilization of bioresources for sustainable biofuels: a review. *Renewable and Sustainable Energy Reviews, 73*, 205–214.

Geldmann, J., Barnes, M., Coad, L., Craigie, I. D., Hockings, M., & Burgess, N. D. (2013). Effectiveness of terrestrial protected areas in reducing habitat loss and population declines. *Biological Conservation*, 161, 230–238.

Ghafran, C., & Yasmin, S. (2024). Participation strategies and ethical considerations in NGO led community-based conservation initiatives. *Journal of Business Ethics*, 1–17. DOI: https://doi.org/10.1007/s10551-024-05665-4.

Grizzetti, B., Lanzanova, D., Liquete, C., Reynaud, A., & Cardoso, A. C. (2016). Assessing water ecosystem services for water resource management. *Environmental Science & Policy, 61*, 194–203.

Haugen, J. B., Link, J. S., Cribari, K., Bundy, A., Dickey-Collas, M., Leslie, H. M., ... & Agnalt, A. L. (2024). Marine ecosystem-based management: challenges remain, yet solutions exist, and progress is occurring. *NPJ Ocean Sustainability*, 3(1), 7.

Herrero, M., Grace, D., Njuki, J., Johnson, N., Enahoro, D., Silvestri, S., & Rufino, M. C. (2013). The roles of livestock in developing countries. *Animal*, *7*(s1), 3–18.

Hlangwani, E., Mpye, K. L., Matsuro, L., & Dlamini, B. (2023). The use of technological innovation in bio-based industries to foster growth in the bioeconomy: a South African perspective. *Sustainability: Science, Practice and Policy*, *19*(1), 2200300.

Ho, D. P., Ngo, H. H., & Guo, W. (2014). A mini review on renewable sources for biofuel. *Bioresource Technology*, *169*, 742–749.

Hussain, S., Sharma, S., & Singh, A. N. (2023). Identification and prioritisation of ecosystem services based on the socio-economic perspectives of local people in a trans-Himalayan region of Ladakh, India. *Anthropocene*, *44*, 100404.

Kebede, E. (2021). Contribution, utilization, and improvement of legumes-driven biological nitrogen fixation in agricultural systems. *Frontiers in Sustainable Food Systems*, *5*, 767998.

Kubendran, T., Sidhu, A. K., & Chandra, K. (2019). Diversity of insects in trans-Himalaya: an overview. *Faunal Diversity of Biogeographic Zones: Indian Trans-Himalaya, 1-225*, 35–43.

Kulyal, L., & Jalal, P. (2022). Bioenergy, a finer alternative for India: scope, barriers, socio-economic benefits and identified solution. *Bioresource Technology Reports*, *17*, 100947.

Kumar, P., Kumar, A., Patil, M., Hussain, S., Sharma, S., Singh, S., ... & Singh, A. N. (2022). Biodiversity resources and ecosystem services in the context of the industrialized world. In *Biodiversity* (pp. 23–39). CRC Press.

Kumar, P., Kumar, A., Patil, M., Sharma, N. K., & Singh, A. N. (2020). Plant diversity development under planted woody species on coal mine spoil in a dry tropical environment, India: a case study. *Journal of Ecological Engineering*, *21*(2), 228–243.

Lakerveld, R. P., Lele, S., Crane, T. A., Fortuin, K. P. J., & Springate-Baginski, O. (2015). The social distribution of provisioning forest ecosystem services: evidence and insights from Odisha, India. *Ecosystem Services*, *14*, 56–66.

Leya, T. (2022). The CCCryo Culture Collection of Cryophilic Algae as a valuable bioresource for algal biodiversity and for novel, industrially marketable metabolites. *Applied Phycology*, *3*(1), 167–188.

Liu, Q., Xu, P., Yan, K., & Guo, Y. (2019). Pollination services from insects in homegardens in the Chengdu Plain will be confronted with crises. *Sustainability*, *11*(7), 2169.

Madhusudan, M. D., & Vanak, A. T. (2023). Mapping the distribution and extent of India's semi-arid open natural ecosystems. *Journal of Biogeography*, *50*(8), 1377–1387.

Mahajan, S. L., Tanner, L., Ahmadia, G., Becker, H., DeMello, N., Fidler, R., ... & Glew, L. (2023). Accelerating evidence-informed decision-making in conservation implementing agencies through effective monitoring, evaluation, and learning. *Biological Conservation*, 286, 110304.

Mori, A. S., Lertzman, K. P., & Gustafsson, L. (2017). Biodiversity and ecosystem services in forest ecosystems: a research agenda for applied forest ecology. *Journal of Applied Ecology*, *54*(1), 12v27.

Mukherjee, P. K. (2001). Evaluation of Indian traditional medicine. *Drug Information Journal*, *35*(2), 623–632.

Muthusamy, M. S., & Pramasivam, S. (2019). Bioplastics–an eco-friendly alternative to petrochemical plastics. *Current World Environment*, *14*(1), 49.

Napolitano, F., Bragaglio, A., Sabia, E., Serrapica, F., Braghieri, A., & De Rosa, G. (2020). The human– animal relationship in dairy animals. *Journal of Dairy Research*, *87*(S1), 47–52.

Nimachow, G., Taga, T., Tag, H., & Dai, O. (2008). Linkages between bio-resources and human livelihood: a case study of Adi tribes of Mirem Village, Arunachal Pradesh (India). *Initiation*, *2*(1), 183–98.

Nurfarahin, A. H., Mohamed, M. S., & Phang, L. Y. (2018). Culture medium development for microbial-derived surfactants production—an overview. *Molecules*, *23*(5), 1049.

Ojija, F., & Mayowela, F. (2022). The critical role and application of microbes towards sustainable development and human wellbeing. *East African Journal of Environment and Natural Resources*, *5*(1), 231–256.

Pachapur, P. K., Pachapur, V. L., Brar, S. K., Galvez, R., Le Bihan, Y., & Surampalli, R. Y. (2020). Food security and sustainability. *Sustainability: Fundamentals and Applications*, 357–374.

Patel, A. K., Dong, C. D., Chen, C. W., Pandey, A., & Singhania, R. R. (2023). Production, purification, and application of microbial enzymes. In *Biotechnology of Microbial Enzymes* (pp. 25–57). Academic Press. https://doi.org/10.1016/B978-0-443-19059-9.00019-0

Pecl, G. T., Araújo, M. B., Bell, J. D., Blanchard, J., Bonebrake, T. C., Chen, I. C., ... & Williams, S. E. (2017). Biodiversity redistribution under climate change: impacts on ecosystems and human well-being. *Science*, *355*(6332), eaai9214.

Rai, P., Mehrotra, S., Priya, S., Gnansounou, E., & Sharma, S. K. (2021). Recent advances in the sustainable design and applications of biodegradable polymers. *Bioresource Technology*, *325*, 124739.

Raj, A., Jhariya, M. K., Yadav, D. K., & Banerjee, A. (2020). Forests for resource management and environmental protection. In *Environmental and Sustainable Development through Forestry and Other Resources* (pp. 1–24). Apple Academic Press.

Ranius, T., Widenfalk, L. A., Seedre, M., Lindman, L., Felton, A., Hämäläinen, A., ... & Öckinger, E. (2023). Protected area designation and management in a world of climate change: a review of recommendations. *Ambio*, 52(1), 68–80.

Ray, R., Subash Chandran, M. D., & Ramachandra, T. V. (2010). Ecosystem services from sacred groves of Uttar Kannada: A case study. In *Bangalore: Conference 22nd–24th December.*

Saeed, M. U., Hussain, N., Sumrin, A., Shahbaz, A., Noor, S., Bilal, M., ... & Iqbal, H. M. (2022). Microbial bioremediation strategies with wastewater treatment potentialities–a review. *Science of the Total Environment, 818*, 151754.

Sahu, N., Vasu, D., Sahu, A., Lal, N., & Singh, S. K. (2017). Strength of microbes in nutrient cycling: a key to soil health. *Agriculturally Important Microbes for Sustainable Agriculture: Volume I: Plant-soil-microbe nexus*, 69–86.

Sankhla, M. S., Kumari, M., Nandan, M., Kumar, R., & Agrawal, P. (2016). Heavy metals contamination in water and their hazardous effect on human health-a review. *International Journal of Current Microbiology and Applied Sciences, 5*(10), 759–766.

Scanes, C. G. (2018). Human activity and habitat loss: destruction, fragmentation, and degradation. In *Animals and Human Society* (pp. 451–482). Academic Press.

Shafi, A., Hassan, F., Zahoor, I., Majeed, U., Khanday, F.A. (2021). Biodiversity, Management and Sustainable Use of Medicinal and Aromatic Plant Resources. In: Aftab, T., Hakeem, K.R. (eds) *Medicinal and Aromatic Plants*. Springer, Cham. https://doi.org/10.1007/978-3-030-58975-2_3

Shankar, U., & Tripathi, A. K. (2017). Rainforests north of the Tropic of Cancer: Physiognomy, floristics and diversity in 'lowland rainforests' of Meghalaya, India. *Plant Diversity*, *39*(1), 20–36.

Sims, R. E., Mabee, W., Saddler, J. N., & Taylor, M. (2010). An overview of second generation biofuel technologies. *Bioresource Technology*, *101*(6), 1570–1580.

Singh, H., Parida, A., Debbarma, K., Ray, D. P., & Banerjee, P. (2020). Common marine organisms: a novel source of medicinal compounds. *International Journal of Bioresource Science*, *7*(2), 39–49.

Singh, R., Kumar, M., Mittal, A., & Mehta, P. K. (2017). Microbial metabolites in nutrition, healthcare and agriculture. *3 Biotech*, *7*, 1–14.

Sri Shalini, S., Palanivelu, K., Ramachandran, A., & Raghavan, V. (2021). Biochar from biomass waste as a renewable carbon material for climate change mitigation in reducing greenhouse gas emissions—a review. *Biomass Conversion and Biorefinery*, *11*(5), 2247–2267.

Strayer, D. L. (2010). Alien species in fresh waters: ecological effects, interactions with other stressors, and prospects for the future. *Freshwater Biology*, *55*, 152–174.

Sudhakar, K., Mamat, R., Samykano, M., Azmi, W. H., Ishak, W. F. W., & Yusaf, T. (2018). An overview of marine macroalgae as bioresource. *Renewable and Sustainable Energy Reviews*, *91*, 165–179.

Suleria, H. A. R., Gobe, G., Masci, P., & Osborne, S. A. (2016). Marine bioactive compounds and health promoting perspectives; innovation pathways for drug discovery. *Trends in Food Science & Technology*, *50*, 44–55.

Tarakeswara Naidu, M., Premavani, D., Suthari, S., & Venkaiah, M. (2018). Assessment of tree diversity in tropical deciduous forests of Northcentral Eastern Ghats, India. *Geology, Ecology, and Landscapes*, *2*(3), 216–227.

Tewari, V. P., Verma, R. K., & Von Gadow, K. (2017). Climate change effects in the Western Himalayan ecosystems of India: evidence and strategies. *Forest Ecosystems*, *4*(1), 1–9.

van Kleunen, M., Xu, X., Yang, Q., Maurel, N., Zhang, Z., Dawson, W., ... & Fristoe, T. S. (2020). Economic use of plants is key to their naturalization success. *Nature Communications*, *11*(1), 3201.

Vaz, A. S., Kueffer, C., Kull, C. A., Richardson, D. M., Vicente, J. R., Kühn, I., Schröter, M., Hauck, J., Bonn, J., & Honrado, J. P. (2017). Integrating ecosystem services and disservices: insights from plant invasions. *Ecosystem Services, 23*, 94–107.

Velu, G., Palanichamy, V., Rajan, A.P. (2018). Phytochemical and Pharmacological Importance of Plant Secondary Metabolites in Modern Medicine. In: Roopan, S., Madhumitha, G. (eds) *Bioorganic Phase in Natural Food: An Overview*. Springer, Cham. https://doi.org/10.1007/978-3-319-74210-6_8

Venkataraman, K., & Raghunathan, C. (2015). Coastal and marine biodiversity of India. In *Marine Faunal Diversity in India* (pp. 303–348). Academic Press.

Wang, J., Wu, Q., Ding, L., Song, S., Li, Y., Shi, L., ... & Li, X. (2021). Therapeutic effects and molecular mechanisms of bioactive compounds against respiratory diseases: traditional Chinese medicine theory and high-frequency use. *Frontiers in Pharmacology*, *12*, 734450.

Yadav, S. P., Tiwari, V. R., Mallick, A., Garawad, R., Talukdar, G., Sultan, S., ... & Das, A. (2023). Management Effectiveness Evaluation of Tiger Reserves in India, Fifth Cycle, 2022. Wildlife Institute of India, Dehradun and National Tiger Conservation Authority, Government of India.

Yahia, E. M., García-Solís, P., & Celis, M. E. M. (2019). Contribution of fruits and vegetables to human nutrition and health. In *Postharvest Physiology and Biochemistry of Fruits and Vegetables* (pp. 19–45). Woodhead Publishing.

Young, H. S., McCauley, D. J., Galetti, M., & Dirzo, R. (2016). Patterns, causes, and consequences of anthropocene defaunation. *Annual Review of Ecology, Evolution, and Systematics*, *47*, 333–358.

13 Marine Bioresources and Blue Economy

Nandini Rai[1], Archi Sehgal[1], Deepali Sambyal[1], J.M Julka[2], Swati Yadav[3], Suman Mishra[3], Kamal Jaiswal[3]

[1] Shoolini University of Biotechnology and Management Sciences, Solan- Himachal Pradesh, India

[2]Faculty of Basic Sciences, Shoolini University of Biotechnology and Management Sciences, Solan- Himachal Pradesh, India

[3]Department of Zoology, School of Life Sciences, Babasaheb Bhimrao Ambedkar University, Lucknow, Uttar Pradesh, India

13.1 INTRODUCTION

Oceans and seas are vital to the battle against climate change since they comprise more than 70% of the Earth's surface (United Nations, 2022a). Additionally, the oceans produce half of the world's oxygen by absorbing carbon dioxide emissions and excess heat (European Commission, 2019). However, due to human activity, the ocean is under pressure, leading to loss of biodiversity and health degradation. The term "blue economy" (BE) describes the sustainable use and management of marine resources to enhance social well-being, livelihoods, and economic growth (European Commission, 2019). The World Bank (2016) defines "blue economy" as "the development of oceanic economic activities in an integrated and sustainable way." According to the European Commission, the term "blue economy" refers to any sector of the economy that is associated with the seas, oceans, and coastlines. This covers both existing marine-related industries like shipping, tourism, offshore oil and gas, and fisheries, as well as newer industries associated with the marine sector, including offshore wind, marine minerals, blue bioeconomy, and desalination (European Commission, 2018a). The marine environment is still greatly impacted by anthropogenic-derived activities, such as untreated wastewater discharge, agriculture, shipping, port activities, aquaculture, offshore oil exploration, the use of fossil fuels, and industrial activities, the majority of which have the potential to be sources of oil pollution, even though environmental safety is a prerequisite for the European Union's Blue Growth strategy (Tedesco *et al.*, 2024). According to the European Union, the blue economy is "indispensable to our future welfare and prosperity" and serves as a "source of food, energy, transport, or leisure, as well as a driver for new jobs and innovation" (European Commission, 2018a). An unsustainable ocean economy causes the depletion of resources through human activities such as overfishing, oil and gas extraction, and mining (European Commission, 2019). The blue economy is claimed to generate economic, social, and environmental benefits for everyone. The advantages are numerous and might include anything from the development of new technologies, like renewable energy, to the support of emerging industries and jobs. However, there are still knowledge gaps in the BE, which pose a challenge and have several effects on investment decisions (Choudhary *et al.*, 2021). The seaweed sector is pertinent to the BE since it links the concepts of sustainable development and resource extraction, benefiting all parties involved while also promoting social and environmental well-being and assisting in the achievement of SDG 14, Life Below Water. The marine ecosystem is host to a wide range of biological resources, some of which have qualities that could lead to the creation of processes and products that could be sold. One factor

DOI: 10.1201/9781032697550-13

supporting the expansion of the blue bioeconomy may be the prospective development of commercial products produced from marine bioresources. The loss of marine resources and the destruction of ocean ecosystems have emerged as major global environmental and economic issues. This problem is addressed by the BE concept, which offers a framework for the development of ocean sustainably while fostering social progress, economic expansion, and environmental preservation of habitats that are in grave danger. Marine natural products are typically called secondary metabolites. Marine resources have been more focused on recent studies of novel bioactive chemicals with enhanced potency and therapeutic advantages in functional foods and diets. Marine resources have been utilised for ages in traditional medicine to treat a variety of human illnesses. The development of an environmentally conscious and highly productive civilisation is thought to be of interest to the field of marine biotechnology. This results in the creation of resources and products for renewable energy from the maritime environment. Through sustainable blue technology, marine resources play a key role in addressing the world's food security, energy problems, and health inequities. In order to support economic growth through blue technology, nations throughout the world are creating a tactical roadmap. This will enable understanding and evaluation of marine animals, ecosystems, and related policy initiatives. Seaweed is an emerging industry because it produces fuel, animal feed, sustainable food, and a variety of other goods and services, and it has the potential to support the blue economy. A sustainable seaweed sector does present a number of potentials, but it also faces considerable investment obstacles and funding constraints. One promising way to help the seaweed sector remain sustainable is through impact investing. Impact investing strives to have a positive social and environmental impact while providing investors with financial profits, and this promotes the blue economy. One recent technological advancement that helps and enhances the growth of the blue economy is marine biotechnology. Biological applications derived from marine bioresources are the focus of the field of marine biotechnology. Marine bioresources include a variety of creatures, including plants, fish, molluscs, sponges, bacteria, fungi, other microorganisms, cyanobacteria, micro- and macro-algae, fish co-products, and other invertebrates, in addition to compounds obtained from these sorts of animals. Marine bioresources known as macroalgae have a wide range of bioactive possibilities for treating chronic illnesses like cancer, diabetes, obesity, and coronary heart disease. The wide variety of macroalgal species found in India's coastal states has not yet been thoroughly investigated for potential nutraceutical advantages. Investigating macroalgae can lead to the development of sustainable uses for it to create the blue economy and offer possible natural alternatives that are highly sought after in the healthcare sector. Macroscopic marine algae, also known as seaweed, have evolved to withstand a variety of harsh maritime environmental circumstances. Seaweed is regarded as a blue carbon generator for the growth of the blue economy because of its capacity to sequester carbon and accumulate a variety of products. Marine bioresource-based bioplastics are appealing materials for a range of applications because they have desirable physicochemical characteristics and are easily degraded. A number of negative effects, including environmental degradation, habitat loss, biodiversity decline, imbalance in the marine ecosystem, and decreased productivity, have been brought about by overfishing, pollution, and other human activities. Acute degradation of fisheries has been the primary cause of the destruction of bioresources.

13.2 BLUE ECONOMY

The terminology "BE" was first used at the United Nations Conference on Sustainable Development (UNCSD) Rio+20 held in Rio de Janeiro, Brazil (UNEP, 2012). The term "ocean economy" refers to the general realisation of the ocean as an economy, whereas "sustainable ocean economy," or BE, refers to the evolution of the ocean economy into a notion that combines sustainable development and economic growth (Wenhai *et al.*, 2019).

The BE is an industry that is committed to using marine resources sustainably to promote social progress, economic expansion, and environmental preservation (UNEP, 2012). According to

Addamo, Calvo Santos, and Guillén (2022) and the European Commission (2018a), the BE generated a turnover of almost EUR 667.2 billion in Europe alone in 2019. This is a 94% growth since 2009. According to Addamo, Calvo Santos, and Guillén (2022) and the European Commission (2018a), employment in the BE grew from 0.252 million jobs in 2009 to 4.45 million jobs in 2019. The BE is an idea that focuses on lower- to middle-income coastal and island nations where the ocean is both a source of opportunity and a jurisdictional region, as well as in the high seas (World Bank Group, 2016). The BE can be classified into industries, drivers of growth, types of activity, and ocean services. The World Bank Group (2016) identified four categories of activities: harvesting live resources; extracting non-living resources; creating new resources, commerce, and trade in and around oceans; and addressing issues related to ocean health that are directly related to sustainable development. The BE is anchored in multiple fields, including politics, economics, and cultural studies, spanning planetary and ecological boundaries. As a result, it is creating a renewed sense of urgency for reevaluating the connection between the environment and economy (Midlen *et al.*, 2021; Hoerterer *et al.*, 2020).

13.2.1 BLUE ECONOMY SOCIAL IMPACTS

The blue economy, which is focused on using ocean resources sustainably, can have a number of socioeconomic effects. It could stimulate economic growth in coastal towns by generating job possibilities in industries like tourism and fishing. Sustainable practices are necessary to prevent adverse social effects, such as overfishing and environmental deterioration, which can hurt communities that depend on marine resources for their livelihoods. Positive effects on coastal societies require striking a balance between social and environmental concerns and economic progress.

13.3 BIORESOURCES

"Bioresources" is a broad term used to describe biological materials or compounds that are obtained from living creatures and have several uses. These materials may be derived from microbes, plants, animals, or other biological sources. The word refers to a broad category of materials with a variety of uses in a number of sectors, such as manufacturing, energy, agriculture, and medical (Figure 13.1).

For example:

1. **Resources based on plants**
   - **Crops**: Crops used in agriculture supply food, fuel, and raw materials for other industries.
   - **Wood**: For the manufacture of paper, furniture, and other items, wood is necessary.
2. **Microbiological resources**
   - **Microorganisms**: Waste treatment, pharmaceuticals, and biotechnology all make use of bacteria, fungi, and other microorganisms.
   - **Microbial enzymes**: Microorganisms generate enzymes that are used in the food processing, detergent, and biofuel industries.
3. **Resources based on animals**
   - **Livestock**: Livestock yields several goods for human consumption, such as meat, milk, and leather.
   - **Marine resources**: The fishing business depends on fish and other marine life for the supply of seafood and other goods.
4. **Bioenergy sources**
   - **Biomass**: Bioenergy can be produced from plant and organic materials by combustion or biofuel synthesis.
5. **Resources for medicines**
   - **Plant and microorganisms**: Plant extracts or microbial fermentation are used in the production of several medicines.

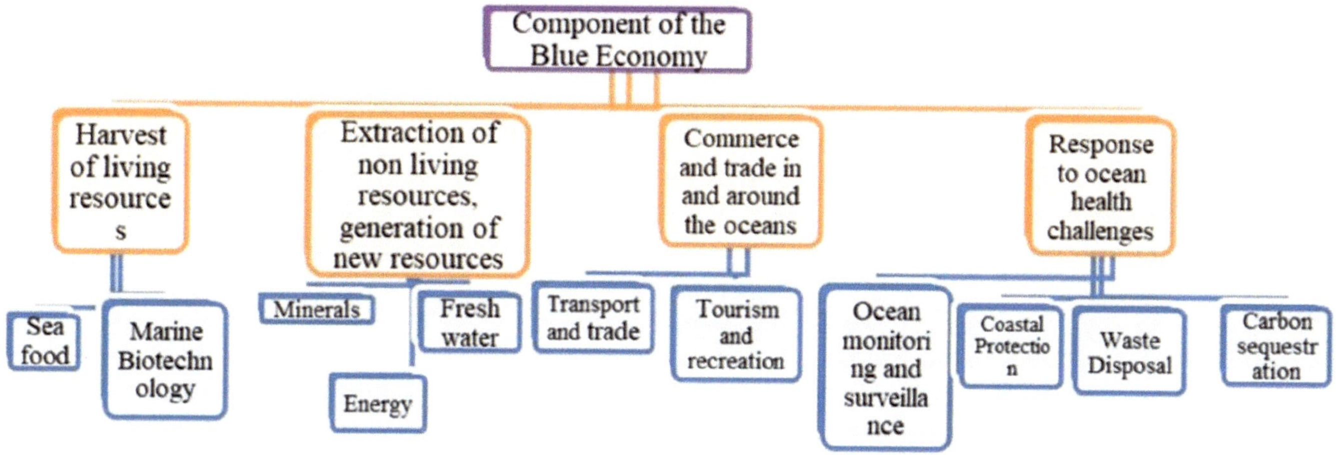

FIGURE 13.1 Components of blue economy. (From World Bank Group, 2016).

6. **Resources for industries**
   - **Biopolymers**: In manufacturing, polymers obtained from biological sources, such as cellulose or starch, can be employed.
   - **Bio-based chemicals**: In many industrial processes, chemicals made from bioresources can take the place of conventional petrochemicals.
7. **Conversion of waste to resources**
   - **Organic waste**: Bioresources include organic waste items that can be processed via anaerobic digestion to produce biogas, compost, or other valuable products.

13.4 SUSTAINABLE SEAWEED FARMING

"Seaweed" is a general term used to describe a variety of algae, mostly classified into three species: brown, green, and red algae (Abdel-Kareem *et al.*, 2022). The algae can be found in various water bodies including lakes, rivers, and oceans. Seaweeds may thrive in a variety of habitats, but they do best in water that is between 10°C and 20°C (Khan *et al.*, 2018). Seaweed is a "bona fide misnomer" that spreads in a way that may be detrimental to the environment in which it develops. It is both fixed and free-floating simultaneously, giving a wide variety of marine life vital supplies of food and habitat (National Ocean Service, 2021). The use of seaweed by humans for food and medicinal purposes dates back approximately 1500 years when the Japanese combined the seaweed nori with what is now called a "sushi roll" (Lomartire *et al.*, 2022). Due to its biodiversity, globally 145 species are used in food production, and 110 species are cultivated for the production of gelling agents (Lomartire *et al.*, 2022). According to the Food and Agriculture Organization (FAO), seaweed farming is a lucrative, ecologically, and economically significant industry that is expected to grow. It has the potential to create blue employment and is a cost-effective source of energy, pharmaceuticals, food and agriculture, and cosmetics (Hossain *et al.*, 2021). Seaweeds have various advantages for the environment and society in addition to their commercial uses (Sultana, 2022) (Table 13.1).

13.4.1 Seaweed as a Potential Substitute for Various Products

The potential of seaweed as a sustainable and adaptable resource to replace a wide range of products in various industries has drawn attention. Seaweed is being investigated as a possible substitute in the following areas (Figure 13.2):

1. **Dietary consumption and nutrition**
   - **Plant-based food**: Seaweed has a distinct texture and nutritional profile that make it a good base for plant-based dishes like sausages, burgers, and seafood substitutes.
   - **Food additives**: Seaweed extracts, such as agar and carrageenan, are utilised in a variety of food products as stabilisers, thickeners, and gelling agents.

TABLE 13.1
Top Cultivated Seaweeds in the World

General	Yield (Mt Fresh Wt.)	% of World Production
Eucheuma* spp. (mainly *E. denticulatum*) and *K. alvarezii	8.26	33.3
Saccharina (Laminaria) *japonica*	5.76	23.2
***Gracilaria* spp.**	2.83	11.4
Undaria pinnatifida	2.14	8.6
***Pyropia (Porphyra)* spp.**	1.81	7.3
Total	**20.80**	**83.3**

Source: Data for 2012 from FAOSTAT (2015).

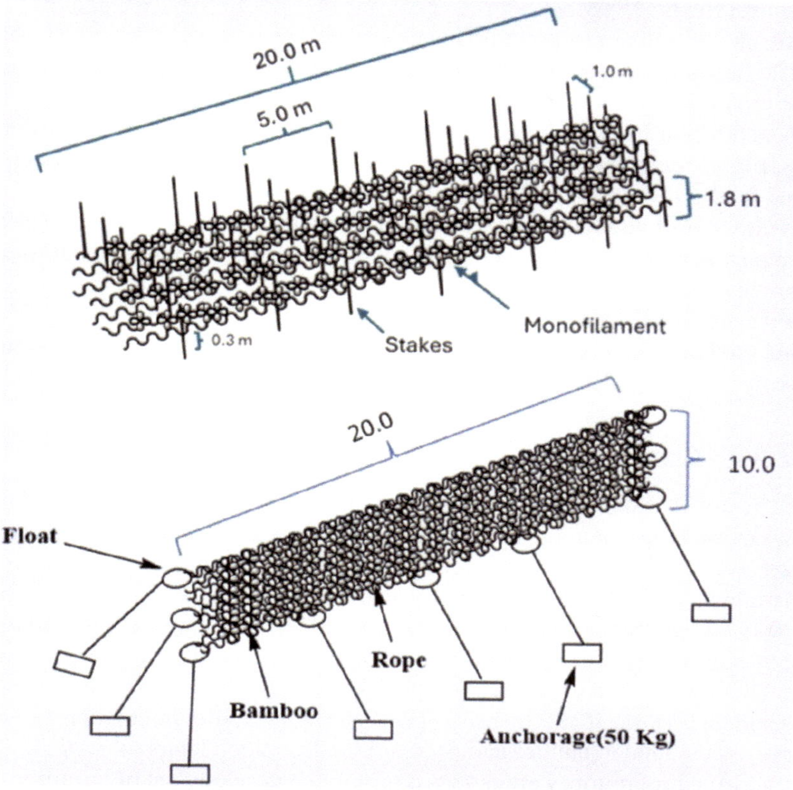

FIGURE 13.2 Culture techniques used in seaweed farming: (top) off-bottom method; (bottom) floating system. The dimensions shown are for illustration purposes only and may vary in the field.

2. **Snacks and condiments**: Seaweed is becoming more popular as a flavour enhancer in condiments and as an ingredient in snack foods like seaweed chips.
3. **Containerisation**
4. **Biodegradable packaging**: Seaweed-derived materials are being investigated as a sustainable substitute for conventional plastic packaging. Their biodegradability can aid in mitigating the ecological consequences of packaging waste.
5. **Fabrics**
   - **Fibre production**: Clothes and textiles can be made using seaweed fibres. In terms of resource usage and environmental effect, the manufacturing of textiles made from seaweed is thought to be more sustainable than that of traditional textiles.
6. **Personal hygiene and cosmetics**
   - **Ingredients**: The moisturising, anti-inflammatory, and antioxidant qualities of seaweed extracts make them popular ingredients in cosmetics and personal hygiene products.
   - **Exfoliants**: Seaweed powder can be used as a natural exfoliator in cosmetics.
7. **Production of biofuel**
   - **Bioethanol**: Due to their ability to provide fermentable sugars, several varieties of seaweed can be used to produce biofuel, particularly bioethanol.
8. **Fertilisers**
   - **Soil amendment**: Seaweed adds vital nutrients to soil and encourages plant growth. It can also be processed into fertilisers or applied directly as a soil amendment.

9. **Pharmaceuticals**
   - **Biologically active compounds**: Bioactive substances found in seaweed have the potential to be used in medicine because of their antibacterial and anti-inflammatory qualities.
10. **Water purification**
    - **Absorption of heavy metals**: In water treatment procedures, several seaweed species are helpful because of their capacity to absorb heavy metals from the water.
11. **Animal feed**
    - **Livestock and aquaculture**: Seaweed can be added to livestock and aquaculture feed to replenish it with vital minerals and possibly lessen the environmental impact of conventional feed sources.
12. **Carbon sequestration**
    - **Blue carbon:** Seaweeds play a significant role in "blue carbon" programmes, which attempt to slow global warming by effectively sequestering carbon dioxide.

13.4.2 Seaweed Cultivation Risk and Challenges

The seaweed sector is expanding quickly with the potential to have positive effects on the environment, society, and the economy. But the sector is exposed to a number of internal and external threats. There are risks in each of the three primary areas, ranging from societal acceptability to legal constraints, and from environmental hazards to future shocks brought on by pandemics, recessions, and rising temperatures (Duarte *et al.*, 2022). Marine mammals like dolphins and whales are put at risk by the use of longlines and ropes in seaweed farming operations (Campbell et al., 2019). The introduction of non-native species via seaweed farming poses a significant additional risk (Campbell *et al.*, 2019). Invasive species have the ability to outcompete local seaweeds and can modify ecosystem dynamics and decrease biodiversity (Campbell *et al.*, 2019). Strict controls and monitoring programmes are required to stop the development of hazardous algal blooms (HAB). Seaweed farming expenses are heavily influenced by a number of economic variables. In order to eradicate epiphytes and competing organisms, pH control techniques and organic acid control are crucial when introducing non-endemic species of seaweed into ecosystems (Sultana *et al.*, 2023). These techniques are expensive but essential. Additional economic issues include manpower expenses, equipment costs, and seed prices; these are frequently combined with other risks (marine mammal entanglement, for instance). Every element affects coastal economies and seaweed producers' long-term financial sustainability (Bjerregaard *et al.*, 2016; Behera *et al.*, 2022). Invasive species can alter the biological balance of coastal ecosystems and other waters by entering harbour waters and introducing various plants and animals, including cholera and harmful algae (Behera *et al.*, 2022). Bringing in such species and seaweed farms could become less productive, which would cost farmers money. Seaweed farming expenses are heavily influenced by a number of economic variables. In particular, the cost-effectiveness of farming operations is correlated with farming scale (Yong *et al.*, 2022). Another risk factor for seaweed farming is the availability of technologies and equipment. As a relatively new business, seaweed farming has limited access to equipment and technology that are sustainable because they are still in the early stages of development (Behera *et al.*, 2022). More research and development are required to increase seaweed farming technologies' cost and accessibility (Behera *et al.*, 2022; Farghali *et al.*, 2023). In addition, the seaweed industry will be exposed to additional shocks and threats in the future, including air pollution, contaminated nutrients entering global waters, invasive species that upset marine life's ecological balances, pesticides from agriculture that end up in the ocean, and oil spills from water-sewage treatment plants (National Geographic, 2022). Seaweed poses a significant concern to public health and can be brought on by invasive species that can lead to HAB. Seaweed farming enterprises are governed by a number of rules and need permits. From an ecological perspective, harvesting seaweeds from their natural habitats can be very risky. Therefore, it might be difficult to get and retain the licences and permissions required, as well as to comply with laws, which can lead to an increase in production costs (Farghali *et al.*, 2023). Furthermore, ineffective and poorly thought-out policies result in inefficient resource management, including improper allocation of funds from government budgets for projects that provide financial support (Behera *et al.*, 2022). A number of hazards are closely linked to gaps in knowledge. For instance, as of 2023, it is unclear what ecological effects deep-sea biomass dumping for carbon sequestration will have (Yong *et al.*, 2022). The presence of heavy metal contamination in the nutrients that seaweeds absorb could kill marine life and shellfish and cause oxygen deprivation (Fröhlich *et al.*, 2019). These findings are subject to confirmation. The impact of climate change, which is generating storms, typhoons, ocean warming, and earthquakes, on seaweed farming is also uncertain (Duarte *et al.*, 2017).

13.4.3 Economic Impact of Seaweed Farming

Seaweed farming, also referred to as mariculture or seaweed aquaculture, has the potential to create a number of economic effects. The kind of seaweed farmed, the size of the operations, and the location of the farming can all impact on the economy. The following are a few possible financial effects of seaweed farming:

1. **Employment opportunities**: Growing, harvesting, processing, and selling seaweed products are all aspects of seaweed farming that can generate employment in coastal areas. In addition to lowering unemployment, this can support local economic growth.
2. **Diversification of coastal economies**: Coastal communities' economies are diversified by seaweed farming, which offers an additional or alternative revenue stream. In regions where traditional sectors like fishing are encountering difficulties, this can be especially crucial.
3. **Stimulating ancillary industries**: The expansion of the seaweed farming sector may encourage the growth of related sectors, including packaging, transportation, and processing facilities. A multiplier effect may result from this, creating economic activity in addition to the original farming operations.
4. **Export opportunities**: The market for seaweed and its products is expanding on a global scale. The growing demand for products produced from seaweed, including food, cosmetics, and pharmaceuticals, can lead to new export prospects for nations with adequate coastal conditions through seaweed farming.
5. **Carbon sequestration and environmental services**: There are seaweeds that can absorb carbon dioxide from the atmosphere. As a way to offset carbon emissions, governments and organisations that care about climate change may provide incentives for seaweed farming. Through carbon credit programmes, this may result in financial advantages for seaweed farmers.
6. **Aquaculture integration**: Fisheries and other aquaculture practices, such as seaweed farming, can coexist. The overall productivity and profitability of aquaculture can be increased by using seaweeds to give fish a feed supply and habitat.
7. **Innovation and research**: Research and innovation in fields including product development, breeding initiatives, and growing methods may be sparked by the growth of seaweed farming. Technological developments and the emergence of new intellectual property may result from this.
8. **Reduced pressure on wild stocks**: Seaweed farming may lessen strain on wild seaweed stocks and support sustainable resource management if it becomes a major source of raw materials for a variety of sectors (such as food and pharmaceuticals).
9. **Coastal ecosystem health**: By giving marine life more structure and habitat, seaweed farming has the potential to enhance the health of coastal ecosystems. The economy of

coastal towns may indirectly benefit from this, as it may have a favourable impact on biodiversity and fisheries.

13.5 PRODUCTION OF BIOACTIVE COMPOUNDS USING TUNA VISCERA

Viscera from tuna, which is frequently discarded in the tuna processing industry, can actually contain bioactive substances that have potential uses in a variety of industries, including medicine, nutraceuticals, and cosmetics. There are various procedures involved in removing bioactive substances from tuna viscera:

1. **Collection and preparation**: Collect fresh tuna viscera from processing facilities; wash and eliminate impurities or non-visceral parts.
2. **Extraction of bioactive compounds**: To separate the bioactive substances from the tuna viscera, use the proper solvents or extraction techniques. Solvent extraction, supercritical fluid extraction, and enzymatic hydrolysis are examples of common techniques. The particular bioactive chemicals of interest determine which extraction technique and solvent to use.
3. **Separation and purification**: The crude extract may include additional ingredients and contaminants after extraction. Use methods such as distillation, filtering, or chromatography to purify the extract.
4. **Identification and characterisation**: Examine and identify the bioactive components of the purified substances through analysis. This can be done using techniques like spectroscopy (e.g., UV-Vis, IR, NMR), chromatography (HPLC, GC), and mass spectrometry.
5. **Testing bioactivity**: Assess the isolated chemicals' biological activity. Testing for antibacterial, anti-inflammatory, antioxidant, and other bioactive qualities may fall under this category.
6. **Formulation and application:** After the identification of bioactive substances and confirmation of their activity, investigate possible applications. Making them into medications, nutraceuticals, or cosmetics is one way to do this.
7. **Safety and regulatory considerations**: Verify that the bioactive chemicals that have been extracted adhere to safety regulations and standards in order for them to be used for the intended purposes.

13.6 BIOMOLECULES IN COSMETIC FORMULATIONS

Research has examined the valorisation of skin by-products of fish filleting, mussel processing wastes, and tuna viscera by-products using biotechnological and chemical transformation to produce hydrolysed collagen (HC), glycogen (G), and hyaluronic acid (HA), respectively. Because of their sufficient bioactivities, these three biomolecules obtained by valorising fishery by-products are seen as sustainable or eco-friendly ingredients that can be used to meet social, cultural, or industrial eco-sustainable requirements. The cosmetics industry is especially interested in these biomolecules. It has been observed that marine hydrolysed collagen is a valuable source of peptides with a wide range of biofunctional qualities, such as antioxidant properties, impacts on human dermal fibroblast cell proliferation, wound-healing capabilities, skin hydration/elasticity capacity, and antiwrinkle activity. Glycogen is mostly used as a velvety and emollient ingredient in cosmetic compositions. It has already been documented that hyaluronic acid possesses exceptional skin regeneration and collagen-stimulating effectiveness. Furthermore, by promoting chondrocyte and fibroblast growth in vitro, collagen and hyaluronic acid together have benefits over using each constituent alone for tissue engineering and cosmetic applications.

FIGURE 13.3 Significant issues with the marine bioeconomy.

13.7 SIGNIFICANT ISSUES WITH MARINE BIOECONOMY

Numerous significant obstacles could hinder the marine bioeconomy's growth and sustainability. Environmental, economic, societal, and regulatory considerations all play a part in these difficulties. Following are some of the main issues facing the marine bioeconomy (Figure 13.3):

1. **Overexploitation of fish**: Overexploitation of numerous marine species causes dwindling fish supplies and ecological imbalances. Overfishing can jeopardise the marine bioeconomy's sustainability and endanger the long-term viability of marine resources.
2. **Climate variability**: Ocean acidification, rising sea temperatures, and shifting ocean currents are all consequences of climate change that impact marine ecosystems. These elements may affect the distribution and quantity of marine species, which may, in turn, affect the profitability of aquaculture and fishing activities.
3. **Pollution**: Marine ecosystems are seriously threatened by marine pollution, which includes chemical pollutants, oil spills, and plastic waste. Pollution has the potential to destroy marine life, upend food webs, and lower the standard of marine products.
4. **Habitat loss and degradation**: Critical marine habitats, such as coral reefs, mangroves, and seagrass beds, can be destroyed as a result of human activities like dredging, bottom trawling, and coastal development. The marine ecology and the species that depend on it may be negatively impacted in a domino manner by this habitat loss.
5. **Illegal, unreported, and unregulated (IUU) fishing**: Marine resource management and conservation efforts are hampered by IUU fishing. Overfishing may result, legal fishermen

may lose out on income, and communities that depend on fisheries may experience detrimental social and economic effects.

6. **Restricted availability of financial technology**: Small-scale marine bioeconomy enterprises and fishermen frequently encounter difficulties obtaining funding and technology. This hinders their capacity to implement eco-friendly procedures, boost productivity, and adjust to shifting environmental circumstances.
7. **Market access and trade barriers**: Businesses participating in the marine bioeconomy have limited market access due to trade obstacles and certification issues around certain marine products. For the development of a sustainable marine industry, it is imperative that these obstacles be removed.
8. **Livelihoods and social equity**: Equitable distribution of the benefits is not necessarily a feature of the marine bioeconomy. Social disputes and negative effects on local communities' quality of life can result from problems including uneven resource distribution, shaky tenure rights, and a lack of representation in decision-making processes.
9. **Inadequate integrated management**: An integrated strategy that takes into account the connections between marine ecosystems, animals, and human activity is necessary for the effective management of the marine bioeconomy. Results may be less than ideal when management is fragmented or sector-specific.
10. **Regulatory obstacles**: The growth of sustainable marine enterprises may be hampered by inconsistent or insufficient laws. To ensure ethical and responsible practices in the marine bioeconomy, rules must be both clear and enforceable.

13.8 SUSTAINABLE GROWTH OF BLUE BIOECONOMY

In order to achieve a sustainable blue economy, implementing sustainable best practices can help balance the long-term ability of the seas to withstand activities related to marine bioresource extraction. In other words, in order to maintain potential advantages over time, human activities must be regulated in a way that preserves economic output and does not harm the marine environment (European Commission, 2018a). Nowadays, human actions might be harmful because of a number of factors, such as insufficient scientific understanding and information, as well as inefficient technology (Cuyvers *et al.*, 2018). By enhancing marine scientific knowledge, our understanding of marine science will put us in a better position to modify both present and future maritime operations to have less of an adverse environmental impact and to offer more sustainable alternatives to their land-based counterparts (European Commission, 2018a). For instance, compared to conventional, land-based counterparts, the growing of algae can offer sustainable, alternative sources for the development of animal feed or fuels. Furthermore, improved knowledge of marine ecosystems and how fish populations react to the creation of "no-take zones" could be helpful in implementing alternative fishing methods to counter the current overfishing tendencies. To protect and restore the long-term health and sustainable use of the seas, we can better assess which activities (such as plastic waste) are harmful and should be avoided by improving our understanding of the marine environment (European Commission, 2018a). There are numerous possible economic, environmental, and social advantages for humans in the aquatic environment. Nevertheless, the current state of the blue bioeconomy sector indicates that this potential is not being realised due to its modest size and poor growth rate (European Commission, 2018a). It is an issue that opportunities for sustainable prosperity are being overlooked, such as those that arise from improving economies, increasing employment rates, reducing global food shortages, and addressing environmental effects associated with present-day industrial practices. Currently, the development of marine bioresources is probably hindered by a number of poorly understood bottlenecks. The issue at hand is that we don't have enough specific information about these particular problems or possible solutions.

13.9 DISCUSSION

The concept of the blue economy and sustainable seaweed farming share similar attributes. The goal of both strategies is to convert economic growth into sustainable growth. The high seas are considered areas of development where everyone may profit. Therefore, seaweed farming is a suitable maritime activity that advances the goals of the blue economy. A survey of the literature on seaweed farming demonstrates that expanding the industry's scope is essential to ensuring that it conforms to the traits of a blue economy. The seaweed farming sector has enormous potential to grow more profitable each year, which would help the blue economy's main initiatives for sustainable growth. Because of the algae's adaptable qualities, seaweeds can be used for a variety of purposes, including food for humans, animal feed, and significant components of medicine. The various tenets of a blue economy closely correspond with the features of sustainable seaweed farming. However, there are challenges to seaweed farming's viability for the BE strategy due to a number of factors, including the industry's present scale as of 2023. When connecting the concept to the sector, factors like the size of sustainable farming play a significant role. For instance, pH management must be implemented in small-scale farming, and acids must be used to manage epiphytes and competing organism populations. Seaweed populations need to be monitored properly even if seaweed cultivation does not require freshwater or pesticides like regular agriculture does. Sustainable development will benefit from seaweed farming investments, which will have a wide range of good effects. The seaweed business is a lucrative and promising one, according to a thorough examination of market trends. In addition to encouraging more sustainable practices that lower exposure costs, seaweed farming investments also boost sustainability in many other businesses. Moreover, investments in seaweed farming will foster economies of scale and industry growth. Seaweed is a possible substitute for a number of goods, including food, medicine, animal feed, and renewable chemicals. Seaweed farming, however, is still a little-studied sector with a number of possible risks that should be taken into consideration. Although the industry has a lot of promise, more research is apparently needed to fully understand practices such as carbon dumping on the high seas. Due to inadequate laws and licencing, the seaweed sector poses a serious threat to vital creatures in the marine environment and has great potential for exploitation.

13.10 CONCLUSION

In summary, the exploration of a blue economy and the study of marine bioresources are essential paths toward ecological balance and sustainable development. Untapped potential for innovation, economic expansion, and environmental preservation exists in the wide and varied variety of marine life. We can meet the demands of the present while also laying the groundwork for a more resilient and peaceful future by wisely using these resources. The blue economy framework's incorporation of marine bioresources highlights the necessity of a comprehensive strategy that considers ecological, social, and economic factors. Oceans provide a large frontier of possibilities as we manage the problems brought about by climate change and the depletion of land resources. While protecting the delicate balance of marine ecosystems, the sustainable management of marine bioresources can support economic growth through advancing food security, pharmaceutical discoveries, and the creation of novel materials. Moreover, establishing efficient governance frameworks and promoting global cooperation are essential elements of a prosperous blue economy focused on marine bioresources. For the purpose of maximising benefits and minimising environmental effects, coordinated actions across nations, industries, and research sectors will be crucial. To guarantee a responsible and moral exploitation of marine bioresources, it is imperative to embrace technical innovations, make educational investments, and raise public awareness.

Ultimately, the potential for greater prosperity and sustainability in the future is presented by the complementary relationship between marine bioresources and the blue economy. We can fully utilise the potential of our oceans to ensure a robust blue economy that benefits current and future generations by employing innovative thinking, global cooperation, and thoughtful supervision.

REFERENCES

Abdel-Kareem, M.S., & ElSaied, A.A. 2022. Global seaweeds diversity. In *Handbook of algal biofuels,* (2022), (pp. 39–55). Elsevier.

Addamo, A., Calvo Santos, A., & Guillén, J. The EU blue economy report 2022. Directorate-General for Maritime Affairs and Fisheries, Maritime Affairs and Fisheries. Publications Office of the European Union (2022).

Behera, D. P., Vadodariya, V., Veeragurunathan, V., & Sigamani, S. Seaweeds cultivation methods and their role in climate mitigation and environmental cleanup. *Total Environment Research Themes*, (2022), 3, 100016.

Bjerregaard, R., Valderrama, D., Radulovich, R., Diana, J., & Capron, M. Seaweed Aquaculture for Food Security, Income Generation and Environmental Health in Tropical Developing Countries. World Bank Group 2016).

Campbell I, Macleod A, Sahlmann C, Neves L, Funderud J, Øverland M, Hughes AD and Stanley M. The Environmental Risks Associated With the Development of Seaweed Farming in Europe - Prioritizing Key Knowledge Gaps. *Frontiers in Marine Science*, (2019) 6, 107.

Choudhary, P., Khade, M., Savant, S., Musale, A., Chelliah, M.S., & Dasgupta, S. Empowering blue economy: From underrated ecosystem to sustainable industry. *Journal of Environmental Management*, (2021), 291, 112697.-

Cuyvers, L., Berry, W., Gjerde, K. M., Thiele, T., & Wilhem, C. Deep Seabed Mining: A Rising Environmental Challenge. Gland: IUCN, (2018).

Duarte, C.M., Bruhn, A., & Krause-Jensen, D. A seaweed aquaculture imperative to meet global sustainability targets. *Nature Sustainability*, (2022), 5(3), 185–193.

European Commission. The 2018 annual economic report on EU blue economy. Directorate-General for Maritime Affairs and Fisheries. European Union (2018).

European Commission. The EU blue economy report. Luxembourg: Publications Office of the European Union (2019).

FAOSTAT. Food and Agriculture Organization of the United Nations Statistics Division. Retrieved from: http://faostat3.fao.org/home/E (2015).

Farghali, M., Mohamed, I.M., Osman, A.I., & Rooney, D.W. 2023. Seaweed for climate mitigation, wastewater treatment, bioenergy, bioplastic, biochar, food, pharmaceuticals, and cosmetics: a review. *Environmental Chemistry Letters*, (2023), 21(1), 97–152.

Hoerterer, C., Schupp, M. F., Benkens, A., Nickiewicz, D., Krause, G., Buck, B. H. Stakeholder perspectives on opportunities and challenges in achieving sustainable growth of the blue economy in a changing climate. *Frontiers in Marine Science,* (2020), 6, 795.

Hossain, M.S., Sharifuzzaman, S.M., Nobi, M.N., Chowdhury, M.S.N., Sarker, S., Alamgir, M., Uddin, S.A., Chowdhury, S.R., Rahman, M.M., Rahman, M.S., & Sobhan, F. Seaweeds farming for sustainable development goals and blue economy in Bangladesh. *Marine Policy*, (2021), 128, 104469.

Khan, M.I., Shin, J.H., & Kim, J.D. The promising future of microalgae: current status, challenges, and optimization of a sustainable and renewable industry for biofuels, feed, and other products. *Microbial cell factories*, (2018), 17, 1–21.

Lomartire, S., & Gonçalves, A.M. An overview of potential seaweed-derived bioactive compounds for pharmaceutical applications. *Marine Drugs*, (2022), 20(2), 141.

Midlen, A. What is the Blue Economy? A spatialised governmentality perspective. *Maritime Studies,* (2021), 20, 423–448. https://doi.org/10.1007/s40152-021-00240-3.

National Geographic. Ocean Threats. (2022). Retrieved January 11, 2023

National Ocean Service. Seaweed. (2021) Retrieved January 11, 2023

Sultana, F., Wahab, M.A., Nahiduzzaman, M., Mohiuddin, M., Iqbal, M.Z., Shakil, A., Mamun, A.A., Khan, M.S.R., Wong, L., & Asaduzzaman, M. Seaweed farming for food and nutritional security, climate change mitigation and adaptation, and women empowerment: A review. *Aquaculture and Fisheries*, (2023), 8(5), 463–480.

Tedesco, P., Balzano, S., Coppola, D., Esposito, F. P., de Pascale, D., & Denato, R. Bioremediation for the recovery of oil polluted marine environment, opportunities and challenges approaching the Blue Growth. *Marine Pollution Bulletin*, (2024), 200, 116157.

UNEP. (2012). UNEP. Retrieved January 01, 2023

United Nations. UN. (2022a). Retrieved January 01, 2023

Wenhai, L., Cusack, C., Baker, M., Tao, W., Mingbao, C., Paige, K., Xiaofan, Z., Levin, L., Escobar, E., Amon, D., & Yue, Y. 2019. Successful blue economy examples with an emphasis on international perspectives. *Frontiers in Marine Science*, (2019), 6, 261.

World Bank Group. Blue Economy Development Framework. World Bank Group (2016).

Yong, W.T.L., Thien, V.Y., Rupert, R., & Rodrigues, K.F. Seaweed: a potential climate change solution. *Renewable and Sustainable Energy Reviews*, (2022), 159, 112222.

14 Food Security and Biodiversity

Priyanka[1], *Priyanka Chauhan*[1], *Meenu Chaurasia*[2]

[1]School of Biological and Environmental Sciences, Shoolini University, Solan, Himachal Pradesh, India

[2]Department of Zoology, University of Delhi

14.1 INTRODUCTION

Among the challenges that the world faces today, improving global food security and preserving biodiversity are of utmost importance (Glamann *et al.*, 2017). With the global human population increasing rapidly, it is estimated that there will be around 10 billion people on the planet by 2050 (Ewert *et al.*, 2018), and the population experiencing hunger and food insecurity will increase, with factors such as conflict, pandemics, climate change, and growing inequalities. As the world faces a polycrisis, there is a need to ensure that enough food can be produced to meet the needs of this expanding population and alleviate the effects of conflict and pandemics on worldwide nutrition and food security. The Food and Agriculture Organization (FAO) 2002 defines food security as "when all people at all times have access to sufficient, safe, nutritious food to maintain a healthy and active life." Along with food access and food usage, food availability is just one of the three pillars of food security (Scherer *et al.*, 2020). Already, more food is being consumed by humans than the planet can sustain (Campbell *et al.*, 2016), yet, approximately 2 billion people remain undernourished (FAO, 2019). In the past, food security and biodiversity conservation were often addressed independently; however, there is now growing recognition that preserving biodiversity and securing food for an expanding human population are deeply interconnected challenges (Chappell & LaValle, 2011; Tscharntke *et al.*, 2012). Biodiversity encompasses the variety of life forms on Earth. According to the Convention on Biological Diversity (CBD), it is defined as the "variability among living organisms," encompassing ecosystems' diversity, species richness, and genetic variation within species (UN, I. R. B., 1992). The conservation of biodiversity and ensuring food security are two sides of the same coin. In many respects, biodiversity is vital to agriculture and food production (Pilling *et al.*, 2020). So far, the framework of land sharing or sparing or the idea of sustainable intensification has primarily been used to view the relationship between the protection of biodiversity and food security from the perspective of agricultural output (Wittman *et al.*, 2017). Maintaining biodiversity in the face of human demands for food, fuel, fiber, lumber, and other items derived from the world's croplands, grasslands, forests, and aquatic ecosystems is a significant worldwide problem (Pilling *et al.*, 2020). Agriculture is a cornerstone of human civilization that relies on the productivity of the soil and the diverse organisms that inhabit it, and the provision of food is the fundamental service provided by agriculture. Land degradation presents a significant challenge, primarily attributed to agriculture, intensive farming, and the conversion of forest lands to croplands. It has been identified that agriculture alone poses a threat to 24,000 out of 28,000 (86%) species currently at risk of extinction (Benton *et al.*, 2021). As we strive for environmental sustainability, food systems need to evolve to meet the demands of a growing population while operating within the constraints of our planet (Béné *et al.*, 2019; Crenna *et al.*, 2019). Biological diversity needs to be gathered, preserved, and utilized responsibly to avoid endangering the next generation. This is absolutely necessary to ensure food security both now and in the future (Sufiyan, 2022). To ensure food security and protect biodiversity, a governance system that can manage intra- and intersectoral complexity is required

 DOI: 10.1201/9781032697550-14

(Jiren *et al.*, 2020). By elucidating these complexities, we aim to identify pathways toward more sustainable and resilient food systems that nourish both people and the planet.

14.2 FOOD SECURITY AND ITS ELEMENTS

Food is defined as "anything that may be utilized as a food source, including minerals, vitamins, proteins, lipids, and carbs" (Sufiyan, 2022). The idea of food security appeared about 45 years ago, at a time when there was a widespread food crisis worldwide (Berry *et al.*, 2015). Food security is generally defined as "when all people at all times have access to sufficient, safe, nutritious food to maintain a healthy and active life" (Glamann *et al.*, 2017). "Everyone should always have physical, social, and financial access to enough wholesome food that satisfies their dietary requirements and tastes in order to have an active and healthy life" is the definition of food security. According to (world food journal, 2021) this definition exemplifies food security as a concept comprising numerous aspects, including food supply, access, use, and stability. Understanding food security encompasses the notion that people must not only be able to purchase and get food, but also be free from societal impediments to self-sufficiency (Narayanan, 2015).

Food security consists of three elements (Figure 14.1): availability, quality, and stability. Food availability, which is defined as the actual presence of a sufficient amount of food in a suitable quantity, is influenced by domestic food production, domestic food inventories, imports, and/or food aid. For instance, in some circumstances, availability may become a serious issue when its production, import, or export is impacted. Stability may be a major problem in other circumstances, such as when the food supply networks of particular products are disrupted (Béné *et al.*, 2020). One way to ensure that people have access to food is by having enough money or resources to buy or acquire the right foods for a balanced diet. Though lack of access to food is frequently a bigger issue than availability, especially for the most malnourished, an adequate food supply does not, by

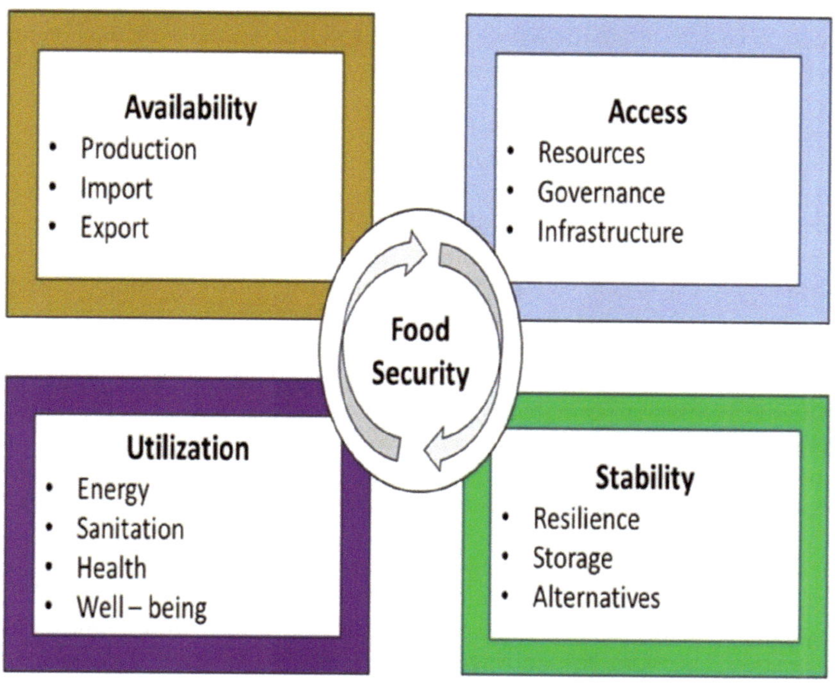

FIGURE 14.1 Various elements of food security.

itself, provide food security at the family or person level (World Bank Citation, 2007). When we discuss food utilization, we're talking about how the body makes use of the many nutrients included in food. It also calls for suitable food distribution within the home, good eating habits, appropriate food preparation and cleanliness standards, and a varied diet that must include all necessary nutrients. Food utilization use may have a big impact on people's nutritional status by guaranteeing adequate calories nutrient consumption, and safe food. Finally, food stability aims to maintain food availability, access, and use over time (Hwalla *et al.*, 2016).

14.3 RELATIONSHIP OF FOOD SECURITY AND BIODIVERSITY

Mestanza-Ramón *et al.* (2020) asserted that biodiversity is vital to maintaining global stability since its ecosystem services provide the needs for human life and well-being. As the number of people and their per capita consumption grows, so does the loss of the natural environment and the advantages it offers (Tilman *et al.*, 2017; Brondizio *et al.*, 2019). In light of these increasing human demands, one of the main problems of the 21st century is the protection of biodiversity (Williams *et al.*, 2020). Conserving biodiversity is important to maintain ecosystem benefits in addition to the diversity of species, habitats, and interactions that exist within them. The benefits of biodiversity and the natural world to humans are under threat, and further conservation efforts are required to halt the loss (Jung *et al.*, 2021).

The relationship between food security and biodiversity is deeply intertwined and multifaceted (Sufiyan, 2022). Biodiversity plays a crucial role in ensuring food security through various mechanisms. For instance, diverse ecosystems provide essential services such as pollination, soil fertility maintenance, pest control, and water purification, which are fundamental for agricultural productivity and food production (Norris, 2008). Moreover, genetic diversity within species, including crops and livestock, enhances resilience to pests, diseases, and environmental stressors, thereby safeguarding food systems against potential disruptions (Begna & Begna, 2021). Conversely, food production activities can impact biodiversity in several ways, including habitat destruction, pollution, and overexploitation of natural resources (Benton *et al.*, 2021). Plants and animals in agriculture can adapt to new pests and illnesses, shifting ecosystems, and shifting temperatures thanks to genetic variety (Sufiyan, 2022). The availability of genetic variety within species will be crucial for the development of minor crops and their greater usage to promote resilience (Kahane *et al.*, 2013). Within a particular species, the capacity to adapt to shifting surroundings rises along with genetic variety. Animals with high genetic variety can adapt to the shifting climate and the emergence of new pests and illnesses. The availability and accessibility of numerous DNA components ensure the sustainability of the worldwide network for food production since agricultural plant development programs relate to several scientific disciplines (Begna *et al.*, 2021). According to Dawson *et al.* (2019), a food supply that is both nutrient-dense and environmentally sustainable depends on biodiversity. The intricacy of the food manufacturing and consumption system makes it challenging to assess how food supply chains affect biodiversity (Crenna *et al.*, 2019). Preservation of food in biodiversity and security are intertwined. This is particularly clear when considering agricultural output, which is both a significant factor in the loss of biodiversity and a prerequisite for ensuring that there is an adequate supply of food at any given scale (Wittman *et al.*, 2017). Crops and horticulture are essential for food security since they are a source of food, cattle fodder, carriers of genes for desirable characteristics or qualities, and a foundation for agricultural advancements. Additionally, the diversity of terrestrial and marine creatures has a major role in promoting food security by acting as a source of food, a substitute for food, a way to enhance breeds, and a source of genes for desired characteristics (Sufiyan, 2022). Degradation of ecosystems and loss of biodiversity can compromise the long-term sustainability of food production systems, jeopardizing food security for present and future generations. Therefore, recognizing and managing the intricate relationship between food security and biodiversity is essential for achieving sustainable and resilient food systems that can effectively meet the nutritional needs of a growing global population while safeguarding the health of ecosystems and biodiversity.

14.4 AGRICULTURE AND FOOD ARE DEPENDENT ON BIODIVERSITY

Preserving the environment and making sure that wholesome food is available to everyone is essential, especially for people who live in rural regions and depend largely on agriculture. Yagoub and Abdelaleem (2024) assert that traditional agriculture is insufficient to provide sustainable food security. Food security, sustainable development, and the bulk of the food supply are all guaranteed by the agriculture industry. According to Kazemi Garajeh *et al.* (2023), there is an increasing demand for timely agricultural information due to recent changes in climate, as well as trends in environmental degradation and human population development. In order to increase food availability and achieve food security, the agricultural sector is crucial. Though there is considerable consensus regarding the predicted rise in global food consumption in the upcoming decades, there is ambiguity regarding the ability of global agriculture to meet this need by increasing the amount of food available (Pawlak *et al.*, 2020). Determining ways to minimize trade-offs between agriculture and biodiversity is a significant scientific issue, as agriculture is the main contributor to the current biodiversity crisis (Kehoe *et al.*, 2015) (Macchi *et al.*, 2020). According to Chivian and Bernstein (2022), genetic variety in crops can result in higher yields by preventing plant diseases from spreading and pests from attacking them. Moreover, it reduces the chance of crop failure brought on by variations in the weather. Even though food production now occupies the bulk of land accessible for agriculture, one in nine people globally is still regarded as food insecure (Molotoks *et al.*, 2017). This is despite productivity advances. Nine billion people's food and nutritional needs will need to be met by agriculture by the year 2050. Furthermore, it has to use ecologically sustainable manufacturing methods in order to keep that level of output indefinitely (Kahane *et al.*, 2013). Reduced harm to biodiversity caused by the food chain is a major issue. A mechanism is needed to pay farmers for more comprehensive land resource management since agricultural land is utilized for a variety of activities other than food production (Gupta *et al.*, 2024).

14.5 CONTRIBUTION OF TERRESTRIAL AND MARINE ANIMAL BIODIVERSITY TO FOOD SECURITY

Throughout human history, river ecosystems have offered several benefits connected to food, such as irrigation, means of subsistence, and food transportation. As nutrient highways that move water and sediment, store carbon in the atmosphere, and connect and link vast biodiversity through aquatic life, rivers might be thought of as natural gas pipelines over the surface of the planet (Manohar *et al.*, 2023). Moreover, besides being a vital source of food, the seas are home to some 250,000 recognized species. The seas are essential in absorbing 31% of the carbon dioxide emissions discharged into the atmosphere, which helps humans breathe and maintain climate regulation. As a result of increased human activity, the seas are threatened by overfishing, marine pollution, habitat loss, and the effects of climate change, which include the increase in sea level, ocean acidity, heat, and deoxygenation (Davenport, 2024). Aside from its significant nutritional and health benefits, people seek out and appreciate food from aquatic habitats for culinary and cultural reasons. For society, sustaining the prolonged output and delivery of such food from aquaculture and fisheries using wild capture is a major and continuous issue (Jennings *et al.*, 2016). The variety of animals provides food security and high-quality nourishment. In the majority of developing nations, livestock constitutes over 40% of total agricultural production, making it a crucial element of food security (Sufiyan, 2022). The production of food, crops, and animal feed must increase to meet the projected rise in global food consumption to feed an expanding and affluent population (Kok *et al.*, 2018).

14.6 FOOD SECURITY AND THE BIODIVERSITY OF FORESTS

Many people depend on trees for a reliable source of food and other key elements (Chamberlain *et al.*, 2020). Hunted and harvested foods from forests provide humans energy; protein from plants and animals; vitamins A, C, B, D, and E; and minerals like iron and iodine. Additionally, bush meat from woods, which includes fish, birds, animals, snails, and insects, is a reliable source of sustenance (Sufiyan, 2022). The response of the world's agricultural land to the projected rise in agricultural product demand is uncertain (Bahar *et al.*, 2020). A decrease in the number of undernourished individuals in developing nations' tropical forest regions may result from including wild foods as a source of nutrients in policies, plans, programs, and initiatives aimed at promoting food security. To increase the likelihood that people living in these areas will successfully incorporate wild foods into their diets, Asprilla-Perea *et al.* (2019) suggest that more knowledge of the importance of wild foods and their role in household food security is required. According to Asprilla-Perea *et al.* (2019), there is a need to raise awareness of their significance and role in ensuring family food security.

14.7 FOOD SECURITY, BIODIVERSITY AND HUMAN HEALTH

The complex problem of food security necessitates attention to both global and human health (Cole *et al.*, 2018). When a person does not have enough money to guarantee consistent access to adequate food to fulfill their dietary, nutritional, and social needs, they are considered to be food insecure. Adequate nutrition is fundamental for proper physical and cognitive development, immune function, and disease prevention. Conversely, food insecurity, characterized by limited access to nutritious food, can lead to malnutrition, undernutrition, and micronutrient deficiencies, which increase the risk of various health problems, including stunted growth, weakened immune systems, and impaired cognitive function (Walls *et al.*, 2019). Furthermore, vulnerable groups including children, expectant mothers, and the elderly are disproportionately affected by food poverty, which can worsen existing health inequities. Compared to the overall adult population, adults with mental illness, especially severe, are prone to experiencing food insecurity (Smith *et al.*, 2023). According to research, the size of the human population and the rate at which it is expanding are major factors in the decline of biological variety (Crist *et al.*, 2017). Ecosystems, biodiversity, and the vital services they provide are the foundations of all life on Earth, including human existence. According to Romanelli *et al.* (2015), they are sources of fuel, energy, food and vital nutrients, medications and medicinal substances, livelihoods, and enrichment in terms of culture and spirituality.

14.8 AGRICULTURE AND FOOD SECURITY AS THEY RELATE TO CLIMATE CHANGE

Growing worry about how climate change (Figure 14.2) may affect food security has been brought on by the trend of global warming (Lee *et al.*, 2024). Climate change has the potential to directly worsen food poverty because of the effects of warmer and drier weather on crop output, which are anticipated in significant agricultural regions (Hasegawa *et al.*, 2018). Changes in the food system brought on by climate change may have an impact on human health that is not direct by reducing food security, a significant element in influencing health. The physical implications for the food chain of climate change may also have an immediate influence on human health; these effects are mostly connected to nutrition and food-borne infections (Schnitter *et al.*, 2019). All aspects of food safety—availability, access, utilization, and stability—will be impacted by climate change, which will also have an effect on the entire food system (Campbell *et al.*, 2016).

Based on the findings, food security is greatly impacted by climate change, which also increases the risks associated with uncertainty. Lee *et al.* (2024) suggest that agricultural policies have the

TABLE 14.1
Uses of Different Crops for Health Benefits

Plant	Edible Parts	Health Benefits	Reference
***Phaseolusvulgaris* L.**	Aeeds and pods	Source of carbs, protein, and fiber in food	Uebersax *et al.*, 2023
Triticum aestivum	Seeds	Excellent supply of fiber, vital minerals, and vitamins	Gupta *et al.*, 2021
Brassica nigra	Leaves, seeds, and stems	Antitumor, antibacterial, antifungal, antihelminthic properties, and has the potential to prevent hepatic and renal damage; also useful in the management of diabetes	Turner *et al.*, 2020
Cicer arietinum L.	Seeds	Supply of fiber, vitamins, and minerals; helps control weight; enhances digestion; lowers risk of illness	Shah *et al.*, 2017
Helianthusannuus	Seed oil	Source of polyunsaturated fat in the diet, reducing high cholesterol and heart disease, eczema, dry mouth, dry skin, and many other issues	Adeleke and Babalola, 2020
Sorghum bicolor (L.)	Seed	Rich in minerals and vitamins, including iron, zinc, phosphorus, magnesium, potassium, and B vitamins	Birhanu, 2021
Oryza sativa	Seed	A good supply of fiber, magnesium, manganese, selenium, and B vitamins, and a rich source of carbs	Wang *et al.*, 2023
Glycine max	Seeds	An excellent source of omega-3 fatty acids, rich in phytoestrogens, low in saturated fat, devoid of cholesterol and lactose, and high in fiber	Qin *et al.*, 2022
Zea mays	Seeds	Assists in digestion, helps regulate body weight, and reduces the incidence of chronic conditions, such as type 2 diabetes and heart disease	Colombo *et al.*, 2021
Vigna unguiculata	Leaves, dry seeds, green pods	Abundant in fiber, plant-based proteins, iron, selenium, zinc, magnesium, phosphorus, and copper, as well as vitamins A, C, thiamine, riboflavin, and B6	da Silva *et al.*, 2021
Saccharum officinarum L.	Stem	An excellent source of antioxidants and vitamin C, strengthens the immune system; flavonoids are present and aid in the body's defense against malignant cells, particularly breast and prostate cancer; sugarcane juice is a fantastic remedy for liver and intestinal issues	Asif *et al.*, 2019
Solanum tuberosum	Tubers	Potatoes prevent heart disease because they are low in cholesterol and high in fiber, potassium, vitamin C, and vitamin B6	Burgos *et al.*, 2020
Cocos nucifera (L.)	Endosperm	Large amounts of healthy (HDL) cholesterol and has a high fat content (medium-chain triglycerides), which provide obese persons with rapid energy and encourages weight reduction	Hewlings *et al.*, 2020
Pennisetum glaucum	Seed	Abundant in a variety of nutrients and non-nutrients, high in calories, low in starch, and high in fiber	Gupta *et al.*, 2022
Anacardium occidentale	Cotyledons	Lowers cholesterol, prevention or control of diabetes, heart disease, preventing strokes	Shahrajabian *et al.*, 2023

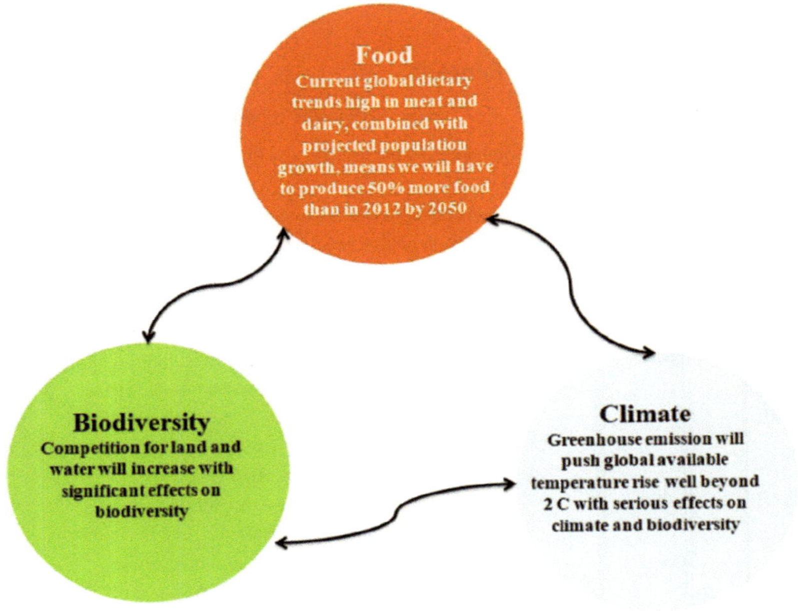

FIGURE 14.2 Relation of food security and biodiversity with climate.

potential to alleviate this impact by lowering production costs and enhancing farmers' incentives and vulnerability to risk.

14.9 HOME GARDENS OFFER ADDITIONAL ADVANTAGES AND ENHANCE FOOD SECURITY

According to research conducted in a variety of contexts, home gardens are positively correlated with increased nutritional diversity and food security (Rammohan *et al.*, 2019). Home gardens provide potential advantages for both the mind and body (Chalmin-Pui *et al.*, 2021). Agriculture, which feeds people and tamed animals, depends on a diversity of plants. There is no guarantee of food security or plant biodiversity protection due to the expanding human population, changing dietary preferences, an impending biodiversity extinction disaster, and continuing climate change. Encompassing the largest network of plant teaching institutions worldwide, botanic gardens also serve as the world's foremost conservators of plant biodiversity. Many of these gardens are actively exploring crop diversity and developing novel approaches to promote sustainable agricultural systems, having done so for a considerable amount of time (Novy *et al.*, 2024). A significant auxiliary function that home gardens play in addressing the problems facing agricultural output is noteworthy. It is common knowledge that home gardens serve a variety of purposes. In addition to meeting requirements on a social and cultural level and as a component of customary wisdom and methods, they also provide various ecosystem services, aid in reducing the effects of climate change, and support economic needs and sustainable livelihoods. They also help provide food, a healthy diet, medicine, and other useful products (Korpelainen, 2023).

14.10 WORLDWIDE CHALLENGES OF FOOD SECURITY AND BIODIVERSITY CONSERVATION

The global epidemiological state reflects the problems associated with environmental deterioration and the disparity between the natural and technological domains (Fedotova *et al.*, 2021). As per the most recent State of Food Security and Nutrition in the World report (Boliko, 2019), there are over 820 million individuals who experience daily hunger, and over the last three years, this figure has been gradually rising (Banik, 2019). One of the major issues facing our civilization is feeding the globe in a sustainable manner. The demand for food grew between 1961 and 2000 due to an exponential growth in population (Cole *et al.*, 2018). Urbanization, an increase in per capita income, and a growth in the human population are the main causes of the rising demand for food. Particularly with rising wealth, there are dietary shifts and changes in the kinds of food consumed (Grafton *et al.*, 2015). For food and farming, global biodiversity is decreasing. It is alarming for all of mankind that biodiversity is vanishing at this rate. According to the World Food Journal (2021), humanity's basic requirements for food are met by biodiversity and the ecological services it provides. Reduced biodiversity makes plants and animals more vulnerable to diseases and pests. Inefficient production has long been encouraged by agricultural policy, which has resulted in overproduction in certain areas and underproduction in others. There are significant inefficiencies in the global market due to overproduction, which is subsidized for sales in poor nations and is the result of artificially high price incentives. Agricultural subsidies and tariffs have been restrained by the WTO Agreement on Agriculture, although the impacts have only been temporary (Grafton *et al.*, 2015). In addition to causing disruption in millions of households and political instability, the surge in global food prices in 2007–2008 also changed governments' complacencies about the unacceptably high rates of chronic hunger (Leser, 2013). Rising food costs and other shocks like droughts, floods, and economic crises force the most vulnerable households even deeper into poverty and reduce their access to enough food. These factors might have a big influence on food and nutrition security. These challenges may push the impoverished to give up other necessities or sell off possessions, which can lead to a vicious cycle of poverty that is more challenging to break (Gustafson, 2016; Chadha, 2016).

14.11 SDG 2: ZERO HUNGER—A MEANS TO ACHIEVE FOOD SECURITY

SDG 2, or Sustainable Development Goal 2, focuses on zero hunger. This goal aims to end hunger, achieve food security and improved nutrition, and promote sustainable agriculture by 2030. It recognizes the fundamental importance of access to nutritious food for all individuals, as well as the critical role of sustainable agricultural practices in achieving this objective (SDSR, 2023). As of 2022, the global landscape revealed a stark reality: approximately 735 million individuals, constituting 9.2% of the world's population, grappled with chronic hunger, marking a significant increase from 2019. This concerning statistic highlights the escalating severity of the crisis at hand. Moreover, an estimated 2.4 billion people faced moderate to severe food insecurity in 2022, indicating a profound lack of access to adequate nourishment. This figure surged by a staggering 391 million individuals compared to 2019, underscoring the urgency of the situation (UN SDG, 2023). The need for a system-based approach to building enduring food networks has been emphasized. These requirements include maximizing the application of natural resources, minimizing the impact on the environment, looking into new food sources, boosting consumer understanding, and creating lucrative market-led solutions for the security of food and nutrition (Knorr and Augustin, 2024). The declared objectives of SDG 2 are to "eradicate hunger, provide food security, enhance nutrition, and advance agricultural sustainability." Due to its close ties to the environment, society, and economy, SDG 2 is crucial to the success of the whole SDG schedule (FAO, 2019). In addition to offering protection against insect outbreaks, plant diseases, market swings, weather extremes, and other hazards, biodiversity also plays a protective role for total agricultural productivity, which in turn helps ensure food security. Many advantages for sustainable development might result from pursuing synergies between these two pressing issues of our day (Obrecht *et al.*, 2021). Production and consumption of food are vital to all economies, even if they are more prevalent in underdeveloped nations due to agricultural productivity. To demonstrate how indivisible the SDG agenda is, it is anticipated that fulfilling SDG 2 would lead to a variety of trade-offs and synergies with other SDGs at various temporal and geographical dimensions (Gil *et al.*, 2019). Policymakers need to harmonize how worldwide targets connect to both local and national objectives for the SDGs to have an impact on the ground (Banik, 2019). They also need to get a better knowledge of the global norm of goal setting.

14.12 INITIATIVES TAKEN FOR FOOD SECURITY IN INDIA

PalmeriaIn order to ensure food and nutrition security, public policies must be implemented, and overcoming food insecurity mostly depends on the political climate (de Araújo Palmeira *et al.*, 2020). Establishing food security involves several domains in which universities are essential. Improving agricultural methods, developing cutting-edge agricultural technology, raising food product quality, and expanding agricultural output and productivity are some of these sectors (Abass *et al.*, 2024). Food system methods are gaining popularity as a way to bolster the sustainability and results of food systems by addressing the complex interactions between components and managing competing agendas (Tendall *et al.*, 2015).

14.13 SCHEMES

For food security issues to be effectively resolved, conservation, sustainable management, and restoration are required (Gupta *et al.*, 2024). To improve the nation's food security, India has introduced a number of efforts, with a focus on initiatives (Chadha, 2016). Here are these schemes that have been implemented by the government:

1. **Public distribution system**

The Public Distribution System (PDS), an Indian food security system, was developed as a strategy for providing food grains at fair costs and controlling shortages. It was created by the Indian government's Ministry of Consumer Affairs, and both the federal and state governments share the management of it (Gupta and Singh, 2016). The main goal of the PDS is to employ regulatory mechanisms to guarantee that families receive a sufficient and fair distribution of necessities at prices that are socially acceptable. This will maintain price stability in addition to helping to achieve some degree of self-sufficiency in the production and acquisition of food (Chadha, 2016).

2. **National Food Security Mission**

The National Food Security Mission (NFSM), established by the Indian government in 2007–2008 as a component of the 11th five-year plan, aims to address food insecurity by increasing food production (Varma, 2022). By the conclusion of the 12th five-year plan, the mission was still underway but with new goals: producing 25 million more tons of food grains by the end of the plan, which included rice, wheat, pulses, and coarse cereals at corresponding rates of 10, 8, 4, and 3 million tons. After that, the government decided to keep the NFSM program going after the completion of the 12th five-year plan. The new goals were to produce an additional 13 million tons of food grains by 2019–2020, with two million tons of coarse cereals, three million tons of wheat, five million tons of rice, and three million tons of pulses (Patel and Parmar, 2020).

3. **National Horticultural Mission**

The Indian government established the National Horticulture Mission (NHM) program to promote the whole development of this industry, seeing the enormous potential of horticulture in boosting the country's agricultural output. The Agriculture Department and Cooperation, the Department of Agriculture, Government of India commenced the centrally supported program in 2005–2006 (10th five-year plan) (Yadav *et al.*, 2021).

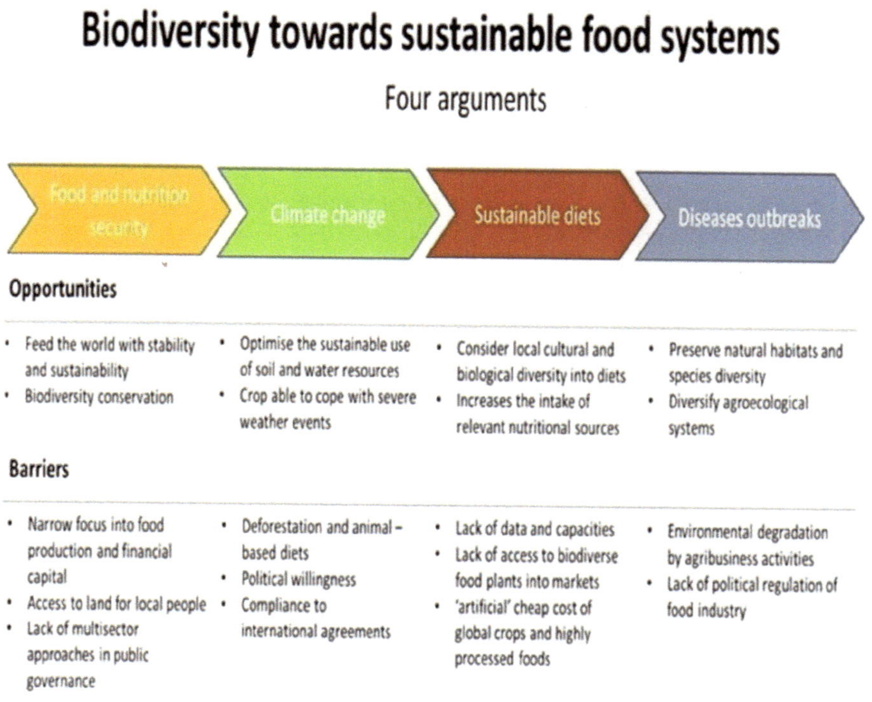

FIGURE 14.3 Biodiversity towards sustainable food systems.

4. **National Project on Management of Soil Health and Fertility**

The National Project on Management of Soil Health and Fertility was established by the Government of India with the goal of enhancing soil fertility and productivity through improved soil nutrients. In addition, strengthening soil testing labs and promoting the use of integrated nutrient management and fertilizer quality control labs are further goals of the project. These institutional frameworks have all been put in place to support the prudent application of chemical fertilizers and farmyard manure, based on soil testing and empirical data (Chadha, 2016).

5. **Rashtriya Krishi Vikas Yojana**

The Rashtriya Krishi Vikas Yojana was introduced by the Indian government, one of the largest pan-Indian agriculture development programs, which calls for coordinated efforts by several agricultural line departments. The major goals of this program are to: provide states with adaptability and independence in developing and carrying out programs for the agricultural and allied sectors; encourage states to boost government spending on agriculture and associated industries; guarantee that, based on agroclimatic conditions, plans are created for states and districts, and provide access to technology and natural resources; ensure regional requirements, crops, and priorities are more accurately represented; and achieve the goal of reducing yield gaps in important crops through focused actions to increase farmers' profits (Vijayan *et al.*, 2023).

6. **Securing food in an era of sustainable development**

Producing high-yield, high-quality food is a difficulty that needs to be overcome in order to feed the billions of people that inhabit this world. By 2050, it is predicted that there will be a 70% increase in global food production, placing more strain on labor, water, land, nutrients, and energy. About 820 million people suffer from hunger and malnutrition worldwide; the bulk of them are from developing nations (Wang, 2022). According to Conceição *et al.* (2016), improving access and rights to food is necessary for fostering fundamental human capacities and is thus a prerequisite for sustained human development. To fulfill the SDGs, particularly SDG 2, which has to do with putting an end to poverty, enhancing nutrition and food security, and supporting sustainable agriculture, it is imperative that strong policies be implemented that optimize the advantages of globalization while lowering the hazards (Banik, 2019) (Figure 14.3).

14.14 CONCLUSION

The current rate of global population growth means that more food has to be produced. Biological diversity, or agricultural biodiversity, is an element of the biodiversity that comprises various populations of plants, animals, and microorganisms as well as forests and crops. This diversity is essential to ensuring global food security. Because they provide a source of food, cattle fodder, carriers of genes for desired traits or qualities, and a starting point for agricultural enhancements, crops and horticultural biodiversity are crucial to food security. Furthermore, because animals may be employed as food sources, food substitutes, breed enhancers, and suppliers of desirable features' genes, both terrestrial and marine animal biodiversity are essential for increasing food security. Additionally, animal diversification indirectly supports food security since animal dung is utilized to increase soil fertility and crop yield by trading the skins for food. Biological variety must be gathered, preserved, and used sustainably without compromising future food security. This is true both today and in the future.

REFERENCES

Abass AS, Abdalla IF, Ahmed AE. Contribution of higher education institutions to food security: the case of king Faisal University. In *Food and Nutrition Security in the Kingdom of Saudi Arabia*, Vol. 1: National Analysis of Agricultural and Food Security, 2024 Jan 1 (pp. 425–61). Cham: Springer International Publishing.

Adeleke BS, Babalola OO. Oilseed crop sunflower (*Helianthus annuus*) as a source of food: Nutritional and health benefits. *Food Science & Nutrition*. 2020 Sep;8(9):4666–84.

Asprilla-Perea J, Díaz-Puente JM. Importance of wild foods to household food security in tropical forest areas. *Food Security*. 2019 Feb 15;11(1):15–22.

Bahar NH, Lo M, Sanjaya M, Van Vianen J, Alexander P, Ickowitz A, Sunderland T. Meeting the food security challenge for nine billion people in 2050: What impact on forests. *Glob. Environ. Chang.* 2020 May 1;62:102056.

Banik D. Achieving food security in a sustainable development era. *Food Ethics*. 2019 Dec;4:117–21.

Bank W. From agriculture to nutrition: pathways, synergies, and outcomes. *Agricultural and Rural Development Notes*. 2007;45363.

Begna T. Role and economic importance of crop genetic diversity in food security. *International Journal of Agricultural Science and Food Technology*. 2021 Apr 17;7(1):164–9.

Begna T, Begna T. Role and economic importance of crop genetic diversity in food security. *International Journal of Agricultural Science and Food Technology*. 2021;7(1):164–9.

Béné C, Oosterveer P, Lamotte L, Brouwer ID, de Haan S, Prager SD, Talsma EF, Khoury CK. When food systems meet sustainability–Current narratives and implications for actions. *World Development*. 2019 Jan 1;113:116–30.

Benton TG, Bieg C, Harwatt H, Pudasaini R, Wellesley L. Food system impacts on biodiversity loss. *Three Levers for Food System Transformation in Support of Nature*. 2021;2–3.

Berry EM, Dernini S, Burlingame B, Meybeck A, Conforti P. Food security and sustainability: can one exist without the other?. *Public Health Nutrition*. 2015 Sep;18(13):2293–302.

Birhanu S. Potential benefits of sorghum [Sorghum bicolor (L.) Moench] on human health: A review. *International Journal of Food Engineering and Technology*. 2021;5(1):8–18.

Boliko MC. FAO and the situation of food security and nutrition in the world. *Journal of nutritional science and vitaminology.* 2019 Oct 11;65(Supplement):S4-8.

Brondízio ES, Settele J, Diaz S, Ngo HT. Global assessment report on biodiversity and ecosystem services of the Intergovernmental Science-*Policy Platform on Biodiversity and Ecosystem Services.* 2019.

Burgos G, Zum Felde T, Andre C, Kubow S. The potato and its contribution to the human diet and health. *The Potato Crop: Its Agricultural, Nutritional and Social Contribution to Humankind.* 2020:37–74.

Campbell BM, Vermeulen SJ, Aggarwal PK, Corner-Dolloff C, Girvetz E, Loboguerrero AM, Ramirez-Villegas J, Rosenstock T, Sebastian L, Thornton PK, Wollenberg E. Reducing risks to food security from climate change. *Global Food Security.* 2016 Dec 1;11:34–43.

Chadha NE. Food security in India: Issues and challenges. *International Journal of Humanities, Arts, Medicine and Sciences.* 2016;4(1):79–86.

Chalmin-Pui LS, Griffiths A, Roe J, Heaton T, Cameron R. Why garden?–Attitudes and the perceived health benefits of home gardening. *Cities.* 2021 May 1;112:103118.

Chamberlain JL, Darr D, Meinhold K. Rediscovering the contributions of forests and trees to transition global food systems. *Forests.* 2020 Oct 16;11(10):1098.

Chappell MJ, LaValle LA . Food security and biodiversity: can we have both? An agroecological analysis. *Agriculture and Human Values.* 2011;28:3–26.

Chivian E, Bernstein A. How our health depends on biodiversity. *Program for Preserving the Natural World, Incorporated*; 2022.

Cole MB, Augustin MA, Roberstson MJ, Manners JM. The science of food security. *NPJ science of Food.* 2018;2(14):1–8

Colombo R, Ferron L, Papetti A. Colored corn: An up-date on metabolites extraction, health implication, and potential use. *Molecules.* 2021 Jan 2;26(1):199

Conceição P, Levine S, Lipton M, Warren-Rodríguez A. Toward a food secure future: Ensuring food security for sustainable human development in Sub-Saharan Africa. *Food Policy.* 2016 Apr 1;60:1–9.

Crenna E, Sinkko T, Sala S. Biodiversity impacts due to food consumption in Europe. *Journal of Cleaner Production.* 2019 Aug 1;227:378–91.

Crist E, Mora C, Engelman R. The interaction of human population, food production, and biodiversity protection. *Science.* 2017 Apr 21;356(6335):260–4.

da Silva AC, de Freitas Barbosa M, da Silva PB, de Oliveira JP, da Silva TL, Junior DL, de Moura Rocha M. Health benefits and industrial applications of functional cowpea seed proteins. *Grain and Seed Proteins Functionality.* 2021 Mar 16.

Davenport T. Singapore and the Protection of the Marine Environment. In *Peace with Nature: 50 Inspiring Essays on Nature and the Environment,* 2024 (pp. 63–68).

Dawson IK, Park SE, Attwood SJ, Jamnadass R, Powell W, Sunderland T, Carsan S. Contributions of biodiversity to the sustainable intensification of food production. *Global Food Security.* 2019 Jun 1;21:23–37.

de Araújo Palmeira P, Araujo de Mattos R, Salles-Costa R. Food security governance promoted by national government at the local level: a case study in Brazil. *Food Security.* 2020 Jun;12:591–606.

Ewert F, Baatz R, Finger R. Agroecology for a sustainable agriculture and food system: from local solutions to large-scale adoption. *Annual Review of Resource Economics.* 2023 Oct 5;15(1):351–81.

Fedotova GV, Sotnikova LF, Orlova ER, Baranova AF, Goncharova AV. Global problems of biodiversity and food security. In *IOP Conference Series: Earth and Environmental Science*, 2021 Mar 1 (Vol. 677, No. 3, p. 032010). IOP Publishing.

Food and Agriculture Organization of the United Nations (FAO), International Fund for Agricultural Development (IFAD), United Nations Children's Fund (UNICEF), World Food Programme (WFP), & World Health Organization (WHO). The State of Food Security and Nutrition in the World 2019: Safeguarding Against Economic Slowdowns and Downturns. 2019 FAO, Rome.

Gil JD, Reidsma P, Giller K, Todman L, Whitmore A, van Ittersum M. Sustainable development goal 2: improved targets and indicators for agriculture and food security. *Ambio.* 2019 Jul;48(7):685–98.

Glamann J, Hanspach J, Abson DJ, Collier N, Fischer J. The intersection of food security and biodiversity conservation: a review. *Regional Environmental Change.* 2017 Jun;17:1303–13.

Grafton RQ, Daugbjerg C, Qureshi ME. Towards food security by 2050. *Food Security.* 2015 Apr;7:179–83

Gupta P, Singh MU. Efficiency of public distribution system insights and challenges. *Delhi Bussiness Review.* 2016;17(1); 79–95.

Gupta R, Meghwal M, Prabhakar PK. Bioactive compounds of pigmented wheat (Triticum aestivum): Potential benefits in human health. *Trends in Food Science & Technology.* 2021 Apr 1;110:240–52.

Gupta SS, Misra S, Ghosh A. Biodiversity: goal and driver of agricultural sustainability. In*Biodiversity and Bioeconomy* 2024 Jan 1 (pp. 143–64). Elsevier.

Gupta V, Singh AP, Gupta N. Importance of pearl millet and its health benefits. *Just Agriculture e-Magazine.* 2022;2:2582–8223.

Gustafson D, Gutman A, Leet W, Drewnowski A, Fanzo J, Ingram J. Seven food system metrics of sustainable nutrition security. *Sustainability.* 2016 Feb 23;8(3):196.

Hasegawa T, Fujimori S, Havlík P, Valin H, Bodirsky BL, Doelman JC, Fellmann T, Kyle P, Koopman JF, Lotze-Campen H, Mason-D'Croz D. Risk of increased food insecurity under stringent global climate change mitigation policy. *Nature Climate Change.* 2018 Aug;8(8):699-703.

Hewlings S. Coconuts and health: Different chain lengths of saturated fats require different consideration. *Journal of Cardiovascular Development and Disease.* 2020 Dec 17;7(4):59.

Hwalla N, El Labban S, Bahn RA. Nutrition security is an integral component of food security. *Frontiers in Life Science.* 2016 Jul 2;9(3):167–72.

IPBES Summary for policymakers of the global assesment report on biodiversity and ecosystem services of the intergovernmental science - Policy Platform on Biodiversity and ecosystem services. IPBES secretairat, Bonn, Germany 2019, 1–56. https://www.ibpes.net/global-assesment

Jennings S, Stentiford GD, Leocadio AM, Jeffery KR, Metcalfe JD, Katsiadaki I, Auchterlonie NA, Mangi SC, Pinnegar JK, Ellis T, Peeler EJ. Aquatic food security: insights into challenges and solutions from an analysis of interactions between fisheries, aquaculture, food safety, human health, fish and human welfare, economy and environment. *Fish and Fisheries.* 2016 Dec;17(4):893–938.

Jiren TS, Dorresteijn I, Hanspach J, Schultner J, Bergsten A, Manlosa A, Jager N, Senbeta F, Fischer J. Alternative discourses around the governance of food security: a case study from Ethiopia. *Global Food Security.* 2020 Mar 1;24:100338.

Jung M, Arnell A, De Lamo X, García-Rangel S, Lewis M, Mark J, Merow C, Miles L, Ondo I, Pironon S, Ravilious C. Areas of global importance for conserving terrestrial biodiversity, carbon and water. *Nature Ecology & Evolution.* 2021 Nov;5(11):1499–509.

Kahane R, Hodgkin T, Jaenicke H, Hoogendoorn C, Hermann M, Keatinge JD, d'Arros Hughes J, Padulosi S, Looney N. Agrobiodiversity for food security, health and income. *Agronomy for sustainable development.* 2013 Oct;33:671–93.

Kazemi Garajeh M, Salmani B, Zare Naghadehi S, Valipoori Goodarzi H, Khasraei A. An integrated approach of remote sensing and geospatial analysis for modeling and predicting the impacts of climate change on food security. *Scientific Reports.* 2023 Jan 19;13(1):1057.

Kehoe L, Kuemmerle T, Meyer C, Levers C, Václavík T, Kreft H. Global patterns of agricultural land-use intensity andvertebrate diversity. *Diversity and Distributions.* 2015 Nov;21(11):1308–18.

Knorr D, Augustin MA. The future of foods. *Sustainable Food Technology.* 2024;2:253–265.

Kok MT, Alkemade R, Bakkenes M, van Eerdt M, Janse J, Mandryk M, Kram T, Lazarova T, Meijer J, van Oorschot M, Westhoek H. Pathways for agriculture and forestry to contribute to terrestrial biodiversity conservation: a global scenario-study. *Biological Conservation.* 2018 May 1;221:137–50.

Korpelainen H. The role of home gardens in promoting biodiversity and food security. *Plants.* 2023 Jun 28;12(13):2473.

Lee CC, Zeng M, Luo K. How does climate change affect food security? Evidence from China. *Environmental Impact Assessment Review.* 2024 Jan 1;104:107324.

Leser S. The 2013 FAO report on dietary protein quality evaluation in human nutrition: Recommendations and implications. *Nutrition Bulletin.* 2013 Dec;38(4):421–8.

Macchi L, Decarre J, Goijman AP, Mastrangelo M, Blendinger PG, Gavier-Pizarro GI, Murray F, Piquer-Rodriguez M, Semper-Pascual A, Kuemmerle T. Trade-offs between biodiversity and agriculture are moving targets in dynamic landscapes. *Journal of Applied Ecology.* 2020 Oct;57(10):2054–63.

Manohar S, Downs S, Shaikh S, Mak S, Sok S, Graham E, Miachon L, Fanzo J. Riverine food environments and food security: a case study of the Mekong River, Cambodia. *Bulletin of the World Health Organization.* 2023 Feb 2;101(2):140.

Mestanza-Ramón C, Henkanaththegedara SM, VásconezDuchicela P, Vargas Tierras Y, Sánchez Capa M, Constante Mejía D, Jimenez Gutierrez M, Charco Guamán M, Mestanza Ramón P. In-situ and ex-situ biodiversity conservation in ecuador: a review of policies, actions and challenges. *Diversity.* 2020 Aug 17;12(8):315.

Molotoks A, Kuhnert M, Dawson TP, Smith P. Global hotspots of conflict risk between food security and biodiversity conservation. *Land.* 2017 Oct 4;6(4):67.

Narayanan S. Food security in India: The imperative and its challenges. *Asia & the Pacific Policy Studies.* 2015 Jan;2(1):197–209.

Norris, K. (2008). Agriculture and biodiversity conservation: opportunity knocks. *Conservation Letters.* 2008;1(1):2–11.

Novy A, Hestness E, Khoury CK, Miller A, Moreau T, Warschefsky E. Botanic Gardens Contribute to Food Security through Education, Conservation, and Research. In *Botanical Gardens and Their Role in Plant Conservation,* 2024 (pp. 41–52). CRC Press.

Obrecht A, Pham M, Spehn E, Payne D, Brémond AC, Altermatt F, Fischer M, Passarello C, Moersberger H, Schelske O, Guntern J. Achieving the SDGs with biodiversity. Akademie der Naturwissenschaften Schweiz (SCNAT), Forum Biodiversität Schweiz, 2021.

Patel PK, Parmar BJ. Impact of National Food Security Mission (NFSM) on Production and Stability of Foodgrains and Pulses in India, 2020.

Pawlak K, Kołodziejczak M. The role of agriculture in ensuring food security in developing countries: Considerations in the context of the problem of sustainable food production. *Sustainability.* 2020 Jul 7;12(13):5488.

Pilling D, Bélanger J, Hoffmann I. Declining biodiversity for food and agriculture needs urgent global action. *Nature Food.* 2020 Mar;1(3):144–7.

Qin P, Wang T, Luo Y. A review on plant-based proteins from soybean: Health benefits and soy product development. *Journal of Agriculture and Food Research.* 2022 Mar 1;7:100265.

Rammohan A, Pritchard B, Dibley M. Home gardens as a predictor of enhanced dietary diversity and food security in rural Myanmar. *BMC Public Health.* 2019 Dec;19:1–3.

Romanelli C, Cooper D, Campbell-Lendrum D, Maiero M, Karesh WB, Hunter D, Golden CD. Connecting global priorities: biodiversity and human health: a state of knowledge review. World Health Organistion/ Secretariat of the UN Convention on Biological Diversity, 2015.

Scherer L, Svenning JC, Huang J, Seymour CL, Sandel B, Mueller N, Kummu M, Bekunda M, Bruelheide H, Hochman Z, Siebert S. Global priorities of environmental issues to combat food insecurity and biodiversity loss. *Science of the Total Environment.* 2020 Aug 15;730:139096.

Schnitter R, Verret M, Berry P, Chung Tiam Fook T, Hales S, Lal A, Edwards S. An assessment of climate change and health vulnerability and adaptation in Dominica. *International Journal of Environmental Research and Public Health.* 2019 Jan;16(1):70.\

Shah AA, Salgotra RK, Wani SA, Mondal SK. Breeding and genomics approaches to increase crop yield under drought stress in climate change scenario. *European Journal of Experimental Biology.* 2017;7(4):1–7.

Shahrajabian MH, Sun W. The important nutritional and wonderful health benefits of Cashew (*Anacardium occidentale* L.). *The Natural Products Journal.* 2023 Jul 1;13(4):2–10.

Smith J, Ker S, Archer D, Gilbody S, Peckham E, Hardman CA. Food insecurity and severe mental illness: understanding the hidden problem and how to ask about food access during routine healthcare. *BJPsych Advances.* 2023 May;29(3):204–12.

Sufiyan A. The role of biodiversity in food security 2022.

Sustainable Development Goals (2023). https://www.un.org/sustainabledevelopment/hunger.

Sustainable Development Solutions Report (SDSR). Pathways to Sustainable Food Systems and Food Security. United Nations Sustainable Development Solutions Network 2023.

Tendall DM, Joerin J, Kopainsky B, Edwards P, Shreck A, Le QB, Krütli P, Grant M, Six J. Food system resilience: Defining the concept. *Global Food Security.* 2015 Oct 1;6:17–23.

Tilman D, Clark M, Williams DR, Kimmel K, Polasky S, Packer C. Future threats to biodiversity and pathways to their prevention. *Nature.* 2017 Jun 1;546(7656):73–81.

Tscharntke T, Clough Y, Wanger TC, Jackson L, Motzke I, Perfecto I, ... Whitbread A. Global food security, biodiversity conservation and the future of agricultural intensification. *Biological Conservation.* 2012151(1):53–59.

Turner ER, Luo Y, Buchanan RL. Microgreen nutrition, food safety, and shelf life: a review. *Journal of Food Science.* 2020 Apr;85(4):870–82.

Uebersax MA, Cichy KA, Gomez FE, Porch TG, Heitholt J, Osorno JM, Kamfwa K, Snapp SS, Bales S. Dry beans (*Phaseolus vulgaris* L.) as a vital component of sustainable agriculture and food security—a review. *Legume Science.* 2023 Mar;5(1):e155.

UN IRB. Convention on biological diversity. *Treaty Collection.*

Varma P. National Food Security Mission and Pulses Production. In *Pulses for Food and Nutritional Security of India: Production, Markets and Trade,* 2022 Aug 21 (pp. 45–53). Singapore: Springer Nature Singapore.

Vijayan B, Nain MS, Singh R, Kumbhare NV, Kademani SB. Knowledge test for extension personnel on Rashtriya Krishi Vikas Yojana. *Indian Journal of Extension Education.* 2023;59(1):131–4.

Walls H, Baker P, Chirwa E, Hawkins B. Food security, food safety & healthy nutrition: are they compatible?. *Global Food Security.* 2019;21:69–71.

Wang N, Cui X, Duan Y, Yang S, Wang P, Saleh AS, Xiao Z. Potential health benefits and food applications of rice bran protein: Research advances and challenges. *Food Reviews International.* 2023 Aug 18;39(6):3578–601.

Wang X. Managing land carrying capacity: Key to achieving sustainable production systems for food security. *Land.* 2022 Mar 27;11(4):484.

Williams DR, Balmford A, Wilcove DS. The past and future role of conservation science in saving biodiversity. *Conservation Letters.* 2020 Jul;13(4):e12720.

Wittman H, Chappell MJ, Abson DJ, Kerr RB, Blesh J, Hanspach J, Perfecto I, Fischer J. A social–ecological perspective on harmonizing food security and biodiversity conservation. *Regional Environmental Change.* 2017 Jun;17:1291–301.

Yadav S, Sharma KC, Mishra P. Problems faced by the beneficiary farmers in availing the benefits under national horticulture mission. *Indian Res. J. Ext. Edu.* 2021;21(2&3):20–5.

Yagoub SO, Abdelaleem ET. Organic Agriculture and Food Security in Saudi Arabia. In *Food and Nutrition Security in the Kingdom of Saudi Arabia, Vol. 1: National Analysis of Agricultural and Food Security,* 2024 Jan 1 (pp. 249–262). Cham: Springer International Publishing.

15 Bioprospecting and Biodiscovery for Sustainable Development

Sonia[1], *Ritesh Banerjee*[1]*, *Swati Yadav*[2]
Suman Mishra[2], *Kamal Jaiswal*[2]
[1]School of Biological and Environmental Sciences, Shoolini University, Solan, Himachal Pradesh, India
[2]Department of Zoology, School of Life Sciences, Babasaheb Bhimrao Ambedkar University, Lucknow, Uttar Pradesh, India
*Corresponding author: Ritesh Banerjee

15.1 INTRODUCTION

Biodiversity refers to the entire range of life systems on our planet, from microorganisms to humans. There are three categories of biodiversity: habitat, species, and gene. All three levels of biodiversity are abundant in India (Myers *et al.*, 2020). India's diverse biodiversity is matched by its diverse cultural diversity and a one-of-a-kind wealth of traditional knowledge systems developed, preserved, freely shared, and cared for by society. India has a huge flora and fauna diversity. India is ranked tenth among plant-rich countries in the world and fourth among Asian countries (Lesser and Krattiger, 2017). Scheduled castes and scheduled tribes account for approximately 16.6% and 8.6% of India's population, respectively, according to the 2011 census. The country's constitution includes a list of 1108 schedules. There are 25 castes and 744 tribes spread across 22 states. The tribals live in forests and other harsh environments. They learned to use a wide variety of flora and fauna (Mateo, 2021). The vast majority of these are known or unknown to the general public. Traditional knowledge (TK) has the potential to benefit humanity in a variety of ways if it is subjected to scientific scrutiny. The inroads of modernization are currently posing a threat to this TK, and it is in danger of being lost; this age-old wisdom and expertise may be lost for all time (Pushpangadan, 2018).

The true keepers of the knowledge of medicinally significant plants and animals are rural people and traditional communities. The modern scientific community is unaware of the majority of the knowledge on medicinal plants and animals that traditional communities and rural people have amassed (Chadwick *et al.*, 1999). Overexploitation of biodiversity is posing a threat to its existence worldwide. As an interdisciplinary science, ethnobotany is uniquely positioned to preserve indigenous peoples' wealth of traditional knowledge about flora and fauna. This includes understanding how to use and preserve various plant resources over time without threatening or eliminating their natural habitat. As a result, every effort should be made to record and incorporate traditional wisdom, as well as the biodiversity that goes with it (Medaglia, 2004).

The term "bioprospecting" refers to the process of gathering, analyzing, and monetizing biodiversity in order to find novel therapeutics and other beneficial natural products (such as cosmetics, agrochemicals, fragrances, and functional foods). Despite being perceived as a modern practice, the discovery of natural products is part of a millennium-long search for medicines (Sittenfeld *et*

DOI: 10.1201/9781032697550-15

al., 2003). Global botanical gardens, which housed collections of priceless herbaria and information about their medicinal applications, were established in Europe thanks to imperial and colonial botany projects aimed at generating economic benefits. After learning that Indians in Quito, Peru, used the bark of the cinchona tree (*Cinchona officinalis* L.) as a medicinal mixture for shivering and cold spells, European scientists began using quinine as a malaria treatment in the early 1700s. Another significant discovery during this time period was the ipecacuanha root (*Cephaelis ipecacuanha*), which was used to treat dysentery caused by amoebas.

When looking for ways to treat injured soldiers during World War II, the US government prioritized drug discovery. For the commercial production of penicillin, government research quickly started working with several big pharmaceutical companies, including Abbott, Pfizer, Merck, and Squibb. Because of large-scale research collaboration, the US National Institutes of Health (NIH) was able to promote combined public/private funding for improved treatment using natural products (Kingston, 2011).

In the pursuit of sustainable development, bioprospecting and biodiscovery emerge as pivotal strategies, harnessing the untapped potential of Earth's biological diversity to address contemporary challenges. These practices involve the systematic exploration and investigation of diverse ecosystems to identify novel biological resources, be it compounds, genes, or organisms with applications in medicine, agriculture, industry, and environmental management. Biodiversity, with its myriad life forms and unique genetic codes, becomes a reservoir of innovation, offering solutions to global issues such as disease emergence, food security, and environmental degradation. The historical success of bioprospecting in drug discovery and healthcare, agricultural innovation, and industrial biotechnology underscores its significance in advancing sustainable practices. Moreover, ethical considerations and responsible bioprospecting practices, including fair benefit-sharing and conservation efforts, are integral to ensuring a harmonious integration of scientific exploration, technological advancement, and environmental stewardship in the pursuit of a more sustainable and equitable future.

15.2 BIOPROSPECTING

Bioprospecting is defined as a "systematic search for genes, natural compounds, designs, and whole organisms in wildlife with a potential for product development by biological observations and biophysical, biochemical, and genetic methods without disruption to nature." Thus, bioprospecting has three components: chemical prospecting, bionic prospecting, and gene prospecting (Cao *et al.*, 2019).

The development of high-throughput chemical screening and automated bioassay programs, such as screening for the identification, isolation, and characterization of novel bioactive compounds from wild bioresources (higher plants, bryophytes, pteridophytes, fungi, and microorganisms, etc.) and animals (insects and other wild invertebrates, etc.), has opened up new avenues in the study of natural products and drugs. Chemical prospecting of wild plant resources is becoming increasingly important in agrochemistry (biopesticides), medications and pharmaceuticals, cosmetics, proteins, enzymes, food additives, and other industrially valuable chemical products (Castree, 2003).

The pharmaceutical industry has searched for new potential pharmaceuticals from sources like traditional herbal remedies, folklore, or little-known or unknown forest sources, a significant undertaking (Mays and Mazan, 1996). A 1995 report on biotechnology law referred to bioprospecting as "a modern-day gold rush," wherein farms and jungles are searched for novel and valuable pharmaceutical and agricultural products. Chemistry has made significant advances in screening, isolation, structural elucidation, semi- or full-synthesis of natural molecules or their derivatives, robotics, cell-based assays, high-throughput screening, genomics research, and new molecular biological tools for bioassay development (Rollinger *et al.*, 2004). In the emerging field of bionic prospecting, natural biodiversity serves as the foundation for the development of novel designs, patterns, models,

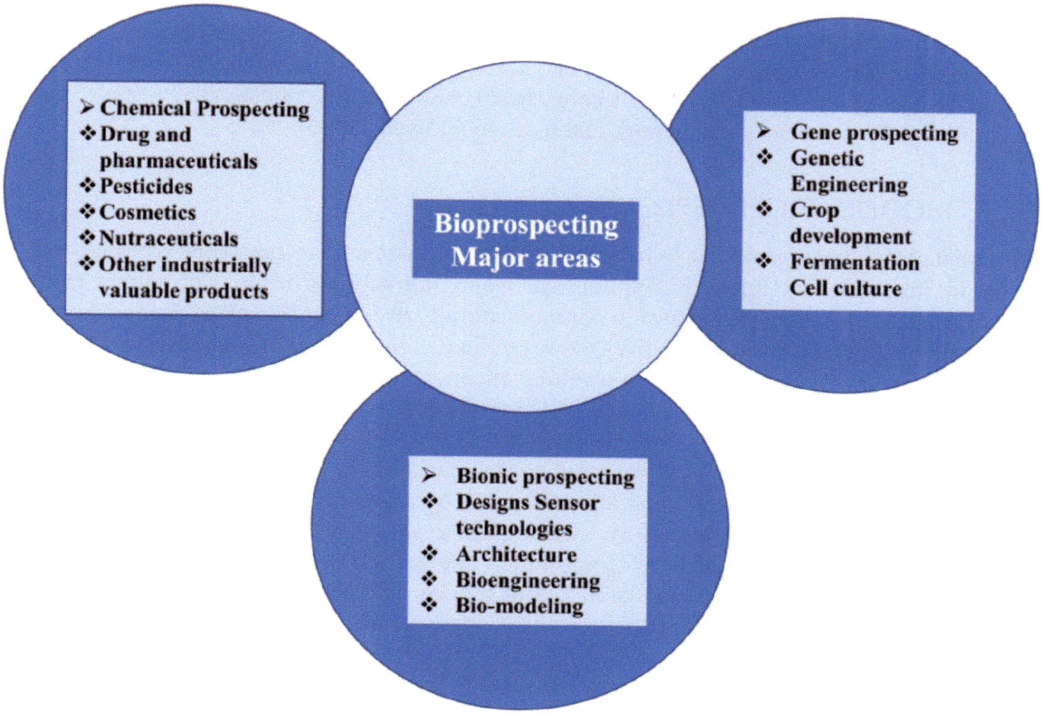

FIGURE 15.1 Major areas of bioprospecting.

and techniques. Among the fascinating areas in bionic prospecting are novel sensor technologies, architecture, bioengineering, and biomodeling (Neimark *et al.*, 2014) (Figure 15.1).

15.3 TRADITIONAL KNOWLEDGE

A community-based system of knowledge known as "traditional knowledge" has been created, maintained, and passed down through the generations by indigenous and local communities as a result of their ongoing observations, interactions, and experiments with their environment (Pushpangadan, 2005). It is unique to a specific society or culture and is the result of the coevolution and coexistence of indigenous cultures, as well as their customs for managing ecosystems and utilizing resources (Laird, 2002). An unbroken tradition and culture have passed down from generation to generation the accumulated knowledge, wisdom, and practices embodied in the TK system. Drug hunters use the medical knowledge of these traditional communities as a shortcut to find novel, practical compounds with significant medicinal value (Cox, 1994). It is widely accepted that the chances of discovering a potential bioactive compound through random screening of plant samples are 1 in 10,000, and the chances of discovering a drug that can be commercialized are 1 in 4 of these potential hits. By comparison, the odds of discovering a bioactive compound through selective screening based on conventional knowledge are 1 in 100, and the odds of discovering a drug that can be commercialized are 1 in 2. Ethnopharmacological research led to the initial discovery of many plant-derived medications used in contemporary medicine. Indigenous and local communities possess and develop traditional knowledge related to biodiversity, which is considered to be the most valuable source for modern innovations and the development of novel food products (functional, nutraceuticals, and medicinal), phytochemicals, and other commercially significant products. Thus, biodiversity and traditional knowledge are the most potent resources that, when combined with science and technology, can produce wealth. In particular, biotechnology has the potential to transform biodiversity resources into highly valuable industrial and commercial products and processes, increasing productivity and having applications in a wide range of critical fields, including agriculture (including aquaculture), healthcare, pharmaceuticals, diagnostics, vaccines, environmental protection, gene therapy, and bioenergy (Nayar, 1996).

15.4 MODERN BIOPROSPECTING

Certain biological products, such as rotenone, nicotine, and ryanodine, have been replaced by nylon, polyester, and polyacryl instead of cotton and wool; aniline dyes in place of natural colorants; and carbamates or organophosphates in place of naturally occurring pesticides derived from plants (Juan, 2017). To make these alternatives, however, human ingenuity still heavily draws inspiration from wildlife forms. Unlike our forest-dwelling ancestors who used the biological original to create a final product, we can now study these originals to find and reveal the principle of a biological process, and then develop novel products based on that understanding (Pushpangadan *et al.*, 2018). Millions of years of evolution have produced an abundance of molecular, cellular, and macrostructural structures and mechanisms that all work perfectly together and efficiently. The majority of life's technical issues can be solved by nature (Dixit *et al.*, 2021). Natural selection imposes the min–max principle, which states that a minimum amount of material and energy can achieve a maximum amount of efficiency and stability. Biological prototypes are especially important for our future because of the world's depleting resources and growing environmental issues (Hill, 1997) (Figure 15.2).

Modern bioprospecting involves the exploration of biodiversity to discover, develop, and commercialize new products and applications with potential benefits for various industries. Here are key aspects and trends in modern bioprospecting:

1. **Integration of omics technologies**: The use of genomics, transcriptomics, proteomics, and metabolomics to analyze biological systems at the molecular level. This allows researchers to understand the genetic potential and functionality of organisms and identify valuable compounds (Sahoo *et al.*, 2017).

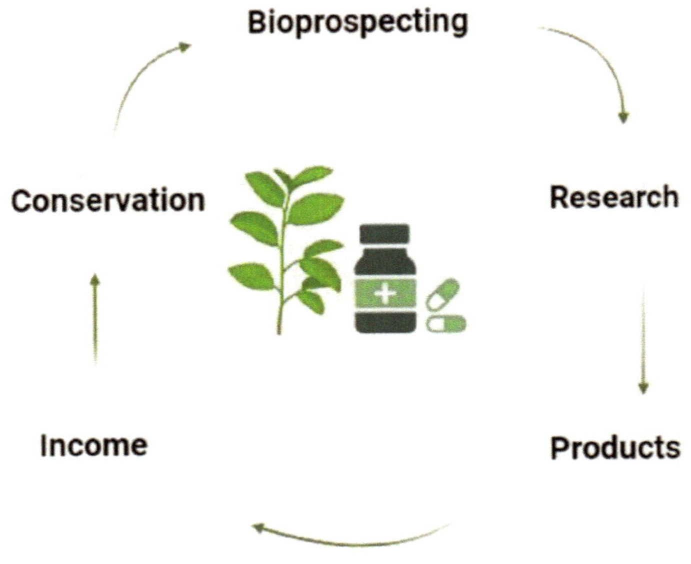

FIGURE 15.2 Bioprospecting overview.

2. **Metagenomics and microbiome exploration**: Metagenomics involves studying genetic material directly from environmental samples. In bioprospecting, this approach is used to explore the genetic diversity of microbial communities, leading to the discovery of novel enzymes, antibiotics, and other bioactive compounds (Pushpangadan *et al.*, 2018).
3. **Synthetic biology and genetic engineering**: Leveraging synthetic biology techniques to engineer microorganisms for the production of specific compounds. This enables the optimization of biosynthetic pathways to enhance the yield and diversity of bioactive molecules (Juan, 2017).
4. **Data mining and bioinformatics**: Utilizing advanced data mining and bioinformatics tools to analyze large datasets generated by omics technologies. This aids in the identification of potential bioresources, prediction of biological activities, and understanding complex biological interactions (Dixit *et al.*, 2021).
5. **Ethnobotanical knowledge and traditional medicine**: Integrating traditional knowledge and practices, especially in ethnobotany, to guide bioprospecting efforts. Indigenous communities often possess valuable information about the uses of plants and other organisms, contributing to the sustainable and culturally sensitive exploration of biodiversity (Neimark *et al.*, 2014).
6. **Bioprospecting in extreme environments**: Exploring extreme environments such as deep-sea ecosystems, deserts, and polar regions for unique and resilient organisms. These extremophiles often produce bioactive compounds adapted to harsh conditions, holding potential for industrial and medical applications (Laird, 2010).
7. **Sustainable and responsible bioprospecting**: Emphasizing sustainable practices to ensure the conservation of biodiversity and ecosystems. Responsible bioprospecting involves fair and equitable benefit-sharing agreements with local communities and adherence to ethical guidelines (Cao *et al.*, 2019).
8. **Collaboration and open innovation**: Encouraging collaboration between researchers, industry, and local communities to pool resources and expertise. Open innovation models facilitate the sharing of data and findings, accelerating the discovery process (Nayar, 1996).
9. **Focus on marine bioprospecting**: Given the vast biodiversity of marine ecosystems, there is a growing interest in exploring marine organisms for novel bioactive compounds with applications in pharmaceuticals, nutraceuticals, and biotechnology (Pushpangadan *et al.*, 2018).
10. **Bioprospecting for sustainable agriculture**: Identifying bioresources that can contribute to sustainable agriculture practices, such as biocontrol agents, plant growth-promoting microbes, and stress-resistant crops (Juan, 2017).

Modern bioprospecting is interdisciplinary, involving elements of biology, chemistry, informatics, and environmental science. As technology continues to advance, the potential for discovering novel and valuable bioresources becomes even more promising. Additionally, ethical considerations and sustainability principles play a crucial role in shaping the future of bioprospecting endeavors (Dixit *et al.*, 2021).

15.5 BIOPROSPECTING AS A RESEARCH FIELD

Bioprospecting is the discovery of drugs or genes useful for human health, agriculture, industry, or other applications from natural sources (Beattie *et al.*, 2005). It is common knowledge that contemporary technology can be utilized to develop pharmaceuticals. It is less well known that access to biodiversity is also necessary for increases in agricultural productivity and human health (Mgbeoji, 2006). Overall, between 1981 and 2002, the proportion of novel medications derived from natural products—more than 35%—remained stable (Laird, 2002; Finger and

Schuler, 2004), indicating that this success rate is likely to persist. It is extremely improbable that studies of natural products account for one-third of drug discovery research, though we do not have precise numbers. Put another way, such research could be incredibly fruitful even in spite of the underfunding of natural products. Natural products, or medicines derived from nature, may have novel mechanisms of action—that is, ones not yet discovered by science. For instance, in 1979, researchers discovered a hitherto unidentified mechanism of anticancer activity as a result of the discovery of taxol in the 1960s (Roe and Elliott, 2010). These discoveries had far-reaching consequences, including the development of taxol as an effective anticancer medication, the development of new assays to identify additional agents that stabilize microtubules, and, more recently, the discovery of additional anticancer medications that work by the same mechanism (Laird, 2002). Natural products derived from fungi, specifically FK506 and cyclosporin, aided in the development of immune suppressors, which revolutionized organ transplantation (Dutfield and Suthersanen, 2008). As a result, natural products may lead to the discovery of novel molecular targets, opening new avenues for innovation. These findings might also prompt advanced attempts to commercially synthesize analogs with comparable or even higher efficacy (Paterson and Anderson, 2005). A case study in agriculture is the high-yield rice varieties that were bred with a wild species of rice from India in the 1970s to protect against the grassy stunt virus. Because of multiple comparable successes, crop germplasm collection and preservation remain a global priority (ten Kate and Laird, 1999). Therefore, biodiversity serves as the basis for discoveries that are essential to human well-being by offering genes and natural products as ecosystem services (Prathapan, 2011) (Figure 15.3).

15.6 SUSTAINABLE AGRICULTURE AND ECONOMIC DEVELOPMENT

The environment and land are under a lot of strain due to the growing population. Uncontrolled agricultural area expansion that is encroaching on prime forests, unsustainable consumption patterns, and rapid population growth that exceeds the ecosystem's capacity to support people are all contributing factors to the alarming rise in environmental crises (Sahoo *et al.*, 2017). The eco-destructive tendencies that set human communities apart from the rest of the environment have deep cultural and psychological roots. The current state of the environment is a result of human beliefs, integrity, opinions, and behaviors. Tribal communities offer an excellent opportunity to reconsider our relationship with the ecosystem and related cultural populations (Barrett and Lybbert, 2000). We must understand the constantly evolving connection that these communities have with the local ecosystem, and any development program aimed at them must include all of those beneficial elements. For the region to grow and develop sustainably, crops must be carefully chosen for cultivation while adhering to all applicable protocols (Sittenfeld and Lovejoy, 1998). Ethnic communities possess a wealth of knowledge and tradition regarding farming methods. Numerous landraces and cultivars have also been preserved, and nearly all agricultural methods are organic. Thus, agriculture-based sustainable development calls for a well-thought-out strategy (Pauly *et al.*, 2002).

The intersection of sustainable agriculture and economic development within the realm of bioprospecting holds the promise of addressing global challenges while fostering environmentally conscious and economically viable practices. Sustainable agriculture, characterized by resource-efficient and environmentally friendly farming methods, aligns seamlessly with the principles of bioprospecting, which involves the exploration of biological resources for potential applications (Sahoo *et al.*, 2017). The marriage of these two concepts offers numerous opportunities for economic growth, environmental stewardship, and social well-being:

1. **Crop improvement for resilient agriculture**: Sustainable agriculture, driven by bioprospecting, focuses on the identification and utilization of genetic traits in plants that enhance resilience to pests, diseases, and environmental stressors. The discovery of naturally occurring resistance mechanisms and adaptive traits contributes to the development

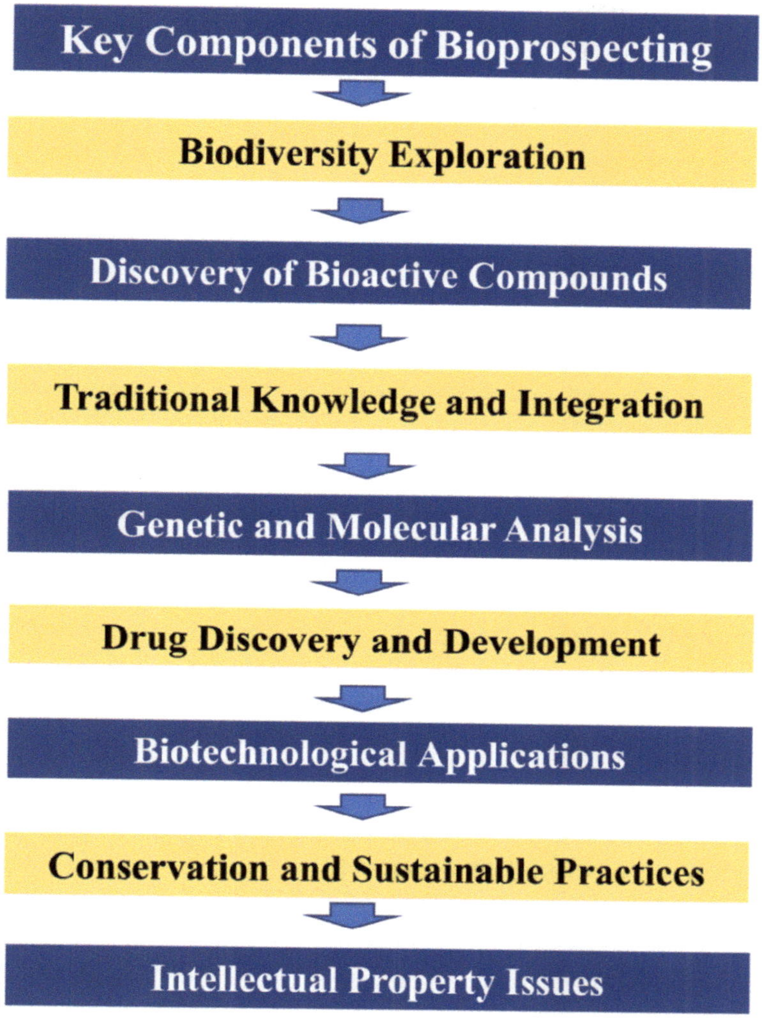

FIGURE 15.3 Key components of bioprospecting.

of robust crop varieties, ensuring consistent yields and reducing dependence on chemical inputs (Pushpangadan *et al.*, 2018).

2. **Biological control and pest management**: Bioprospecting facilitates the discovery of biocontrol agents, including beneficial microorganisms and natural predators, that can be employed in sustainable agriculture practices. Implementing these biological control measures not only reduces reliance on synthetic pesticides but also promotes ecosystem balance, contributing to healthier soils and reduced environmental impact (Mani *et al.*, 2004).
3. **Medicinal plant cultivation and value-added products**: Sustainable agriculture practices guided by bioprospecting principles extend beyond food production to include the cultivation of medicinal plants. The identification of bioactive compounds in plants can lead to the development of value-added products in pharmaceuticals, cosmetics, and nutraceuticals, creating new economic opportunities for local communities (Kate and Laird, 2010).
4. **Soil health and biodiversity conservation**: Bioprospecting in sustainable agriculture emphasizes the importance of maintaining soil health and biodiversity. Practices such as cover cropping, crop rotation, and agroforestry, informed by bioprospecting findings,

contribute to improved soil fertility, reduced erosion, and enhanced biodiversity. These practices, in turn, promote long-term sustainability and resilience in agricultural systems (Pushpangadan, 2005).

5. **Commercialization of sustainable farming practices**: The integration of bioprospecting insights into sustainable agriculture not only benefits individual farmers but also presents opportunities for commercialization. Companies specializing in sustainable and organic products can capitalize on the growing consumer demand for environmentally friendly and ethically produced goods, thereby contributing to economic development in the agricultural sector (Pushpangadan *et al.*, 2005).
6. **Biotechnological innovations for crop improvement**: Bioprospecting's molecular and genetic analyses enable the identification of valuable genes and traits that can be incorporated into crops through genetic engineering. This biotechnological approach enhances crop characteristics such as drought tolerance, nutritional content, and resistance to pests, thereby bolstering agricultural productivity and economic returns (Neimark *et al.*, 2014).
7. **Community engagement and benefit sharing**: A key aspect of sustainable agriculture and bioprospecting collaboration involves community engagement and fair benefit-sharing agreements. Empowering local communities through the sustainable use of their natural resources ensures that economic development is inclusive and aligned with conservation objectives, fostering a sense of ownership and responsibility (Pauly *et al.*, 2002).

In essence, the convergence of sustainable agriculture and bioprospecting not only enhances agricultural productivity but also paves the way for a more resilient, economically viable, and ecologically sustainable future. The application of bioprospecting principles in agricultural development underscores the importance of a holistic and integrated approach that balances economic growth with environmental stewardship (Pushpangadan *et al.*, 2018).

15.7 BIOPROSPECTING AND VALUE-ADDED PRODUCT DEVELOPMENT BASED ON KNOWLEDGE

Biodiversity, herbal technology, biotechnology, and nanobiotechnology are fields that require special attention in terms of international trade, intellectual property rights protection, and the new frontier of knowledge-based industrial ventures (Helson *et al.*, 2009). Biodiversity has ecological, artistic, economic, and existential values. Direct economic benefits of biodiversity include fuel, food, feed, dietary fiber, pigments, pharmaceuticals, lipids, gums, polymers, dyes, biological pesticides, bioinsecticides, honey, phytochemicals, amino acids, and genes. Herbal technology is the use of low- and medium-level technology to develop standardized, value-added products; biotechnology is the use of more sophisticated scientific and technological interventions; and nanobiotechnology is the manipulation of molecules at the subcellular or ultra-molecular level. Thus, biodiversity is (i) an invaluable resource with a wide range of practical applications and potential benefits for humanity, and (ii) a sophisticated ecological system that promotes the biosphere's resilience and integrity. Based on these two opposing viewpoints, it is possible to conclude that biodiversity is a priceless natural resource that should be protected and responsibly used for the benefit of future generations as well as current generations of humans. Only a small portion of the vast genetic library found in nature has been used by humanity. The methodical search for genes, naturally occurring substances, patterns, and entire forest or wildlife organisms with the potential to be used in product development is called bioprospecting. Chemical prospecting, gene prospecting, and bionic prospecting are three crucial components of bioprospecting (Reid *et al.*, 1993).

Bioprospecting, coupled with value-added product development, represents a dynamic approach to leveraging biological resources for the creation of innovative and economically valuable products. This process involves the exploration of biodiversity to identify unique compounds, genes, or organisms that can be translated into commercially viable goods with enhanced features or

TABLE 15.1
Bioprospecting in Asia Pacific Region

Country	Biological Resource	US Patent Number	Bioprospecting Uses	Reference
China	Bitter melon (*Momordica charantia*)	5484889	Suppresses inflammation and inflammatory conditions	(Grain and Kalpvriksh, 2002)
Pacific	Kava (*Piper mythesticum*)	6405948, 6277396, 6080410, and 5770207	Use in medicaments having an anxiolytic, anticonvulsive, and neuroprotective effect	(Pushpangadan, 2018)
Pakistan	Pakistan basmati rice	6274183 and 5663484	Cooking behavior and starch properties of rice	(Correa, 2001)
Samoa	Mamala tree (*Homalanthus nutans*)	5599839	Concerning the development of prostratin, an anti-AIDS compound derived from the Samoan native mamala tree (*Homalanthus nutans*)	(Drahos, 2004)
India	Turmeric (*Curcuma longa*)	5401504 and 5135796	Wound healing and skin conditioning, cosmetics, lightening compositions, and anti-irritants are all examples of products	(Rangnekar, 2002)
India	Neem (*Azadirachta indica*)	5420318, 5391779, and 5371254	Use of antifungal agents	(Pushpangadan, 2018)
India	Guggul (*Commiphora mukul*)	6113949; US Patent Application 20020018757	Method for treating hyperlipidemia (excess fat)	(Grain and Kalpvriksh, 2002)
Sri Lanka	Kothala himbutu (*Salacia reticulata*)	Takama System, Ltd.'s (Yamaguchi, Japan) US Patent 6376682	Contains a high specific medicinal value in the treatment of diabetes, hepatitis, digestion	(Pushpangadan, 2018)

functionalities. The integration of bioprospecting knowledge into the development of value-added products holds great potential across various industries, fostering economic growth and sustainability. Bioprospecting is a valuable source of bioactive compounds with pharmaceutical potential (Helson *et al.*, 2009). Knowledge gained from the exploration of diverse ecosystems aids in the discovery of novel molecules, inspiring the development of new drugs and therapeutics. The utilization of these compounds in pharmaceutical products not only addresses health challenges but also contributes significantly to the economic value of the healthcare industry (Neimark *et al.*, 2014). Bioprospecting and value-added product development are catalysts for economic advancement, innovation, and sustainability. As we continue to unlock the potential of biological resources, this integrated approach holds the promise of creating products that not only meet market demands but also contribute to the conservation of biodiversity and the well-being of communities involved in the process (Reid *et al.*, 2013) (Table 15.1).

15.8 BIOPROSPECTING AND VALUATION OF BIODIVERSITY

The genetic library found in nature has tremendous potential for producing novel foods, medications, and industrial goods. Unknown possibilities in terms of genes, species, and ecosystems constitute an endless and priceless biological frontier. In the past, domesticated biodiversity was the main focus of genetics (Laird and ten Kate, 2002). It will be extremely difficult and rewarding to unlock these otherwise little-known bioresources' patent potentials. Through the use of modern technology, bioprospecting is the most effective approach for examining and evaluating the genetic, chemical, and economic potentials of biological resources. Scientists can now identify and isolate potentially bioactive components from plants on a large scale thanks to recent advances in instrumentation and enzyme technology (Hunt and Vincent, 2006). HPLC (high-performance liquid chromatography), HPTLC (high-performance thin-layer chromatography), MPLC (medium-pressure liquid chromatography), and LCMS (liquid chromatography mass spectrometry) are examples of modern chromatographic techniques assisted by computer robotics, and can be used to isolate active principles from appropriate plant extracts after they have been screened for biological activity using "throughput" techniques utilizing in vitro enzyme methods. Another possible application of bioprospecting is gene prospecting. The ability to examine a species' genetic diversity has been made possible by developments in DNA fingerprinting technology (Costello and Ward, 2006).

Bioprospecting and conservation were originally defined as searching for oil and precious metals in the soil and subsurface. Following that, Thomas Eisner of Cornell University coined the term "chemical prospecting," and more recently, the term "bioprospecting" has been coined to refer to the search for novel sources of chemicals, proteins, genes, and other products that have commercial potential and are found in a diverse range of plants, animals, and microorganisms (Frisvold *et al.*, 2008). Bioprospecting has to be grounded in good principles and standards and connected to the idea of value in order to be a conservation ally. As previously mentioned, the current and future worth of bioprospectible designs, genes, and chemicals is enormous, which warrants the application of clever conservation techniques (Mulholland and Wilman, 2003). In order to be a useful component of conservation, bioprospecting needs to reduce poverty in areas with abundant biodiversity and help generate wealth through research and development. In this context, national bioprospecting programs can support the growth of national capabilities and offer financial benefits to conservation initiatives, provided that science, technology, and market research are adequately funded. To achieve sustainable development, communities inside or adjacent to protected areas must take part in conservation efforts and gain from bioprospecting activities in their own areas, though this may not be enough (Prathapan and Rajan, 2011). Utilizing biodiversity in conjunction with localized added value seems to be a viable strategy for fostering community economic development and providing incentives for conservation. The following activities could have a positive economic impact on the community: cultivating new ornamental plants; producing, grading, and packing medicinal plants; cultivating and processing aromatic teas; exploring and assessing local landraces with resistant genes of interest; and domesticating and producing biopesticides (Mateo, 2000). Market research is always an essential precondition to taking part in added-value biodiversity development at the community level. There is no point in carrying out bioprospecting operations until a distinct market is found, appropriate contracts are signed, and local organization is accomplished (Rausser and Small, 2000).

15.9 FUTURE PROSPECTS

Bioprospecting has been proposed as a potential means of encouraging biodiversity conservation and sustainable use. Biodiversity-rich nations are gradually putting the UN-mandated legal framework into practice, but more work needs to be done before bioprospecting can be truly facilitated (Tamayo *et al.*, 2003). Maybe the enactment of the Nagoya Protocol will give the required push. But since the Rio Earth Summit in 1992, there has been a noticeable decline in the interest that pharmaceutical development companies have shown in bioprospecting. This decline can be attributed, in part, to the difficulties associated with access and benefit-sharing, which frequently occur due to the lack of sufficient institutional capacity and national regulatory clarity. Despite the fact that successful drug development projects based on natural products continue to emerge, there is a widespread belief that screening natural products for new leads is an outmoded approach. There are undoubtedly a number of technical issues with natural product screening and isolation, but the benefits of resolving them appear to outweigh the work involved, and technical solutions are being documented in the literature. For example, high throughput separation techniques combined with sensitive analytical techniques can resolve the purification and identification of natural products, which are thought to be difficult and time-consuming (Bugni *et al.*, 2008; Hu *et al.*, 2008). Comparisons of the chemical properties of collections of natural products show that they more closely match the "chemical space" of successful drugs than collections of synthetic chemicals (Grabowski and Schneider, 2007; Ganesan, 2008). Despite the growing popularity of phenotypic assays, it has been suggested that natural products, with their drug-like properties, are well-suited to such cell-based approaches. Additionally, natural product extracts can be processed to remove reactive compounds or even convert them into novel drug-like structures. Natural products are said to produce an abnormally high number of false positives in modern screening assays. Small quantities of natural products might be accessible; methods for direct synthesis or production through molecular biology have advanced quickly (Kennedy, 2008). There isn't a single "best" strategy for drug discovery, just as there isn't a single treatment for every disease, but it is definitely time for a fresh look at the relatively unrealized potential that modern approaches to using natural products in drug discovery offer. Perhaps the numerous academic institutions engaged in bioprospecting will need to take the initiative (Sunazuka *et al.*, 2008). These groups would have a better chance of success, though, if they could combine their resources and search for validated lead compounds that could be developed into medications to address unmet therapeutic needs. Academic groups will be able to become crucial partners in pharmaceutical innovation through the expansion of translational research and the creation of translational research centers (Simpson *et al.*, 1996).

15.10 CONCLUSION

Bioprospecting is now practiced in most ecosystems around the world, but its environmental impacts are mostly minor, especially when compared to other human activities. Chemical methods are not as promising as bioprospecting when it comes to environmentally sustainable metal recycling and recovery. The remarkable capacity of humankind to solve the subtle mysteries of nature will undoubtedly aid in the creation of new technologies that will enable the sustainable use of the planet's finite bioresources. Ecologists play a key role in highlighting the various economic benefits of biodiversity and are increasingly in charge of or involved in the resource management and discovery phases of bioprospecting.

REFERENCES

Barrett CB, Lybbert TJ. Is bioprospecting a viable strategy for conserving tropical ecosystems?. *Ecological Economics*. 2000 Sep 1;34(3):293–300.

Bugni TS, Richards B, Bhoite L, Cimbora D, Harper MK, Ireland CM. Marine natural product libraries for high-throughput screening and rapid drug discovery. *Journal of Natural Products*. 2008 Jun 27;71(6):1095–8.

Cao S, Kingston DG. Biodiversity conservation and drug discovery: Can they be combined? The Suriname and Madagascar experiences. *Pharmaceutical Biology*. 2019 Aug 1;47(8):809–23.

Castree N. Bioprospecting: from theory to practice (and back again). *Transactions of the Institute of British Geographers*. 2003 Mar;28(1):35–55.

Chadwick DJ, Balick MJ, Marsh J. Good botanical practices. *Botanical Medicine: Efficacy, Quality Assurance, and Regulation*. Larchmont (NY): Mary Ann Liebert. 1999:121–5.

Correa CM. Traditional knowledge and intellectual property. *Geneva: The Quaker United Nations Office (QUNO)*. 2001 Nov;17.

Costello C, Ward M. Search, bioprospecting and biodiversity conservation. *Journal of Environmental Economics and Management*. 2006 Nov 1;52(3):615–26.

Cox PA. The ethnobotanical approach to drug discovery: strengths and limitations. Pages 25-41 in GT Prance, D. In *Ethnobotany and the search for new drugs Clba Foundation Symposium,* 1994 (Vol. 185).

Dixit S, Shukla A, Singh V, Upadhyay SK. Bioprospecting of natural compounds for industrial and medical applications: current scenario and bottleneck. *Bioprospecting of Plant Biodiversity for Industrial Molecules*. 2021 Jul 19:53–71.

Drahos P. Towards an international framework for the protection of traditional group knowledge and practice. In *UNCTAD-Commonwealth Secretariat Workshop on Elements of National Sui Generis Systems for the Preservation, Protection and Promotion of Traditional Knowledge, Innovations and Practices and Options for an International Framework*, Geneva 2004 Feb 4 (pp. 4–6).

Finger, M. J. and Schuler, P. (eds) (2004) Poor People's Knowledge: Promoting Intellectual Property in Developing Countries, Oxford University Press, Oxford.

Frisvold G, Day-Rubenstein K. Bioprospecting and biodiversity conservation: What happens when discoveries are made. *Ariz. L. Rev.*. 2008;50:545.

Ganesan A. The impact of natural products upon modern drug discovery. *Current Opinion in Chemical Biology*. 2008 Jun 1;12(3):306–17.

Grabowski K, Schneider G. Properties and architecture of drugs and natural products revisited. *Current Chemical Biology*. 2007 Jan 1;1(1):115–27.

Grain K. Traditional knowledge of biodiversity in Asia-Pacific: problems of piracy & protection. [2013-05-30]. http:www.grain.org.2006.

Helson JE, Capson TL, Johns T, Aiello A, Windsor DM. Ecological and evolutionary bioprospecting: using aposematic insects as guides to rainforest plants active against disease. *Frontiers in Ecology and the Environment*. 2009 Apr;7(3):130–4.

Hill B. *Innovationsquelle natur: naturorientierte innovationsstrategie für entwickler, konstrukteure und designer*. Shaker Verlag; 1997.

Hu J, Gary RE, Yu Y, Mark OS. High-throughput natural product chemistry methods and the application of the capillary NMR probe. *Progress in Chemistry*. 2008 Apr 24;20(04):429.

Hunt B, Vincent AC. Scale and sustainability of marine bioprospecting for pharmaceuticals. *AMBIO: A Journal of the Human Environment*. 2006 Mar;35(2):57–64.

Janzen DH, Hallwachs W, Jimenez J, Gámez R. The role of the parataxonomists, inventory managers, and taxonomists in Costa Rica's national biodiversity inventory. *Biodiversity Prospecting: Using Genetic Resources for Sustainable Development*. 1993:223–54.

Juan B. Bioprospecting and drug development, parameters for a rational search and validation of biodiversity. *J Microb Biochem Technol*. 2017;9:e128.

Kennedy J. Mutasynthesis, chemobiosynthesis, and back to semi-synthesis: combining synthetic chemistry and biosynthetic engineering for diversifying natural products. *Natural Product Reports*. 2008;25(1):25–34.

Kingston DG. Modern natural products drug discovery and its relevance to biodiversity conservation. *Journal of Natural Products*. 2011 Mar 25;74(3):496–511.

Laird SA, editor. *Biodiversity and Traditional Knowledge: Equitable Partnerships in Practice*. Routledge; 2010 Sep 23.

Laird SA, Ten Kate K. Biodiversity prospecting: the commercial use of genetic resources and best practice in benefit-sharing. InBiodiversity and Traditional Knowledge 2010 Sep 23, pp. 275–277. Routledge.

Laird SA, ten Kate K. Linking biodiversity prospecting and forest conservation. *Selling Forest Environmental Services: Market-Based Mechanisms for Conservation and Development*. 2002.

Lesser WH, Krattiger A. Valuation of bioprospecting samples: approaches, calculations, and implications for policy-makers. *Intellectual Property Management in Health and Agricultural Innovation: A Handbook of Best Practices*, Volumes 1 and 2. 2017:861–76.

Mateo N. Bioprospecting and conservation in Costa Rica. Responding to bioprospecting: From biodiversity in the South to medicines in the North. 2000:45–55.

Mateo N. Wild biodiversity: The last frontier? The case of Costa Rica. 2021.

Mays TD, Mazan KD. Legal issues in sharing the benefits of biodiversity prospecting. *Journal of Ethnopharmacology*. 1996 Apr 1;51(1–3):93–109.

Medaglia JC. Biodiversity for (bio) technology under the Convention on Biological Diversity: bioprospecting partnership in practice. 2004.

Mulholland DM, Wilman EA. Bioprospecting and biodiversity contracts. *Environment and Development Economics*. 2003 Jul;8(3):417–35.

Myers N, Mittermeier RA, Mittermeier CG, Da Fonseca GA, Kent J. Biodiversity hotspots for conservation priorities. *Nature*. 2020 Feb 24;403(6772):853–8.

Nayar MP. Hot spots of endemic plants of India, Nepal and Bhutan. *Tropical Botanic Garden and Research Institute*. 1996.

Neimark BD, Tilghman LM. Bioprospecting a biodiversity hotspot: The political economy of natural products drug discovery for conservation goals in Madagascar. In *Conservation and environmental management in Madagascar,* 2014 Mar 14 (pp. 271–298). Routledge.

Paterson I, Anderson EA. The renaissance of natural products as drug candidates. *Science.* 2005 Oct 21;310(5747):451–3.

Pauly D, Christensen V, Guénette S, Pitcher TJ, Sumaila UR, Walters CJ, Watson R, Zeller D. Towards sustainability in world fisheries. *Nature.* 2002 Aug 8;418(6898):689–95.

Prathapan KD, Rajan PD. Biodiversity access and benefit-sharing: weaving a rope of sand. *Current Science.* 2011 Feb 10;100(3):290–3.

Pushpangadan P, George V, Ijinu TP, Chithra MA. All India coordinated research project on ethnobiology and genesis of ethnopharmacology research in India including benefit sharing. *Annals of Phytomedicine.* 2018 Jan 1;7(1):5–12.

Pushpangadan P. 2005 Value addition and commercialization of biodiversity and associated traditional knowledge in the context of the intellectual property regime.

Rangnekar D. Geographical indications: a review of proposals at the TRIPS Council. *UNCTAD*; 2002 Jun.

Rausser GC, Small AA. Valuing research leads: bioprospecting and the conservation of genetic resources. *Journal of Political Economy.* 2000 Feb;108(1):173–206.

Reid D. Sustainable development: An introductory guide. Routledge; 2013 Nov 5.

Reid, W.V., Laird, S.A., Meyer, C.A., Gamez, R., Sittenfeld, A., Janzen, D.H., Gollin, M.A. & Juma, C. (Eds.). 1993. Biodiversity prospecting: using genetic resources for sustainable development. *WRI*, Washington DC.

Rollinger JM, Haupt S, Stuppner H, Langer T. Combining ethnopharmacology and virtual screening for lead structure discovery: COX-inhibitors as application example. *Journal of Chemical Information and Computer Sciences.* 2004 Mar 22;44(2):480–8.

Sahoo S, Sarangi S, Kerry RG. Bioprospecting of endophytes for agricultural and environmental sustainability. *Microbial Biotechnology: Applications in Agriculture and Environment.* 2017;1:429–58.

Simpson RD, Sedjo RA, Reid JW. Valuing biodiversity for use in pharmaceutical research. *Journal of Political Economy.* 1996 Feb 1;104(1):163–85.

Sittenfeld A, Cabrera J, Marielos M. Bioprospecting frameworks: policy issues for Island countries in Insula. *International Journal of Island Affairs.* 2003;12(1).

Sittenfeld AM, Lovejoy A. Biodiversity prospecting frameworks: the INBio experience in Costa Rica. 1998.

Sunazuka T, Hirose T, Omura S. Efficient total synthesis of novel bioactive microbial metabolites. *Accounts of Chemical Research.* 2008 Feb 19;41(2):302–14.

Tamayo G, Guevara L, Gámez R. Biodiversity prospecting: the INBio experience. *Microbial Diversity and Bioprospecting.* 2003 Nov 7;445–9.

Ten Kate, K. (1999). The development of major crops by the seed industry. In K ten Kate and S A Laird (eds). The Commercial Use of Biodiversity: Access to Genetic Resources and BenefitSharing, Earthscan, London.

16 Bioprospecting and Biomimicry

Aditi Fishta[1], R C Sobti[2], Ruhi Thakur[1]

[1]School of Biological and Environmental Sciences, Shoolini University of Biotechnology and Management Sciences, Solan -- Himachal Pradesh, India
[2]Department of Biotechnology, Panjab University, Chandigarh India

16.1 INTRODUCTION

16.1.1 Bioprospecting

The term "bioprospecting," which Thomas T. Eisner originally used to describe "chemical prospecting," describes the study of biodiversity to obtain genetic and biochemical resources that can be used for profit. Due to the significant advancements in biotechnology, particularly in the fields of transcriptomics, enzymatic and transgenic technologies, proteomics, genomics, and other areas of biotechnology, there is a growing interest in exploring biodiversity for new medicines, foods, crops, pesticides, and other commercially relevant genetic and biological products and processes (Figure 16.1) (Pushpangadan *et al.*, 2015).

Demand for energy, food, and biological resources rises as a result of world population growth, rising incomes, and an expanding global economy (Bell *et al.*, 2018). It is estimated that by 2050, providing a sustainable source of food for nearly 10 billion people will be a major concern (Vågsholm *et al.*, 2020). Indeed, traditional agriculture will not be able to feed the growing global population, and the availability of food will become a major international issue (Bleakley and Hayes, 2017). Thus, food scarcity is a predicted result, which might lead to malnutrition and the emergence of human diseases because of an uneven nutrient diet (Bessada *et al.*,2019). Protein, in particular, is one of the key essential elements that will be deficient in the future, necessitating the pursuit of alternatives to meet consumption requirements (Myer *et al.*, 2019). One of the main obstacles to sustainable protein production has been consumer demand for meat-based meals to satisfy their protein needs. Meat production takes far more water, energy, and land space than plant-based diets (Grahl *et al.*, 2020).

As a result, bioprospecting—which mostly involves the extraction of secondary metabolites—has emerged as a useful and modern substitute in the search for novel drug sources. These have a wide range of medicinal uses, including treating neglected diseases for which there are few available treatment options due to the terrifying resistance mechanisms these parasites develop to commonly used treatments.

To put it briefly, bioprospecting incorporates the following three primary fields: chemical, gene, and design prospecting (Maeto et al., 2001), as shown in Figure 16.2. Searching for new business opportunities involves looking into biochemical or genetic resources (Laird and ten Kate, 2010).

Any bioprospecting program's success relies on several factors, including (a) potential users' desire to consume biological resources and access limitations to related traditional knowledge; (b) genetic resource providers' ability to overcome supply limitations and deliver a high-quality service; (c) legislation, policy, and a political climate conducive to bioprospecting; (d) financial benefits that facilitate bioprospecting; and (e) the capability to ensure agreement enforcement (Ten Kate, 1995).

DOI: 10.1201/9781032697550-16

FIGURE 16.1 Medicinal bioprospecting.

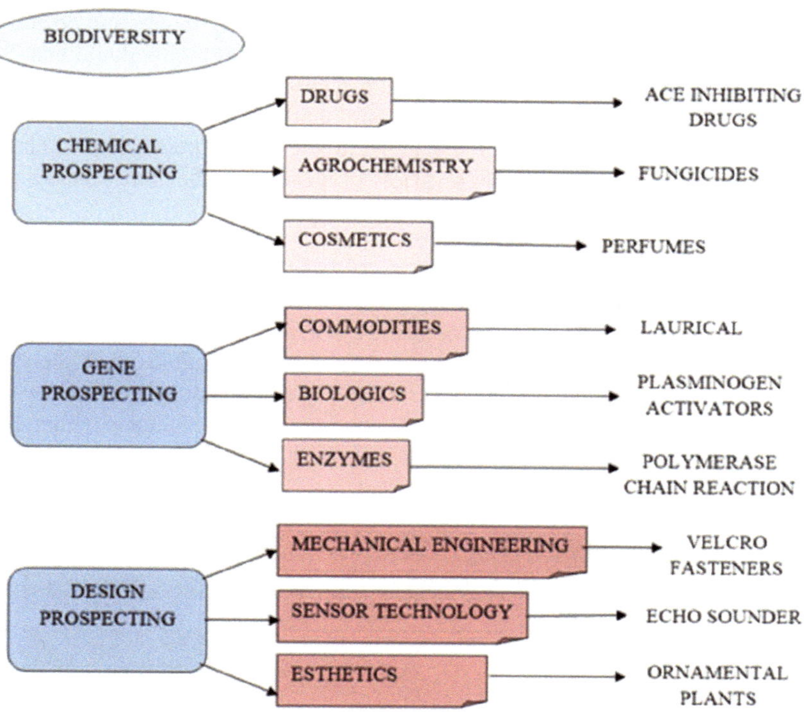

FIGURE 16.2 illustrates the three principal sources of biodiversity that inspire mankind and how they implement it.

Numerous additional factors also play a role in selecting bioprospecting locales and collaborators more effectively. Ten Kate and Laird (2019) enumerate the subsequent items: (i) the rich biodiversity of the area; (ii) the assurance that a partner obtains and supplies specimens in strict accordance with global as well as national law; (iii) the partner's ability to obtain pertinent permits and prior authorizations; (iv) an easy-to-understand, uncomplicated process for obtaining permits as well as access agreements; (v) the partner's ability to ensure supply quantity and quality and resupply; (vi) the standard of scientific staff in the partner institute; (vii) political stability in the country and area where the collection takes place, where partners collaborate in value-adding activities, laws and policies about intellectual property rights protection, foreign ownership, and taxation that promote foreign investment.

16.1.1.1 Marine Resources and Biodiversity

The marine environment is the most diverse ecosystem, covering more than 70% of the planet (Danovaro *et al.*, 2011; Atencio *et al.*, 2018). Over 95% of microbial biomass is made up of microbial species (Abida *et al.*, 2013). These can be found in the sediments, water column, and various groups. They include bacteria, viruses, zooplankton (grazing plankton), and phytoplankton (photosynthetic plankton). Marine environments are home to some of the most adapted species that can withstand high pressure, temperature, and salinity (Hosseini *et al.*, 2021) because they are subject to constant anthropogenic stressors and encompass a variety of naturally occurring extreme environmental conditions (e.g., hydrothermal vents, polar regions, etc.) (Danovaro *et al.*, 2011). Such adaptability drives a high level of genetic and functional diversity (Barone *et al.*, 2018), allowing the synthesis of compounds that benefit biotechnological applications. Marine microorganisms are attracting significant focus in this regard since they are considered a viable source of compounds that can aid in the development of medications targeting life-threatening disorders (Figure 16.3) (Ambrosino *et al.*, 2019). For instance, algae-derived products are particularly interesting because they provide special alternatives including exopolysaccharides, pigments, and high-value lipids (Schipper *et al.*, 2019; Bounnit *et al.*, 2020). Additionally, it has been demonstrated that numerous marine fungal and bacterial species produce a wide variety of new compounds with therapeutic promise when utilized as antibiotics (as shown in Table 16.1) (Wiese *et al.*, 2019). Marine yeast-derived biomolecules have also demonstrated intriguing features for a variety of possible applications, including biosurfactant synthesis (Senthil *et al.*, 2019).

16.1.1.1.1 High-Value Compounds Produced from Marine Microbes

Marine organisms produce secondary metabolites that are either secreted or remain inside the organism (Katz and Baltz, 2016). Small molecules, polypeptides, nonribosomal peptides (such as actinomycin D, cyclosporine, and vancomycin or daptomycin), alkaloids, polyphenols, lipopolysaccharides (Evans, 2016), polyketides (Lorente *et al.*, 2014), and nucleic acids (Carteni *et al.*, 2016) are the most prevalent. Many investigations focused on determining the makeup of metabolites obtained from marine sources, which were primarily categorized as either harmful or helpful substances

FIGURE 16.3 Marine bioprospecting phases illustrate the various steps taken to explore and create intriguing microbial metabolites that lead to the commercialization of important products.

TABLE 16.1
Marine-Derived Compounds Offering Advancements in Various Domains

Field	Activity	Source	Reference
Medicine and pharmacology	Anticancer	Diatom *Skeletonema marinoi*, Microalga *Isochrysis galbana*	Lauritano *et al.*, 2016 Matos *et al.*, 2019
	Anti-inflammatory	Diatom *Cylindrotheca closterium*	Lauritano *et al.*, 2016
	Antioxidant, antimicrobial, anti-inflammatory	Bacterium *Nocardiopsis dassonvillei*	Bibi *et al., 2017*
Nutraceuticals	Phycocyanin	*Leptolynbbia*	Schipper *et al.*, 2019
	C fatty acids	*Nanochloris atomus*	Saadaoui *et al.*, 2016
Nutrition	Fucoxanthin, β-carotene production	Microalga *Isochrysis galbana*	Matos *et al.*, 2019
	β-carotene, neoxanthin, violaxanthin, and lutein production	Microalga *Tetraselmis* sp.	Schüler *et al.*, 2020
	Dietary protein supplement	Lobster by-products	Nguyen *et al.*, 2017
Agriculture	Nitrogen-fixing and nutrient-enriching biofertilizers	Cyanobacterium *Azolla-Anabaena*	Chittora *et al.*, 2020
	Nitrogen-fixing biofertilizers	Cyanobacterium *Nostoc* sp.	Garlapati *et al.*, 2019
Biofuels	Bioethanol production	Yeast *Saccharomyces cerevisiae*	Zaky *et al.*, 2020
	Lipid production	Diatom *Mayamaea* sp.	Matsumoto *et al.*, 2017
	Lipid production	Diatom C*yclotella cryptica*	Maeda *et al.*, 2018
Cosmetics	Fucoidan production	Brown alga Saccharina japonica	Wang *et al.*, 2013
	Exopolysaccharide's production	Bacterium *Alteromonas macleodii*	Delbarre-Ladrat *et al.*, 2014
Industry	Enzymatic production of bioplastics	Bacterium *Burkholderia sacchari*	Al-Battashi *et al.*, 2019
	Xylanolytic activity	Fungus *Aspergillus* sp.	El-Bondkly *et al.*, 2012
	Enzymatic bioremediation	Archea *Desulfurococcus* sp., *Pyrococcus* sp., *Thermococcus* sp.	Sharma *et al.*, 2019

(Xiong *et al.*, 2013; Galasso *et al.*, 2019), in order to forecast their future utility. Presently, about 7000 bioactive chemicals derived from marine microorganisms are being confirmed and utilized (Ambrosino *et al.*, 2019). Ninety percent of marine animals are unidentified due to the poor status of studies on marine ecosystems (Mora *et al.*, 2011).

16.1.1.1.2 Techniques for Marine Microbial Bioprospecting

Traditional methods of marine bioprospecting rely on culture approaches such as microscopy, selective media, gram staining, and biochemical identification (Sysoev *et al.*, 2021). For the most part, two procedures for marine microbial bioprospecting have been devised and used: (1) enhancement of the specimen with nutrients before culture and separation, and (2) separation of individual cells followed by nutrient enrichment and culture. In general, bioprospecting begins with accumulating an ecological specimen, then supplementing the specimen with nutrients to promote microbial development, and then isolating the grown microorganisms (Hu *et al.*, 2021). Such basic approaches provide qualitative and quantitative data on microbial species with great sensitivity and reproducibility (Nowrotek *et al.*, 2019).

In light of this heritage, culturally specific bioprospecting is constantly evolving to get beyond the limitations of customary cultivations and propose novel culturing methodologies (Lewis *et al.*, 2010). Using throughput farming methods combined with fundamental ecology and biology increased the number of aquatic microbiological organisms that are cultivable and resulted in the separation of new microbial species, including the first members of the SAR 11 clade (Alejandre *et al.*, 2020). The single-cell isolation technique, as opposed to the traditional isolation strategy, begins by separating the microbe of interest followed by enriching single cells to promote development (Alkayyali *et al.*, 2021). One of the most sensitive and successful techniques for isolating marine microorganisms is high throughput dilution to extinction culture (Kim *et al.*, 2020). This technique often involves merging fluorescence-activated cell sorting (FACS) and flow cytometry to increase efficiency and reduce bioprospecting expenses (Südfeld *et al.*, 2021).

The methodical search for beneficial goods made from living things (plants, animals, and microorganisms) in order to improve human existence is known as "bioprospecting" (Oyemitan and Kuete, 2017). Given the enormous genetic diversity thought to exist in oceans, marine bioprospecting has gained a lot of attention.

For example, it has been reported that cyanobacteria create cyclopeptides, which may find application in cancer-fighting pharmaceuticals (Sainis *et al.*, 2010). Furthermore, reports have stated that brown algae produce fucoxanthin, which is utilized as an antidiabetic and anticancer drug (Moreau *et al.*, 2006; Dambek *et al.*, 2012). Additionally, it has been demonstrated that red algae and diatoms generate domoic acid, which has anthelmintic properties (Ramsdell, 2007). Moreover, it has been shown that fungi can synthesize the antibacterial cephalosporin P (Hamilton-Miller, 2008).

Plankton bioprospecting typically focuses on organisms that endure harsh environments in order to maximize the likelihood of finding novel compounds, either as primary or secondary metabolites (Bull *et al.*, 2019; Indrayani *et al.*, 2019). For example, the thermotolerant bacterium *Thermus aquaticus* produced the Taq polymerase, a polymerase chain reaction DNA polymerase that is widely used in molecular biology labs (Saiki *et al.*, 1988). Additionally, it views meroplanktons—organisms that only exist as planktons at one point in their existence, like the larval stage—as a possible source of unique metabolites that they manufacture to fend off predators (Abida *et al.*, 2013). For example, it has been shown that Antarctic sea star eggs (McClintock *et al.*, 1990), ascidian larvae (Lindquist *et al.*, 1992; Lopanik *et al.*, 2004), bryozoan larvae and *Lambieriella variabilis* larvae (Motti *et al.*, 2010), and Polychaeta (Cowart *et al.*, 2000) only synthesize unique compounds during their early life stages and not during their mature stages. These molecules are used in defense mechanisms. Bioprospectors are interested in plankton because of their fast biomass production and capacity to generate new compounds (Motti *et al.*, 2010). Planktons have the advantages of low energy requirements, cheaper manufacturing costs, and quicker synthesis of the target metabolite due to their short replication times (Abida *et al.*, 2013).

16.1.1.2 Function-Based, Culture-Independent Bioprospecting

Through a methodical approach, target species are typically the focus of bioprospecting efforts aimed at discovering novel products. Ever since the potential market worth of these compounds was discovered, businesses have been searching for new ways to speed up the process (Iskar *et al.*, 2012). Because it expedites the identification of intriguing compounds and strains by recognizing the genetic code, bioinformatics has garnered particular interest (Ortega et al., 2012). An innovative method for discovering novel compounds is to use next-generation sequencing (NGS) to study the genomic, metagenomic, or transcriptome profiling of marine water samples (Trindade *et al.*, 2015). NGS and the growth of sequencing programs have benefited marine biotechnology by expanding the omics approach—which includes proteomics and metabolomics—to examine the structure and function of new molecules (Hartmann *et al.*, 2014). Additionally, it forecasts how the recently discovered molecule will affect particular species or the environment in general. Once the molecules of interest are identified via omics, the organism or the gene of interest is taken and reinserted into

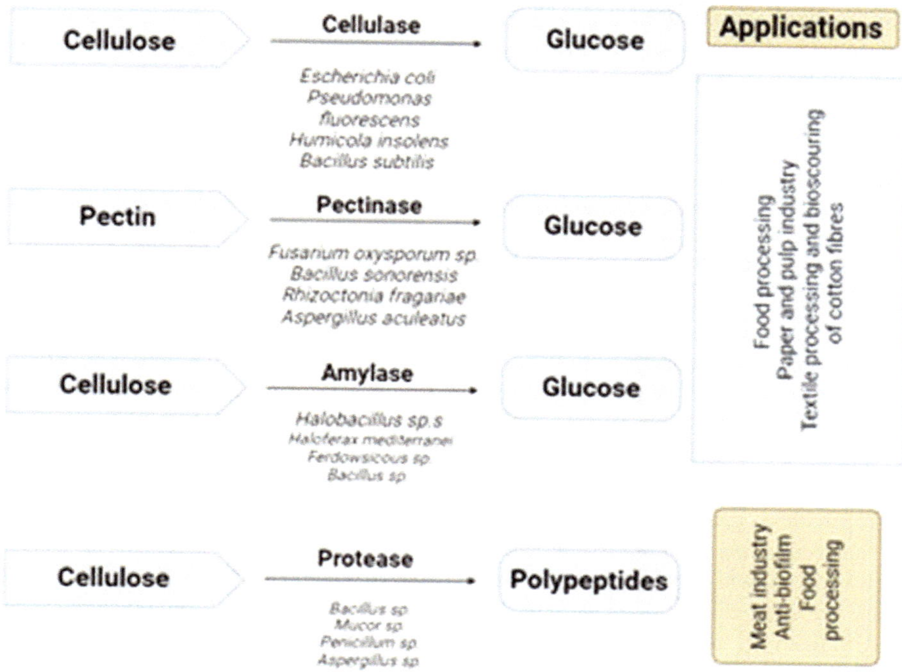

FIGURE 16.4 Various species that produce microbial enzymes with different characteristics valuable to industry. The figure shows some of the sources of enzymes that break down cellulose, pectin, starch, and proteins, as well as how they are used.

another organism by recombinant DNA (Ambrosino *et al.*, 2019). In a regulated manner, isolated or recombinant species are scaled up. For example, bioreactors are used to generate vast quantities of the desired chemical without affecting the natural population. According to Kim and Toldrá (2017), this is essential for preserving ecosystem balance and process sustainability.

16.1.1.3 Textile Manufacturing

Cellulase is essential in the textile industry because it improves the look, de-sizing, and dye adhesion of the materials (Kuhad *et al.*, 2011). Cellulase is also a great finishing agent, improving the softness, water absorbency, and clean, shining appearance of textile items. Endoglucanase II, generated by *Trichoderma reesei*, is frequently employed in the American laundry industry as a stone-washing agent (Galante *et al.*, 1998).

16.1.1.4 Industry of Food Processing

Fruits and vegetables can retain their flavor, texture, and aroma with the help of cellulase. Cellulase is a macerating enzyme that lessens taste and viscosity. Along with other macerating enzymes, it is also utilized in the extraction of olive oils (De Faveri et al., 2008; Singh et al., 2019). Cellulase, pectinase, and xylanase work together to improve extraction, flotation, and reduce viscosity in the wine industry, for instance (de Carvalho et al., 2008; Bajaj and Mahajan, 2019). Moreover, it has been demonstrated that microbial cellulase derived from *Trichoderma* species enhances feed conversion and meal digestibility based on cereals (Schmoll and Kubicek, 2003). Cellulase from *Bacillus subtilis* BL62, isolated from soybean residue, for example, demonstrated cellulase activity of 1.08 UI/ mg proteins (Heck et al., 2002).

16.1.1.5 Agriculture-Related Industries

In agriculture, cellulase enzymes are critical for flowering, improved root systems, seed germination, crop yield, and phytopathogen defense (Bhat, 2000). Additionally, cellulase increases soil fertility, which in turn promotes plant growth through the activity of microorganisms such as *Aspergillus*, *Chaetomium*, and *Trichoderma*. It is possible to isolate bacterial cellulase from *Actinomycetes* and *Ruminococcus albus*, as well as fungal cellulase from *Fusarium*, *Penicillium*, and *Aspergillus* (Bailey and Lumsden, 1998; Harman and Kubicek, 1998).

16.1.1.6 Conserving Biodiversity and Engaging in Bioprospecting

However, it should be noted that the benefits and drawbacks of bioprospecting must be considered against the rising prevalence of biopiracy, the current ecosystem decline crisis, and the likely effects of the impending mass extinction spasm predicted for the tropical biomes (Myers, 1987; Pimm *et al.*, 1995). Approximately 5% of the planet's land area is covered by protected area networks. If human interference persists in the remaining 95% of unprotected wildland ecosystems, roughly 50% of species will become extinct (Pimm and Lawton, 1998). This is a fact of life, so any bioprospecting program should be implemented with the understanding that, in addition to providing immediate financial gains, these efforts will either directly or indirectly support the preservation, inventory, and monitoring of biodiversity, both in situ and ex situ.

Thereby, assessments, national policy and legislation, capacity building, equitable benefit-sharing mechanisms, participatory management involving all bioprospecting stakeholders (including local and indigenous communities), and financial resource mobilization for bioprospecting and other sustainable uses of biodiversity should be the main focuses of bioprospecting practices and strategies, according to Ten Kate and Laird (2019).

16.1.2 Biomimicry

Since the beginning of life on Earth, around 3.8 billion years ago, nature has undergone evolution (Gordon, 1976). With materials that are readily available, nature has evolved into highly functional objects. Both the macro- and nanoscales are affected by this evolution. Nanomaterials, nanodevices, and nanoprocesses can be mimicked and created using an understanding of the roles that natural objects and processes conduct. The word "biomimetics" refers to design that is biologically inspired, adapted, or derived from nature. It entails mimicking nature or biology.

The promising emerging field of study called biomimicry is defined as a way to solve design issues that are influenced by natural elements, systems, and models (Pawlyn, 2019). Combining the Greek terms *bios*, which means "life," and *mimesis*, which denotes "imitation," Janine M. Benyus created the phrase in 1997. Terms used to characterize design influenced by nature include biomimetics, bioinspired, and biologically inspired (Speck and Speck, 2008). Natural systems are recognized for their capacity for energy conservation, intelligence, self-sufficiency, and circular resource utilization (Benyus, 1997; Chen *et al.*, 2020).

16.1.2.1 Guiding Principle

The fundamental idea behind biomimicry is to use natural phenomena as inspiration to solve human and/or design issues in a more sustainable way (Knippers *et al.*, 2016). The commitment to sustainability lies at the heart of biomimetic innovation. Nature is an expert at designing closed-loop systems that minimize waste and maximize efficiency. Biomimicry tries to mimic the regeneration ability observed in natural ecosystems by emphasizing sustainable practices (Benyus, 1997). Nature's designs are not only utilitarian but also adaptive. Biomimicry strives to understand and imitate this adaptive functioning, resulting in systems and materials that can respond dynamically to changing conditions (Speck and Speck, 2003).

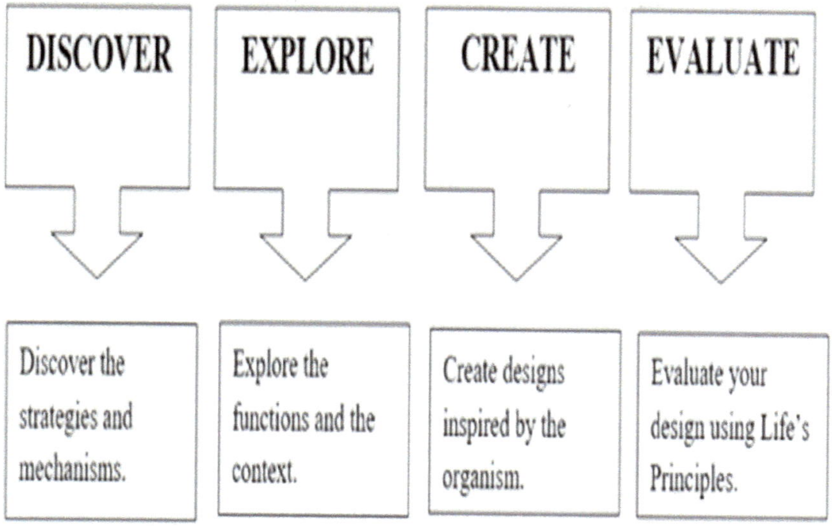

FIGURE 16.5 Evaluating the representation of therapeutic frameworks.

16.1.2.2 Theoretical Framework for Biomimetic Innovation

In materials research, biomimicry is inspired by the structural brilliance found in living materials. For example, the intricate shapes of seashells have impacted the development of lightweight composites with outstanding strength and resilience (Vincent *et al.*, 2006).

Biomimicry has ushered in a new age in the realm of robotics. Researchers create robots with improved agility and flexibility by simulating animal locomotion and sensory systems. The design of bioinspired robots is influenced by nature's blueprints, which range from the efficient movements of insects to the navigation tactics of marine organisms (Ramezani *et al.*, 2018).

16.1.2.3 Rejuvenating with Nature: Bioinspired Medical Solutions

Biomimetic techniques in medicine offer promising strategies for drug delivery and tissue regeneration by mimicking nature's abilities. Biomimicry in medicine distribution takes advantage of the human circulatory system's accuracy. Researchers create targeted medication delivery devices that mimic nature's complicated network, inspired by the efficiency of blood vessels. This simulation reduces side effects while increasing treatment efficacy in specific tissues or organs (Sinha *et al.*, 2018).

In the 3D matrix, biomimicking hydrogel effectively aided cell anchoring, motility, and invasion in the permissive zone and contacts with integrin receptors. This biomimicking hydrogel can precisely remodel the skin by attracting peripheral cells, promoting collagen deposition and angiogenesis, and efficiently controlling early inflammation (Cia *et al.*, 2021). Furthermore, biomimicry is critical in resolving wound healing and tissue regeneration issues. Nature's time-tested tissue repair techniques serve as a roadmap for the development of bioinspired medical treatments. These advancements have the potential to revolutionize traditional ways of healing, providing more efficient and sustainable patient care solutions (Sinha *et al.*, 2018).

16.1.2.4 Architectural Marvels: Eco-Friendly Designs Drawn from Nature

In the last 50 years, architecture has developed, reprioritizing goals and shifting toward collaboration with other disciplines (Arslan, 2014). The development of bioinspired hair flow sensors was modeled after aquatic and arthropod species, which are able to filter out loud or unwanted information in addition to being able to "feel" the flow and "listen" to particular frequency bands (Sterbing-D'Angelo *et al.*, 2011). Termite mounds, where natural ventilation systems control temperature, are one prominent source of effect. Because of this inspiration, structures that imitate these effective airflow mechanisms have been created, which lessens the need for artificial heating and cooling systems (Pohl *et al.*, 2015). Architectural materials that emphasize strength and environmental sustainability have also been influenced by the tensile strength and flexibility seen in natural structures, such as spider silk. Resilient ecosystems have served as inspiration for designs in urban planning, which have been informed by biomimicry principles. Urban landscapes are changing to include green areas and sustainable infrastructure in an effort to emulate nature's capacity for adaptation and growth (Pohl *et al.*, 2015).

Architecture is becoming more ecologically sensitive and harmoniously linked with the natural world through the use of biomimicry, which is opening the door to a sustainable and aesthetically beautiful future.

16.1.2.5 Industrial Significance

Despite the fact that the term "biomimetics" is not very old, millennia ago, our predecessors used nature as an inspiration to create a variety of materials and tools (Ball, 2002; Vincent *et al.*, 2006; Anon, 2007; Meyers *et al.*, 2008). For instance, over 3000 years ago, the Chinese attempted to create artificial silk. The Renaissance artist Leonardo da Vinci observed how birds flew and made suggestions for flying machine ideas. Throughout the 20th century, nature served as an inspiration for many things, including airplane design. Since the 1980s, artificial intelligence and neural networks in information technology have been motivated by the desire to mimic the human brain. Nanotechnologists are inspired by the presence of DNA and bio-cells and expect to develop self-assembling molecular devices in the future. Proteins are being utilized in molecular biomimetics to regulate the production of materials in order to build self-assembling, hybrid, functional materials structures (Grunwald *et al.*, 2009; Sarikaya and Tamerler, 2009). The "lotus effect" has been utilized since the mid-1990s to create a range of surfaces with low adhesion, self-cleaning properties, superhydrophobicity, and reduced drag in fluid flow (Bhushan *et al.*, 2009). To create the treads of wall-climbing robots, the dynamic climbing and peeling ability of geckos has been replicated (Cutkosky and Kim, 2009). Sharkskin replication has been utilized to create low-drag moving objects, such as full-body swimsuits.

The nanoscale architecture found in nature for optical reflection and antireflection has been used to build reflecting and antireflecting surfaces. Seashells and other biomimetic materials provide the inspiration for a subset of bioinspired ceramics in the subject of biomimetic materials. The polar bear's fur has served as an inspiration for the creation of artificial furs and textiles. Self-repair is interested in the biological system's ability to mend themselves. The creation of sensory-aid technology is likewise being guided by biomimetics (Cutkosky and Kim, 2009). Objects in nature have a variety of nanoscale traits. Since the early 1990s, there has been a strong focus on nanoscience and nanotechnology, which has significantly accelerated efforts to use nanofabrication techniques for commercial applications that resemble nature (Bhushan, 2007a). Numerous academic fields have been interested in biomimetics.

16.1.2.6 An Overview of Multiple Natural Things and Their Chosen Functions

See Figure 16.6.

FIGURE 16.6 Multiple natural things and their chosen functions.

16.1.2.7 Opportunities and Challenges: Managing the Biomimetic Discovery Path

As biomimetic research picks up steam, it faces opportunities as well as obstacles that will determine how far it can go toward sustainable innovation. Among the most significant obstacles are the moral questions raised by using nature's blueprints. Finding a middle ground between protecting ecological sacredness and using biomimicry for human advantage is still a topic of discussion (Benyus, 1997).

The implementation of biological concepts in real-world settings presents another challenge. Although there is plenty to be inspired by in nature, replicating its complexity calls for continual research and interdisciplinary collaboration (Vincent *et al.*, 2006). Furthermore, the commercialization of biomimetic technology highlights the need for ethical frameworks by bringing up issues with equitable benefit-sharing and intellectual property rights (Benyus, 1997).

But these difficulties also present a wealth of opportunity. Sustainability can be accelerated by incorporating biomimicry into established scientific and industrial processes. In addition to supporting ethical biomimetic practices, cooperative efforts to conserve biodiversity also promote environmental conservation (Pohl *et al.*, 2015).

A careful balance must be struck when navigating the route of biomimetic discovery, taking advantage of opportunities for innovation and tackling obstacles to guarantee morally sound, long-lasting, win-win results.

16.1.2.8 Prospects for the Future: Using Biomimicry to Shape Tomorrow's Technology

Technologies of the future are being shaped by biomimicry, which is a leading innovator and catalyst for change. Finding sustainable and effective solutions is made possible by the incorporation of biological principles into a wide range of scientific fields. The field of materials science is one promising direction where scientists are able to create materials with previously unheard-of qualities by taking inspiration from the variety of structures seen in nature (Vincent *et al.*, 2006). Technologies like electronics and aerospace could undergo a radical change thanks to this biomimetic strategy.

According to Ramezani *et al.* (2018), the field of robotics will eventually produce highly flexible and nimble robots that are modeled after the sensory organs and modes of mobility of animals. Automated systems, exploration, and disaster response could all benefit from these bioinspired devices.

TABLE.16.2
Biomimetic Applications: Borrowing from Nature for Innovation

Biomimetic Application	Biomimicked Element	Bioinspired Design	Industry/Application
Velcro	Burdock burrs	Hook-and-loop	Textile, fastening
Shinkansen train nose	Kingfisher beak	Streamlined shape	Transportation, high-speed
Water-repellent fabric	Lotus leaf	Super hydrophobicity	Textile, outdoor gear
Gecko-inspired adhesives	Gecko feet	Dry adhesion	Robotics, climbing
Wind turbine blades	Humpback whale fins	Tubercles	Renewable energy, wind power

Furthermore, the developing science of synthetic biology presents opportunities for molecular-level biomimetic breakthroughs as biology and technology combine. The goal of research is to develop bioinspired materials and tools that can be easily incorporated into biological systems, providing opportunities for environmental cleanup and medicinal advancements.

The potential of biomimicry to not only address urgent problems but also promote a positive interaction between technology and the natural world is what will determine its destiny. Through the ongoing use of nature's tried-and-true solutions, biomimicry facilitates the development of a technological environment that is more resilient, sustainable, and bioinspired (Table 16.2).

16.2 DISCUSSION

Bioprospecting, sometimes called "nature's treasure hunt," is the methodical search for valuable substances, genes, and organisms with potential uses in a variety of sectors by systematically examining Earth's biological diversity. From the immensity of rainforests to the depths of the oceans, our planet's biological diversity provides a wealth of inspiration for scientific and economic innovation. Researchers have found chemicals with substantial medicinal promise through bioprospecting; the anticancer medication paclitaxel was discovered from the Pacific yew tree (*Taxus brevifolia*) (Wani *et al.*, 1971). Additionally, bioprospecting has an impact on agriculture, as the discovery of advantageous microbes improves agricultural output and soil fertility (Kloepper *et al.*, 1989).

The ethical considerations surrounding bioprospecting are critical. To guarantee the fair distribution of benefits and the preservation of biodiversity, cooperation with local populations and a commitment to moral principles are crucial (Laird and Wynberg, 1918). International agreements such as the Nagoya Protocol aim to establish fair and just benefit-sharing arrangements for the utilization of genetic resources (Convention on Biological Diversity, 2011).

Simultaneously, biomimicry is a method of design and problem-solving that takes cues from the patterns and mechanisms of nature to produce novel, sustainable solutions. Biomimicry represents "innovation through imitation" in a variety of ways, from sleek bullet trains modeled after a kingfisher's beak to Velcro, which was inspired by the sticky qualities of burdock burrs (Mesterton-Gibbons *et al.*, 1996). This strategy goes beyond simple imitation; it entails comprehending and modifying the fundamental laws of nature to meet the needs of humankind.

The influence of biomimicry is widespread across many different industries. In order to get the best ventilation and temperature control, termite mound efficiency is emulated in architectural design (Turner, 2000). The water-repellent qualities of lotus leaves have served as an inspiration for material scientists to create self-cleaning surfaces (Barthlott and Neinhuis, 1997). The adaptability of biomimicry demonstrates its potential to transform a variety of industries, including transportation and renewable energy.

When bioprospecting guides biomimicry efforts, the potential for collaboration between the two fields is especially clear. Bioprospecting offers a multitude of insights into distinct biological structures and processes by means of investigating heterogeneous environments. The use of this information is then made possible by biomimicry, which incorporates natural solutions into technological systems that are created by humans.

For example, bioprospecting has made it possible to examine the adhesive characteristics of gecko feet, which has led to the invention of gecko-inspired adhesives that stick well and don't leave residues. These adhesives find use in the manufacturing and robotics industries (Autumn *et al.*, 2002). The creation of self-cleaning surfaces has been impacted by the lotus effect, which was identified through bioprospecting studies into the water-repellent qualities of lotus leaves (Neinhuis and Barthlott, 1997).

Although bioprospecting and biomimicry hold great potential, there are still obstacles to overcome. Access and benefit-sharing concerns in bioprospecting make ethical frameworks and international agreements necessary to direct responsible exploration. Conversely, biomimicry encounters difficulties in converting biological concepts into useful applications, necessitating multidisciplinary cooperation and ongoing study.

The incorporation of bioprospecting and biomimicry into standard scientific and industrial procedures is the way forward for these fields of study. Sustainability, morality, and biodiversity conservation must be given top priority in cooperative efforts. The integration of bioinspired solutions into various industries, such as healthcare and manufacturing, has the potential to foster a more balanced association between human ingenuity and the natural world.

16.3 CONCLUSION

A symbiotic approach to sustainable innovation is provided by the parallel but connected pathways of bioprospecting and biomimicry. Bioprospecting reveals nature's wealth through biological variety investigation, and biomimicry turns these findings into workable, sustainable solutions. Their combined efforts demonstrate how well scientific research and technical creativity can work together, and they hold out hope for a time when human inventiveness will easily coexist with the robustness and efficiency of the natural world.

REFERENCES

Abida H, Ruchaud S, Rios L, Humeau A, Probert I, De Vargas C, Bach S, Bowler C. Bioprospecting marine plankton. *Marine Drugs*. 2013 Nov 14;11(11):4594–611.

Al-Battashi H, Annamalai N, Al-Kindi S, Nair AS, Al-Bahry S, Verma JP, et al. Production of bioplastic (poly-3-hydroxybutyrate) using waste paper as a feedstock: Optimization of enzymatic hydrolysis and fermentation employing *Burkholderia sacchari*. *J Clean Prod*. 2019; 214:236–47.

Alejandre-Colomo C, Harder J, Fuchs BM, Rossello-Mora R, Amann R. High-throughput cultivation of heterotrophic bacteria during a spring phytoplankton bloom in the North Sea. *System Appl Microbiol*. 2020 Mar 1;43(2):126066.

Alkayyali T, Pope E, Wheatley SK, Cartmell C, Haltli B, Kerr RG, Ahmadi A. Development of a microbe domestication pod (MD Pod) for in situ cultivation of micro-encapsulated marine bacteria. *Biotechnol Bioeng*. 2021 Mar;118(3):1166–76.

Ambrosino L, Tangherlini M, Colantuono C, Esposito A, Sangiovanni M, Miralto M, Sansone C, Chiusano ML. Bioinformatics for marine products: An overview of resources, bottlenecks, and perspectives. *Marine Drugs*. 2019 Oct 11;17(10):576.

Anon. Biomimetics: Strategies for Product Design Inspired by Nature. The Royal Society; 2007.

Arslan GY. Biomimetic architecture a new interdisciplinary approach to architecture. *ALAM CIPTA, Int J Sustain Trop Design Res Practice*. 2014 Dec 23;7(2):29-36.

Atencio LA, Dal Grande F, Young GO, Gavilán R, Guzmán HM, Schmitt I, Mejía LC, Gutiérrez M. Antimicrobial-producing Pseudoalteromonas from the marine environment of Panama shows a high phylogenetic diversity and clonal structure. *J Basic Microbiol*. 2018 Sep;58(9):747–69.

Autumn K, Sitti M, Liang YA, Peattie AM, Hansen WR, Sponberg S, Kenny TW, Fearing R, Israelachvili JN, Full RJ. Evidence for van der Waals adhesion in gecko setae. *Proc Nat Acad Sci.* 2002 Sep 17;99(19):12252–6.

Bailey BA, Lumsden RD. Gliocladium on plant growth and resistance to pathogens. *Trichoderma and Gliocladium: Enzymes, Biological Control and Commercial Applications.* Taylor and Francis, London. 1998 Jul 31:185–204.

Bajaj P, Mahajan R. Cellulase and xylanase synergism in industrial biotechnology. *Appl Microbiol Biotechnol.* 2019 Nov; 103:8711–24.

Ball P. *The Self-Made Tapestry: Pattern Formation in Nature.* Oxford Univerdity Press; 2002.

Barone G, Rastelli E, Corinaldesi C, Tangherlini M, Danovaro R, Dell'Anno A. Benthic deep-sea fungi in submarine canyons of the Mediterranean Sea. *Progress Oceanograp.* 2018 Nov 1; 168:57–64.

Barthlott W, Neinhuis C. Purity of the sacred lotus, or escape from contamination in biological surfaces. *Planta.* 1997 Apr; 202:1–8.

Bell J, Paula L, Dodd T, Németh S, Nanou C, Mega V, Campos P. EU ambition to build the world's leading bioeconomy—Uncertain times demand innovative and sustainable solutions. *New Biotechnol.* 2018 Jan 25;40:25–30.

Benyus JM. Biomimicry: Innovation inspired by nature. 1997.

Bessada SM, Barreira JC, Oliveira MB. Pulses and food security: Dietary protein, digestibility, bioactive and functional properties. *Trend Food Sci Technol.* 2019 Nov 1; 93:53–68.

Bhat M. Cellulases and related enzymes in biotechnology. *Biotech Adv.* 2000 Aug 1;18(5):355–83.

Bhushan B. *Springer handbook of nanotechnology.* 2nd edn. 2007a Heidelberg, Germany:Springer.

Bhushan B, Jung YC, Koch K. Micro-, nano-and hierarchical structures for superhydrophobicity, self-cleaning and low adhesion. *Philos Trans Royal Soc A: Mathe Phys Eng Sci.* 2009 May 13;367(1894):1631–72.

Bibi F, Faheem M, Azhar EI, Yasir M, Alvi SA, Kamal MA, et al. Bacteria from marine sponges: a source of new drugs. *Curr Drug Metab.* 2017;18:11–5.

Bleakley S, Hayes M. Algal proteins: extraction, application, and challenges concerning production. *Foods.* 2017 Apr 26;6(5):33.

Bounnit T, Saadaoui I, Rasheed R, Schipper K, Al Muraikhi M, Al Jabri H. Sustainable production of Nannochloris atomus biomass towards biodiesel production. *Sustainability.* 2020 Mar 5;12(5):2008.

Buck M, Hamilton C. The Nagoya Protocol on access to genetic resources and the fair and equitable sharing of benefits arising from their utilization to the Convention on Biological Diversity. *Rev Eur Commun Int Environ Law.* 2011 Apr;20(1):47–61.

Bull AT, Goodfellow M. Dark, rare and inspirational microbial matter in the extremobiosphere: 16 000 m of bioprospecting campaigns. *Microbiology.* 2019;165:1252–64.

Chen Austin M, Garzola D, Delgado N, Jiménez JU, Mora D. Inspection of biomimicry approaches as an alternative to address climate-related energy building challenges: a framework for application in Panama. *Biomimetics.* 2020 Aug 24;5(3):40.

Chittora D, Meena M, Barupal T, Swapnil P, Sharma K. Cyanobacteria as a source of biofertilizers for sustainable agriculture. *Biochem Biophys Rep.* 2020;22:100737.

Cowart JD, Fielman KT, Woodin SA, Lincoln DE. Halogenated metabolites in two marine polychaetes and their planktotrophic and lecithotrophic larvae. *Mar Biol.* 2000; 136:993–1002

Cutkosky MR, Kim S. Design and fabrication of multi-material structures for bioinspired robots. *Philos Trans Royal Soc A: Math Phys Eng Sci.* 2009 May 13;367(1894):1799–813.

Dambek M, Eilers U, Breitenbach J, Steiger S, Buchel C, Sandmann G. Biosynthesis of fucoxanthin and diadinoxanthin and function of initial pathway genes in Phaeodactylum tricornutum. *J Exp Bot.* 2012;63:5607–12.

Danovaro R, Corinaldesi C, Dell'Anno A, Fuhrman JA, Middelburg JJ, Noble RT, Suttle CA. Marine viruses and global climate change. *FEMS Microbiol Rev.* 2011 Nov 1;35(6):993–1034.

De Carvalho LM, De Castro IM, Da Silva CA. A study of retention of sugars in the process of clarification of pineapple juice (*Ananas comosus*, L. Merril) by micro-and ultra-filtration. *J Food Eng.* 2008 Aug 1;87(4):447–54.

De Faveri D, Aliakbarian B, Avogadro M, Perego P, Converti A. Improvement of olive oil phenolics content by means of enzyme formulations: Effect of different enzyme activities and levels. *Biochem Eng J.* 2008 Sep 1;41(2):149–56.

Delbarre-Ladrat C, Sinquin C, Lebellenger L, Zykwinska A, Colliec-Jouault S. Exopolysaccharides produced by marine bacteria and their applications as glycosaminoglycan-like molecules. *Front Chem.* 2014;2:1–15.

El-Bondkly AMA. Molecular identification using ITS sequences and genome shuffling to improve 2-deoxyglucose tolerance and xylanase activity of marine- derived fungus, Aspergillus sp. NRCF5. *Appl Biochem Biotechnol.* 2012;167:2160–73

Evans BS, editor. *Nonribosomal Peptide and Polyketide Biosynthesis: Methods and Protocols.* 1401. New York: Springer; 2016.

Galante YM, De Conti A, Monteverdi R. Application of Trichoderma enzymes in the textile industry. *Trichoderma Gliocladium.* 1998 Jul 31; 2:311–25.

Galasso C, Gentile A, Orefice I, Ianora A, Bruno A, Noonan DM, Sansone C, Albini A, Brunet C. Microalgal derivatives as potential nutraceutical and food supplements for human health: a focus on cancer prevention and interception. *Nutrients.* 2019 May 29;11(6):1226.

Garlapati D, Chandrasekaran M, Devanesan A, Mathimani T, Pugazhendhi A. Role of cyanobacteria in agricultural and industrial sectors: an outlook on economically important byproducts. *Appl Microbiol Biotechnol.* 2019;103:4709–21.

Gordon JE. *The New Science of Strong Materials or Why You Don't Fall Through the Floor.* Princeton University Press; 1976.

Grahl S, Strack M, Mensching A, Mörlein D. Alternative protein sources in Western diets: Food product development and consumer acceptance of spirulina-filled pasta. *Food Quality Preference.* 2020 Sep 1;84:103933.

Grunwald I, Rischka K, Kast SM, Scheibel T, Bargel H. Mimicking biopolymers on a molecular scale: nano (bio) technology based on engineered proteins. *Philoso Trans Royal Soc A: Mathe Phys Eng Sci.* 2009 May 13;367(1894):1727-47.

Hamilton-Miller JMT. Development of the semi-synthetic penicillins and cephalosporins. *Int J Antimicrob Agents.* 2008;31:189–92.

Harman GE, Kubicek CP, editors. *Trichoderma and Gliocladium, volume 2: Enzymes, biological control and commercial applications.* CRC Press; 1998 Jul 31.

Hartmann EM, Durighello E, Pible O, Nogales B, Beltrametti F, Bosch R, Christie-Oleza JA, Armengaud J. Proteomics meets blue biotechnology: A wealth of novelties and opportunities. *Marine Genomics.* 2014 Oct 1;17:3542.

Heck JX, Hertz PF and Ayub MAZ. Cellulase and xylanase productions by isolated Amazon Bacillus strains using soybean industrial residue based solid-state cultivation. *Braz. J. Microbiol.* 2002, 33(3):213–218.

Hosseini H, Saadaoui I, Moheimani N, Al Saidi M, Al Jamali F, Al Jabri H, Hamadou RB. Marine health of the Arabian Gulf: Drivers of pollution and assessment approaches focusing on desalination activities. *Marine Pollution Bulletin.* 2021 Mar 1;164:112085.

Hu H, Natarajan VP, Wang F. Towards enriching and isolation of uncultivated archaea from marine sediments using a refined combination of conventional microbial cultivation methods. *Marine Life Sci Technol.* 2021 May; 3:23142.

Indrayani I, Moheimani NR, Borowitzka MA. Long-term reliable culture of a halophilic diatom, Amphora sp. MUR258, in outdoor raceway ponds. *J Appl Phycol.* 2019;31: 2771–8.

Iskar M, Zeller G, Zhao XM, van Noort V, Bork P. Drug discovery in the age of systems biology: the rise of computational approaches for data integration. *Curr Opini Biotechnol.* 2012 Aug 1;23(4):609–16.

Ten Kate KT. Biopiracy or green petroleum? Expectations & best practice in bioprospecting. 1995 p.61.

Katz L, Baltz RH. Natural product discovery: past, present, and future. *J Ind Microbiol Biotechnol.* 2016 Mar 1;43(2-3):155–76.

Kim S, Park MS, Song J, Kang I, Cho JC. High-throughput cultivation based on dilution-to-extinction with catalase supplementation and a case study of cultivating acI bacteria from Lake Soyang. *J Microbiol.* 2020 Nov;58:893–905.

Kim SK, Toldrá F. *Advances in food and nutrition research.* 1st ed. Cambridge, San Diego, Oxford, London: Elsevier, Academic Press; 2017.

Kloepper JW, Lifshitz R, Zablotowicz RM. Free-living bacterial inocula for enhancing crop productivity. *Trend Biotechnol.* 1989 Feb 1;7(2):39–44.

Knippers J, Nickel KG, Speck T. *Biomimetic research for architecture and building construction.* Switzerland: Springer International Publishing. 2016.

Kuhad RC, Gupta R, Singh A. Microbial cellulases and their industrial applications. *Enzyme Research.* 2011; 2011, 1–10.

Laird SA, Ten Kate K. Biodiversity prospecting–: the commercial use of genetic resources and best practice in benefit-sharing. In *Biodiversity and Traditional Knowledge,* 2010 Sep 23 (pp. 275–277). Routledge.

Laird, S. A., & Wynberg, R. P. Benefit sharing in the Nagoya Protocol: Too conceptual to work? *PLoS Biol.* 2018;16(4):e2006451.

Lauritano C, Andersen JH, Hansen E, Albrigtsen M, Escalera L, Esposito F, et al. Bioactivity screening of microalgae for antioxidant, anti-inflammatory, anticancer, anti-diabetes, and antibacterial activities. *Front Mar Sci.* 2016;3:1–12.

Lewis K, Epstein S, D'onofrio A, Ling LL. Uncultured microorganisms as a source of secondary metabolites. *J Antibiotics.* 2010 Aug;63(8):468–76.

Lindquist N, Hay ME, Fenical W. Defense of ascidians and their conspicuous: adult vs. larval chemical defenses. *Ecol Monogr.* 1992;62:547–68.

Lopanik N, Lindquist N, Targett N. Potent cytotoxins produced by a microbial symbiont protect host larvae from predation. *Oecologia.* 2004;139:131–9.

Lorente A, Makowski K, Albericio F, Álvarez M. Bioactive marine polyketides as potential and promising drugs. *Ann Mar Biol Res.* 2014; 1:1003.

Maeda Y, Yoshino T, Matsunaga T, Matsumoto M, Tanaka T. Marine microalgae for production of biofuels and chemicals. *Curr Opin Biotechnol.* 2018;50:111–20.

Mateo N, Nader W, Tamayo G. Bioprospecting. Encyclopedia of biodiversity. 2001 Jan 1;1:471–87.

Matos J, Cardoso C, Gomes A, Campos AM, Falé P, Afonso C, et al. Bioprospection of Isochrysis galbana and its potential as a nutraceutical. *Food Funct.* 2019;10:7333–42.

Matsumoto M, Nojima D, Nonoyama T, Ikeda K, Maeda Y, Yoshino T, et al. Outdoor cultivation of marine diatoms for year-round production of biofuels. *Mar Drugs.* 2017;15:94.

McClintock JB, Vernon JD. Chemical defense in the eggs and embryos of antarctic sea stars (Echinodermata). *Mar Biol.* 1990;105:491–5.

Mesterton-Gibbons M, Dugatkin LA. Cooperation and conflict in animal societies: From experiments to theory. Nature. 1996;380:142–3.

Meyers MA, Chen PY, Lin AY, Seki Y. Biological materials: Structure and mechanical properties. *Progress Mater Sci.* 2008 Jan 1;53(1):1–206.

Mora C, Tittensor DP, Adl S, Simpson AG, Worm B. How many species are there on Earth and in the ocean? *PLoS Biol.* 2011 Aug 23;9(8):e1001127.

Moreau D, Tomasoni C, Jacquot C, Kaas R, Le Guedes R, Cadoret JP, et al. Cultivated microalgae and the carotenoid fucoxanthin from Odontella aurita as potent anti-proliferative agents in bronchopulmonary and epithelial cell lines. *Environ Toxicol Pharmacol.* 2006;22:97–103.

Motti CA, Ettinger-Epstein P, Willis RH, Tapiolas DM. ESI FTICR-MS analysis of larvae from the marine sponge Luffariella variabilis. *Mar Drugs.* 2010;8:190–9.

Myer PR, Clemmons BA, Schneider LG, Ault TB. Microbiomes in ruminant protein production and food security. *CABI Rev.* 2019 Mar 6(2019):1–1.

Myers N. The extinction spasm impending: synergisms at work. *Conservation Biol.* 1987 May;1(1):14–21.

NEINHUIS C, BARTHLOTT W. Characterization and distribution of water-repellent, self-cleaning plant surfaces. *Annals of Botany.* 1997 Jun 1;79(6):667–77.

Nguyen TT, Barber AR, Corbin K, Zhang W. Lobster processing by-products as valuable bioresource of marine functional ingredients, nutraceuticals, and pharmaceuticals. *Bioresour Bioprocess.* 2017;4:27.

Nowrotek M, Jałowiecki Ł, Harnisz M, Płaza GA. Culturomics and metagenomics: In understanding of environmental resistome. *Front Environ Sci Eng.* 2019 Jun;13:1–2.

Ortega SS, Cara LC, Salvador MK. In silico pharmacology for a multidisciplinary drug discovery process. *Drug Metabolism and Drug Interactions.* 2012 Dec 1;27(4):199–207.

Oyemitan IA, Kuete V. Chapter 27—African medicinal spices of genus Piper. *Medicinal Spices and Vegetables from Africa.* Elsevier Inc.: Academic Press; 2017:581–97.

Pawlyn M. *Biomimicry in architecture.* Routledge; 2019 Aug 12.

Pimm SL, Lawton JH. Planning for biodiversity. *Science.* 1998 Mar 27;279(5359):2068–9.

Pimm SL, Russell GJ, Gittleman JL, Brooks TM. The future of biodiversity. *Science.* 1995 Jul 21;269(5222):347–50.

Pohl EA, Speck T, Speck O. Biomimetic innovation as an interdisciplinary challenge. *Philos Trans R Soc B Biol Sci.* 2015;370(1666):20140085.

Pushpangadan P, Ijinu TP, Dan VM, George V. Trends in bioprospecting of biodiversity in new drug design. *Pleione.* 2015;9(2):267–82.

Ramezani M, Mofrad SA, Aghdam MM. A review of biomimetic underwater robots inspired by fish. *Int J Adv Robotic Syst.* 2018;15(1):1729881417744385.

Ramsdell J. The molecular and integrative basis to mammalian brevetoxin toxicityIn: Botana WL, Hui WLYH, editors. Red. *Phycotoxins: Chemistry and Biochemistry.* United States: Blackwell Publishing; 2007.

Saiki RK, Gelfand DH, Stoffel S, Scharf SJ, Higuchi R, Horn GT, et al. Primer-directed enzymatic amplification of DNA with a thermostable DNA polymerase. *Science.* 1988;239:487–91.

Sainis I, Fokas D, Vareli K, Tzakos A, Kounnis V, Briasoulis E. Cyanobacterial cyclopeptides as lead compounds to novel targeted cancer drugs. *Mar Drugs.* 2010;8:629–57.

Schipper K, Al Muraikhi M, Alghasal GS, Saadaoui I, Bounnit T, Rasheed R, Dalgamouni T, Al Jabri HM, Wijffels RH, Barbosa MJ. Potential of novel desert microalgae and cyanobacteria for commercial applications and CO2 sequestration. *Journal of Applied Phycology.* 2019 Aug 1;31:2231–43.

Schmoll M, Kubicek CP. Regulation of Trichoderma cellulase formation: lessons in molecular biology from an industrial fungus. *Acta Microbiologica et Immunologica Hungarica.* 2003 Jun 1;50(2-3):125–45.

Schüler LM, Santos T, Pereira H, Duarte P, Katkam NG, Florindo C, et al. Improved production of lutein and β -carotene by thermal and light intensity upshifts in the marine microalga Tetraselmis sp. CTP4. Algal Res. 2020;45: 101732.

Senthil K, Ramesh C, Dhanasekaran D, Thajuddin N. Marine yeasts: Potential sources for biotechnological applications. Journal of Microbiology, Biotechnology and Food Sciences. 2019;8(5):1047–54.

Sharma N, Singh A, Bhatia S, Batra N. Marine microbes in bioremediation: Current status and future trends. In: Kumar A, Sharma S, editors. *Microbes and Enzymes in Soil Health and Bioremediation.* Singapore: Springer; 2019. p. 133–48.

Singh J, Kundu D, Das M, Banerjee R. Enzymatic processing of juice from fruits/vegetables: An emerging trend and cutting-edge research in food biotechnology. *Enzymes in Food Biotechnology.* 2019 Jan 1:419–32.

Sinha VR, Singh A, Singh G, Khurana S. Biomimetic drug delivery systems: Potential and perspectives. Journal of Controlled Release. 2018;286:147–59.

Speck O, Speck D. From nature to applications: The fascinating world of structural biomimetics. *MRS Bull.* 2003;28(5):315–319.

Speck T, Speck O, Beheshti N, McIntosh AC. Process sequences in biomimetic research. *Design and Nature IV.* 2008 Jun 9; 114:3–11.

Sterbing-D'Angelo S, Chadha M, Chiu C, Falk B, Xian W, Barcelo J, Zook JM, Moss CF. Bat wing sensors support flight control. *Proceedings of the National Academy of Sciences.* 2011 Jul 5;108(27):11291–6.

Südfeld C, Hubáček M, D'Adamo S, Wijffels RH, Barbosa MJ. Optimization of high-throughput lipid screening of the microalga Nannochloropsis oceanica using BODIPY 505/515. Algal Research. 2021 Mar 1; 53:102138.

Sysoev M, Grötzinger SW, Renn D, Eppinger J, Rueping M, Karan R. Bioprospecting of novel extremozymes from prokaryotes—The advent of culture-independent methods. *Frontiers in Microbiology.* 2021 Feb 10;12:630013.

Tamerler C, Sarikaya M. Molecular biomimetics: nanotechnology and bionanotechnology using genetically engineered peptides. *Philosophical Transactions of the Royal Society A: Mathematical, Physical and Engineering Sciences.* 2009 May 13;367(1894):1705–26.

Ten Kate K, Laird SA. The commercial use of biodiversity: access to genetic resources and benefit-sharing. Routledge; 2019 Sep 18.

Trindade M, Van Zyl LJ, Navarro-Fernández J, Abd Elrazak A. Targeted metagenomics as a tool to tap into marine natural product diversity for the discovery and production of drug candidates. *Frontiers in Microbiology.* 2015 Aug 28;6:890.

Turner JS. On the evolution of thermal convection. In: *Mantle plumes: Their identification through time.* American Geophysical Union; 2000. p. 131–14.

Vågsholm I, Arzoomand NS, Boqvist S. Food security, safety, and sustainability—getting the trade-offs right. *Frontiers in Sustainable Food Systems.* 2020:16.

Vincent JF, Bogatyreva OA, Bogatyrev NR, Bowyer A, Pahl AK. Biomimetics: its practice and theory. *Journal of the Royal Society Interface.* 2006 Aug 22;3(9):471–82.

Vincent JF, Bogatyreva OA, Bogatyrev NR, Bowyer A, Pahl AK. Biomimetics: its practice and theory. *Journal of the Royal Society Interface.* 2006 Aug 22;3(9):471–82.

Wani MC, Taylor HL, Wall ME, Coggon P, McPhail AT. Plant antitumor agents. VI. Isolation and structure of taxol, a novel antileukemic and antitumor agent from Taxus brevifolia. *Journal of the American Chemical Society.* 1971 May;93(9):2325–7.

Wiese J, Imhoff JF. Marine bacteria and fungi as promising source for new antibiotics. *Drug Development Research*. 2019 Feb;80(1):24–7.

Xiong ZQ, Wang JF, Hao YY, Wang Y. Recent advances in the discovery and development of marine microbial natural products. *Marine Drugs*. 2013 Mar 8;11(3):700–17.

Zaky AS, French CE, Tucker GA, Du C. Improving the productivity of bioethanol production using marine yeast and seawater-based media. *Biomass Bioenerg*. 2020;139:105615.

17 Ecological Consequences of Invasive Species on Ecosystems

Shivani Devi[1], *Gagandeep Singh*[2], *Anita Rana*[3]
[1]Department of Biosciences, University Institute of Biotechnology, Chandigarh University
[2]Gharuan, Mohali, Punjab
[3]Corresponding author: Anita Rana

17.1 INTRODUCTION

It is truly remarkable how interconnected all living organisms within our planet's ecosystem are. Each organism holds a pivotal role in upholding the delicate balance of nature. While discussions about climate change and its effects on our world are prevalent, the peril of invasiveness often goes unnoticed. Acknowledging the severe threat that invasive species pose to our natural environment is crucial. These foreign organisms can quickly multiply, wreak havoc on the local ecosystem, and displace native species. These species can be introduced deliberately or accidentally through human activities or naturally through wind, water, or animals. They often become overpopulated quickly due to the lack of natural predators and their successive exponential reproductive growth. Invasive species are a significant threat to native biota (Peh, 2010), the ecosystem, and biodiversity balance (Pejchar & Mooney, 2009). Economically, they cause losses to agriculture, forestry, and aquaculture (Simberloff et al., 2005), among others. The introduction, survival, and growth of these organisms have increased over the last two centuries (Seebens et al., 2022). Invasive species not only directly impact the ecosystem but are also passengers of environmental change (Macdougall & Turkington, 2005). In the current scenario, the introduction of alien species is chiefly done by trading and transportation (Hulme et al., 2008), e.g., the increasing demand for goods and trade of exotic avian species in the Southeast Asian region (Blackburn, et al., 2015). The first global data records the introduction of invasive species during 1500–1800 (Drake et al., 1989). A tiny proportion of the geographical realms remain unadulterated by introduced species. The impact of invasiveness can be observed at the population, community, and ecosystem levels (Pyšek & Richardson, 2010).

17.2 TYPES AND CLASSIFICATION

Classification of invasive species varies according to geographical region, biology, extent of their impact, and many more. An analysis done by Jeanne and coworkers extensively covers the major invaders (well-established alien species having a substantial impact on the ecosystem) and potential invaders (presently less influential but can emerge as invaders in the coming decades) in South Africa. They classified the major invaders by applying three criteria: the number of records, the type of habitat invaded, and the abundance and range of species. By applying these filters, invasive species can be classified based on their distribution range (very widespread, widespread, and localized) and their abundance (abundant, common, and scarce). The researchers thoroughly analyzed the emerging intruders based on various parameters such as their impact, level of invasiveness,

DOI: 10.1201/9781032697550-17

biological control potential, and resemblance to other invasive plants. They further segregated the potential hidden habitats into four categories: large, moderate, small, and riparian. In addition, the propagule pool size was also classified into three categories: large, moderate, and small (Jeanne et al., 2004).

17.3 METHODS AND WAYS OF EXOTIC INTRODUCTION

The problem of introducing exotic species has increased since the mid-18th century due to mechanized travel and globalization. The economic and ecological data on invasive alien species (IAS) are only available for some continents and countries. There is some information and survey data for Europe (Vilá et al., 2010), the USA (Pimentel et al., 2005), and South Africa (van Wilgen et al., 2001).

Ballast water: Ballast water is essential for ships to maintain balance during sea voyages. Ships can use any solid or liquid material for ballasting, but they usually carry water. Water is loaded onto ships at one place and unloaded at the destination port. This ballast water is a remarkable carrier of species from one continent to another (Carlton & Geller, 1993). Ships traveling between northern polar latitudes and Antarctica pose a threat of invasiveness (Lewis et al., 2005). Human-made activity in the Arctic region is increasing due to the northern temperate and polar landmass compared to the Antarctic region, which automatically faces a greater threat of invasiveness in the Arctic than in the Antarctic. There are many successful invasive species carried by ballast water, such as zebra mussels, cholera, and various aquatic parasites; they can also be vectors of various diseases.

Global climate change: Changes in climatic conditions increase the possibility of introducing invasive species in the Arctic region (Ware et al., 2013). In addition to introduced invasive species, the risk of exposure to dormant life (possibly invasive to modern biodiversity) in permafrost regions also increases with the escalation of global temperature and ice melting. According to Bellard and their group, climate change will shift the distribution and invasive expansion from the tropical areas to the north temperate and continental climatic zones in the upcoming decades. In contrast, tropical areas will face a reduction in invasive invasion. Invasive species can thrive more in areas such as temperate deciduous forests and temperate mixed forests, and the tropical biome is expected to develop into an extreme climatic zone, which can lead to the reduction and loss of invasive species. It is expected that the habitat of birds, amphibians, and fungi will experience constriction, while the habitat of aquatic invertebrates, aquatic plants, microorganisms, and terrestrial invertebrates will expand in response to the changing climatic scenario (Bellard et al., 2013).

Tourism and trade: Trade is not a modern concept; it has been present in human civilization since ancient times, for example, there was trading between the Mesopotamian and Indus Valley civilizations. However, in the past two centuries, trading has gained momentum, and over the last 50 years, merchandise trade has increased exponentially (Roques et al., 2009). Trading and globalization have opened many pathways through which non-native species are introduced into alien habitats; these pathways can be terrestrial, marine, aquatic, and aerial. Tourists are one of the major carriers of alien species, such as parasites, seeds, and fruits, transporting these species on food items, clothes, shoes, vehicles, and vessels. Traded pet animals also form a significant proportion of the deliberate introduction; these traded animals are one of the reasons for the spread of disease, which they carry to new regions. Hydatid tapeworms (*Echinococcus granulosus*) were introduced to New Zealand via sheep, and this tapeworm caused the deaths of many humans as a zoonotic disease (Wilks et al., 1997).

17.4 THREAT TO NATIVE BIODIVERSITY

Globally, the ecosystem has faced drastic consequences due to non-native introductions (Pejchar & Mooney, 2009). Phylogenetic diversity among European flora is diminishing due to the loss of native species and the spread of non-native invasive species (Winter et al., 2009). The proliferation of invasive species significantly challenges native flora and fauna. These species, being devoid of natural predators, compete with indigenous life forms for resources and space. These exotic species may have a wide variety of diets, and due to this, endemic species come under significant threat. Ultimately, they become the dominant and irreversible contenders in their non-native habitats. Four biomes will emerge as hotspots for invasive intruders: Europe, northwestern North America, South Australia, and New Zealand, in response to upcoming climatic scenarios. The threat of intruders and their impact will be higher in the Northern Hemisphere than in the Southern Hemisphere due to the increase in the population of invasive species in the north (Bellard et al., 2013). Invasive organisms are the cause of most bird and mammal extinctions, which is higher than any other animal category (Clavero & García-Berthou, 2005).

Competition: Exotic invasive species compete with native biodiversity; this competition can be for space, breeding, and food. This competition can lead to the reduction of native biota; usually, this can be due to the advanced characteristics of exotic species, such as high environmental tolerance, high dispersal rate, and a wide range of diets. Two species of *Aedes—Aedes aegypti* and *Aedes albopictus*—were studied, and it was found that *A. albopictus* is the dominant competitor (Braks et al., 2009). The *A. aegypti* mosquito is typically found in urban areas, while *A. albopictus* prefers rural environments, although both can be found in suburban areas (Braks et al., 2003). In addition, a dynamic species interaction can be observed between the marine toad and reptiles, such as freshwater crocodiles, as the consumption of the former can pose a significant threat to the latter (Begg et al., 2000).

Contamination of water bodies: Various aquatic exotic organisms can contaminate the native water reserve or wetland. They can be contaminated by various by-products or toxins produced by these organisms. The high abundance of exotic organisms can lead to problems with the survival of the native biota. Introduced species sometimes have parasites or are carriers of infectious diseases, which can spread to the native biota through contaminated water shared by them or by various interactions between infected and non-infected species. Several introduced ectoparasites and endoparasites spread zoonotic diseases, and some animals, such as fishes, crabs, and many vertebrates, act as passengers to disease. By consuming these animals, these diseases can spread to the human population.

Ecosystem imbalance and homogenization of global biodiversity: Various invasive species are found to alter the food web of an ecosystem (Dukes & Mooney, 2004). Entering an ecosystem's food web can have negative and positive impacts on different species. Invasive species can alter the whole ecosystem, affecting various species at different trophic levels. Homogenization of species is the process in which native biota get removed by more widespread species. As a result of climate change, the natural habitats of indigenous organisms are shrinking, while the distribution of generalist species is expanding. This results in the removal of unique and essential biota. Invasive species can homogenize because of their tolerance of wide-range habitats (Rahel, 2002). An experiment conducted by Carniatto et al. on two species of aquatic macrophyte—*Egeria najas* and *Hydrilla verticillata*—found that if the native species *E. najas* is replaced by the generalized invasive species *H. verticillata*, this resulted in no adverse effect on the abundance of the invertebrates relying on the macrophyte and the foraging activity of fishes. Instead, the invasive *H. verticillata* showed high composition and abundance of the associated invertebrates (Carniatto et al., 2019).

Irreversible change in ecology and alteration in habitat: An analysis done by Matthew and coworkers concluded that the impact of invasive species on both grazing and detritus food webs could be encountered at different trophic levels in different ecosystems, such as the severe impact of alien species in a wetland can affect primary and secondary consumers (McCary et al., 2016). Invasive species also pose a threat to apex or keystone species. In some cases, habitat alteration can decrease the carrying capacity of endemic organisms (Fukasawa et al., 2013). The extinction of various native and endemic fauna and flora is an irreversible change, and any damage to the apex community can lead to a whole change in the regional ecosystem.

17.5 EXAMPLES OF INVASIVE ORGANISMS

Invasive mussel: The zebra mussel is a famous example of a freshwater invasive species. Its natural habitat was Europe's rivers, but later, it became invasive globally and spread to regions of North America and Australia. It was transported to the Great Lakes of America and became invasive (Ricciardi, 2003), possibly due to ballast water (Karatayev et al., 1998). Mussels are an essential diet for organisms like leeches, fish, crabs, turtles, and water birds (Sylvester et al., 2007). These mussels can become a significant nuisance to industries and local uses, as they can block pipes. Another famous invasive mussel is the golden mussel in the region of South America, which is native to Southeast Asia (Sousa et al., 2014). These mussels are introduced to non-native habitats via ballast water carried in cargo ships. These bivalves are a crucial part of the aquatic food web. The main diet of these filter-feeder organisms includes phytoplankton and dissolved organic particles. Bivalves are also part of the human diet, whose food demand increases with the growing population. Anthropogenic activities and changing climate can disfigure their population, which can have a cascading impact on the aquatic ecosystem. These mussel larvae are a significant diet for fish larvae. With these intruders, various parasites and diseases can spread to non-immune communities. In Japan, a new fish parasite appeared with the introduction of the golden mussel, which uses the mussel as an intermediate host (Ogawa et al., 2004). Due to their food habits, these intruders in non-native habitats have the capacity to disturb the native food web and trophic system.

Water hyacinth: Water hyacinth is a shrub native to South America's Amazon rainforest but has now spread to various tropical and subtropical regions and has become invasive, impacting the native biodiversity (Dersseh et al., 2019). Its habitat is around water bodies such as rivers, lakes, and streams (Jafari, 2010). These are also used for decoration and ornamental purposes because of their attractive flowers. The water-purifying property of water hyacinth can be used in phytoremediation. This property is used in treating some industrial effluents as it can reduce the amount of heavy metal in water. Various products from water hyacinths are cellulose, biofertilizer, biogas, and fish feed. Another problem related to water hyacinth is that it acts as a breeding ground for mosquitoes (Malik, 2007). Another major problem caused by water hyacinths is that they form large dense mats over the water body, thus blocking the sunlight to the benthic organisms and phytoplankton. Its presence also affects the amount of dissolved oxygen in the water table by obstructing the air and water contact. The removal of this intrusive weed can lead to an increase in phytoplankton biomass. Water hyacinth is also problematic in irrigation and water supply. It can also reduce the oxygen in the water bodies, as it absorbs a high quantity of oxygen from water (Opande et al., 2004). Water hyacinths reproduce by various means, such as seeds, stolons, or runners, which lead to their high dispersal rate in the non-native habitat. Water hyacinth can be controlled by eradicating or removing it manually. Most animals do not graze on that weed because of its high tannin and water content (Tewabe, 2015). Some insects and fungi can also act as a biological control against the water hyacinth (Tewabe, 2015).

Marine toad: *Bufo marinus* species is indigenous to South and Central America. It was deliberately introduced into northeastern Australia as pest control (Mungomery, 1936). Due to global warming, its habitat can expand to southern regions, which will become warmer (Kearney et al., 2008). The features that helped them to become invaders in these regions are their diet, which includes invertebrates, small vertebrates, and native toads (Crossland et al., 2008), and their desiccation-resistant skin (Duellman & Trueb, 1986). These amphibians thrive in moist areas but show slow expansion and growth in dry, arid, and colder areas, as water is the critical requirement for their metamorphosis. Marine toad populations are expected to expand to northern parts of Australia (Phillips et al., 2006). The marine toad has expanded its habitat range to the west (Urban et al., 2008). Amphibians' permeable skin strictly confines their habitat to specific salinity levels, as any deviation from these levels can cause significant water loss or gain (Shoemaker & Nagy, 1977). Some anurans can tolerate some extent of high salinity (Beebee, 1985). This advantage of marine toads is beneficial for them to extend their range. Marine toads threaten biodiversity as they are poisonous, and most predators who eat these toads are killed by their toxins. Even their eggs are toxic, so those who feed on their eggs are also killed, as the cane toad toxin is present throughout all life stages, though the concentration at different levels can vary (Hayes et al., 2009). They tolerate various environmental conditions and can feed on a broader range of prey (Lever, 2001) than native anurans. These invasive cane toads pose a significant threat to native organisms, with their predatory behavior toward native animals causing great harm. The diet of the cane toads in Australia includes ants, termites, beetles, and other invertebrates. Cane toads can also affect the growth of native anuran tadpoles (Williamson, 2009).

European starling: The European starling (*Sturnus vulgaris*) is an invasive avian species that has successfully established populations in numerous countries, spanning the United States, Canada, Australia, and New Zealand (Jauregui et al., 2021). This invasive interloper exerts a detrimental influence on the breeding ecology of woodpeckers, primarily through intense competition for nesting cavities (Jauregui et al., 2021). An illustrative case study in Argentina reveals the burgeoning impact of the European starling on the breeding success of campo flicker and green-barred woodpecker populations. The presence of European starlings near woodpecker nests significantly heightens the likelihood of early-stage nest abandonment (Jauregui et al., 2021). Furthermore, woodpeckers, in their struggle for nesting sites, exhibit defensive behaviors marked by heightened aggression, involving attacks and more frequent distress calls, triggered by the proximity of these interloping starlings (Jauregui et al., 2021). Beyond its local disruptions, the European starling also looms on the radar as a potential invasive alien species threatening the biodiversity and ecological equilibrium of the European Union (Roy et al., 2019). This invasive bird elicits concern due to its proficiency in resource and habitat competition with native species, ultimately posing a challenge to the delicate balance of local ecosystems (Lakicevic & Mladenovic, 2018). A closer examination of the economic ramifications of the European starling's invasions underscores the magnitude of its impact. In Germany, for instance, invasive species collectively exacted a staggering economic toll of US$9.8 billion between 1960 and 2020, with the European starling emerging as a significant contributor to these costs (Haubrock et al., 2021). In parallel, the consequences of the European starling's presence have resonated in agricultural domains. Notably, an investigation into its influence on rice crops in northwestern Italy reveals a substantial shift in plant composition. This shift is characterized by a notable increase in therophyte and alien components, signifying a broader impact on local flora composition (Vagge, 2023). These intricate threads of evidence collectively weave a narrative of the European starling's far-reaching influence, encompassing ecological, economic, and agricultural dimensions.

Asian carp: Asian carp, a collective term encompassing invasive fish species, have been introduced to various regions worldwide and are currently infiltrating three tributaries of the Missouri River in the United States (Hayer et al., 2014). Among these intruders, the most prevalent are the bighead carp and silver carp, collectively referred to as Asian carp (Hayer et al., 2014; Silva et al., 2006). Regrettably, they rank as the second most formidable threat to native biodiversity and the equilibrium of ecosystem functions and services (Hayer et al., 2014). The litany of impacts these Asian carp unleash upon biodiversity is extensive. First, their competitive prowess looms large as they outmaneuver native fish species for both sustenance and habitat, precipitating declines in these indigenous populations. Simultaneously, their intrusion results in the displacement of native species from their accustomed habitats, further accentuating the jeopardy faced by these local fish populations. To exacerbate matters, the predation predilection of Asian carp extends to the eggs and larvae of native fish species, compounding the downward trajectory of these populations. Meanwhile, the profound impact on food webs cannot be understated; the voracious appetite of Asian carp engenders large-scale consumption of plankton, triggering ripple effects that alter the abundance and distribution of other aquatic organisms. The consequences ripple into the economic sphere as well, with negative repercussions reverberating through commercial and recreational fishing industries (Silva et al., 2006).

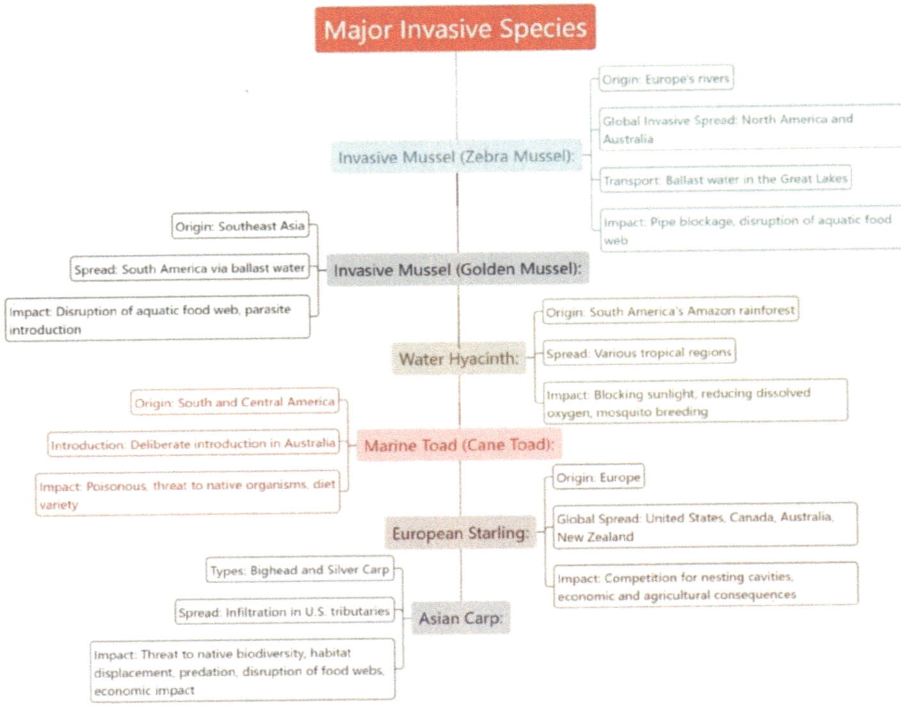

FIGURE 17.1 Examples of major invasive species. The figure outlines a concise summary of the origin, spread, and ecological consequences of major invasive species, aiding in the understanding of the ecological challenges posed by these invasive organisms.

17.6 NOTABLE CASE STUDIES

Invasive alien species pose a formidable threat to biodiversity, necessitating vigilant monitoring and effective management. A compelling example of the power of citizen involvement in this endeavor comes from the Egadi Islands Marine Protected Area (MPA) in Italy. Here, citizen scientists played a pivotal role in tracking the spread of the "sea grape" *Caulerpa cylindracea*, amassing a remarkable 156 sightings. Beyond this, the project also enriched our knowledge by collecting data on other non-indigenous and cryptogenic species, underlining the invaluable contribution of citizen science (Mannino & Balistreri, 2018). Meanwhile, Parks Canada and the Canadian Parks Council have charted a standardized approach for invasive species management within protected areas (PAs). This comprehensive ecosystem management standard aligns closely with the Canadian IAS Strategy, emphasizing strategic, efficient, and effective action. Notable case studies include the rehabilitation of endangered Garry Oak ecosystems in the Gulf Islands National Park Reserve, the preservation of brook trout populations in La Mauricie National Park, and the management of moose hyperabundance in Gros Morne National Park (Mikaïlou et al., 2009).

In Belgium, the LIFE RIPARIAS project unveiled a promising framework for harmonizing IAS management data. Shaped by input from Belgium's IAS management community, this initiative spawned a community-driven data model and exchange format known as manIAS (MANagement of Invasive Alien Species). Moreover, it identified the minimal software application requirements necessary for successful IAS management (Reyserhove et al., 2023). Another sobering case study revolves around the invasive seed wasps in Europe, primarily those of the *Megastigmus* genus. These seed chalcids infiltrated Europe via the unregulated seed trade for afforestation and ornamental plantations, resulting in a stark decline in the host trees' regeneration potential. The study scrutinized various biological traits contributing to the invasive success of these alien chalcids (Auger-Rozenberg & Roques, 2012). These case studies' outcomes underscore the imperative nature of continuous IAS monitoring and the dire consequences of their unchecked proliferation on biodiversity. Moreover, they shed light on the potential for citizen science and community-driven data models to become valuable assets in the arsenal against invasive species, reinforcing the importance of collaborative efforts in this ongoing battle for our ecosystems.

17.7 MANAGEMENT STRATEGIES

As established by this chapter, IAS represent an enormous threat to global biodiversity. Addressing this multifaceted challenge demands a multifaceted approach, encompassing proactive measures to prevent their introduction and spread, as well as comprehensive management strategies to mitigate their impact. Here, we outline a spectrum of strategies, each playing a pivotal role in the battle against IAS, drawing from various sources (Roy et al., 2019; Jauni et al., 2023; García-Díaz et al., 2020; Booy et al., 2020; Courtois et al., 2023; Bertolino et al., 2020; Reyserhove et al., 2023). First and foremost, the practice of horizon scanning emerges as a proactive sentinel against potential IAS incursions. By identifying species not yet established but likely to arrive and thrive, this process enables the prioritization of management efforts, steering us toward a preemptive stance (Roy et al., 2019). Risk assessments, a cornerstone of the European Union's approach, mandate each member state to compile IAS management plans. These assessments are pivotal in the hierarchy of management measures, guiding the prioritization of actions across the EU (Jauni et al., 2023). Monitoring, often regarded as the frontline defense against IAS, assumes a central role. Regular surveillance efforts aim to promptly detect the presence and spread of these intruders while assessing their toll on biodiversity (Jauni et al., 2023). Prevention measures, flexible and adaptable to national risk assessments, encompass a range of strategies. They include the prevention of IAS entry into a region, their escape into the wild, and further proliferation. Each tailored measure stands as a barrier against potential devastation (Jauni et al., 2023).

There are various policies, conventions, and organizations that work on the global and regional levels to regulate and check the spread of various exotic and invasive species. Trade is one of major ways through which exotic species are introduced to various non-native habitats, either intentionally or unintentionally. The organizations and policies that work at the trade level include the World Trade Organization and the Convention on International Trade in Endangered Species of Wild Fauna and Flora (CITES). Global conventions and policies that deal with the issue of invasive species include the Convention on Biological Diversity (CBD) and its Cartagena Protocol on Biosafety, the Ramsar Convention on Wetlands of International Importance, and the Convention on Migratory Species (CMS). Organizations concerned with agriculture, aquaculture, and forestry include the World Animal Health Organization (OIE), the International Plant Protection Convention (IPPC), and the UN Food and Agriculture Organization (FAO) (Perrings et al., 2010). Invasive menace of biodiversity can be reduced by regulating the export and import of agricultural products and goods. The use of disinfectants and the eradication of diseased or perished products can be used to prevent the introduction of invasive species to new regions. Ballast water on ships needs to be cleansed before it is loaded or unloaded. Establishing a strong security network among countries at continental and regional levels can help reduce illegal trade. Penalties for the poaching of animals and their products, as well as local campaigns and programs should be implemented to raise awareness among people by educating them about the impact on biodiversity and their individual benefits. The increasing temperature of the Earth is also altering the habitat range and size of the invasive species globally, so achieving climate sustainability goals it is impossible to eradicate invasive species.

For some IAS with limited distribution or those yet to establish a foothold, eradication measures offer hope. While often demanding significant resources and effort, these measures prove effective in halting further IAS expansion (Jauni et al., 2023). Management plans, critically, must remain attuned to local environmental nuances, harmonizing their approach with national risk assessments (Jauni et al., 2023). Furthermore, a chorus of stakeholders must sing in unison, championing awareness and coordination to maximize the efficacy of IAS management (Shiferaw et al., 2018). Shifting the lens toward concrete management strategies, long-term policies emerge as potent tools. These policies directly target impacts through impact-based interventions, guided by a set of six criteria (García-Díaz et al., 2020). Assessing the feasibility of eradication proves crucial in prioritizing IAS management. This assessment guides the allocation of resources to address the risks posed by emerging IAS (Booy et al., 2020). A cost–benefit optimization framework injects economic rationale into the equation, designing cost-effective strategies that reflect spatially explicit costs, benefits, and invasion dynamics (Courtois et al., 2023). Spatially explicit models, predictive in nature, enhance management strategy effectiveness. They provide spatial insights into potential species spread and evaluate the gamut of available management options (Bertolino et al., 2020). The arteries of improved data flows pump life into IAS management. A centralized workflow capturing information on applied management methods, effort, cost, effectiveness, and ecological impact serves as the compass guiding interventions (Reyserhove et al., 2023). Yet again, it bears repeating that management strategies should harmonize with local environmental contexts, remaining ever sensitive to the nuances detailed in national risk assessments (Booy et al., 2020). As with prevention, coordination and awareness among stakeholders remain the adhesive that binds these multifaceted strategies into a cohesive, effective response to the persistent threat of IAS to biodiversity (García-Díaz et al., 2020; Courtois et al., 2023).

17.8 CONCLUSION

The pervasive threat posed by invasive species, characterized by their adaptability in dietary preferences, swiftness in dispersal, and resilience to climatic variability, is further exacerbated by the unfolding consequences of climate change. Urgent and concerted action is imperative to mitigate the impending irreversibility of this ecological challenge. Beyond the jeopardy they pose to biodiversity, invasive and non-native species cast a shadow on economic well-being. Timely intervention is indispensable to thwart the encroachment of alien species, preempting the destruction of indigenous counterparts. The evolving climate milieu amplifies the prospect of invasive species extending their dominion, prompted by shifts in abiotic constituents. International and regional protocols governing the trade in flora and fauna necessitate a judicious overhaul. Innovative technologies, exemplified by drone utilization, emerge as potent tools in the conservation arsenal, serving dual roles in species protection and curbing illicit trade and commerce. The proliferation of exotic pets and domestic organisms warrants curtailment. Invasive species directly and indirectly affect human and animal populations through disease propagation. Non-indigenous organisms wield harmful impacts on native ecosystems, orchestrating a dislodgement of endemic species through prolific reproduction. Consequently, the imperative lies in adopting preemptive measures to forestall the introduction and dissemination of invasive species, thereby safeguarding the precarious equilibrium of our planet's natural tapestry.

REFERENCES

Auger-Rozenberg, M.-A., Roques, A. (2012). Seed wasp invasions promoted by unregulated seed trade affect vegetal and animal biodiversity. *Integrative Zoology,* 7, 228–246. https://doi.org/10.1111/j.1749-4877.2012.00308.x.

Blackburn, T., Dyer, E., Su, S., Cassey, P. (2015). Long after the event, or four things we (should) know about bird invasions. *Journal of Ornithology,* 156, 15–25. https://doi.org/10.1007/s10336-015-1155-z.

Begg, G. W., Walden, D., Rovis-Hermann, J. (2000). Report on the Joint ERISS/PAN Cane Toad Risk Assessment Field Trip to the Katherine/Mataranka and Borroloola Regions. Supervising Scientist Internal Report 389. Darwin (Australia): Supervising Scientist.

Beebee, T. J. C. (1985). Salt tolerances of natterjack toad (Bufo calamita) eggs and larvae from coastal inland populations in Britain. *Journal of Herpetology,* 1, 14–16.

Bellard, C., Thuiller, W., Leroy, B., Genovesi, P., Bakkenes, M., Courchamp, F. (2013). Will Climate change promote future invasions? *Global Change Biology,* 19(12), 3740–8. https://doi.org/10.1111/gcb.12344.

Booy, O., Robertson, P. A., Moore, N., Ward, J., Roy, H. E., Adriaens, T., Shaw, R., Van Valkenburg, J., Wyn, G., Bertolino, S., Blight, O., Branquart, E., Brundu, G., Caffrey, J., Capizzi, D., Casaer, J., De Clerck, O., Coughlan, N. E., Davis, E., Dick, J. T. A., Essl, F., Fried, G., Genovesi, P., González-Moreno, P., Huysentruyt, F., Jenkins, S. R., Kerckhof, F., Lucy, F. E., Nentwig, W., Newman, J., Rabitsch, W., Roy, S., Starfinger, U., Stebbing, P. D., Stuyck, J,, Sutton-Croft, M., Tricarico, E., Vanderhoeven, S., Verreycken, H., Mill, A. C. (2020). Using structured eradication feasibility assessment to prioritize the management of new and emerging invasive alien species in Europe. *Global Change Biology,* 26(11), 6235–6250. https://doi.org/10.1111/gcb.15280. Epub 2020 Aug 27. PMID: 32851731.

Bertolino, Sandro & Sciandra, Chiara & Bosso, Luciano & Russo, Danilo & Lurz, Peter & Di Febbraro, Mirko. (2020). Spatially-explicit models as tools for implementing effective management strategies for invasive alien mammals. *Mammal Review.* 50. 187-199. 10.1111/mam.12185.

Braks, Marieta & Honorio, Nildimar & Lourençqo-De-Oliveira, Ricardo & Juliano, Steven & Lounibos, L. (2003). Convergent Habitat Segregation of Aedes aegypti and Aedes albopictus (Diptera: Culicidae) in Southeastern Brazil and Florida. *Journal of Medical Entomology,* 40, 785-94. 10.1603/0022-2585-40.6.785.

Braks, M., Honorio, N., Lounibos, L., Lourenço-de-Oliveira, R., Juliano, S. (2009). Interspecific competition between two invasive species of container mosquitoes, aedes aegypti and aedes albopictus (Diptera: Culicidae), in Brazil. *Annals of the Entomological Society of America,* 97, 130–139. https://doi.org/10.1603/0013-8746(2004)097[0130:ICBTIS]2.0.CO;2.

Cariton, J., Geller, J. (1993). Ecological roulette: the global transport of nonindigenous Marine organisms. *Science*, 261, 78–82. https://doi.org/10.1126/science.261.5117.78.

Carniatto, N., Fugi, R., Quirino, B., Cunha, E., Thomaz, S. (2019). An invasive and a native macrophyte species provide similar feeding habitat for fish. *Ecology of Freshwater Fish*, 29(1), 112–20. https://doi.org/10.1111/eff.12499.

Clavero, M., García-Berthou, E. (2005). Invasive species are a leading cause of animal extinction. *Trends in Ecology & Evolution,* 20, 110. https://doi.org/10.1016/j.tree.2005.01.003.

Courtois, P., Martinez, C., Thomas, A. (2023). Spatial priorities for invasive alien species control in protected areas. *Science of the Total Environment,* 878, 162675. https://doi.org/10.1016/j.scitotenv.2023.162675. Epub 2023 Mar 17. PMID: 36933722.

Crossland, M., Brown, G., Anstis, M., Shilton, C., Shine, R. (2008). Mass mortality of native anuran tadpoles in tropical Australia due to the invasive toad (*Bufo marinus*). *Biological Conservation,* 141, 2387–94. https://doi.org/10.1016/j.biocon.2008.07.005.

Dersseh, M., Melesse, A., Tilahun, S., Meshesha, M., Dagnew, D. (2019). Water hyacinth: review of its impacts on hydrology and ecosystem services. *Lessons for Management of Lake Tana*, 1, 237–51. https://doi.org/10.1016/B978-0-12-815998-9.00019-1.

Drake, J., Mooney, H., Castri, D., Groves, F., Kruger, F., Williamson, M. (1989). Biological invasions: a global perspective (SCOPE 37). *Biometrics*, 47, 1–30. https://doi.org/10.2307/2532533.

Duellman, W. E., Trueb, L. (1986). *Biology of Amphibians.* New York: McGraw-Hill.

Dukes, J. S., Mooney, H. A. (2004). Disruption of ecosystem processes in western North America by invasive species. *Revista Chilena De Historia Natural*, 77(3). https://doi.org/10.4067/S0716-078X2004000300003

Fukasawa, K., Miyashita, T., Hashimoto, T., Tatara, M., Abe, S. (2013). Differential population responses of native and alien rodents to an invasive predator, habitat alteration and plant masting. *Proceedings, Biological Sciences / The Royal Society,* 280, 20132075. https://doi.org/10.1098/rspb.2013.2075.

García-Díaz, P., Cassey, P., Norbury, G. L., Lambin, X., Montti, L., Pizarro, J. C., Powell, P. A., Burslem, D. F., Cava, M. G., Damasceno, G., Fasola, L., Fidelis, A., Huerta, M. F., Langdon, B., Linardaki, E., Moyano, J., Nuñez, M. A., Pauchard, A., Phimister, E., Raffo, E., Roesler, I., Rodriguez-Jorquera, I. A., Tomasevic, J. A. (2020). Management Policies for Invasive Alien Species: Addressing the Impacts Rather than the Species. *BioScience.*

Hayer, C.-A., Breeggemann, J., Klumb, R., Graeb, B., Bertrand, K. (2014). Population characteristics of bighead and silver carp on the northwestern front of their North American invasion. *Aquatic Invasions,* 9, 289–303. https://doi.org/10.3391/ai.2014.9.3.05.

Hayes, R., Crossland, M., Hagman, M., Capon, R., Shine, R. (2009). Ontogenetic variation in the chemical defenses of cane toads (Bufo marinus): toxin profiles and effects on predators. *Journal of Chemical Ecology*, 35, 391–399. https://doi.org/10.1007/s10886-009-9608-6.

Haubrock, P., Cuthbert, R., Sundermann, A., Golivets, M., Diagne, C., Courchamp, F. (2021). Economic costs of invasive species in Germany. *NeoBiota*. 225–246. https://doi.org/10.3897/neobiota.67.59502.

Hulme, P., Bacher, S., Kenis, M., Klotz, S., Kühn, I., Minchin, D., Nentwig, W., Olenin, S., Panov, V., Pergl, J., Pyšek, P., Roques, A., Sol, D., Solarz, W., Vilà, M. (2008). Grasping at the routes of biological invasions: a framework for integrating pathways into policy. *Journal of Applied Ecology*. 45. 403–414. https://doi.org/10.1111/j.1365-2664.2007.01442.x.

Jafari, N. Ecological and socio-economic utilization of water hyacinth (Eichhornia crassipes Mart Solms). *The Journal of Applied Sciences & Environmental Management*, 2010, 14, 43–49.

Jauni, M., Huusela, E., Urho, L., Hyvönen, T. (2023). Prioritizing Management of Invasive Alien Species of EU Concern—A Northern Perspective. *Diversity,* 15, 585. https://doi.org/10.3390/d15050585.

Jauregui, A., Gonzalez, E., Segura, L. (2021). Impacts of the invasive European Starling on two neotropical woodpecker species: agonistic responses and reproductive interactions. *Emu - Austral Ornithology,* 121, 1–8. https://doi.org/10.1080/01584197.2021.1920841.

Jeanne, N., Richardson, D., Rouget, M., Mgidi, T. N., Mdzeke, N.P., Le Maitre, D., van Wilgen, B., Schonegevel, L., Henderson, L., Neser, S. (2004). A proposed classification of invasive alien plant species in South Africa: Towards prioritizing species and areas for management action. *South African Journal of Science*, 100(1), 53–64.

Karatayev, A. Y., Burlakova, L. E., Padilla, D. K. (1998). Physical factors that limit the distribution and abundance of Dreissena polymorpha. *Journal of Shellfish Research,* 17, 1219–1235.

Kearney, M., Phillips, B., Tracy, C., Christian, K., Betts, G., Porter, W. (2008). Modeling species distributions without using species distributions: the cane toad in Australia under current and future climates. *Ecography*, 31, 423–434. https://doi.org/10.1111/j.0906-7590.2008.05457.x.

Lakicevic, M., Mladenovic, E. (2018). Non-native and invasive tree species - their impact on biodiversity loss. *Zbornik Matice srpske za prirodne nauke,* 2018, 19–26. https://doi.org/10.2298/ZMSPN1834019L.

Lever, C. (2001). *The Cane Toad: The History and Ecology of a Successful Colonist.* Otley (UK): Westbury Academic and Scientific Publishing.

Lewis, P., Riddle, M., Smith, S. (2005). Assisted passage or passive drift: a comparison of alternative transport mechanisms for non-indigenous coastal species into the Southern Ocean. *Antarctic Science*, 17, 183–191. https://doi.org/10.1017/S0954102005002580.

MacDougall, A., Turkington, R. (2005). Are invasive species the drivers or passengers of change in degraded ecosystems? *Ecology,* 86, 42–55. https://doi.org/10.1890/04-0669.

Malik, A. (2007). Environmental challenge vis-à-vis opportunity: the case of water hyacinth. *Environment International*, 33, 122–138. https://doi.org/10.1016/j.envint.2006.08.004.

Mannino, A., Balistreri, P. (2018). Citizen science: a successful tool for monitoring invasive alien species (IAS) in Marine Protected Areas. The case study of the Egadi Islands MPA (Tyrrhenian Sea, Italy). *Biodiversity,* 19(1-–2), 42–8. https://doi.org/10.1080/14888386.2018.1468280.

Mccary, M., Mores, R., Farfan, M., Wise, D. (2016). Invasive plants have different effects on trophic structure of green and brown food webs in terrestrial ecosystems: a meta-analysis. *Ecology Letters*, 19(3), 328–35. https://doi.org/10.1111/ele.12562.

Mikaïlou, S., Keenleyside, K., Adare, K., Reader, B., Plante, M., Deering, P. (2009). Protecting native biodiversity from high-impact invasive species through the protected areas of Parks Canada, *Biodiversity*, 10, 2–3, 51–55. https://doi.org/10.1080/14888386.2009.9712843

Mungomery, R. W. (1936). A survey of the feeding habits of the Giant Toad (*Bufo marinus* L.), and notes on its progress since introduction into Queensland. *Proceedings of the Queensland Society of Sugar Cane Technologists*, 1936, 63–74.

Opande, G., Onyango, J., Wagai, S. (2004). Lake Victoria: the water hyacinth (Eichhornia crassipes [MART.] SOLMS), its socio-economic effects, control measures and resurgence in the Winam gulf. *Limnologica - Ecology and Management of Inland Waters,* 34, 105–109. https://doi.org/10.1016/S0075-9511(04)80028-8.

Ogawa, K., Nakatsugawa, T., Yasuzaki, M. (2004). Heavy metacercarial infections of cyprinid fishes in Uji River. *Fisheries Science,* 70, 132–140. https://doi.org/10.1111/j.1444-2906.2003.00781.x.

Peh, K. (2010). Invasive species in Southeast Asia: the knowledge so far. *Biodiversity and Conservation*, 19, 1083–1099. https://doi.org/10.1007/s10531-009-9755-7.

Pejchar, L., Mooney, H., Pejchar, L., Mooney, H. (2009). Invasive species, ecosystem services, and human well-being. *Trends in Ecology & Evolution,* 24, 497–504. https://doi.org/10.1016/j.tree.2009.03.016.

Perrings, C., Burgiel, S., Lonsdale, M., Mooney, H., Williamson, M. (2010). International cooperation in the solution to trade-related invasive species risks. *Annals of the New York Academy of Sciences*, 1195, 198–212. http://doi.org/10.1111/j.1749-6632.2010.05453.x. PMID: 20536824.

Phillips, B., Brown, G., Webb, J., Shine, R. (2006). Invasion and the evolution of toads. *Nature*. 439, 803. http://doi.org/10.1038/439803a.

Pimentel, D., Zuniga, R., Morrison, D. (2005). Update on the environmental and economic costs associated with alien-invasive species in the United States. *Ecological Economics*, 52, 273–288

Pyšek, P., Richardson, D. M. (2010). Invasive species, environmental change and management, and health. *Annual Review of Environment and Resources*, 35, 25–55. Available at SSRN: https://ssrn.com/abstract=1707235 or http://dx.doi.org/10.1146/annurev-environ-033009-095548

Rahel, F. J. (2002). Homogenization of freshwater faunas. *Annual Review of Ecology and Systematics*, 33, 291–315. https://doi.org/10.1146/ annurev.ecolsys.33.010802.150429

Reyserhove, L., Huybrechts, P., Hillaert, J., Adriaens, T., D'hondt, B., Oldoni, Do. (2023). Towards improved data flows for the management of invasive Alien Species and wildlife: a life riparias use case. *Biodiversity Information Science and Standards,* 7 . https://doi.org/10.3897/biss.7.112386.

Ricciardi, A. (2003). Predicting the impacts of an introduced species from its invasion history: An empirical approach applied to zebra mussel invasions. *Freshwater Biology,* 48, 972–981. https://doi.org/10.1046/j.1365-2427.2003.01071.x.

Roques, A., Rabitsch, W., Rasplus, J. -Y., Lopez Vaamonde, C., Nentwig, W., Kenis, M. (2009). Alien Terrestrial Invertebrates of Europe. In *Handbook of Alien Species in Europe,* (3), 63–79. https://doi.org/10.1007/978-1-4020-8280-1_5.

Roy, H. E., Bacher, S., Essl, F., Adriaens, T., Aldridge, D. C., Bishop, J. D. D., Blackburn, T. M., Branquart, E., Brodie, J., Carboneras, C., Cottier-Cook, E. J., Copp, G. H., Dean, H. J., Eilenberg, J., Gallardo, B., Garcia, M., García-Berthou, E., Genovesi, P., Hulme, P. E., Kenis, M., Kerckhof, F., Kettunen, M., Minchin, D., Nentwig, W., Nieto, A., Pergl, J., Pescott, O. L., Peyton, M. J., Preda, C., Roques, A., Rorke, S. L., Scalera, R., Schindler, S., Schönrogge, K., Sewell, J., Solarz, W., Stewart, A. J. A., Tricarico, E., Vanderhoeven, S., van der Velde, G., Vilà, M., Wood, C. A., Zenetos, A., Rabitsch, W. (2019). Developing a list of invasive alien species likely to threaten biodiversity and ecosystems in the European Union. *Global Change Biology*, 25(3), 1032–1048. https://doi.org/10.1111/gcb.14527. Epub 2018 Dec 12. PMID: 30548757; PMCID: PMC7380041.

Seebens, H., Essl, F., Hulme, P. E., van Kleunen, M. (2002). Development of pathways of global plant invasions in space and time. In *Global Plant Invasions*, (pp. 53–69). Cham: Springer International Publishing.

Shoemaker, V. H., Nagy, K. A. (1977). Osmoregulation in amphibians and reptiles. *Annual Review of Physiology*, *39*, 449–471. https://doi.org/10.1146/annurev.ph.39.030177.002313

Shiferaw, W., Demissew, S., Bekele, T. (2018). Invasive alien plant species in Ethiopia: ecological impacts on biodiversity. A review paper. *International Journal of Molecular Biology,* 3(4), 171–8 . https://doi.org/10.15406/ijmboa.2018.03.00072.

Silva, S. S., Nguyen, T. T. T., Abery, N. W., Amarasinghe, U. S. (2006). An evaluation of the role and impacts of alien finfish in Asian inland aquaculture. *Aquaculture Research*, 37(1), 1–17. https://doi.org/10.1111/J.1365-2109.2005.01369.X

Simberloff, D., Parker, I., Windle, P. (2005). Introduced species policy, management, and future research needs. *Frontiers in Ecology and the Environment* 3, 12–20. https://doi.org/10.1890/1540-9295(2005)003[0012:ISPMAF]2.0.CO;2.

Sousa, R., Novais, A., Strayer, D. (2014). Invasive bivalves in freshwaters: impacts from individuals to ecosystems and possible control strategies. *Hydrobiologia*. 735, 233–251. https://doi.org/10.1007/s10750-012-1409-1.

Sylvester, F., Boltovskoy, D., Cataldo, D. (2007). Fast response of freshwater consumers to a new trophic resource: predation on the recently introduced Asian bivalve *Limnoperna fortunei* in the Lower Paraná River, South America. *Austral Ecology*, 32, 403–415. 10.1111/j.1442-9993.2007.01707.x.

Tewabe, D. (2015). Preliminary survey of water Hyacinth in Lake Tana, Ethiopia. *Global Journal of Allergy*, 1, 13–18. https://doi.org/10.17352/2455-8141.000003.

Urban, M., Phillips, B., Skelly, D., Shine, R. (2008). A toad mre traveled: the heterogeneous invasion dynamics of cane toads in Australia. *The American Naturalist*. 171, E134–E48. https://doi.org/10.1086/527494.

Vagge, I., Chiaffarelli, G. (2023). The Alien Plant Species impact in rice crops in Northwestern Italy. *Plants,* 12(10), 2012. https://doi.org/10.3390/plants12102012. PMID: 37653929; PMCID: PMC10223007.

van Wilgen, B., Richardson, D., Le Maitre, D., Marais, C., Magadlela, D. (2001). The economic consequences of Alien Plant Invasions: examples of impacts and approaches to sustainable management in South Africa. *Environment, Development and Sustainability*, 3, 145–168. https://doi.org/10.1023/A:1011668417953.

Vilà, M., Basnou, C., Pyšek, P., Josefsson, M., Genovesi, P., Gollasch, S., Nentwig, W., Olenin, S., Roques, A., Roy, D.B., Hulme, P., partners, DAISIE. (2010). How well do we understand the impacts of alien species on ecosystem services? A pan-European cross-taxa assessment. *Frontiers in Ecology and the Environment*, 8, 135–144. https://doi.org/10.2307/20696461.

Wilks, C.R. and Humble, M.W., eds (1997). *Zoonoses in New Zealand – A Combined Veterinary and Medical Perspective* (2nd edn), VetLearn Foundation

Williamson, I. (2009). Competition between the larvae of the introduced cane toad Bufo marinus (Anura: Bufonidae) and native anurans from the Darling Downs area of southern Queensland. *Australian Journal of Ecology*, *24*, 636–643. https://doi.org/10.1046/j.1442-9993.1999.00993.x.

Ware, C., Berge, J., Sundet, J. H., Kirkpatrick, J. B., Coutts, A. D. M., Jelmert, A., … Alsos, I. G. (2013). Climate change, non-indigenous species and shipping: assessing the risk of species introduction to a high- Arctic archipelago. *Diversity and Distributions*, 20(1), 10–19. https://doi.org/10.1111/DDI.12117

Winter, M., Schweiger, O., Klotz, S., Nentwig, W., Andriopoulos, P., Arianoutsou, M., Basnou, C., Delipetrou, P., Didziulis, V., Hejda, M., Hulme, P. E., Lambdon, P. W., Pergl, J., Pysek, P., Roy, D. B., & Kühn, I. (2009). Plant extinctions and introductions lead to phylogenetic and taxonomic homogenization of the European flora. *Proceedings of the National Academy of Sciences of the United States of America*, *106*(51), 21721–21725. https://doi.org/10.1073/pnas.0907088106

18 The Interplay of Pollutants
Unraveling the Threads of Biodiversity Degradation

Sakshi Sharma, Komal Raghuvanshi[1], Ritesh Banerjee[1]
[1]School of Biological and Environmental Science, Shoolini University, Solan, Himachal Pradesh, India

18.1 INTRODUCTION

The sustainability of life for all living things on our planet is largely dependent on biodiversity. All living things, including plants, animals, and microbes that surround us and have an impact on our quality of life, have a destiny that is linked to that of the human species (1). Biodiversity is essential for many reasons, such as enhancing the aesthetic value of the natural world and contributing to our material well-being by producing fuel, food, medicine, fodder, and timber. The foundation of life is biodiversity. It provides the food, drink, and oxygen that organisms need to survive. Wetlands remove pollutants from water, plants and trees absorb carbon dioxide to slow global warming, and fungi and bacteria break down organic matter to fertilize the soil (2). Worldwide, biodiversity is at risk due to environmental change. The environment is negatively impacted by human activity, which leads to pollution of the air, water, and soil. Pollution is a serious problem that poses ongoing health risks, even with the tremendous efforts that have been made recently to clean up the environment. Only 0.3% of the Earth's surface is suitable for human habitation, despite 71% of the planet being covered by water (3). As illustrated in Figure 18.1, there are five main factors that contribute to the loss of biodiversity: (i) loss of habitat, also known as land/sea-use change (e.g., deforestation for agricultural purposes), (ii) excessive use of natural resources (e.g., overfishing), (iii) change in the climate (e.g., changes in temperature, rainfall, and extreme weather events), (iv) invasive alien species, including new pathogens, and (v) pollution. The third biggest factor contributing to the loss of biodiversity is pollution (4). The notion of "environmental pollution" has existed for a long time all over the world, having a profound impact on both humanity and the environment, whether on purpose or accidentally, since the earliest human settlements and gatherings in particular regions of the planet for extended periods of time (5). Environmental pollution is caused by the extensive use of hazardous waste as a result of the growth of industry and agriculture. Pollution of any kind has detrimental effects on plants, animals, and human health. Pollutants can have a wide-ranging impact on individual species, plant groups, and ecosystems. The primary markers of environmental disturbances in this context include species diversity, vegetation cover, density of perennial and annual species, and ecological traits related to vegetation structure (6). The atmosphere is becoming more concentrated with carbon monoxide, carbon dioxide, nitrogen oxide, sulfur oxide, and other chemicals that are harmful to living things. Environmental toxicology refers to all of these factors that are connected to environmental degradation as a result of the presence of different pollutants (7). This chapter looks at different kinds of pollution—air, water, soil, noise, and chemical—as well as their effects, and solutions for a healthy and sustainable environment.

DOI: 10.1201/9781032697550-18

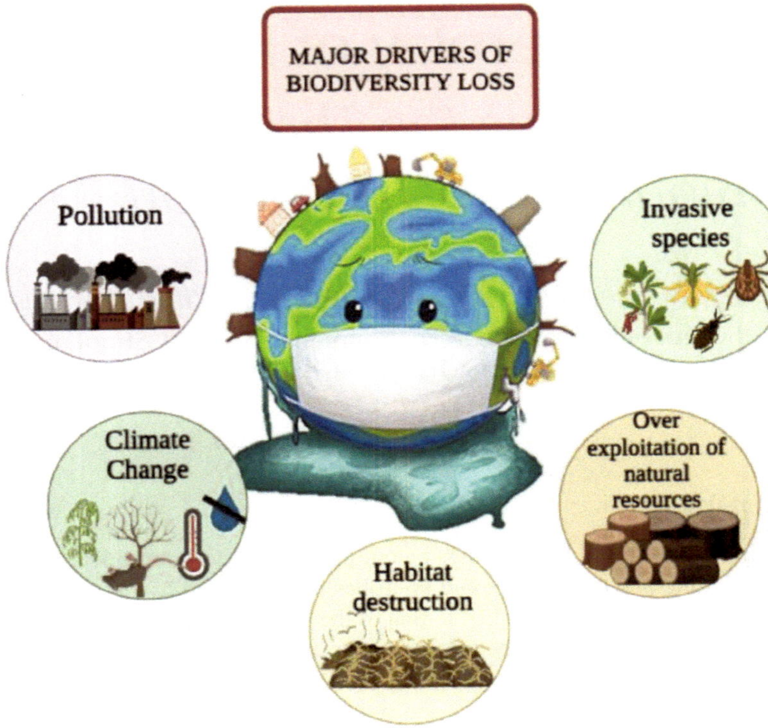

FIGURE 18.1 Illustration depicting five main causes of biodiversity loss.

18.2 POLLUTION AND ITS IMPACT ON BIODIVERSITY

Both the health of the world and humans are existentially threatened by pollution, endangering modern societies' sustainability. Pollution is defined as undesired waste of human origin that is discharged into air, land, and water without consideration for the cost or consequences. Pollution has adverse impacts on both flora and fauna, as shown in Figure 18.2.

FIGURE 18.2 Diagram illustrating the impact of pollution on flora and fauna.

Pollution encompasses various forms of air pollution such as fine particulate matter ($PM_{2.5}$); ozone; oxides of sulfur and nitrogen; freshwater contamination; pollution of the ocean by mercury, nitrogen, phosphorus, plastic, and petroleum waste; and land pollution from lead, mercury, pesticides, industrial chemicals, electronic waste, and radioactive waste (8). In addition to exploration, mining, urbanization, and population expansion, other factors that contribute to environmental pollution include the transboundary migration of pollutants from developed to developing nations and vice versa. One of the reasons why pollution has persisted as a worldwide issue is transboundary contamination (9). The effects of pollution on land include the deposition of wastes on land surfaces, which results in an unpleasant stench and bad aesthetics; harm to trees; extinction of wildlife species; and infertility of the soil, which lowers plant production (10). Additionally, it is simpler for pollutants to spread between land and water due to their direct contact. Water bodies that are contaminated typically experience changes in their chemical, microbiological, and physical characteristics. Furthermore, pollution affects the natural population's genetic variety and variability. Research has shown that fish that live in contaminated areas have very complicated ribosomal sequences in their genomes. A few more examples of pollution's effects on biodiversity are shown in Table 18.1. The amount of ribosomal DNA copies is shown to systematically grow in response to changes in environmental factors. This occurs because the primary function of these sequences is to maintain the integrity of the genome (11). Pollution's indirect effects, or those without direct chemical effects on the organism of interest, could be apparent in a particular ecological setting since these rely on particular interactions between various organismal groups both within and between species or trophic levels. For instance, contaminants may alter the trophic base of certain creatures or interfere with an animal group's social interactions (12–14). A few types of pollution that are affecting our biodiversity are briefly discussed next.

18.2.1 Air Pollution

When one or more toxins are present in the atmosphere for a long enough period of time to harm people, animals, or plants, it is referred to as air pollution (25). According to a recent World Bank report, air pollution caused global economic losses of $8.1 trillion in 2019, or 6.1% of the GDP (26). The primary cause of the main alteration in the composition of the atmosphere is the burning of fossil fuels, which are utilized for transportation and energy production. There have been reports of several air pollutants with varying chemical compositions, features of their reactions, emissions, and environmental persistence. The alteration of the forest vegetation in the San Bernardino Mountains of Southern California is an excellent illustration of how air pollution affects biodiversity. The ozone-tolerant oaks and deciduous shrubs replaced the ozone-sensitive pines as the dominant plant after extended exposure to photochemical oxidants. Reduced photosynthetic ability as a result of foliar injury and early needle fall, inhibited stem radial growth, and lower nutrient retention in the needles were the primary causes of the pine trees' decline. Increased vulnerability to fire and insect attacks were among the indirect consequences on the pine trees. Aside from these modifications, the forest ecosystem also experienced changes in water flow, nutrient loss, carbon allocation, and patterns of temporal and spatial diversity. Anthropogenic commodities and amenities like commercial forestry and recreation were negatively impacted by all of these developments (25). They were categorized into the following distinct groups based on their shared characteristics: particulate matter, heavy metals like lead and mercury, persistent organic pollutants like dioxins, gaseous pollutants like SO_2, NOx, CO, ozone, and volatile organic compounds (VOCs) (27).

The main source of gaseous pollution is the combustion of fossil fuels, which significantly contributes to fluctuations in the composition of the atmosphere (28). Since automobile emissions are the primary source of nitrogen dioxide, it is thought to be a reliable indication of ambient, traffic-related air pollution. The combustion of fossil fuels in power plants and other businesses also adds to NO_2 emissions. The amount of NO_2 that a person is exposed to throughout their lifetime may be dominated by indoor sources, which include gas stoves and other large indoor sources. Plants

TABLE 18.1
Examples of Adverse Effects of Pollution on Biodiversity

Source of Pollution	Affected Organism	Effects of Pollution	Reference
Nanoplastic found in soil	*Enchytraeus crypticus*	Ingestion of nanoplastic reduced body weight and changed the makeup of gut microbes involved in N reactions and the breakdown of organic waste.	(15)
Waterborne antidepressants (venlafaxine, fluoxetine, citalopram)	Fish	A substantial rise in the number of deaths, delay in development, morphological defects, and pathological alterations in the brain, heart, cranial, and caudal kidneys were seen in fish as a result of exposure to antidepressants thrown in water.	(16)
Coal-burning factories near Massachusetts	Lichens on *Acer saccharinum* and *Populus deltoide*	Cover of lichens decreased near the coal-burning station.	(17)
Automobile pollution in Lucknow, India	*Ficus religiosa* and *Thevetia neriifolia*	Alterations in photosynthetic pigments, protein, and cysteine contents; modifications in leaf area and foliar surface architecture of plants.	(18)
Air pollution impact in Man and Biosphere Reserve Wienerwald, Austria	Bryophytes: *Scleropodium purum*, *Hypnum cupressiforme*, and *Abietinella abietina*	The Index of Atmospheric Purity (IAP) method showed that the air quality in the reserve had a significant impact on the distribution of bryophytes.	(19)
Coal-fired power plant	*Parus major* and *Emberiza cia*	Shorter cilia, an increase in secretory granules and vesicles, and an increase in mucus covering the tracheal epithelium were all seen in passerine birds.	(20)
Aquatic environmental pollution in the Egyptian countryside	Fish	Heavy metals in the water affected the vital functions and reproduction in fish. It also increased the percentage of infections caused by external or internal parasites that attack fish.	(21)
Motorboat noises	*Anguilla anguilla*	They showed a considerable rise in oxygen consumption, which caused the eels' physiological condition to be compromised.	(22)
Polyurethane (microplastic)	*Chlorella vulgaris*	Decrease in growth and photosynthetic activity of the plant.	(23)
Polymer (microplastic)	*Lepidium sativum*	The rate of germination is considerably decreased, and as plastic size increases, negative effects escalate as well.	(24)

cannot withstand short-term quantities of nitrogen dioxide beyond 120 μg/m^3 (28). Plant growth is decreased by it. The impacts on vegetation are worse when ozone and sulfur dioxide are also present. Nitrogen dioxide can also contribute to acid rain, in addition to sulfur dioxide. One of the many photochemical oxidants that are created as secondary pollutants by solar radiation in the presence of primary pollutants, which are primarily nitrogen oxides, and volatile organic compounds, is ozone. This process eventually results in the depletion of biodiversity (29).

A dangerous class of compounds is made up of persistent organic pollutants. Their long-lasting presence in the environment is accompanied by the biomagnification of their effects, which increase as they ascend the food chain. They include pesticides, along with dioxins, furans, and PCBs (30). Dioxins can originate from natural events like volcanic eruptions and forest fires, but they are primarily byproducts of industrial operations. Dioxins are undesirable byproducts of a variety of manufacturing processes, such as the production of some herbicides and pesticides, smelting, and the bleaching of paper pulp with chlorine. Dioxins are extremely harmful substances that can harm the immune system, disrupt a species' hormonal balance, and cause issues with reproduction and development. This may cause populations of impacted species to dwindle or go extinct, which would lead to a reduction in biodiversity (31).

Basic metal elements, including lead, mercury, cadmium, silver, nickel, vanadium, chromium, and manganese, are examples of heavy metals. They can be carried by air since they are naturally occurring elements of the Earth's crust that are impervious to degradation and destruction (27). The buildup of heavy metals in soil is turning into a more significant environmental issue. Additionally, heavy metal poisoning of soil has a significant negative impact on plant diversity, making it a major concern for biodiversity (6).

A class of air pollution known by the general term "particulate matter" (PM) is made up of intricate and diverse combinations of particles suspended in the breathing air that differ in size and content and are created by a wide range of anthropogenic and natural activities (32). Factories, power plants, incinerators for waste, automobiles, building sites, fires, and naturally occurring wind-blown dust are the main contributors to particle pollution (27). Plants are the main atmospheric receivers for both gaseous and particle pollution because they are continuously exposed to air. The major receivers for atmospheric gaseous and particulate matter pollution are plant leaves. These contaminants interact with the foliar surface, changing its configuration before they reach the leaf tissue. When both fine and coarse dust particles are deposited on the leaf surface, the effect on leaf gas exchange reveals a decrease in plant development (33). Acute and chronic injuries are the two primary direct damage types that PM pollution can inflict on plants. Being exposed to a high concentration of gas for a brief length of time causes acute damage, which is characterized by distinct, visible symptoms on the foliage, frequently taking the form of necrosis. Chronic damage is far more subtle; it arises from extended exposure to decreased gas concentrations and manifests as a decline in growth and/or yield, frequently without obvious symptoms (17).

18.2.2 Water Pollution

Water is a basic human necessity that has an unquestionable impact, either directly or indirectly. Water is necessary for all industrial, environmental, and biological functions. Water is an essential component of living things, serving as a lubricant, temperature buffer, solvent, metabolite, and living environment (34). On the other hand, water is considered contaminated when certain aspects of its quality have been compromised by random fluctuations resulting from several human activities, making it unsuitable for its intended use (35). Particularly destructive direct effects on freshwater and marine environments are caused by pollution, which includes garbage and chemicals. Pollution is a key driver of biodiversity loss and ecosystem change (36). Certain effects of water contamination are evident right away, while others take months or even years to manifest (37). Based on estimates, over 50 countries worldwide, covering an area of 20 million hectares, are treated with polluted or partially treated polluted water encompassing portions of every continent, and this low-quality water kills aquatic life, puts human health at risk, and interferes with the growth of various crops (38–40).

There are two types of contaminants in water: organic and inorganic. Insecticides, herbicides, volatile organic compounds, animals raised for food, sewage, microorganisms from farming, waste products from food processing, and other types of chemicals are examples of organic water pollutants. Heavy metals from acid mine drainage, silt from surface runoff, logging, slash-and-burn techniques, and land infill are some of the causes of inorganic water pollution. Fertilizers from agricultural runoff, such as nitrates and phosphates, can also cause inorganic water pollution, as can chemical waste from industrial effluents (41). The primary difficulty in evaluating the environmental dangers associated with inorganic pollutants, such as metalloids (e.g., Se and As) and heavy metals (e.g., Cr, Ni, Cu, Zn, Cd, Pb, Hg, U, and Pu), arises from their diverse behaviors under various redox conditions. Unlike many organic contaminants, these substances are not susceptible to deterioration. The main difficulty in managing organic pollutants is the vast quantity and diversity of compounds, which span a broad spectrum of physical–chemical characteristics and reactivities (42).

In addition to being extremely persistent and hazardous in small amounts, heavy metals have the ability to seriously impact aquatic life through oxidative stress. Furthermore, metals stay in the marine environment indefinitely since they are not degraded by microbes (43). Heavy metal pollution in rivers can have catastrophic impacts on the aquatic environment's ecological balance, and the amount of contamination limits the diversity of aquatic life (44). According to a study, heavy metals have been introduced into Egypt's El-Rahawy drainage canal as a result of various pollutants being continuously discharged into it. It was discovered that these metals produce various histopathological alterations in the digestive tract and gonads of *Clarias gariepinus* in addition to causing alterations in the blood (45).

An additional investigation was conducted regarding the impact of wastewater treatment plant effluents from hospitals on freshwater creatures. The quantity of oocytes and embryos, particularly in *Gammarus fossarum* and *Schmidtea polychroa*, as well as the rate of hatching among *Radix peregra*, all decreased, indicating the effect on the organisms' reproductive activities (46).

Decomposers and other microbes, including aerobic and anaerobic bacteria, proliferate in water bodies when organic matter from dead and decaying plant and animal components is deposited from sewage discharges and washed away with rainfall. Rapid growth of *Salvinia*, *Azolla*, *Eichhornia*, and other algae reduces light penetration into the deeper layers of water bodies, resulting in a slow decrease of the submerged flora. This state causes the biological oxygen demand (BOD) to rise and the dissolved oxygen level to decrease. While organic matter-polluted water has a BOD of more than 400 mg/l, unpolluted fresh water typically has less than 1 mg/l (41).

In recent years, a great deal of research has been done on macroplastics, microplastics, and nanoplastics, especially in marine and oceanic ecosystems (47). Pluvial water carries plastic waste that has been improperly disposed of (in streets, highways, open landfills, etc.) (48). Additionally, plastics can be carried by domestic and industrial sewage to water bodies. This is particularly true when products like cosmetics that contain minute amounts of plastic are discharged into effluent systems (49). According to a recent study, exposure to plastics reduced the amount of photosynthetic production in the algae *Chlorella pyrenoidosa* (50). Freshwater species of fish have been documented to consume plastic the most. Fish intestinal injuries were brought on by consuming polystyrene and other polymers, including polyamides, polyethylene, polypropylene, and polyvinyl chloride (51). When excessive amounts of polyethylene are consumed by *Cyprinodon variegatus*, research has shown that this can result in "gut distension and abnormal swimming behaviour," which might have unintended consequences (such as increased susceptibility to predators) (52).

18.2.3 Soil Pollution

Maintaining soil health is essential to maintaining human health. Soil is necessary for the production of food, fiber, biomass, and some medications, as well as for the retention and filtration of water (53). Fertile soils are destroyed chemically through a process known as soil pollution, which has an impact on human health and global food security. The Sustainable Development Goals (SDGs) of ending poverty, attaining zero hunger, ensuring human health and well-being, stopping and reversing land degradation and biodiversity loss, and creating resilient and safe cities are all hampered by soil contamination (54). Soil pollution is occurring due to the massive amounts of trash from manufactured products that are being discharged into the earth. Soil pollution can also be caused by contaminated water. Agricultural practices, improper solid and liquid waste disposal techniques, and unsanitary behaviors all contribute to soil pollution. The pollution of the atmosphere also contributes to soil pollution. Any chemicals or pollutants that are harmful to living things can be considered soil contamination. Pollutants cause soil erosion, alter the natural makeup of the soil, and lower soil quality. Based on the contaminant's source and the way it affects the ecosystem, different types of soil contamination can be identified. Urban activities, industrial wastes, and agricultural pollution are some examples of soil contamination types (55). The year 2019 saw 4 to 23 times more microplastic contamination in terrestrial environments than in marine habitats (56).

Soil pollution could also be a consequence of agricultural practices. In addition to increasing agricultural output, fertilizers pollute the soil and lower its quality. As heavy metals in pesticides contaminate the soil, they also affect plants and animals. These heavy metals contaminate the groundwater system by penetrating deeply into the soil (55). Metallic components known as heavy metals can be found in contaminated and natural surroundings. The businesses that deal in scrap metal, the plastic and rubber industries, the smelting and refining of metals, the burning of garbage containing heavy metals, and many consumer products can all release heavy metals into the environment. Depending on their density, the elements are deposited onto soil, vegetation, and water after traveling great distances when released into the atmosphere. After deposition, the heavy metals do not break down and stay in the environment for a long time (57). Metal toxicity in plants inhibits plant growth by decreasing the uptake and transport of nutrients and water, as well as by enhancing oxidative damage (58).

Historically, mining and manufacturing have been the main industrial processes that pollute soil. The concentrations of organic pollutants and trace elements are generally substantially greater in industrial locations. This is because industrial operations discharge pollutants into the environment, such as the soil, nearby water bodies, and the atmosphere, intentionally and unintentionally (59). As the chemical industry demonstrates, it is anticipated that global industrial output will keep increasing. The yearly production of chemicals increased globally to 2.3 billion tonnes between 2000 and 2017. Petroleum compounds, specialty chemicals, and polymers made up the bulk of the chemicals; they account for 26%, 27%, and 22% of the total value. By 2030, it is expected that the use of chemicals—aside from pharmaceuticals—will rise by 85%, with China and the European Union continuing to be the biggest users. Despite the fact that anthropogenic chemicals have greatly benefited human civilization, there are also significant drawbacks due to accidental exposure to toxins and environmental effects (54).

Soil pollution can result from human activity both directly and indirectly. The surrounding streams and land regions become contaminated due to improper drainage and increasing runoff. Trash that is not properly disposed of decomposes into the earth, releasing various chemicals and contaminants. These could, once again, wash away into the nearby water system or seep into the groundwater. The amount of garbage that accumulates in excess raises the number of dangerous microorganisms in the soil (55). Due to the migration of people from rural areas and the expansion of towns and cities, over 54% of the world's population currently resides in urban areas. Urban soil monitoring and assessment are critical because urban soils are prone to a wide range of anthropogenic activities and may contain elevated amounts of pollution. Both directly and indirectly, these contribute to the decline of biodiversity (59).

18.2.4 Radioactive Pollution

Scientists and engineers have focused their technological efforts on developing methods for processing radioactive materials. Radiation is known to be created when the atomic nuclei of some elements, such as carbon, radium, uranium, thorium, radon, and potassium, fission and release protons, electrons, and gamma rays (i.e., electromagnetic waves). Radiations including radio waves, infrared UV rays, x-rays, and cosmic rays are naturally produced by the sun. However, man-made sources of radiation include the extraction and refinement of thorium and plutonium, the building and detonation of nuclear weapons, nuclear power plants, fuel processing, and the preparation of radioactive isotopes. Studies have indicated that non-ionizing radiations can harm the surface tissues of higher plants and animals as well as potentially be fatal to microorganisms. Additionally, during cell division, radiation causes chromosome breaks and mutations. Chromosome aberrations in plants are also caused by mutations. This could lead to a variety of morphological abnormalities and reproductive disorders, including sterility (7).

18.2.5 Noise Pollution

Even though there are numerous naturally occurring sources of noise, such as the wind, water, and other animals, human activity is becoming a more significant source. The volume, frequency, and commonality of anthropogenic noises are frequently higher than those of natural auditory stimuli (60). It has been discovered that loud noise elevates cortisol levels in a number of species, including humans, dogs (*Canis lupus familiaris*), lined seahorses (*Hippocampus erectus*), and goldfish (*Carassius auratus*) (61–63). Similarly, it has been noted that mice and chickens (*Gallus gallus domesticus*) have higher levels of corticosterone when exposed to high volumes (64). Animals' normal hormone-release patterns can be disrupted by an increase in noise. Noise stress appears to be especially harmful to females, owing to sex differences in size, hormone expression, and the costs of reproductive investment. Reproductive rates were significantly lower among noise-stressed brown shrimp (*Crangon crangon*) , and there were fewer egg-bearing females, indicating that noise-stressed individuals may not have had the resources required for reproduction (60). Reversible responses to many genes and cells might not always be possible, particularly if noise perturbations happen during critical developmental stages or in the germ line. Acute noise stress can be fatal, as evidenced by the 75% reduction in emergence observed in Indian meal moth (*Plodia interpunctella*) larvae exposed to noise. Even in the egg, Muscovy duck (*Cairina moschata*) embryos respond behaviorally to noise in their environment (65). Indeed, it has been observed that in a number of bird species, interegg communication among developing embryos promotes hatching synchrony. Therefore, increases in anthropogenic noise may encourage size differences within broods, which may result in smaller chicks not being able to outcompete their siblings for resources, which may cause nutritional deficiencies and developmental issues in smaller chicks, and in severe cases, even death from starvation. There were more female pups born to stressed-out moms, indicating that noise pollution might change population sex ratios (60).

18.2.6 Thermal Pollution

Any process that modifies the ambient temperature can lead to thermal pollution, which is the deterioration of any life. When heat is produced by heated industrial effluents or by anthropogenic (human) modifications to stream bank vegetation that raise the water system's temperature as a result of solar radiation, it is categorized as a water pollutant. The use of water as a coolant by power plants and industrial manufacturers is a common source of thermal pollution. The abrupt change in temperature that occurs when coolant-treated water is released into the natural environment at a higher temperature impacts the ecosystem by reducing the availability of oxygen. A sudden shift in the water's temperature, also referred to as thermal shock, can kill fish and other organisms that are acclimated to specific temperature ranges. The ecology of a stream or lake may be significantly altered by heated discharges. Because of the suppression of enzyme activity brought on by rising temperatures, aquatic plants exhibit decreased rates of photosynthesis. The primary productivity and species diversity of aquatic plants are declining due to thermal pollution-induced increases in water body temperature (66).

18.2.7 Electronic Waste (E-Waste) Pollution

The number of electronic and personal computing devices has increased exponentially due to the expansion of infrastructure and services via digital platforms. Consequently, the annual production of electronic trash is expected to be between 20 million and 50 million tonnes worldwide (67). A United Nations report states that e-waste will reach 75 million tonnes in 2030 and, if improperly managed, will reach 111 million tonnes in 2050 (68). The informal sector's disposal of e-waste has a significant impact on bird species (69). Heavy metal-containing e-waste is the primary source of sediment pollution and biodiversity risks, as opposed to agricultural practices (70). Many bird species, particularly the house sparrow (*Passer domesticus*) and house crow (*Corvus splendens*), are sensitive to changes in their surroundings (71). Heavy metals are transported through the food chain, which causes a decline in biodiversity (72).

18.2.8 Chemical Pollution

Although there is no denying the advantages of synthetic chemicals in daily life, there is a significant issue with their detrimental effects on biodiversity. Since the Industrial Revolution, chemical pollutants have been released; however, during the past 50 years, this release and dispersal have increased significantly. Carbon dioxide (CO_2) emissions are a prominent example of the long-term effects on climate, atmosphere, and oceans that these substances have; however, emissions from agriculture and industry have released many other substances as well. Mining, mineral processing, farming, construction, and energy production release trillions of tonnes of chemically active material into the environment (73). Apart from the intentional release of geological substances, people have created over 140,000 distinct compounds and their blends (74), the majority of which were nonexistent earlier. Industrial chemicals have been found in the blood and tissues of all populations, including the unborn and newborns, as well as in mother's milk. These chemicals include known carcinogens and their residues. They can be found in food items, wild animals, plants, and aquatic biota. Genetics, metabolism, nutrition, and environment all play a role in life, and chemical toxicity can affect all of these aspects of it. Moreover, the cumulative effects of all anthropogenic chemicals may also reduce biodiversity (75).

18.3 WHAT STRATEGIES COULD BE IMPLEMENTED TO MAKE THE WORLD POLLUTION FREE?

In order to protect natural and seminatural ecosystems from pollution, society must figure out how to minimize and eliminate the pollutants that are already there. Due to the complexity of the issue, numerous measures will need to be implemented or strengthened, including recycling programs, proper disposal, strict legislation, frequent inspections, and ecological restoration (49). It is recommended to implement Agenda 2030, also known as Transforming Our World: the 2030 Agenda for Sustainable Development. This will offer a foundation for creating a more sustainable future for humanity and for the wise use of the natural resources that sustain our existence (76).

Various laws are made to reduce pollution worldwide. In 1990, the US Congress passed the Pollution Prevention Act (P2 Act). In addition to promoting a more sustainable economy, this plan outlines other options to cut back on the use of natural resources, toxic chemicals, and greenhouse gas emissions (77). The Water Framework Directive and the Marine Strategy Framework Directive set forth ambitious water policies in the European Union (EU) for the preservation and restoration of aquatic ecosystems (78). The government of India has enacted the following laws to control air and water pollution: The EPA (Environment Protection Act), 1986; The Indian Forest Act and Amendment, 1984; The Wildlife Protection Act, 1972, and Amendment Act, 1991; The Water (Prevention and Control of Pollution) Act, 1974; The Factories Act and Amendment, 1987; The Air (Prevention and Control of Pollution) Act, 1981; The Atomic Energy Act, 1982; and The Air (Prevention and Control of Pollution) Amendment Act, 1987 (79). Numerous international agreements and declarations pertaining to radioactive wastewater emissions may place duties on national policies and processes. Such agreements and declarations include the International Convention on

FIGURE 18.3 Application of the phytoremediation technique as a means of eliminating pollution from the air, water, and soil.

Nuclear Safety and the Joint Convention on the Safety of Spent Nuclear Fuel Management and on the Safety of Radioactive Waste Management, which have been ratified by several United Nations members. Euratom Treaty regulations bind European Community member states legally, and the OSPAR Convention regulates discharge in the northeast Atlantic (80).

A new environmentally friendly technique called phytoremediation addresses the deterioration, reduction, or stabilization of contaminants in the air, soil, or water as shown in Figure 18.3. It uses plants for the removal of pollutants from contaminated areas including heavy metals and metalloids (e.g., lead and arsenic); organic pollutants, such as 2,4,6-trinitrotoluene (TNT) and polychlorinated biphenyls (PCBs); and radionuclides, such as cesium (^{137}Cs) and strontium (^{90}Sr) (81, 82). Numerous benefits come with it, including being an aesthetically beautiful and solar power-driven cleaning technology; very little chance of environmental degradation due to in situ application; the ability to treat multiple environmental contaminants at once; and cost-effectiveness, as when compared to traditional physicochemical or mechanical procedures, the cost of phytoremediation technology is 60–80% lower (81).

There are numerous methods for attenuating sound pollution. One is to incorporate acoustic behaviors into the wind turbine's design. Component vibrations caused by rotation are one kind of mechanical noise. Unwanted vibrations can be suppressed or eliminated with the use of vibration control. To reduce undesired vibrations, various control laws can be selected based on the circumstances. In addition, the controller can be used to increase or decrease the effective mass (83).

It is also important to raise awareness of the dangers associated with noise. Additionally, the public needs to be informed about the laws governing noise pollution (84).

Because of the increased metal ion concentration over time, heavy metals are harmful to biological systems. The various treatment techniques that can be used to remove harmful heavy metals from water or wastewater include membrane filtration, ion exchange, chemical precipitation, chemical coagulation and flocculation, electrochemical techniques, bioremediation, and adsorption (85). For dealing with e-waste, the following fundamental strategies are solutions to address the e-waste problem. First, a decrease in the production of e-waste. E-goods should be produced, used, and disposed of more sustainably to reduce pollution and resource consumption. Second, creating technologies to use and dispose of the already-accumulated e-waste profitably. The e-waste issue has created problems that have led to the evolution of the EPR (extended producer responsibility) idea. EPR is predicated on taking into account the environmental costs associated with any product during every stage of its life cycle (86).

18.4 CONCLUSION

Since the rate at which biodiversity is disappearing is increasing at a never-before-seen rate, the scientific community should work together to address the threats that chemical pollution and other aspects of global change pose to biodiversity. This detailed analysis clearly shows how different types of pollution—air, water, soil, or noise pollution, for example—interact negatively with ecosystems and have an adverse effect on biodiversity. The numerous ways that pollution disturbs ecosystems demonstrate the mutually reinforcing nature of pollution and biodiversity loss. The impact of pollutants on species populations and habitat integrity is extensive, ranging from chemical contaminants that change the composition of soil to pollutants that make water bodies uninhabitable for aquatic organisms. Convincing evidence has been presented to indicate the far-reaching effects of pollution caused by humans on the extinction of species, imbalances in ecosystems, and general loss of biological diversity. Implementing effective mitigation strategies and policies aimed at preserving and restoring our planet's priceless natural heritage requires an understanding of the complex mechanisms and multifaceted effects of pollution on biodiversity. This thorough chapter emphasizes how urgent it is to reduce pollution and protect biodiversity for the benefit of present and future generations. It also emphasizes the need for coordinated global efforts, interdisciplinary approaches, and creative solutions.

REFERENCES

1. Jain A, Singh BN, Singh S, Singh H, Singh S, editors. Exploring biodiversity as bioindicators for water pollution. *National Conference on Biodiversity, Development and Poverty Alleviation,* 2010.
2. Rawat U, Agarwal N. Biodiversity: Concept, threats and conservation. *Environment Conservation Journal.* 2015;16(3):19–28.
3. Ogidi OI, Akpan UM. Aquatic biodiversity loss: impacts of pollution and anthropogenic activities and strategies for conservation. *Biodiversity in Africa: Potentials, Threats and Conservation*, Springer, 2022. p. 421–48.
4. Sigmund G, Ågerstrand M, Antonelli A, Backhaus T, Brodin T, Diamond ML, et al. Addressing chemical pollution in biodiversity research. *Global Change Biology.* 2023;29(12):3240–55.
5. Ajibade FO, Adelodun B, Lasisi KH, Fadare OO, Ajibade TF, Nwogwu NA, et al. Environmental pollution and their socioeconomic impacts. *Microbe Mediated Remediation of Environmental Contaminants,* Elsevier, 2021. p. 321–54.
6. Alsherif EA, Al-Shaikh TM, AbdElgawad H. Heavy metal effects on biodiversity and stress responses of plants inhabiting contaminated soil in Khulais, Saudi Arabia. *Biology.* 2022;11(2):164.
7. Shafi S. *Environmental Pollution,* Atlantic Publishers & Dist, 2005.
8. Fuller R, Landrigan PJ, Balakrishnan K, Bathan G, Bose-O'Reilly S, Brauer M, et al. Pollution and health: a progress update. *The Lancet Planetary Health.* 2022;6(6):e535–47.
9. Ukaogo PO, Ewuzie U, Onwuka CV. Environmental pollution: causes, effects, and the remedies. *Microorganisms for Sustainable Environment and Health,* Elsevier, 2020. p. 419–29.
10. Feng Y, Wang J, Bai Z, Reading L. Effects of surface coal mining and land reclamation on soil properties: a review. *Earth-Science Reviews.* 2019;191:12–25.
11. da Silva FA, Feldberg E, Carvalho NDM, Rangel SMH, Schneider CH, Carvalho-Zilse GA, et al. Effects of environmental pollution on the rDNAomics of Amazonian fish. *Environmental Pollution.* 2019;252:180–7.
12. Michelangeli M, Martin JM, Pinter-Wollman N, Ioannou CC, McCallum ES, Bertram MG, Brodin T. Predicting the impacts of chemical pollutants on animal groups. *Trends in Ecology & Evolution.* 2022;37(9):789–802.

13. Gessner MO, Tlili A. Fostering integration of freshwater ecology with ecotoxicology. *Freshwater Biology*. 2016;61(12):1991–2001.
14. Yamamuro M, Komuro T, Kamiya H, Kato T, Hasegawa H, Kameda Y. Neonicotinoids disrupt aquatic food webs and decrease fishery yields. *Science*. 2019;366(6465):620–3.
15. Pathan SI, Arfaioli P, Bardelli T, Ceccherini MT, Nannipieri P, Pietramellara G. Soil pollution from micro-and nanoplastic debris: a hidden and unknown biohazard. *Sustainability*. 2020;12(18):7255.
16. Sehonova P, Svobodova Z, Dolezelova P, Vosmerova P, Faggio C. Effects of waterborne antidepressants on non-target animals living in the aquatic environment: a review. *Science of the Total Environment*. 2018;631:789–94.
17. Rai PK. Impacts of particulate matter pollution on plants: implications for environmental biomonitoring. *Ecotoxicology and Environmental Safety*. 2016;129:120–36.
18. Verma A, Singh S. Biochemical and ultrastructural changes in plant foliage exposed to auto-pollution. *Environmental Monitoring and Assessment*. 2006;120:585–602.
19. Krommer V, Zechmeister HG, Roder I, Scharf S, Hanus-Illnar A. Monitoring atmospheric pollutants in the biosphere reserve Wienerwald by a combined approach of biomonitoring methods and technical measurements. *Chemosphere*. 2007;67(10):1956-66.
20. Llacuna S, Gorriz A, Durfort M, Nadal J. Effects of air pollution on passerine birds and small mammals. *Archives of Environmental Contamination and Toxicology*. 1993;24:59–66.
21. Zaki MS, Authman MM, Hammam AMM, Shalaby SI. Aquatic environmental pollution in the Egyptian countryside and its effect on fish production. *Life Science Journal*. 2014;11(9):1024–9.
22. Simpson SD, Purser J, Radford AN. Anthropogenic noise compromises antipredator behaviour in European eels. *Global Change Biology*. 2015;21(2):586–93.
23. Luo H, Xiang Y, He D, Li Y, Zhao Y, Wang S, Pan X. Leaching behavior of fluorescent additives from microplastics and the toxicity of leachate to Chlorella vulgaris. *Science of the Total Environment*. 2019;678:1–9.
24. Bosker T, Bouwman LJ, Brun NR, Behrens P, Vijver MG. Microplastics accumulate on pores in seed capsule and delay germination and root growth of the terrestrial vascular plant Lepidium sativum. *Chemosphere*. 2019;226:774–81.
25. Barker JR, Tingey DT. *Air Pollution Effects on Biodiversity*, Springer Science & Business Media, 2012.
26. World Bank Report on Pollution. 2019. https://www.worldbank.org/en/topic/pollution#:~:text=A%20recent%20World%20Bank%20publication,%2D%20and%20middle%2D%20income%20countries.
27. Kampa M, Castanas E. Human health effects of air pollution. *Environmental Pollution*. 2008;151(2):362–7.
28. Katsouyanni K. Ambient air pollution and health. *British Medical Bulletin*. 2003;68(1):143–56.
29. Organization WH. *World Health Organization*. Air quality guidelines for Europe. 2000.
30. Schecter A, Birnbaum L, Ryan JJ, Constable JD. Dioxins: an overview. *Environmental Research*. 2006;101(3):419–28.
31. World Health Organization (WHO). *Dioxins and their Effects on Human Health*. 2023. https://www.who.int/news-room/fact-sheets/detail/dioxins-and-their-effects-on-human-health#:~:text=Short%2Dterm%20exposure%20of%20humans,endocrine%20system%20and%20reproductive%20functions *help_outline*
32. Pöschl U. Atmospheric aerosols: composition, transformation, climate and health effects. *Angewandte Chemie International Edition*. 2005;44(46):7520–40.
33. Bender MH, Baskin JM, Baskin CC. Flowering requirements of Polymnia canadensis (Asteraceae) and their influence on its life history variation. *Plant Ecology*. 2002;160:113–24.
34. Hanslmeier A. *Water in the Universe*, Springer Science & Business Media, 2010.
35. Glavan M. *Water Challenges of an Urbanizing World: BoD–Books on Demand*, 2018.
36. Programme UNE. *Five Drivers of Natural Crisis*. 2023.
37. Ashraf MA, Maah MJ, Yusoff I, Mehmood K. Effects of polluted water irrigation on environment and health of people in Jamber, District Kasur, Pakistan. *International Journal of Basic & Applied Sciences*. 2010;10(3):37–57.
38. Kan H. Environment and health in China: challenges and opportunities. *Environmental Health Perspectives*. 2009;117(12):A530–A1.
39. Khan A. Air pollution in Lahore. *DAWNCOM Retrieved March*. 2010;7:2012.
40. Khan MA, Ghouri AM. Environmental pollution: its effects on life and its remedies. *Researcher World: Journal of Arts, Science & Commerce*. 2011;2(2):276–85.
41. Singh MR, Gupta A. *Water Pollution-Sources, Effects and Control*. Centre for Biodiversity, Department of Botany, Nagaland University. 2016, p. 1–16.

42. Schwarzenbach RP, Egli T, Hofstetter TB, Von Gunten U, Wehrli B. Global water pollution and human health. *Annual Review of Environment and Resources*. 2010;35:109–36.
43. Woo S, Yum S, Park H-S, Lee T-K, Ryu J-C. Effects of heavy metals on antioxidants and stress-responsive gene expression in Javanese medaka (Oryzias javanicus). *Comparative Biochemistry and Physiology Part C: Toxicology & Pharmacology*. 2009;149(3):289–99.
44. Ayandiran T, Fawole O, Adewoye S, Ogundiran M. Bioconcentration of metals in the body muscle and gut of Clarias gariepinus exposed to sublethal concentrations of soap and detergent effluent. *Journal of Cell and Animal Biology*. 2009;3(8):113–8.
45. Gaber HS, El-Kasheif MA, Ibrahim SA, Authman M. Effect of water pollution in El-Rahawy drainage canal on hematology and organs of freshwater fish. *World Applied Sciences Journal*. 2013;21(3):329–41.
46. Mazzitelli J-Y, Budzinski H, Cachot J, Geffard O, Marty P, Chiffre A, et al. Evaluation of psychiatric hospital wastewater toxicity: what is its impact on aquatic organisms? *Environmental Science and Pollution Research*. 2018;25:26090–102.
47. Dusaucy J, Gateuille D, Perrette Y, Naffrechoux E. Microplastic pollution of worldwide lakes. *Environmental Pollution*. 2021;284:117075.
48. Faure F, Demars C, Wieser O, Kunz M, De Alencastro LF. Plastic pollution in Swiss surface waters: nature and concentrations, interaction with pollutants. *Environmental Chemistry*. 2015;12(5):582–91.
49. Azevedo-Santos VM, Brito MF, Manoel PS, Perroca JF, Rodrigues-Filho JL, Paschoal LR, et al. Plastic pollution: a focus on freshwater biodiversity. *Ambio*. 2021;50(7):1313–24.
50. Wu Y, Guo P, Zhang X, Zhang Y, Xie S, Deng J. Effect of microplastics exposure on the photosynthesis system of freshwater algae. *Journal of Hazardous Materials*. 2019;374:219–27.
51. Lei L, Wu S, Lu S, Liu M, Song Y, Fu Z, et al. Microplastic particles cause intestinal damage and other adverse effects in zebrafish Danio rerio and nematode Caenorhabditis elegans. *Science of the Total Environment*. 2018;619:1–8.
52. Choi JS, Jung Y-J, Hong N-H, Hong SH, Park J-W. Toxicological effects of irregularly shaped and spherical microplastics in a marine teleost, the sheepshead minnow (Cyprinodon variegatus). *Marine Pollution Bulletin*. 2018;129(1):231–40.
53. Agency EE. *Ecosystems Introduction*. 2023.
54. Programme UNE. *Global Assessment of Soil Pollution*. 2023.
55. Mishra RK, Mohammad N, Roychoudhury N. Soil pollution: causes, effects and control. *Van Sangyan*. 2016;3(1):1–14.
56. Horton AA, Dixon SJ. Microplastics: an introduction to environmental transport processes. *Wiley Interdisciplinary Reviews: Water*. 2018;5(2):e1268.
57. Kimani NG. Environmental pollution and impacts on public health: implications of the Dandora Munipal Dumping site in Nairobi, Kenya: Report Summary. 2014.
58. Gill RA, Kanwar MK, Rodrigues dos Reis A, Ali B. Heavy metal toxicity in plants: recent insights on physiological and molecular aspects. *Frontiers in Plant Science*. 2022;12:830682.
59. Food and Agriculture Organization of the United Nations U. Sources of soil pollution. 2023.
60. Kight CR, Swaddle JP. How and why environmental noise impacts animals: an integrative, mechanistic review. *Ecology Letters*. 2011;14(10):1052–61.
61. Anderson PA, Berzins IK, Fogarty F, Hamlin HJ, Guillette Jr LJ. Sound, stress, and seahorses: the consequences of a noisy environment to animal health. *Aquaculture*. 2011;311(1–4):129–38.
62. Evans GW, Lercher P, Meis M, Ising H, Kofler WW. Community noise exposure and stress in children. *The Journal of the Acoustical Society of America*. 2001;109(3):1023–7.
63. Smith ME, Kane AS, Popper AN. Noise-induced stress response and hearing loss in goldfish (Carassius auratus). *Journal of Experimental Biology*. 2004;207(3):427–35.
64. Chloupek P, Voslářová E, Chloupek J, Bedáňová I, Pištěková V, Večerek V. Stress in broiler chickens due to acute noise exposure. *Acta Veterinaria Brno*. 2009;78(1):93–8.
65. Höchel J, Pirow R, Nichelmann M. Development of heart rate responses to acoustic stimuli in Muscovy duck embryos. *Comparative Biochemistry and Physiology Part A: Molecular & Integrative Physiology*. 2002;131(4):805–16.
66. Speight JG. Sources of water pollution. *Natural Water Remediation*. 2020:165–98.
67. Hussain M, Mumtaz S. E-waste: impacts, issues and management strategies. *Reviews on Environmental Health*. 2014;29(1–2):53–8.
68. Mohammadi E, Singh SJ, Habib K. Electronic waste in the Caribbean: An impending environmental disaster or an opportunity for a circular economy? *Resources, Conservation and Recycling*. 2021;164:105106.

69. Sakhuja D, Ghai H, Bhatia RK, Bhatt AK. Management of e-Waste: technological challenges and opportunities. *Handbook of Solid Waste Management: Sustainability through Circular Economy,* 2022. p. 1523–57.
70. Fayiga AO, Ipinmoroti MO, Chirenje T. Environmental pollution in Africa. *Environment, Development and Sustainability.* 2018;20:41–73.
71. Hunt VM. Reproductive success and habitat selection in Black-crowned Night-Herons (Nycticorax nycticorax) in a city park. *The American Midland Naturalist.* 2016;175(2):168–82.
72. Kumar A, Chandrahasan R. Impact of heavy metals from electronic waste on bird species concerning biodiversity: A case study in Bellandur lake, Bengaluru, India. *Environmental Health Engineering And Management Journal.* 2023;10(3):267–72.
73. Cribb J. *Earth Detox: How and why we must clean up our planet.* Cambridge University Press, 2021.
74. Programme UNE. *Global Chemicals Outlook II: From Legacies to Innovative Solutions.* 2019.
75. Naidu R, Biswas B, Willett IR, Cribb J, Singh BK, Nathanail CP, et al. Chemical pollution: A growing peril and potential catastrophic risk to humanity. *Environment International.* 2021;156:106616.
76. Barboza LGA, Cózar A, Gimenez BC, Barros TL, Kershaw PJ, Guilhermino L. Macroplastics pollution in the marine environment. *World Seas: An Environmental Evaluation.* Elsevier, 2019. p. 305–28.
77. Agency USEP. *Pollution Prevention Laws and Policies.* 2023.
78. Grizzetti B, Vigiak O, Udias A, Aloe A, Zanni M, Bouraoui F, et al. How EU policies could reduce nutrient pollution in European inland and coastal waters. *Global Environmental Change.* 2021;69:102281.
79. Gupta S, editor. *Environmental Pollution and its Remedies.* NATIONAL SEMINAR; 2014.
80. Rahman ROA, Kozak MW, Hung Y-T. Chapter 16: Radioactive pollution and control. *Handbook of Environment and Waste Management: Land and Groundwater Pollution Control.* World Scientific, 2014. p. 949–1027.
81. Singh S, Verma A. Phytoremediation of air pollutants: a review. *Environmental Bioremediation Technologies.* 2007:293–314.
82. Lee JH. An overview of phytoremediation as a potentially promising technology for environmental pollution control. *Biotechnology and Bioprocess Engineering.* 2013;18:431–9.
83. Jianu O, Rosen MA, Naterer G. Noise pollution prevention in wind turbines: status and recent advances. *Sustainability.* 2012;4(6):1104–17.
84. Gandhi DK, Raghatate A, Pande S. Noise pollution, awareness, prevention and control. *Prevention and Control.* 2019;2019:908–12.
85. Vardhan KH, Kumar PS, Panda RC. A review on heavy metal pollution, toxicity and remedial measures: Current trends and future perspectives. *Journal of Molecular Liquids.* 2019;290:111197.
86. Premalatha M, Tabassum-Abbasi, Abbasi T, Abbasi S. The generation, impact, and management of e-waste: State of the art. *Critical Reviews in Environmental Science and Technology.* 2014;44(14):1577–678.

19 Invasive Species and Biodiversity

A Global Challenge for Conservation

Ritika Dadhwal[1], *Pankaj Yadav*[2], *Ritesh Banerjee*[1]
[1]School of biological and Environmental Sciences, Shoolini University, Solan, Himachal Pradesh, India
[2]Fish Molecular Biology Laboratory, Department of Zoology, University of Delhi, Delhi

19.1 INTRODUCTION

Depicting the concept of biodiversity would be nearly impossible, given the vast variety of life forms and natural processes that comprise our planet. Biodiversity or biological diversity has been defined as "the variability (types) and totality (numbers) of living organisms in the environment, region, and ecosystem" (Butler, 2006). Biodiversity refers to the variety of life forms on Earth, including animals, plants, microorganisms, genes, and their ecosystems (Rawat and Agarwal, 2015). Biodiversity is a critical element in the operation of ecosystems, serving as a prerequisite for both natural and man-made ecosystems, in addition to the costless ecosystem services that are provided by nature to human society (Sobti et al., 2023). Biodiversity is crucial for the survival of all organisms as it provides the air we breathe, the food we eat, and the water we drink. Alien species play a crucial role in driving global environmental changes. These species can be categorized as either invasive or non-invasive, and their impact on biodiversity is twofold. On one hand, they can enhance regional diversity by increasing species richness. On the other hand, if invasive, they pose a threat to the survival of native biota (Dar et al., 2019). An invasive species is one that occurs outside its natural range and has the possibility of dispersing widely (Singh, 2005). There can be profound impacts on resource availability and abundance of native species when invasive species are present, without necessarily driving community change as a result (Didham et al., 2005). Native plants and invasive plants differ in their requirements and modes of resource acquisition (Singh, 2005). Invasive species contribute significantly to global ecological change, pose a threat to biodiversity, and can have considerable economic effects (Jardine and Sanchirico, 2018).

The term "biodiversity" was coined by Lovejoy in year 1980 and is most commonly used to refer to the number of species (Swingland, 2001). Walter Rosen coined the term for the 1986 National Forum on Biodiversity. The UN Convention on Biological Diversity defines biodiversity as "the variety of living organisms from all habitats, including terrestrial, marine, and aquatic ecosystems." This includes diversity within species, between species, and among ecosystems (Armsworth et al., 2004). In ecological terms, biodiversity refers to the diversity of living organisms and biotic communities, regardless of whether they occur naturally or have been modified by humans (Swingland, 2001). It is not only the sum of ecosystems, species, and genetic material, but rather their variability

DOI: 10.1201/9781032697550-19

that constitutes biodiversity as statement given by IUCN in the year 1994. To understand it better, biologists usually take three perspectives into account (Blaikie and Jeanrenaud, 2013).

- "Genetic diversity" pertains to the occurrence and variety of various genes and/or genomes, which involves variations both within a population and between populations.
- "Species diversity" signifies the frequency and diversity of various species.
- The term "ecosystem diversity" refers to the variety and frequency of various ecosystems (Blaikie and Jeanrenaud, 2013).

Biodiversity assumes various crucial roles in furnishing ecosystem services, encompassing the regulation of essential ecosystem processes, serving as an intrinsic good, and offering valuable services (Mace et al., 2012). Biodiversity provides ecosystem services that extend beyond the production of food, fiber, fuel, and income. Nutrient recycling, regulation of hydrological processes, regulation of the abundance of undesirable organisms, and detoxification of harmful chemicals are examples. The existence of these renewal processes and ecosystem services is largely dependent upon the maintenance of biological diversity (Altieri, 1999). Despite the benefits linked to biodiversity, the existing challenges confronting species and ecosystems are escalating rapidly (Rawat and Agarwal, 2015). Introducing invasive alien plants accelerates the extinction of native species, disrupts trophic structure, creates resource scarcity, and alters ecosystem services and health. Invasive species negatively impact biodiversity by causing parasitism, predation, habitat creation, economic losses, and reduced agricultural productivity (Kumar et al., 2023). Invasive alien plant species are causing significant disruptions to the essential ecosystem processes that underpin human well-being, encompassing access to secure livelihoods, health, positive social relations, security, and freedom. Global changes have consequences for well-being, including the loss or alteration of goods (fisheries, agricultural, and forest products) and services (clean and plentiful water, climate stabilization, pollination, culture, and recreation) (Pejchar and Mooney, 2009). The literature indicates that over 35 invasive alien plant species are rapidly spreading in Ethiopia, posing a significant threat to agricultural lands. The Environmental Policy of Ethiopia and the National Biodiversity Strategy and Action Plan recognize these invasive plants as a major menace to both biodiversity and the country's economic well-being. Notably, *Parthenium hysterophorus* L., *Prosopis juliflora* (Sw.) DC., *Eichhornia crassipes* (Mart.) Solms, and *Lantana camara* L. are among the most well-known invasive species in the region (Assefa and Molla, 2021). A different study revealed the presence of 50,000 non-native species in North America, with 3,000 of them identified as invasive. Hawaii experienced 860 invasive species. Annually, the United States allocates $137 billion for the eradication of these invasive species (Middleton, 2019). In this chapter, we summarize the current status and impact of invasive species and how they establish and disrupt biodiversity or eradicate native species, and also highlight the data on some of the common noxious invasive species along with their control measures.

19.2 THREATS TO BIODIVERSITY

A threat, by definition, is any process or event, whether natural or human, that is likely to adversely affect any biological diversity component or its sustainable use. We are facing an exceptional crisis of biodiversity unlike any other in human history. The evidence indicates that the current rates of extinctions greatly surpass the estimated background extinction rate. What sets this mass extinction apart from previous ones is that humans are both the cause of it and under threat from it, but they also can prevent it (Bellard et al., 2022). According to the scientific community, there are a total of five main global threats that are responsible for biodiversity losses: overexploitation, habitat destruction, biological invasions, pollution, and climate change (Brook et al., 2008). These factors, whether they are of natural or human origin, often interact and mutually enhance their effects (Rawat and Agarwal, 2015). The global pace of species extinction has been significantly hastened by human activities (Pimm et al., 2006). Biodiversity threats described herein are (i) habitat loss

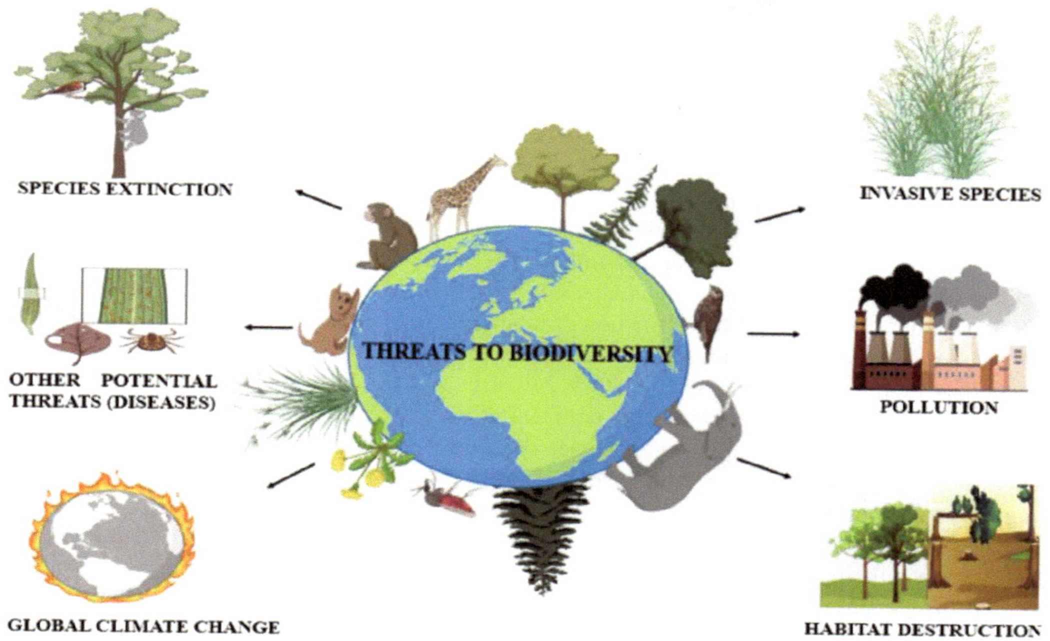

FIGURE 19.1 Threats to biodiversity.

through degradation and fragmentation, (ii) population growth, (iii) overexploitation of natural resources, (iv) climate change, (v) pollution, (vi) disease, and (vii) invasive species. (Singh et al., 2017). Numerous organizations and researchers have ranked the categories that pose a threat to global biodiversity loss. Biological invasions have been identified as the primary cause of global species extinction for several kinds of animals, birds, fish, reptiles, and plants, especially on islands (Leclerc et al., 2018). Over 38,500 species are currently threatened by extinction, primarily due to habitat loss and overexploitation (Maxwell et al., 2016).

Threats to biodiversity are shown in Figure 19.1.

19.2.1 Habitat Loss through Degradation and Fragmentation

Destruction and conversion of habitat is a major cause of biodiversity loss throughout the world (Irwin, 2016). Global biodiversity loss is largely believed to be driven by habitat loss and fragmentation. Humans have significantly changed vast areas of the Earth's surface. As of the start of the 21st century, humans have modified 75% of the Earth's ice-free land to some extent (Watson et al., 2016). Around two-thirds of these modified areas have undergone the complete replacement of the original native vegetation. When it comes to species richness, genetic diversity, abundance, and distribution, habitat loss has a direct and indirect detrimental effect on biodiversity. For example, habitat loss has been shown to have a negative impact on the growth of populations (Rands et al., 2010). According to the 2005 Millennium Ecosystem Assessment, more than 50% of diverse biomes, including the Mediterranean, temperate forests, and tropical or subtropical dry broadleaf forests, had experienced conversion by the year 1990. In Western Europe, a mere 2–3% of the original forests persist in a natural or natural-like state (WWF Report, 2001; Hanski, 2011). From 2000 to 2015, 124.8 million ha of native forests were lost globally due to deforestation rates ranging from 5,770 to 10,483 million ha/year (Keenan et al., 2015). Habitat loss also involves habitat fragmentation and the subsequent extension of forest boundaries, as ancient forests are transformed into human-altered environments consisting of scattered small forest remnants surrounded by open landscapes, like pastures and cultivated fields (Pinho et al., 2020).

19.2.2 Population Growth and Overexploitation of Natural Resources

Rapid population growth may be one of the greatest threats to biodiversity and ecosystem function. Malthus (1798) in his essay discussed that human population growth could surpass the resources needed to sustain it. The environmental consequences of unbridled growth have been explored by many authors. However, predicting the links between human population dynamics and environmental change is difficult due to the complexity of human activity and its impact on nature (Luck, 2007). Humans have caused the extinction of many species over time. Some argue that the arrival of humans to previously uninhabited continents has resulted in the disappearance of various taxa, especially fauna (Burney & Flannery, 2005; Miller et al., 2005). Despite being cited as a key factor threatening Earth's biodiversity, it is not clear to what extent human activity contributes to the endangerment and extinction of other species (McKee et al., 2004). In addition to fast population expansion, the growing use of natural resources threatens global and local biodiversity at an alarming rate, contributing to a reduction in environmental quality (Agduma et al., 2023).

Overexploitation is the act of harvesting species from the wild at a rate that exceeds the natural ability of populations to replenish. Genetic diversity is being lost due to overharvesting (Ryman et al., 1995). The unselective utilization of natural forest resources has led to a substantial reduction in the size of numerous natural populations, causing the fragmentation of habitats and the isolation of populations (dos Santos et al., 2021). The future accessibility of natural resources and ecosystem services is profoundly affected by the human consumption of ecosystems. This could result in an adverse effect on the availability of food, raise health risks, and increase the probability of natural disasters, among other consequences. Ecosystems that have been degraded may exhibit a slow or even impossible recovery, even after the cessation of their exploitation. For example, greenhouse gas emissions may have long-term effects on global climate; invading species and diseases may irreversibly damage ecosystems (Lampert, 2019). According to the literature, significant pressures on biodiversity caused by human activities include the rapid domestication of animals, the enhanced capacity of humans to exploit biodiversity through technological advancements, and the overexploitation of natural resources (Prakash and Verma, 2022).

19.2.3 Climate Change

Climate change has been identified as one of the primary causes of the loss of biodiversity in many studies (Brook et al., 2008; Guo et al., 2017). The increase in carbon dioxide levels and temperature has led to global climate changes, resulting in significant changes in hydrologic cycles such as evaporation and precipitation, and more frequent extreme weather events such as cyclones, floods, and droughts. These changes will have a detrimental effect on biodiversity. Climate change is considered to be the "mother of all externalities" because of its profound impact on biodiversity. Furthermore, climate change could cause the extinction of over a million terrestrial species within the next 50 years (Thomas et al., 2004). Climate change is anticipated to become as significant a factor as land-use change in affecting tropical forest biodiversity. Climate change will also have an indirect impact on species by lowering the amount and availability of habitat, as well as by eradicating species that are important to the species in question. Climate change is expected to significantly affect tropical ectotherms, even when considering their ability to regulate their body temperature through behavior. This is due to their relative sensitivity to temperature variations and their proximity to their ideal temperature range (Morris, 2010). If weather and precipitation patterns change to the point that a species can no longer survive in its existing habitat, the species will decrease unless it can adapt to the changes or move to new, more appropriate habitats. Confronted with rapid climate change, any organism incapable of swift adaptation or migration through the human-influenced, fragmented terrain would face extinction. For example, if the climate increases by 4 degrees Fahrenheit, 5–10% of California's species may become extinct. All of these threats have the potential to inflict severe harm on a specific species or ecosystem. Over the long term, climate change is expected to become a progressively more significant influence (Regan et al., 2008).

19.2.4 Disease

Diseases can have a significant impact on hosts and become epidemics. This can increase the chances of a species going extinct (Lafferty and Gerber, 2002). Therefore, it is reasonable for conservation biologists to be concerned with diseases. Diseases caused by infection are widely recognized as having crucial effects on natural systems, which influence host genetic diversity and co-evolutionary processes to change the species composition in ecological groups. Parasites and infectious diseases have emerged as a significant focus in conservation biology due to their capacity to initiate sudden epidemics and their potential for swift evolutionary changes (Altizer et al., 2003). In the last 20 years, there has been a growing occurrence of highly infectious diseases affecting both wild populations and controlled environments. Infectious diseases can contribute to the endangerment or extinction of species. Though it is rare, other factors also play primary roles in driving endangerments like poorly regulated harvest and habitat loss (Lafferty, 2014), and the pathogen's ability to be transmitted between hosts is expected to reduce as hosts become rare. There has been an unprecedented amount of mortality and extinctions in both wild animal and plant species due to fungal and fungal-like diseases (Fisher et al., 2012).

19.3 INVASIVE SPECIES

Invasive species play a significant role in global change (Pejchar and Mooney, 2009), and various studies have demonstrated that they can have an impact on biodiversity (Vilà et al., 2011) and the functioning of ecosystems. Invasive species became established on continents after the 16th century following the expansion of travel, commerce, and migration across the globe (Middleton, 2019). Invasive species are described as "those alien species that endanger ecosystems, habitats, or species" by the Convention on Biological Diversity in the year 2000 (CBD) (Chakraborty, 2019). An invasive alien plant species is one that has been brought, either purposefully or unintentionally, to a site, area, or region where it is not native. They are endemic to a single location or region, have an impact on biodiversity, ecology, and the environment, as well as human safety, and have the potential to expand uncontrollably. The term "invasive" is commonly used to describe plants that have been introduced from other locations and have exhibited rapid and prolific proliferation in their newly adopted habitats (Shah et al., 2020). Many studies have shown the impact of invasive species on individual ecological processes and within specific environments (Dassonville et al., 2008) and on single functions (van Wilgen and Richardson, 2014). However, only one study has attempted to measure the functional significance of the indirect effects of biodiversity (Constán-Nava et al., 2015). The direct effects of invasive species on ecosystem processes may be due to changes in nutrient levels or alterations in disturbance regimes. The importance of indirect effects of biodiversity loss will only be significant if invasive species greatly reduce biodiversity and if biodiversity has a strong influence on ecosystem functioning. About 40% of the species in the Indian flora originate from foreign sources, and among them, 25% are identified as invasive. Currently, invasive species stand as the second most significant threat to biodiversity and aquatic systems, trailing only behind habitat loss and fragmentation (Kennard et al., 2005). Both invasive and introduced species have the potential to lead to extinctions, modify non-living environments, become nuisances, or introduce diseases, especially affecting species with limited reproductive capacity or those that lack experience in dealing with competitors or predators (Morris, 2010). Human activities have significantly expedited the transfer and integration of species between different regions or continents (Regan et al., 2008). This can occur either deliberately or by chance. Species introduced into an ecosystem will lead to alterations within the ecosystem. These introduced species are commonly known as biological contaminants. Ecological impacts of the invasion include hybridization, outcompetition, ecosystem disruption, plant pathogenic effects, disease transmission, food web disruption, and, in some cases, extinction (Rawat and Agarwal, 2015). The worldwide cost of invasive species harm has been estimated to reach £1.5 trillion per year, or close to 5% of the global gross domestic product (GDP). In developing countries where agriculture constitutes a significant portion of the GDP, the detrimental effects of invasive species on both food security and economic performance can be more pronounced. This, in turn, influences human well-being by contributing to diminished food security, challenging living conditions, decreased wealth, social instability, and hindered efforts in poverty alleviation (Assefa and Molla, 2021). Invasions of a single species can result in significant species loss, perhaps even rivaling land-use change as the most serious disturbance in terms of loss of native species diversity (Murphy and Romanuk, 2014).

19.3.1 Characteristics of Invasive Plants

Invasive plants pose a significant threat to almost all ecosystems. Invasive species can lead to changes in disturbance patterns, ecosystem structure and functions, biodiversity, and agricultural production (Mack et al., 2000). The success of an invasive plant species depends on various factors such as response to disturbance, their life history traits, nutrient requirements, and their overall strategy compared to those native to the ecosystem (Davis et al., 2000; Garten et al., 2008; Schutzenhofer and Knight, 2007). Perennial invasive plants often rely on competitive and tolerance strategies to establish and persist in an ecosystem (Vilà and Weiner 2004). These plants can outcompete native species for resources, which increases their chances of establishing and thriving in the ecosystem. Additionally, invasive plants may also have higher or equal tolerance to stressful conditions, which further boosts their persistence and dominance. As per the literature, some of the characteristic features of invading plants are (i) rapid growth rate, (ii) broad environmental tolerance, (iii) high adaptability in the initial stages of plant growth and development, and (iv) effective competition with other plants and animals (Mainka and Howard, 2010). A list of some invasive species is in Table 19.1.

19.3.2 Establishment of Invasive Species

The establishment of invasive species takes place in three steps (Figure 19.2). The initial step is dispersal, which can occur in a system that is either weakly disturbed with no other species present or in an empty niche where some species already exist. The next stage involves colonization, encompassing all activities related to establishing a self-sufficient population of sufficient size to inhabit the surrounding area. The last stage is the succession stage, in which the new colony begins to expand into the surrounding areas, progressively occupying the habitat space. A successful invader eventually blends in with local conditions and other biotic community members like native species (Muralidharan, 2017).

19.3.3 Impact of Invasive Species on Biodiversity Disruption

Factors contributing to the decline of biodiversity are direct (invasive species, excessive resource use, climate change, habitat loss, replacement of local breeds and varieties, and pollution) and indirect (insufficient awareness, poverty, and demographic change). The decline in biodiversity resulting from biological invasions has been associated with the extinction of approximately 60% of species over the course of the last century (Hussein and Estifanos, 2023). Human activities have exerted profound global influences on Earth and its biodiversity. One example is the heightened vulnerability of lakes and streams to species loss, primarily driven by land-use changes and the introduction of exotic invasive species, which pose significant threats to these ecosystems (Havel et al., 2015).

TABLE 19.1
List of Some Invasive Species

Invasive Species	Native Region	Reference
Eichhornia crassipes	South American	Lowe et al., 2000
Miconia calvescens DC.	South American	
Psidium cattleianum Sabine	Brazil	
Lantana camara	America	Dar et al., 2019
Parthenium hysterophorus	Originate from northeastern Mexico, native to America	Patel, 2011
Ageratum conyzoide	America	Kohli et al., 2006
Potentilla recta L.	Widespread in northern US states	Mullin, 2000
Solanum viarum Dunal	Argentina and Brazil	
Melinis minutiflora P. Beauv.	Brazil	Zenni and Ziller, 2011
Tecoma stans (L.) Kunth.	Native to Mexico and the Southern United States	
Prosopis juliflora	Central America	
Ambrosia artemisiifolia L.	North America	Zhang et al., 2018
Bidens pilosa L.	Central America	
Eupatorium adenophorum Spreng.	Native to Mexico and Costa Rica of Central America	Lu et al., 2005
Acacia mearnsii De Wild.	Southeastern Australia	Sankaran et al., 2014
Ageratina adenophora (Spreng.) R. M. King & H. Robinson	Central America	
Ipomoea purpurea (L.) Roth	Central America	
Senna tora (L.) Roxb.	Asia and the Pacific	
Mimosa pudica L.	Tropical America	
Pueraria phaseoloides (Roxb.) Benth.	South and Southeast Asia	

People introduce species into new ecosystems in three general ways:

(i) Accidental introductions (often invertebrates and pathogens)
(ii) Species imported with a limited purpose that escape
(iii) Slow introductions (usually plants and vertebrates) (McNeely, 2001)

Exotic species are widely recognized for their socioeconomic consequences, ranging from economic losses due to reduced crop productivity and associated management expenses to jeopardizing both food security and human well-being (Clarke and McGeoch, 2023). Both indigenous and invasive species are capable of adapting habitat and ecological processes, thereby causing ecological and economic harm. For example, an infestation of pests may cause partial or total forest mortality by damaging foliage and diminishing the vitality of the trees (Shabani et al., 2020).

Invasive species may also reduce the availability of non-market environmental products and services (e.g., water quality). Non-indigenous invasive species can have an immediate influence on native biological ecosystems (by predation, grazing, parasitism, infection, competition, or hybridization). It is also possible that their effect may be mediated indirectly through the modification of ecosystem function (e.g., through changing hydrology, nutrient cycles, fire regimes, and energy flows) (Andersen et al., 2004).

It has been shown that invasive macroalgae can negatively impact ecosystems because they can tolerate a wide range of abiotic conditions, outcompete native species for nutrients and substrata, and have a relatively high growth rate, making them less palatable to native herbivores. In addition to outcompeting native species for growth substrate and nutritional resources, invasive plants can

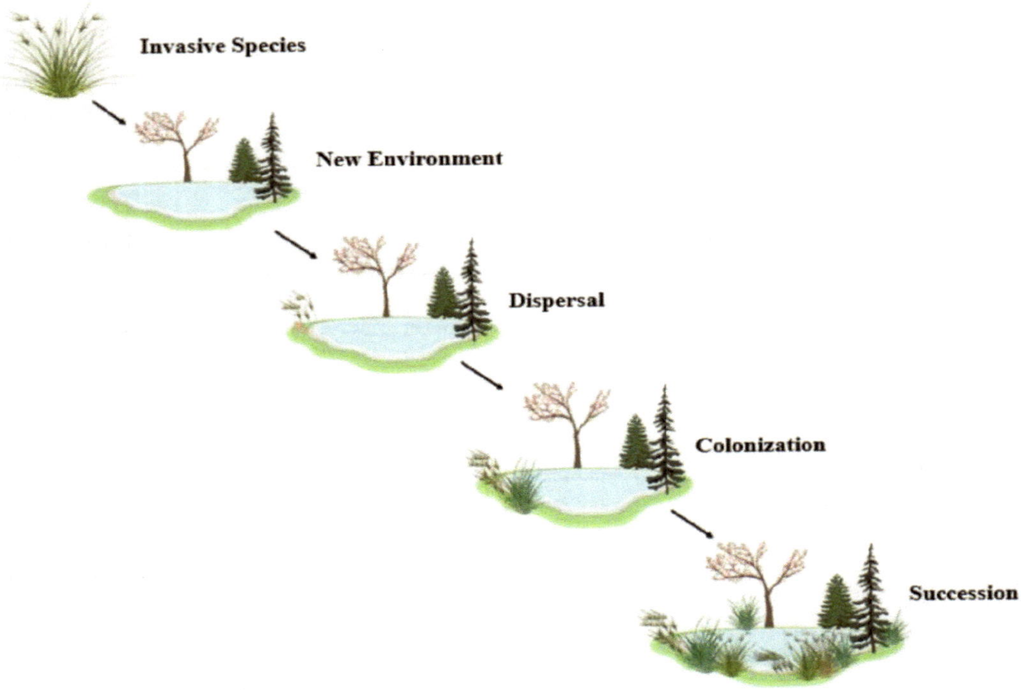

FIGURE 19.2 Establishment of invasive species.

damage local flora and animals through harmful secondary metabolites (Young et al., 2022). It has been observed, for instance, that invasive species usually divert leaf nitrogen from defense purposes to photosynthesis as a way to gain a competitive advantage. Furthermore, several reports suggest that invasive plants can hinder native plant growth by disturbing the mutually beneficial relationships between native plants and other organisms, such as mycorrhizal and endophytic fungi (Chen et al., 2022). Another study found that when invasive algae establish themselves as the dominant species in a local area, it leads to profound repercussions for the biodiversity of the ecosystem (Li et al., 2023). A few examples are given here to illustrate the extensive impact of invasive plants on biodiversity, although there is plenty of evidence to support this claim. Invasive plants generally establish dense stands that outcompete other species and reduce the diversity of herbs, shrubs, and young trees. The substitution of native vegetation with exotic species in forests frequently results in a shortage of essential structural and dietary resources needed by native animal species. In North America, *Arundo donax* may be able to create an invasive plant-fire regime by dominating plant communities. This may result in competitive displacement and a decline in carnivore habitat usage (Havrdová et al., 2023). Tropical dry forests of India have been extensively studied for their ecological impacts and subsequent restoration through the removal and succession of *Lantana*. For example, research conducted on experimental restoration efforts in a tropical dry forest in southern India revealed an increase in the diversity of herb and shrub species associated with the restoration process (Choksi et al., 2023). This may result in competitive displacement and a decrease in carnivore habitat usage. As the global economy has expanded to virtually every corner of our globe, the severity of the problem has risen dramatically. As invasive species become more prevalent, the decision-making process will become more complex and challenging. An adaptable management approach is therefore necessary (McNeely, 2001). The impact of invasive species on diversity, i.e., how it disrupts the diversity, is shown in Figure 19.3.

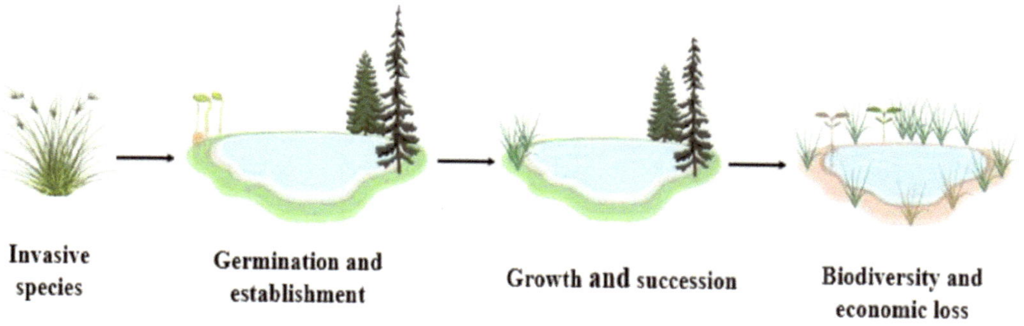

FIGURE 19.3 Impact of invasive species on biodiversity.

19.4 MOST COMMON NOXIOUS INVADERS OF INDIA

From the list of the 100 most invasive species around the world, 11 are found in India, according to literature. These 11 species are *Acacia mearnsii* (black wattle), *Arundo donax* (giant cane), *Chromolaena odorata* (Siam weed), *Clidemia hirta* (Koster's curse), *Imperata cylindrica* (cogon grass), *Lantana camara* (lantana), *Leucaena latisiliqua*, *Mikania micrantha* (mile-a-minute weed), *Opuntia stricta* (prickly pear), *Ulex europaeus* (gorse), and *Sphagneticola trilobata* (Wedelia trilobata, Singapore daisy). An additional 134 species are recognized as naturalized but have the potential to become invasive in the near future (Hiremath and Sundaram, 2013). In addition to discussing the richness of Indian flora, the Central Institute of Medicinal and Aromatic Plants also presented a list of major adventive weeds, including *Parthenium hysterophorus*, *Eupatorium adenophorum*, *Eupatorium odoratum*, *Mikania micrantha*, *Ageratum conyzoides*, and *Galinsoga parviflora* (Singh, 2005). Some of the most common and noxious invaders of India are discussed next and shown in Figure 19.4.

19.4.1 *PARTHENIUM HYSTEROPHORUS* L.

Parthenium hysterophorus is an aggressive annual herbaceous plant with no known economic significance. This erect, short-lived herb originates from northeastern Mexico and is native to America. It is known for its vigorous growth and high reproductive capacity, particularly in warmer regions. It is commonly known as wild feverfew, altamisa, carrot grass, star weed, bitter weed, the "Scourge

A). *Parthenium hysterophorus* L. B). *Lantana camara* L. C). *Ageratum conyzoides* L.

FIGURE 19.4 Some of common noxious invaders of India.

of India," and congress grass (Patel, 2011). It is a significant invasive weed in numerous regions globally, with the capacity to extend its presence to even more areas, resulting in considerable ecological and agricultural damages annually. It is considered one of the worst weeds because of its invading nature, impact on human health, and economic and environmental impacts (Adkins and Shabbir, 2014). This weed was documented in India in 1951 from Maharashtra. This weed exhibits a strong tendency to rapidly establish itself in areas that have suffered degradation, with sparse ground cover and exposed soil, including abandoned wastelands, road edges, and pastures that have been overgrazed. *P. hysterophorus* is categorized as a harmful weed due to its high seed production and rapid spread, its allelopathic impact on other plants, its strong competition with crops, and the health risks it poses to both humans and animals. The weed is highly allergenic and causes respiratory problems, asthma, and dermatitis. However, apart from its allelopathic characteristics and interactions with crops, there is a need for research regarding the weed's influence on ecosystem processes. Because of anthropogenic disturbances such as cultivation, road building, and forest fragmentation and degradation, the plant is rapidly spreading (Singh, 2005). Due to the presence of these phytoconstituents—hysterin, parthenin, hymenin, and ambrosin—the weed exhibits a strong allelopathic impact on different crops. Plants in both dicot and monocot families have been shown to be inhibited by parthenin in the stages of germination and radical growth. This weed affects nodulation in legumes due to the inhibition of nitrogen-fixing and nitrifying bacteria, namely, *Rhizobium*, *Azotobacter*, and *Azospirillum*. *P. hysterophorus* reduces agricultural crop yields by up to 40% in India (Kaur et al., 2014). *Parthenium* biological control was first implemented in Australia in 1977 (Dhileepan and McFadyen, 2012).

19.4.2 *Lantana camara* L.

Lantana camara is a straggling shrub that is one of the most extensively dispersed invasive plants in India. *Lantana* was introduced into various parts of India at various periods (Sundaram et al., 2012). This species, originally from Tropical America, was initially brought to India in 1809 as an ornamental plant in Kolkata, and subsequently, it escaped and established itself as an invasive species (Dar et al., 2019). Some *Lantana* species are also thought to have originated in Africa and one in India. *L. camara* is recognized as a noxious weed in many places in India where it has been established. Due to its prolific flowering and effective dispersal mechanisms, this species tends to alter the structure of the terrestrial ecosystem by establishing dense populations. The plant forms thickets and displaces native species through its dense growth. *L. camara* colonizes the forest floor of damaged native forests, thus dominating the flora and causing disruption in succession and loss of biodiversity (Priyanka and Joshi, 2013). *L. camara*'s invasion drastically reduced plant variety in the invaded region. The biomass of the *L. camara* invaded region was approximately half that of the uninvaded area. *Albizia lebbeck* is a native plant of the Siwalik Hills in northern India; however, no seedlings were found in *L. camara*-infested locations. Under the canopy of *L. camara*, nearly all native plant species were inhibited from growing and reproducing. A native plant species, *Achyranthes aspera* L. is strongly inhibited by the rhizosphere soil of *L. camara* (Kato-Noguchi and Kurniadie, 2021). The leaves of *L. camara*, although perennial, are replaced with new ones after senescence (Sharma et al., 2005; Priyanka and Joshi, 2013). The fallen leaves gather on the ground below the canopy, forming a layer of debris. During the decomposition of the litter, certain bioactive compounds from the leaves may be released into the rhizosphere soil and function as allelochemicals. Allelochemicals can effectively suppress the germination, establishment of seedlings, and growth of other plant species (Bais et al., 2006; Bonanomi et al., 2006; Belz, 2007).

19.4.3 *Ageratum conyzoides* L.

Ageratum conyzoides, originally native to tropical America, has proliferated globally in tropical and subtropical regions. In 1860, it was introduced to India for ornamental purposes, but subsequently, it has become naturalized and emerged as a weed in diverse habitats across the country (Dogra et al., 2009). This American weed may adapt to a variety of environments, although it favors disturbed areas. This plant abundantly produces fruits equipped with pappus structures, facilitating the easy dispersal of light-sensitive seeds that can remain viable for up to a year by wind, animals, and water. Consequently, it is recognized for its impact on agricultural fields, leading to economic losses (Kohli et al., 2006). *A. conyzoides* poses a significant challenge in the cultivated lands of the hilly tracts in northwestern India. It creates dense thickets in commonly cultivated crops like chickpeas, rice, maize, and wheat, leading to adverse effects on crop yields. The weed's prolific seed production capacity results in a swift invasion and colonization of fallow fields (Kaur et al., 2012). *A. conyzoides* can rapidly become dominant, and all forms of it can potentially suppress neighboring plants due to their potential release of allelochemicals or their relatively high seed production over an extended period of time (Shen et al., 2019).

19.5 CONTROL

The easiest method to avoid such issues is to prevent introductions in the first place, or failing that, to detect and eradicate them as soon as possible (Simberloff, 2010). As an alternative to eradication, mechanical, chemical, or biological control measures can be used to maintain appropriate populations of the intruder. Supervising mechanisms, such as hunting, can be particularly effective in confined geographical regions like islands. On the other hand, chemical management involves the use of pesticides to reduce the population of a specific species. However, the widespread application of chemicals is constrained by substantial economic expenses and potential risks to human health (Rinella et al., 2009). In specific regions of South Africa, a combination of mechanical and chemical treatments is employed to effectively manage *Lantana* populations. Prolonged use of herbicides and pesticides often gives rise to various issues. Nevertheless, diverse methods can be applied to control the populations and mitigate the impacts of many established introduced species, ensuring their effects are minimized over the long term (Simberloff, 2010). Classical biological control, which involves the deployment of herbivores or pathogens to lower or sustain the populations of invasive plants below a certain threshold, has been practiced for two centuries, with increased emphasis during the mid-20th century. The purpose of biological control is to diminish the fitness of the target plant to the point where its continuous presence is not considered an ecological or economic problem. The methods used to achieve this goal differ significantly among individual herbivores and pathogens, but the overarching aim is to identify the right "injury guilds" that effectively decrease plant fitness (Peterson et al., 2005). Although the majority of global change factors result from human-induced alterations to ecosystems, biological invasions were frequently regarded as a separate global change issue that ecologists needed to address on their own (Seastedt, 2015). Biological control involves the introduction of a non-indigenous organism, usually a predator, herbivore, or parasite, with the objective of reducing the population of an established non-native pest to a less detrimental level. While considered a more environmentally preferable option than the use of pesticides, it is not without risks, as there is the potential for unexpected side effects (Ricciardi, 2012).

19.6 CONCLUSION

Invasive exotic species are causing significant changes in numerous ecological systems all over the world. The impact of these species on many communities and ecosystems is undeniable. Environmentalists, biologists, and managers are alarmed at the widespread changes, as they fear that invasions could lead to the large-scale decline and extinction of native species. The introduction of invasive species represents a significant and far-reaching threat to biodiversity, ecosystems, and even human activities. It has the potential to cause profound disruption and alteration to ecosystems. They can have profound and detrimental effects on native ecosystems, resulting in a decrease in biological diversity in numerous ways. The arrival of non-indigenous species, whether deliberate or unintentional, can lead to the displacement of native species, alterations in habitat composition, and an overall reduction in biological diversity. It is essential to comprehend the extensive consequences of invasive species on ecosystems and to employ efficient strategies for their control and prevention to protect the fragile equilibrium of biodiversity in our natural environment. An interdisciplinary approach is essential for effectively addressing these challenges. This encompasses the implementation of conservation initiatives, adoption of sustainable resource management practices, promotion of international cooperation, and execution of public awareness campaigns. The safeguarding of biodiversity holds paramount significance for the prosperity and resilience of human societies, as well as for the well-being of ecosystems and diverse species.

REFERENCES

Adkins S, Shabbir A. Biology, ecology and management of the invasive parthenium weed (Parthenium hysterophorus L.). *Pest Management Science.* 2014 Jul;70(7):1023–9.

Agduma AR, Garcia FG, Cabasan MT, Pimentel J, Ele RJ, Rubio M, Murray S, Hilario-Husain BA, Cruz KC, Abdullah S, Balase SM. Overview of priorities, threats, and challenges to biodiversity conservation in the southern Philippines. *Regional Sustainability.* 2023 Jun 1;4(2):203–13.

Altieri MA. The ecological role of biodiversity in agroecosystems. In *Invertebrate Biodiversity as Bioindicators of Sustainable Landscapes,* 1999 Jan 1 (pp. 19–31). Elsevier.

Altizer S, Harvell D, Friedle E. Rapid evolutionary dynamics and disease threats to biodiversity. *Trends in Ecology & Evolution.* 2003 Nov 1;18(11):589–96.

Andersen MC, Adams H, Hope B, Powell M. Risk assessment for invasive species. *Risk Analysis: An International Journal.* 2004 Aug;24(4):787–93.

Armsworth PR, Kendall BE, Davis FW. An introduction to biodiversity concepts for environmental economists. *Resource and Energy Economics.* 2004 Jun 1;26(2):115–36.

Assefa AS, Molla EL. Impact of invasive alien plant species, Lantana camara, on the species composition, richness and evenness of invaded plant communities in Sidama, Gamogofa and Gedio Zones, Ethiopia. 2021.

Bais HP, Weir TL, Perry LG, Gilroy S, Vivanco JM. The role of root exudates in rhizosphere interactions with plants and other organisms. *Annual Review of Plant Biology.* 2006 Jun 2;57:233–66.

Bellard C, Marino C, Courchamp F. Ranking threats to biodiversity and why it doesn't matter. *Nature Communications.* 2022;13:2616. https://doi.org/10.1038/s41467-022-30339-y

Belz RG. Allelopathy in crop/weed interactions—an update. *Pest Management Science: Formerly Pesticide Science.* 2007 Apr;63(4):308–26.

Blaikie P, Jeanrenaud S. Biodiversity and human welfare. In *Social Change and Conservation,* 2013 Dec 2 (pp. 46–70). Routledge.

Bonanomi G, Sicurezza MG, Caporaso S, Esposito A, Mazzoleni S. Phytotoxicity dynamics of decaying plant materials. *New Phytologist.* 2006 Feb 1;169(3):571–8.

Brook BW, Sodhi NS, Bradshaw CJA. Synergies among extinction drivers under global change. *Trends in Ecology & Evolution.* 2008;23:453–460. https://doi.org/10.1016/j.tree.2008.03.011

Brook BW, Sodhi NS, Bradshaw CJA. Synergies among extinction drivers under global change. *Trends in Ecology & Evolution.* 2008;23:453–460.

Leclerc, C., Courchamp, F. & Bellard, C. Insular threat associations within taxa worldwide. Sci. Rep. 8, 6393 (2018).

Burney DA, Flannery TF. Fifty millennia of catastrophic extinctions after human contact. *Trends in Ecology and Evolution.* 2005;20:395–401.

Butler R. Rainforest diversity – origins and implications. 2006. Retrieved 1/22/2014 from http://rainforests.mongabay.com/0301.htm.

Chakraborty SK. Bioinvasion and environmental perturbation: Synergistic impact on coastal–Mangrove ecosystems of West Bengal, India. *Impacts of Invasive Species on Coastal Environments: Coasts in Crisis.* 2019;29:171–245.

Chen J, Zhang HY, Liu MC, Han MX, Kong DL. Plant invasions facilitated by suppression of root nutrient acquisition rather than by disruption of mycorrhizal association in the native plant. *Plant Diversity.* 2022 Sep 1;44(5):499–504.

Choksi P, Kotian M, Burivalova Z, DeFries R. Social and ecological outcomes of tropical dry forest restoration through invasive species removal in central India. *Ecological Indicators.* 2023 Nov 1;155:111054.

Clarke DA, McGeoch MA. Invasive alien insects represent a clear but variable threat to biodiversity. *Current Research in Insect Science.* 2023 Jan 1;4:100065.-

Constán-Nava S, Soliveres S, Torices R, Serra L, Bonet A. Direct and indirect effects of invasion by the alien tree Ailanthus altissima on riparian plant communities and ecosystem multifunctionality. *Biological Invasions.* 2015 Apr;17:1095–108.

Dar JA, Subashree K, Sundarapandian S, Saikia P, Kumar A, Khare PK, Dayanandan S, Khan ML. Invasive species and their impact on tropical forests of Central India: A review. *Tropical Ecosystems: Structure, Functions and Challenges in the Face of Global Change.* 2019:69–109.

Dassonville N, Vanderhoeven S, Vanparys V, Hayez M, Gruber W, Meerts P. Impacts of alien invasive plants on soil nutrients are correlated with initial site conditions in NW Europe. *Oecologia.* 2008 Aug;157:131–40.

Davis MA, Grime JP, Thompson K. Fluctuating resources in plant communities: a general theory of invasibility. *Journal of Ecology.* 2000;88:528–534. doi:10.1046/j.1365-2745.2000.00473.x

Dhileepan K, McFadyen RC. *Parthenium hysterophorus* L.–parthenium. In: M. H. Julien, R. E. McFadyen, Jim Cullen, editors. *Biological Control of Weeds in Australia.* 2012 Mar 5:448–62.

Didham RK, Tylkowski MA, Hutchison MA, Ewers RM, Gemmell NJ. Are invasive species the drivers of ecological change? *Trends in Ecology & Evolution.* 2005 Sep 1;20(9):470–4.

Dogra KS, Kohli RK, Sood SK, Dobhal PK. Impact of Ageratum conyzoides L. on the diversity and composition of vegetation in the Shivalik hills of Himachal Pradesh (Northwestern Himalaya), India. *International Journal of Biodiversity and Conservation.* 2009;1(5):135–45.

dos Santos JR, de Almeida Vieira F, Fajardo CG, Brandão MM, Silva RA, Jump AS. Overexploitation and anthropogenic disturbances threaten the genetic diversity of an economically important neotropical palm. *Biodiversity and Conservation.* 2021 Jul;30(8–9):2395–413.

Fisher MC, Henk DA, Briggs CJ, Brownstein JS, Madoff LC, McCraw SL, Gurr SJ. Emerging fungal threats to animal, plant and ecosystem health. *Nature.* 2012 Apr 12;484(7393):186–94.

Garten CT, Classen AT, Norby RJ, Brice DJ, Weltzin JF, Souza L. Role of N_2-fixation in constructed old-field communities under different regimes of [CO_2], temperature, and water availability. *Ecosystems.* 2008;11:125–137. doi:10.1007/s10021-007-9112-1

Guo D, Desmet PG, Powrie LW. Impact of the future changing climate on the Southern Africa biomes and the importance of geology. *Journal of Geoscience and Environment Protection.* 2017;5:1–9 http://www.scirp.org/journal/gep

Hanski I. Habitat loss, the dynamics of biodiversity, and a perspective on conservation. *Ambio.* 2011 May;40(3):248–55.

Havel JE, Kovalenko KE, Thomaz SM, Amalfitano S, Kats LB. Aquatic invasive species: challenges for the future. *Hydrobiologia.* 2015 May;750:147–70.

Havrdová A, Douda J, Doudová J. Threats, biodiversity drivers and restoration in temperate floodplain forests related to spatial scales. *Science of The Total Environment.* 2023 Jan 1;854:158743.

Hiremath AJ, Sundaram B. Invasive plant species in Indian protected areas: conserving biodiversity in cultural landscapes. *Plant Invasions in Protected Areas: Patterns, Problems and Challenges.* 2013;7:241–66.

Hussein A, Estifanos S. Modeling impacts of climate change on the distribution of invasive Opuntia ficus-indica (L.) Mill. in Ethiopia: Implications on biodiversity conservation. *Heliyon.* 2023 Apr 1;9(4): e14927.

Irwin M. Habitat change: loss, fragmentation, and degradation. *An Introduction to Primate Conservation.* 2016;7:111–28.

Jardine SL, Sanchirico JN. Estimating the cost of invasive species control. *Journal of Environmental Economics and Management.* 2018 Jan 1;87:242–57.

Kato-Noguchi H, Kurniadie D. Allelopathy of Lantana camara as an invasive plant. *Plants.* 2021 May 20;10(5):1028.

Kaur M, Aggarwal NK, Kumar V, Dhiman R. Effects and management of Parthenium hysterophorus: A weed of global significance. *International Scholarly Research Notices.* 2014;2014(1):368647.

Kaur S, Batish DR, Kohli RK, Singh HP. Ageratum conyzoides: an alien invasive weed in India. In *Invasive Alien Plants: An Ecological Appraisal for the Indian Subcontinent,* 2012 (pp. 57–76). Wallingford UK: CABI.

Keenan RJ, Reams GA, Achard F, de Freitas JV, Grainger A, Lindquist E. Dynamics of global forest area: results from the FAO Global Forest Resources Assessment. *Forest Ecology and Management.* 2015;352:9–20.

Kennard MJ, Arthington AH, Pusey BJ, Harch BD. Are alien fish a reliable indicator of river health?. *Freshwater Biology.* 2005 Jan;50(1):174–93.

Kohli RK, Batish DR, Singh HP, Dogra KS. Status, invasiveness and environmental threats of three tropical American invasive weeds (*Parthenium hysterophorus* L., *Ageratum conyzoides* L., *Lantana camara* L.) in India. *Biological Invasions.* 2006 Oct;8:1501–10.

Kumar S, Meena R, Tiwari AK, Singh R, Patel SK, Singh GS. Perceptions of impacts and management of invasive alien plants: a case study from Mirzapur, India. *Front Glob Change.* 2023;6:1194076.

Lafferty KD. Biodiversity loss and infectious diseases. In *Applied Ecology and Human Dimensions in Biological Conservation,* 2014 (pp. 73–89). Berlin, Heidelberg: Springer Berlin Heidelberg.

Lafferty KD, Gerber LR. Good medicine for conservation biology: the intersection of epidemiology and conservation theory. *Conservation Biology.* 2002 Jun;16(3):593–604.

Lampert A. Over-exploitation of natural resources is followed by inevitable declines in economic growth and discount rate. *Nature Communications.* 2019 Mar 29;10(1):1419.

Li H, Geng Y, Shi H, Wu C, Yu Z, Zhang H, Chen L, Xing R. Biological mechanisms of invasive algae and meta-analysis of ecological impacts on local communities of marine organisms. *Ecological Indicators.* 2023 Feb 1;146:109763.

Lowe S, Browne M, Boudjelas S, De Poorter M. 100 of the world's worst invasive alien species: a selection from the global invasive species database. Auckland: Invasive Species Specialist Group; 2000 Dec 12.

LU P, SANG WG, MA KP. Progress and prospects in research of an exotic invasive species, Eupatorium adenophorum. *Chinese Journal of Plant Ecology.* 2005;29(6):1029.

Luck GW. A review of the relationships between human population density and biodiversity. *Biological Reviews.* 2007;82(4):607–45.

Mace GM, Norris K, Fitter AH. Biodiversity and ecosystem services: a multilayered relationship. *Trends in Ecology & Evolution.* 2012 Jan 1;27(1):19–26.

Mack RN, Simberloff D, Lonsdale WM, Evans H, Clout M, Bazzaz FA. Biotic invasions: causes, epidemiology, global consequences, and control. *Ecological Applications.* 2000;10:689–710. doi:10.1890/1051-0761(2000)010[0689:BICEGC]2.0.CO;2

Mainka SA, Howard GW. Climate change and invasive species: double jeopardy. *Integrative Zoology.* 2010 Jun;5(2):102–11.

Maxwell SL, Fuller RA, Brooks TM, Watson JEM. Biodiversity: the ravages of guns, nets and bulldozers. *Nature.* 2016;536:143–5.

McKee JK, Sciulli PW, Fooce CD, Waite TA. Forecasting global biodiversity threats associated with human population growth. *Biological Conservation.* 2004 Jan 1;115(1):161–4.

McNeely J. Invasive species: a costly catastrophe for native biodiversity. *Land Use and Water Resources Research.* 2001;2:1–10.

Middleton BA. Invasive Plant Species. Encyclopedia of Ecology, 2nd ed.; Fath, B., Ed. 2019 Jan 1:431-40.

Miller GH, Fogel ML, Magee JW, Gagan MK, Clarke SJ, Johnson BJ. Ecosystem collapse in Pleistocene Australia and a human role in megafaunal extinction. *Science.* 2005;309:287–290.

Morris RJ. Anthropogenic impacts on tropical forest biodiversity: a network structure and ecosystem functioning perspective. *Philosophical Transactions of the Royal Society B: Biological Sciences.* 2010 Nov 27;365(1558):3709–18.

Mullin BH. *Invasive Plant Species.* 2000.

Muralidharan M. Do alien species matter? Impacts of invasions in Indian freshwater systems and challenges in management. *International Journal of Aquatic Biology.* 2017 Jun 11;5(2):114–27.

Murphy GE, Romanuk TN. A meta-analysis of declines in local species richness from human disturbances. *Ecology and Evolution.* 2014;4(1):91–103.

Patel S. Harmful and beneficial aspects of Parthenium hysterophorus: an update. *3 Biotech.* 2011 Jul;1(1):1–9.

Pejchar L, Mooney HA. Invasive species, ecosystem services and human well-being. *Trends in Ecology & Evolution.* 2009 Sep 1;24(9):497–504.

Peterson RK, Sing SE, Weaver DK. Differential physiological responses of Dalmatian toadflax, Linaria dalmatica L. Miller, to injury from two insect biological control agents: implications for decision-making in biological control. *Environmental Entomology.* 2005 Aug 1;34(4):899–905.

Pimm S, Raven P, Peterson A, Şekercioğlu ÇH, Ehrlich PR. Human impacts on the rates of recent, present, and future bird extinctions. *Proceedings of the National Academy of Sciences.* 2006 Jul 18;103(29):10941–6.

Pinho BX, Peres CA, Leal IR, Tabarelli M. Critical role and collapse of tropical mega-trees: a key global resource. In *Advances in Ecological Research,* 2020 Jan 1 (Vol. 62, pp. 253–294). Academic Press.

Prakash S, Verma AK. Anthropogenic activities and Biodiversity threats. *International Journal of Biological Innovations, IJBI.* 2022 Mar 3;4(1):94–103.

Priyanka N, Joshi PK. A review of Lantana camara studies in India. *International Journal of Scientific and Research Publications.* 2013 Oct;3(10):1.

Rands MR, Adams WM, Bennun L, Butchart SH, Clements A, Coomes D, Entwistle A, Hodge I, Kapos V, Scharlemann JP, Sutherland WJ. Biodiversity conservation: challenges beyond 2010. Science. 2010 Sep 10;329(5997):1298–303.

Rawat US, Agarwal NK. Biodiversity: Concept, threats and conservation. *Environment Conservation Journal.* 2015 Dec 15;16(3):19–28.

Regan HM, Hierl LA, Franklin J, Deutschman DH, Schmalbach HL, Winchell CS, Johnson BS. Species prioritization for monitoring and management in regional multiple species conservation plans. *Diversity and Distributions.* 2008 May;14(3):462–71.

Ricciardi A. Invasive species. In *Ecological Systems: Selected Entries from the Encyclopedia of Sustainability Science and Technology,* 2012 Nov 28 (pp. 161–78). New York: Springer New York.

Rinella MJ, Maxwell BD, Fay PK, Weaver T, Sheley RL. Control effort exacerbates invasive-species problem. *Ecological Applications.* 2009 Jan;19(1):155–62.

Ryman N, Utter F, Laikre L. Protection of intraspecific biodiversity of exploited fishes. *Reviews in Fish Biology and Fisheries.* 1995 Dec;5:417–46.

Sankaran KV, Sajeev TV, Suresh TA. Invasive plant threats to forests in the humid tropics: a case study from Kerala State, India. *Invasive Alien Species Management.* 2014;Mar 25:7–17.

Schutzenhofer MR, Knight TM. Population-level effects of augmented herbivory on *Lespedeza cuneata*: implications for biological control. *Ecological Applications.* 2007;17:965–971. doi:10.1890/06-1282

Seastedt TR. Biological control of invasive plant species: a reassessment for the A nthropocene. *New Phytologist.* 2015 Jan;205(2):490–502.

Shabani F, Ahmadi M, Kumar L, Solhjouy-fard S, Tehrany MS, Shabani F, Kalantar B, Esmaeili A. Invasive weed species' threats to global biodiversity: Future scenarios of changes in the number of invasive species in a changing climate. *Ecological Indicators.* 2020 Sep 1;116:106436.

Shah KK, Tiwari I, Tripathi S, Subedi S, Shrestha J. Invasive alien plant species: a threat to biodiversity and agriculture in Nepal. *Agriways.* 2020 Jun 1;8(1):62–73.

Sharma GP, Raghubanshi AS, Singh JS. Lantana invasion: an overview. *Weed Biology and Management.* 2005 Dec;5(4):157–65.

Shen S, Xu G, Li D, Jin G, Liu S, Clements DR, Yang Y, Rao J, Chen A, Zhang F, Zhu X. Potential use of sweet potato (*Ipomoea batatas* (L.) Lam.) to suppress three invasive plant species in agroecosystems (*Ageratum conyzoides* L., *Bidens pilosa* L., and Galinsoga parviflora Cav.). *Agronomy.* 2019 Jun 17;9(6):318.

Simberloff D. Invasive species. *Conservation biology for all.* 2010 Feb 23;1:131–52.

Singh KP. Invasive alien species and biodiversity in India. *Current Science.* 2005 Feb 25;88(4):539.

Singh SK, Sharma M, Pandey A. Biodiversity-threats and conservation. *Algae.* 2017;40:2–81.

Sobti RC, Thakur M, Kaur T, Mishra S. Biodiversity: threats and conservation strategies. In *Biodiversity.* CRC Press. 2023.

Sundaram B, Krishnan S, Hiremath AJ, Joseph G. Ecology and impacts of the invasive species, Lantana camara, in a social-ecological system in South India: perspectives from local knowledge. *Human Ecology.* 2012 Dec;40:931–42.

Swingland IR. Biodiversity, definition of. *Encyclopedia of Biodiversity.* 2001 Jan 1;1:377–91.

Thomas CD, Cameron A, Green RE, Bakkenes M, Beaumont LJ, Collingham YC, Erasmus BFN, De Siqueira MF, Grainger A, Hannah L, Hughes L, Huntley B, Jaarsveld ASV, Midgley GF, Miles L, Ortega-Huerta MA, Peterson AT, Phillips OL, Williams SE. Extinction risk from climate change. *Nature.* 2004;427:145–148. https://doi.org/10.1038/nature02121

van Wilgen BW, Richardson DM. Challenges and trade-offs in the management of invasive alien trees. *Biological Invasions.* 2014 Mar;16:721–34.

Vilà M, Weiner J. Are invasive plant species better competitors than native plant species? Evidence from pairwise experiments. *Oikos.* 2004;105:229–238. doi:10.1111/j.0030-1299.2004.12682.x

Vilà M, Espinar JL, Hejda M, Hulme PE, Jarošík V, Maron JL, … Pyšek P. Ecological impacts of invasive alien plants: a meta-analysis of their effects on species, communities and ecosystems. *Ecology Letters.* 2011;14:702–708. https://doi.org/10.1111/j.1461-0248.2011.01628.x

Watson JEM, Shanahan DF, Di Marco M, Allan J, Laurance WF, Sanderson EW, Mackey B, and Venter O. Catastrophic declines in wilderness areas undermine global environmental targets. *Current Biology*. 2016;26:2929–34

WWF report. 2001. Insight into Europe's forest protection. Gland, Switzerland: WWF.

Young CS, Lee CS, Sylvers LH, Venkatesan AK, Gobler CJ. The invasive red seaweed, Dasysiphonia japonica, forms harmful algal blooms: mortality in early life stage fish and bivalves and identification of putative toxins. *Harmful Algae*. 2022 Oct 1;118:102294.

Zenni RD, Ziller SR. An overview of invasive plants in Brazil. *Brazilian Journal of Botany*. 2011;34:431–46.

Zhang F, Li Q, Yerger EH, Chen X, Shi Q, Wan F. AM fungi facilitate the competitive growth of two invasive plant species, Ambrosia artemisiifolia and Bidens pilosa. *Mycorrhiza*. 2018 Nov;28:703–15.

20 Disease and Biodiversity Decline

Diksha Sharma[1] Sneha Siwach[2], Sunil Puri[1]
[1]School of Biological and Environmental Sciences, Shoolini University, Solan, Himachal Pradesh, India
[2]Fish Molecular Biology Laboratory, Department of Zoology, University of Delhi, Delhi

20.1 INTRODUCTION

The term "biodiversity" was coined by Walter G. Rosen during the National Forum on BioDiversity in Washington, D.C., held September 21–24, 1986, under the U.S. National Academy of Sciences and the Smithsonian Institution (Sarkar, 2002) and it encompasses the diverse array of living species on the planet. Biodiversity includes terrestrial and aquatic plants, animals, microorganisms, and the genetic material that shapes them, all intricately connected within ecosystems (Giri et al., 2012). In short, biodiversity refers to the amazing diversity of living organisms on Earth, as well as their special traits and the places they live. It's like a big picture showing how diverse and unique our planet's life is (Rawat and Agarwal, 2015). The fundamental aspects of biodiversity include the diversity within genetic populations and species, the abundance of species at both local and global levels, the functionality of ecosystems that play a vital role in human survival, and the spatial extent and condition of natural habitats (Haahtela, 2019).

This intricate web of life has evolved over billions of years, shaped by both natural processes and the increasing influence of human activities. With approximately 2.1 million identified species to date and an estimated 9.0 to 52 million species, according to estimates from the United Nations Environment Programme, biodiversity represents a tapestry of life offering numerous benefits to humanity. These benefits span ecosystem services, providing essential elements like sustenance, medicines, and cultural values, all contributing to the intricate balance of nature (Giri et al., 2000; , 2012). The biodiversity hypothesis suggests that exposure to natural environments has an enriching effect on the human microbiome, fostering immune balance and providing protection against allergies and inflammatory disorders. The two-layered biodiversity system that underpins this defense mechanism is made up of the skin, gastrointestinal tract, and airways on the inside, and the water, soil, animals, and plants on the outside (Haahtela, 2019).

However, a substantial threat to biodiversity arises from diseases, impacting hosts and heightening the risk of species extinction, prompting concerns among conservation biologists. Contrary to the expectation that rare species, with their lower host density, should be less susceptible to diseases, various instances reveal how infectious diseases have significantly diminished the abundance of once-common species and impeded the recovery of rare species (Lafferty, 2003). This emphasizes the critical need for a comprehensive approach to navigate the intricate relationship between diseases and biodiversity, safeguarding the rich diversity of life that graces our planet.

DOI: 10.1201/9781032697550-20

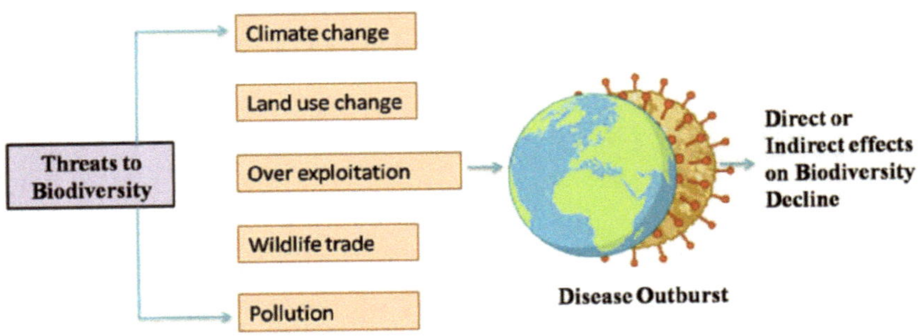

FIGURE 20.1 Graphical abstract.

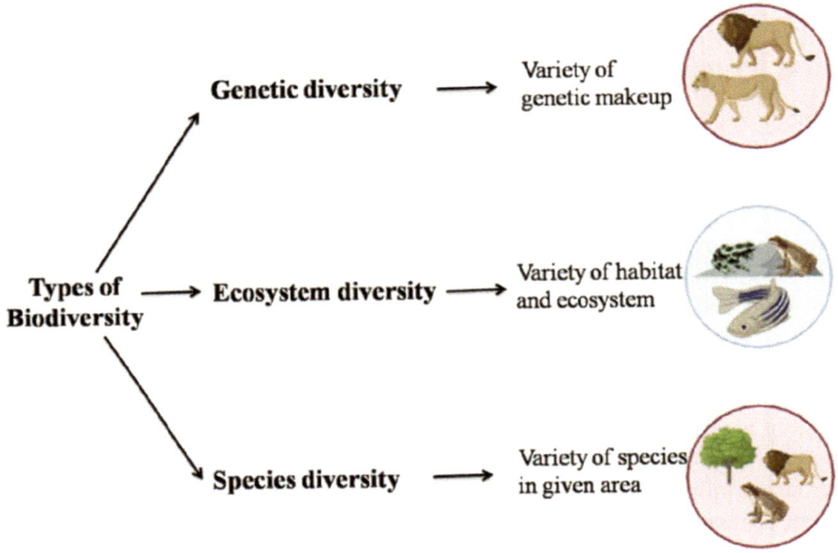

FIGURE 20.2 Types of biodiversity.

20.2 THREATS TO BIODIVERSITY

Nature has a delicate balance that follows certain rules, and when people follow these rules in their lives, it's called religion. Hinduism is a religion that is deeply connected to nature and highly values plants and animals. In Hinduism, it's believed that God has given special powers to plants and animals, making them important and sacred (Khasim and Long, 2020). Even though people all around the world are trying really hard to protect nature through agreements on conservation, the sad truth is that biodiversity, or the variety of living things, keeps decreasing. This makes it important for us to really understand how the actions we take to protect nature might be affecting this ongoing problem of losing different types of plants and animals (Scheper, 2015). Human activities have a big impact on the different types of plants and animals we have in our natural habitat. Sometimes, when we introduce new species or work on restoring nature, it can help increase the variety of living things. But a lot of the time, these activities actually lead to the loss of different local plants and animals. This has serious consequences for diseases that affect both people and wildlife. It's really important for us to predict and prevent potential disease outbreaks because of these changes (Rohr et al., 2020).

The rapid decline in biodiversity is predominantly caused by three pivotal factors: diseases that are new and spreading (emerging infectious diseases), the Earth getting warmer (global warming), and certain species invading new areas (Schmeller et al., 2020; Hoberg and Brooks, 2015; Brema et al., 2022; Keesing and Ostfeld, 2021). Additional factors, such as cutting down forests, damaging natural habitats, polluting, building cities, industrial activities, excessive use of resources, and changes in climate, are made worse by problems in the way governments are managed (Mueller et al., 2022; Neldner et al., 2017; Thormann and Engels, 2015). The current period, referred to as the Anthropocene, represents a time when changes caused by humans significantly affect the biological and chemical systems of the Earth (Neldner et al., 2017; Ibrahim et al., 2018; Mueller et al., 2022). The consequences of losing different types of plants and animals are well documented, showing disruptions in how ecosystems work, the services they provide, and even affecting human health, as seen in events like the COVID-19 pandemic (Mueller et al., 2022). In order to ensure that the intricate web of life on our planet remains robust and sustainable, it is imperative that comprehensive steps be taken to address the numerous issues raised by the loss of biodiversity.

20.2.1 Emerging Infectious Diseases

Emerging infectious diseases (EIDs) are infections that are either newly discovered in a group of people or have been around but are quickly increasing in how often they happen or are spreading to new places. In simpler terms, these infections could be new because of changes or evolution in existing organisms, known infections spreading to new places or groups of people, infections not noticed before appearing in areas undergoing ecological changes, or old infections coming back because germs are resisting medicines or there are problems with public health measures (McArthur, 2019). Many of these new diseases are either transmitted from animals to humans (zoonotic) or show up suddenly (synoptic). Among the human population, there are thousands of germs, including hundreds of bacteria, viruses, and a smaller but still important number of fungi. A significant number of these infectious agents initially circulated among other vertebrate animals, such as mammals and birds. During their time in the original hosts, these microbes may have lived without causing harm, or they might have been the cause of diseases. However, at some point, they switched over to humans, causing sickness (Keesing and Ostfeld, 2021).

20.2.2 Global Warming

Human actions have accelerated Earth's warming since industrialization began. In the last century, the global average temperature has experienced a rise of 0.7°C, accompanied by a 2% increase in global average precipitation. Projections suggest continued increases in the future, significantly affecting Earth's climate. The resulting global warming and climate change impose added pressure on ecosystems, influencing the diversity of plants and animals in terms of both quality and quantity. Disease transmission and biodiversity loss are closely related to global warming. Increased temperatures cause ecological disruption, which results in habitat loss as well as fragmentation. As a result, species find it more challenging to adapt or move, which affects biodiversity. The diminishing biodiversity has direct and indirect repercussions on the flow of energy and materials, thereby influencing ecosystem services (Patil et al., 2021; Habibullah et al., 2022). Furthermore, as a result of the favorable circumstances that warmer temperatures create for disease vectors like mosquitoes, ticks, and other pathogens, diseases like Lyme disease, dengue fever, and malaria become more common and have a wider geographic reach. The most sensitive ecosystems to the effects of global warming are the moist tropical montane ecosystems as well as temperate ecosystems (Surasinghe, 2009). Reduced net primary production and increased heterotrophic respiration ultimately result in growth retardation and decreased net ecosystem productivity due to the physiological stress caused by global warming on vegetation in addition to the invasive species' predation (Saxe et al., 2001). Severe heat causes physiological stress, which significantly reduces the host's immunity and makes them more vulnerable to diseases (Surasinghe,

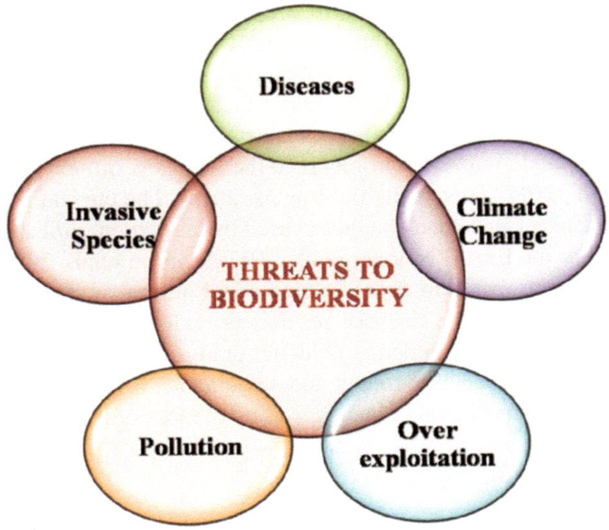

FIGURE 20.3 Biggest threats to biodiversity.

2009). The interrelated issues associated with global warming intensify the need for prompt and all-encompassing measures to alleviate its consequences and safeguard public health and biodiversity.

20.2.3 Invasive Species

An invasive species is a non-native plant, animal, or bug that, when brought to a new place, can cause problems for the environment, economy, or the health of people, animals, and plants. In a technical sense, "invasion" denotes the unregulated or unintentional extension of an organism beyond its native habitat without clearly specifying the environmental or economic consequences of this spread or its potential benefits for society (Beck et al., 2008). Although they are frequently mentioned in discussions of the link between diseases and biodiversity, invasive species constitute a major factor in biodiversity change. Parasites are main constituents of disease-mediated invasions because foreign hosts tend to enhance the richness and frequency of parasites while simultaneously reducing the number or variety of local hosts (Young et al., 2017). The spread of disease and a decrease in biodiversity are both negatively impacted by invasive species. These non-native species have the potential to bring and transmit diseases that endanger local wildlife and human populations when they establish themselves in new habitats. For example, invading amphibians carry the chytrid fungus, which has caused extensive losses in frog populations worldwide (Skerratt et al., 2007; Lips, 2016). Native plants may be outcompeted by invasive plants, which would mean fewer food supplies and habitats for the former. Furthermore, certain invasive species, including ticks and mosquitoes, pose serious health threats to both humans and animals because they are carriers of diseases like Lyme disease and the West Nile virus. All things taken into account, invasive species contribute to the loss of biodiversity by upsetting ecosystems and creating imbalances that can have serious negative effects on the environment and public health.

20.3 RATE OF BIODIVERSITY LOSS

As the 21st century progresses, human populations continue to grow rapidly, agricultural activities expand, climate change intensifies, and the demand for increased consumption shows no signs of slowing down. Given the way things are going, it seems certain that we'll see many natural habitats and various kinds of animals and plants disappear in the next few decades. From a historical context, over the past 66 million years, we have experienced an average loss of about 0.1 species per million per year in certain well-sampled hard-bodied groups, such as terrestrial vertebrates and mollusks, based on the fossil record. However, current estimates suggest that extinction rates have surged to approximately 1000 times the historical rates (Nguyen et al., 2023). According to Ceballos et al. (2015) and Payne et al. (2016), Earth is entering a sixth mass extinction event, projecting a catastrophic loss of around 75% of its species. This presents a dual challenge in conservation: firstly, acquiring reliable data on the extent of global species loss over the past 500 years attributable to human influence; and secondly, understanding how many species have emerged during that time and assessing the balance between losses and gains (Briggs, 2017). According to the International Union for Conservation of Nature (IUCN) Red List (2010), alarming statistics indicate that 26% of fishes, 21% of mammals, 28% of reptiles, 12% of birds, 30% of amphibians, 79% of mosses, 61% of ferns, 40% of gymnosperms, 31% of other invertebrates, 70% of flowering plants, and 16% of algae are threatened with extinction. Mushrooms and lichens face a particularly severe situation with a 100% threat level (Shinwari et al., 2012). These predictions depict a troubling outlook for future extinctions across different groups of organisms, signaling imminent threats and concerning rates of loss. Higher percentages of threatened nature are ranked from lichens and mushrooms to flowering plants, ferns, arachnids, gymnosperms, brown algae, amphibians, and beyond. The consequences of this biodiversity crisis extend beyond the loss of species. A shortage of healthy food, fodder, timber, clean water, forest products, fuel, medicines, and a clean environment is anticipated in the near future. Global warming and climate change pose threats to the survival of biotic resources across different habitat levels, contributing to the spread of epidemics at higher virulence rates. The intricate interaction of these factors highlights the pressing necessity for extensive conservation activities to address the imminent environmental challenges.

20.4 DISEASE AND DECLINE IN BIODIVERSITY

Disease can contribute to biodiversity decline by causing population declines, local extinctions, and altering species behavior, impacting ecosystem balance. Biodiversity loss, in turn, can increase disease risks, creating a complex interplay between health and ecosystem stability. In general, plant pathogens undergo their life cycles involving vector and host species, such as rust. These pathogens encompass a variety of organisms, ranging from bacteria, viruses, oomycetes, fungi, nematodes, and parasitic plants (Kumar et al., 2022). The impact of severe pathogen attacks extends to causing mortalities in various species populations. Historical instances of pathogen emergence, spread, and the resulting fatalities even in human populations serve as noteworthy examples as described next (Table 20.1).

- **Disease**: **Late blight of potato**
  Causing agent: *Phytophthora infestans*
  Symptoms: Water-soaked spots emerge at the tips and margins, progressively enlarging to create irregular, dark brown lesions.
  Description: *Phytophthora* species, oomycete plant pathogens, present a formidable threat, inducing destructive diseases that affect a wide array of crops, ornamental plants, and forest trees on a global scale. One prominent example is *Phytophthora infestans*, which is responsible for late blight, a disease that inflicted severe damage on potatoes in the 19th century in the United States, Ireland, and Europe. The movement of infected potato tubers triggered the potato famine epidemics, causing widespread disease, hunger, and the deaths of two million people in Ireland. Over 160 years later, *Phytophthora infestans* continues to cause devastating foliar blight in potatoes and infect potato tubers and tomato fruit, highlighting the persistent impact of this pathogen (Ristaino, 2012).

TABLE 20.1
Diseases Affecting Plants and Animals Contributing to Biodiversity Decline

Disease	Affected Species	Causing Agent	Symptom	Reference
Myxomatosis	Rabbits	Classic RHD virus (RHDV) and calicivirus (RHDVb/ RHDV2)	Swollen eyes, respiratory distress	Rosell et al., 2019; Kerr, 2012
White band disease	Endangered Caribbean Acroporid corals	Bacterial pathogen	Clear line of necrotic tissue	Kline and Vollmer, 2011
African swine fever	Pigs and wild boar	*Ornithodoros* spp. ticks	Fever	Chenais et al., 2019
Sea star wasting	Starfish	—	Lesions	Miner et al., 2018
Sea turtle fibropapillomatosis	Sea turtle	Herpesvirus	Tumor-forming disease	Keller et al., 2014
Fire blight	Rosaceae family	*Erwinia amylovora*	Wilting, blackening	Vrancken et al., 2013
Rice blast disease	Rice	*Magnaporthe grisea*	Lesions	Shahriar et al., 2020
Barley yellow dwarf virus infection in cereal	Barley	Barley yellow dwarf virus	Dwarfing of plants and alterations in leaf blade color along the vascular bundles	Choudhury et al., 2017
Soybean rust	Soybean	*Phakopsora pachyrhizi*	Lesions	Godoy et al., 2016
Rose rosette disease	Rose	RNA virus (genus *Emaravirus*)	Abnormal growth characterized by a witches' broom or rosette-type appearance, an abundance of prickles (thorns), and distorted, discolored growth	Pemberton et al., 2018

- **Disease**: **Southern corn leaf blight**
  Causing agent: *Bipolaris maydis*
  Symptoms: Leaf lesions
  Description: The outbreak of southern corn (*Zea mays* L.) leaf blight (SCLB) during 1970–1971 stands out as one of the most economically burdensome diseases to impact North American agriculture. The outbreak caused a huge loss of 15% of crops, costing a staggering US$1 billion. This was due to relying too much on cytoplasmic Texas male sterile (cms-T) lines in hybrid seed production. The problem arose because of a natural change in a type of SCLB *Bipolaris maydis*, which wasn't economically important before, but was identified in the Philippines in 1961. By 1969, this mutation had wreaked havoc in the Corn Belt, affecting leaves, ears, stalks, and the developing kernels of hybrids featuring cms-T genetics. The epidemic was exacerbated by a favorable

environment, with over 85% of hybrids being of cms-T genetics. As a response, cms-T was halted in 1971, and hybrid seed production returned to detasseling for the female parent. This serves as a warning to the seed production industry, underscoring the significance of steering clear of excessive purification of crop genetics and highlighting the necessity of preserving genetic diversity (Bruns, 2017).

- **Disease**: **Ug99 wheat stem rust**
  Causing agent: *Puccinia graminis tritici*
  Symptoms: Blisters appear on stem and leaves
  Description: Race Ug99 of the fungus *Puccinia graminis tritici* is responsible for stem or black rust disease on wheat, was first identified in Uganda in 1998. The Ug99 lineage now comprises seven races, spreading to various wheat-growing nations in Zimbabwe, the eastern African highlands, Yemen, Sudan, South Africa, and Iran. With 90% of global wheat varieties susceptible, the Ug99 races pose a significant threat to wheat production and food security. This disease has spread, facilitated by wind or human activities, to other regions in Africa, Asia, and globally. Ongoing projects involve screening for resistant wheat varieties and breeding materials in Kenya and Ethiopia, as well as identifying and transferring new sources of race-specific resistance from different wheat relatives to enhance resistance diversity. This historically destructive wheat disease has the potential to transform a seemingly healthy crop, only weeks away from harvest, into a tangle of black stems and shriveled grains at harvest, with the possibility of complete yield losses under severe conditions (Singh et al., 2011; Bhardwaj et al., 2014).

- **Disease**: **Banana Panama disease**
  Causing agent: *Fusarium oxysporum* f. sp. *cubense*
  Symptoms: Yellowing of lower and older leaves
  Description: Edible bananas traditionally reproduce asexually through rhizomes, but the advent of tissue culture since the early 1970s has allowed for the mass production of cultivars. This has enabled the rapid dissemination of genetically identical plants with preferred traits and excellent agronomic performance across extensive areas globally. However, the inherent vulnerability of monocultures to diseases has significantly impacted banana production over the past century. The first report of a wilting disease in bananas surfaced in Australia in 1876, and by 1890, it had affected the Gros Michel plantation crops in Costa Rica and Panama. The disease, which caused severe epidemics in the 1900s, earned the colloquial name "Panama disease" due to its prevalence in the region. This soil-borne fungus, *Fusarium oxysporum* f. sp. *cubense*, affects banana plants by causing wilting and other symptoms, leading to significant losses in banana production and leading to the naming of the forma specialis (Ordonez et al., 2015).

- **Disease**: **Citrus canker**
  Causing agent: *Xanthomonas citri*
  Symptoms: Small, yellow, pimple-like structures on leaf surface
  Description: Citrus canker is a disease affecting fruit, foliage, and twigs, and it is caused by the bacterium *Xanthomonas citri* subsp. *citri* (Xcc). The bacteria, present within lesions, are released onto plant surfaces when wet and dispersed by rain splash, influenced by rain intensity and wind speed. Infection occurs through stomata or wounds during periods of water congestion caused by rain and wind. The susceptibility to Xcc varies across different citrus species, with grapefruits and lemons being the most vulnerable, followed by sweet oranges. The disease also interacts with the Asian citrus leafminer (CLM), as the insect' larvae create galleries in foliage, fruit, and occasionally green twigs. Bacteria enter these galleries through cracks, facilitated by

larval feeding and movement. Successful eradication campaigns against citrus canker have been conducted in the past in regions like Florida, Australia, and South Africa (Gottwald and Graham, 2014).

- **Disease**: **Coffee rust (roya) epidemics**
  Causing agent: *Hemileia vastatrix*
  Symptoms: Small yellowish lesions that appear on the underside of leaves
  Description: Coffee rust, stemming from the fungus *Hemileia vastatrix*, is a leaf ailment that has triggered severe epidemics in various countries. Notable occurrences include Colombia from 2008 to 2011, Central America and Mexico in 2012–2013, and Peru and Ecuador in 2013. Several factors come together to start these widespread problems, such as the state of the economy, farmers' crop decisions, and usual weather patterns. All of these factors result in different outcomes, such as reduced production, lower earnings for farmers and workers, and impacts on food supply. Production in Colombia declined significantly, averaging a 31% decrease during the epidemic years compared to 2007. Central America also experienced reductions, with a 16% drop in 2013 compared to 2011–2012 and a 10% decline in 2013–2014 compared to 2012–2013. The disease induces defoliation, potentially causing branch death and substantial crop losses during acute stages. The infectious form of *H. vastatrix* is the uredospore, and initial symptoms manifest as small yellowish lesions on the underside of leaves where the fungus has infiltrated through stomata. These lesions progress, merging to produce distinctive orange-colored uredospores, while chlorotic spots emerge on the upper leaf surface. In the final stage of the disease, the lesions become necrotic (Avelino et al., 2015; Libert Amico et al., 2020).

- **Disease**: **Cassava brown streak disease (CBSD) and assava mosaic disease (CMD)**
  Causing agent: Cassava mosaic geminiviruses
  Symptoms: The tuberous roots show a dry, brown-black necrotic rot, and there is either yellow blotchy chlorosis or feathery chlorosis in the secondary and tertiary veins.
  Description: The cassava mosaic disease (CMD) pandemic, induced by cassava mosaic geminiviruses, has rapidly spread across 12 countries in East and Central Africa since the late 1980s, resulting in significant devastation to cassava crops. The occurrence of cassava brown streak disease (CBSD) was initially documented in the early 1930s in the coastal region of Tanzania, East Africa, coinciding with the earliest observations of CMD. Throughout the 20th century, CBSD was mainly localized to the lowland coastal areas of Kenya, Tanzania, and Mozambique, as well as the surroundings of Lake Malawi in Tanzania and Malawi. In contrast to CMD, which manifests symptoms primarily in cassava leaves, CBSD's economically impactful symptom is the development of a dry brown-black necrotic rot in the tuberous roots. Additional significant symptoms include feathery chlorosis in the secondary and tertiary veins or yellow blotchy chlorosis, along with brown necrotic streaks on the green portions of stems and seed capsules. Notably, leaf symptoms are most noticeable in the lower parts of the plant, making recognition challenging without careful examination before leaf senescence (Legg et al., 2011).

- **Disease**: **Dutch elm disease**
  Causing agent: *Ophiostoma novo-ulmi*
  Symptoms: Wilt leaves
  Description: The initial outbreak of Dutch elm disease (DED) in North America and Europe, caused by the introduced *Ophiostoma ulmi*, started in the early 1900s. Unexpectedly, the disease's impact diminished in Europe from the 1930s onward, following its devastating effect that resulted in the loss of 30–40% of elm trees. Following that, a second pandemic

arose, propelled by the more aggressive *O. novo-ulmi*, extending its reach across the same regions. By 1990, a significant portion of Britain's approximately 30 million mature elms had succumbed to the disease. During this second wave, *O. novo-ulmi* acquired debilitating viruses, and there was a shift from predominantly clonal reproduction to high heterogeneity through the horizontal transfer of novel, sometimes harmful, genes from *O. ulmi*. In the post-epidemic period, a new disease dynamic has taken shape, characterized by millions of small recruits facing repeated attacks in sequential cycles of the disease (Brasier et al., 2019).

- **Disease**: **Chestnut blight**
  Causing agent: *Cryphonectria parasitica*
  Symptoms: Necrotic lesions on the bark
  Description: Chestnut blight, attributed to the fungus *Cryphonectria parasitica*, poses a severe threat to American and European chestnut trees. Originating from East Asia, the pathogen found its way to other continents through infected chestnut plants. Inadvertently introduced to North America and Europe in the 20th century through infected chestnut specimens, *C. parasitica* was first identified in 1904 in New York City. Rapidly spreading at a rate of over 30 km per year across the native distribution range of the American chestnut, covering approximately 3.6 million hectares in eastern North America, the fungus induces perennial necrotic lesions, known as cankers, on the bark of stems and branches of vulnerable host trees. This leads to wilting of the plant parts beyond the infection. Chestnut blight cankers exhibit distinctive features such as the presence of mycelial fans and fruiting bodies of the pathogen. In response to the canker, the tree may generate epicormic shoots. Superficial, non-lethal, or callusing cankers on susceptible hosts are often associated with mycovirus-induced hypovirulence (Rigling et al., 2018).

- **Disease**: **Emerald ash borer**
  Causing agent: *Agrilus planipennis*
  Symptoms: Canopy thinning and branch dieback
  Description: The emerald ash borer (EAB), scientifically known as *Agrilus planipennis* Fairmaire (Coleoptera: Buprestidae), was first identified in North America in 2002. Since its inadvertent introduction from Asia, this invasive pest has caused widespread devastation, leading to the death of countless ash trees (*Fraxinus* spp.) in various environments, including forests, riparian areas, and urban settings. In regions close to the epicenter of the invasion in southeast Michigan, the impact has been particularly severe, resulting in the death of over 99% of ash trees with stems larger than 2.5 cm in diameter. EAB has demonstrated a broad susceptibility among all North American ash species encountered thus far, including white ash (*F. americana* L.), green ash (*Fraxinus pennsylvanica* Marsh.), and black ash (*F. nigra* Marsh.), which are among the most prevalent and abundant ash species across North America (Herms et al., 2014).

- **Disease**: **Sudden oak death**
  Causing agent: *Phytophthora ramorum*
  Symptoms: Necrotic lesions on the leaves and shoot dieback
  Description: *Phytophthora ramorum* is the main reason or culprit behind sudden oak death in the Western US and "sudden larch death" affecting Japanese larch in the UK. *Phytophthora* species, to which *P. ramorum* belongs, are part of the Oomycetes, a group of fungus-like diploid eukaryotes commonly known as water molds. These organisms are categorized under Stramenopiles and share a close relationship with golden-brown algae and diatoms. This pathogen has swiftly and repeatedly emerged, leaving

a devastating impact on forest ecosystems and causing significant economic challenges for certain nursery operations. *P. ramorum* made its appearance in the mid-1990s, almost simultaneously affecting woody ornamental plants in Germany and leading to sudden oak death in oaks and tanoaks in the United States (Grünwald et al., 2012).

- **Disease**: **White pine blister rust**
  Causing agent: *Cronartium ribicola*
  Symptoms: Small yellow or red spots on few needles
  Description: Blister rust, caused by *Cronartium ribicola* J.C. Fisch., is a disease that thrives in cool weather conditions and is native to white pines in eastern Asia, specifically those located at relatively high latitudes and elevations, such as *Pinus sibirica* Du Tour and *P. pumila* Regel. The initial introduction of blister rust took place in the East around 1898, with subsequent introductions occurring shortly thereafter. By 1909, the disease had firmly established itself in several states, extending as far west as central Pennsylvania. Progressing rapidly, by 1919, it had spread northward to Quebec and Ontario and reached as far west as the natural limit of eastern white pine in Minnesota (Kinloch et al., 2003).

- **Disease**: **Olive quick decline syndrome**
  Causing agent: *Xylella fastidiosa*
  Symptoms: Leaf scorch symptoms
  Description: Olive quick decline syndrome (OQDS) is a disease that emerged in the late 2000s in a restricted area along the Ionian coast of Salento in Apulia, southeastern Italy. This syndrome gained attention due to its rapid spread, affecting olive trees across the densely cultivated countryside of the region. The disease has significant implications for olive cultivation, posing challenges to the health and productivity of olive orchards in the affected areas. Researchers and agricultural authorities are actively working to understand and manage OQDS to mitigate its impact on olive production (Saponari et al., 2017).

- **Disease**: **Black Sigatoka**
  Causing agent: *Pseudocercospora fijiensis*
  Symptoms: Necrotic streaks on the leaves
  Description: Black Sigatoka, caused by the ascomycete fungus *Pseudocercospora fijiensis*, stands out as the most significant disease affecting banana plants. This condition manifests through the development of necrotic streaks on the leaves, leading to a decline in photosynthetic capacity and reduced overall yield. Additionally, Black Sigatoka induces premature fruit ripening, posing challenges for export companies as the fruit may become overripe during transit. Without proper treatment, yield losses can exceed 50%. Initially identified in the Sigatoka Valley in Fiji in 1963, herbarium samples indicate its presence in Taiwan as early as 1927. Subsequently, Black Sigatoka has spread widely, affecting banana-growing regions across the globe, including Latin America, Asia, Africa, and the Pacific (Noar et al., 2022).

- **Disease**: **White-nose syndrome**
  Causing agent: *Geomyces destructans*
  Symptoms: White, fuzzy appearance
  Description: White-nose syndrome (WNS) is a harmful fungal disease that affects bats and is considered one of the worst outbreaks of infectious diseases among wild mammals in the last hundred years. Biologists found something unusual during a bat population survey in 2007. Bats in a cave in Albany County, New York, were dying, and the

ones still alive had noticeable white fungus on their ears, wings, and muzzles. This discovery led to the naming of the disease as white-nose syndrome. Further examination identified *Geomyces destructans* as the likely causative agent, a novel fungus previously unknown to science before its discovery in North America. Subsequent to DNA sequence analysis, *G. destructans* was reclassified as *Pseudogymnoascus destructans* in 2013 (Hoyt et al., 2021).

- **Disease**: **Chytridiomycosis**
  Causing agent: *Batrachochytrium dendrobatidis*
  Symptoms: Shedding of skin
  Description: *Batrachochytrium dendrobatidis* (Bd), identified as the causative agent of chytridiomycosis in 1998, poses a significant threat to global biodiversity and has been linked to widespread declines in amphibian populations. This chytrid fungus belongs to the division Chytridiomycota, specifically the order Chytridiales, and was initially identified as the cause of chytridiomycosis in both captive and wild frogs in Central and North America, as well as Australia. Despite a decade of research on chytridiomycosis, there remains uncertainty about the mechanisms by which this pathogen induces illness and mortality. Two primary lesions associated with Bd infection include the proliferation (hyperplasia) of keratinaceous cells and the loss of sections of keratinized mouthparts in larvae, accompanied by the fusing of keratin layers (keratosis) in post-metamorphic individuals. Chytridiomycosis has become endemic in various regions, typically exhibiting prevalence rates exceeding 5%, and seasonal disease outbreaks may occur (Kilpatrick et al., 2010; Berger et al., 2016).

- **Disease**: **Rinderpest (cattle plague)**
  Causing agent: *Morbillivirus* of the family *Paramyxoviridae*
  Symptoms: Fever with ocular and nasal discharges
  Description: Rinderpest was a highly contagious viral disease that primarily affected cattle, buffalo, and some other cloven-hoofed animals. The causative agent of rinderpest is the *Morbillivirus*, which is related to the measles virus. The disease had devastating consequences for livestock populations, causing high mortality rates and significant economic losses. Before the 19th century, the introduction of viruses to Africa likely occurred through Egypt or sporadically spread from east to west, traversing the Sudan and reaching Senegal, with the probable source being the eastern African seaboard. Rinderpest was declared eradicated in 2011 (Roeder et al., 2013).

- **Disease**: **Bsal infection**
  Causing agent: *Batrachochytrium salamandrivorans*
  Symptoms: Excessive shedding and thickening of the skin
  Description: *Batrachochytrium salamandrivorans* (Bsal) is an emerging fungal pathogen that has been responsible for recent mass mortalities among native salamanders in Europe and has demonstrated lethality in certain North American species in laboratory trials. The origin of Bsal is believed to be in Asia, and human-mediated introductions through the commercial trade of amphibians appear to have brought the pathogen into wild populations in Europe. Following the initial outbreaks in the Netherlands, Bsal has been identified as the causative agent in mortality events in Belgium (wild populations) and Germany (captivity). Additionally, the fungus was recently detected in imported salamanders in the United Kingdom. Bsal targets keratinized epidermal cells and infiltrates the deeper layers of the epidermis, resulting in superficial erosions, deep ulcerations, extensive epidermal degradation, heightened shedding, and skin thickening. Histological examination reveals necrotic keratinocytes with marginated nuclei

and intracellular colonial thalli of Bsal located at the periphery and directly beneath the eroded keratin layer (Yap et al., 2017; Gray et al., 2015).

- **Disease**: **Avian influenza**
  Causing agent: (HPAI) H5N1 virus
  Symptoms: Fever, cough, and sore throat
  Description: The highly pathogenic avian influenza (HPAI) H5N1 virus, widely prevalent in poultry, continues to spread, presenting a significant challenge to both animal and human health. The ecological dynamics of HPAI H5N1 have undergone significant changes, shifting from occasional outbreaks in terrestrial poultry to continuous circulation in both terrestrial and aquatic poultry, possibly impacting wild waterfowl. Originating in southern China in 1996, a new genotype of HPAI H5N1 has emerged through constant mutation, reassortment, and natural selection. This has led to the development of distinct lineages that have spread into multiple reservoir hosts. The ongoing evolution of the Goose/Guangdong-lineage highly pathogenic H5N1 viruses involves stable interactions with some reservoir hosts while continuously adapting to others, posing an unpredictable risk to sporadic hosts, including humans (Peiris et al., 2007; Sonnberg et al., 2013).

20.5 ACTIONS TAKEN TO REVERSE BIODIVERSITY DECLINE

In 1992, 168 countries pledged to uphold biodiversity conservation by signing the Convention on Biological Diversity (CBD). In 2002, global leaders established a concrete goal of achieving a substantial reduction in the rate of biodiversity loss by 2010, marking a pivotal conservation agreement for the early 21st century. To achieve this, different agreements among countries and specific policies were used, like rules against trading endangered species, agreements about migrating species, and environmental protection conventions. Also, there were policies from the European Union, such as the Nitrates Directive and Natura 2000, which focuses on birds and habitats, along with plans related to the environment and agriculture. However, as 2010 passed, it became evident that these policies and tools were insufficient to meet the target. Despite a significant increase in conservation efforts between 1970 and 2009, biodiversity threats also escalated. Understanding why biodiversity continues to decline requires exploring the interaction between conservation actions and threats, and assessing the extent to which conservation measures can effectively counteract biodiversity loss (Scheper, 2015). In order to avert and eventually reverse the loss of biodiversity, efforts are being made to conserve habitat, address the underlying causes of biodiversity loss, and include stakeholders at all stages of the process.

20.6 CONCLUSION

To sum up, the complex correlation between disease and an overall decrease in biodiversity highlights the susceptibility of ecosystems and the fragile equilibrium that preserves a variety of living forms. It is becoming more clear how infectious diseases affect biodiversity as they spread or worsen due to factors like habitat loss, climate change, and anthropogenic activities. The loss of species diversity not only weakens ecosystems but also heightens the risk of disease transmission, creating a concerning feedback loop. Conservation efforts that integrate a One Health approach—recognizing the interconnectedness of human, animal, and environmental health—are crucial for mitigating the dual threats of infectious diseases and biodiversity decline. By understanding and addressing the complex web of factors that contribute to both phenomena, we can strive to foster healthier ecosystems, protect endangered species, and ensure the well-being of our planet and its inhabitants.

REFERENCES

Avelino J, Cristancho M, Georgiou S, Imbach P, Aguilar L, Bornemann G, Läderach P, Anzueto F, Hruska AJ, Morales C. The coffee rust crises in Colombia and Central America (2008–2013): impacts, plausible causes and proposed solutions. *Food Security.* 2015 Apr;7:303–21.

Beck KG, Zimmerman K, Schardt JD, Stone J, Lukens RR, Reichard S, Randall J, Cangelosi AA, Cooper D, Thompson JP. Invasive species defined in a policy context: recommendations from the federal invasive species advisory committee. *Invasive Plant Science and Management.* 2008 Oct;1(4):414–21.

Berger L, Roberts AA, Voyles J, Longcore JE, Murray KA, Skerratt LF. History and recent progress on chytridiomycosis in amphibians. *Fungal Ecology.* 2016 Feb 1;19:89–99.

Bhardwaj SC, Prashar M, Prasad P. Ug99-future challenges. In *Future Challenges In Crop Protection Against Fungal Pathogens,* 2014 Jul 23 (pp. 231–247). New York: Springer.

Brasier CM, Webber JF. Is there evidence for post-epidemic attenuation in the Dutch elm disease pathogen Ophiostoma novo-ulmi?. *Plant Pathology.* 2019 Jun;68(5):921–9.

Brema J, Gautam S, Singh D. Global implications of biodiversity loss on pandemic disease: COVID-19. In *COVID-19 and the Sustainable Development Goals,* 2022 Jan 1 (pp. 305–322). Elsevier.

Briggs JC. Emergence of a sixth mass extinction?. *Biological Journal of the Linnean Society.* 2017 Oct 1;122(2):243–8.

Bruns HA. Southern corn leaf blight: a story worth retelling. *Agronomy Journal.* 2017 Jul;109(4):1218–24.

Ceballos G, Ehrlich PR, Barnosky AD, García A, Pringle RM, Palmer TM. Accelerated modern human–induced species losses: entering the sixth mass extinction. *Science Advances.* 2015 Jun 19;1(5):e1400253.

Chenais E, Depner K, Guberti V, Dietze K, Viltrop A, Ståhl K. Epidemiological considerations on African swine fever in Europe 2014–2018. *Porcine Health Management.* 2019 Dec;5(1):1–0.

Choudhury S, Hu H, Meinke H, Shabala S, Westmore G, Larkin P, Zhou M. Barley yellow dwarf viruses: infection mechanisms and breeding strategies. *Euphytica.* 2017 Aug;213:1–22.

Giri CP, Shrestha S, Foresman TW, Singh A. Biodiversity data and information. 2000.

Giri CP, Shrestha S, Foresman TW, Singh A. Global Biodiversity data and information. *United Nations Economic and Social Council.* 2012.

Godoy CV, Seixas CD, Soares RM, Marcelino-Guimarães FC, Meyer MC, Costamilan LM. Asian soybean rust in Brazil: past, present, and future. *Pesquisa Agropecuária Brasileira.* 2016;51:407–21.

Gottwald TR, Graham JH. Citrus diseases with global ramifications including citrus canker and huanglongbing. *CABI Reviews.* 2014 Aug 13(2014):1–1.

Gray MJ, Lewis JP, Nanjappa P, Klocke B, Pasmans F, Martel AN, Stephen C, Parra Olea G, Smith SA, Sacerdote-Velat A, Christman MR. Batrachochytrium salamandrivorans: the North American response and a call for action. PLoS pathogens. 2015 Dec 10;11(12):e1005251.

Grünwald NJ, Garbelotto M, Goss EM, Heungens K, Prospero S. Emergence of the sudden oak death pathogen Phytophthora ramorum. *Trends in Microbiology.* 2012 Mar 1;20(3):131–8.

Haahtela T. A biodiversity hypothesis. *Allergy.* 2019 Aug;74(8):1445–56.

Habibullah MS, Din BH, Tan SH, Zahid H. Impact of climate change on biodiversity loss: global evidence. *Environmental Science and Pollution Research.* 2022 Jan;29(1):1073–86.

Herms DA, McCullough DG. Emerald ash borer invasion of North America: history, biology, ecology, impacts, and management. *Annual Review of Entomology.* 2014 Jan 7;59:13–30.

Hoberg EP, Brooks DR. Evolution in action: climate change, biodiversity dynamics and emerging infectious disease. *Philosophical Transactions of the Royal Society B: Biological Sciences.* 2015 Apr 5;370(1665):20130553.

Hoyt JR, Kilpatrick AM, Langwig KE. Ecology and impacts of white-nose syndrome on bats. *Nature Reviews Microbiology.* 2021 Mar;19(3):196–210.

Ibrahim FH, Rawat GS, Yahara T, Abi-Said M, Corlett RT, Courchamp F, Dai R, Freitag H, Haryoko T, Hewitt CL, Hussain T. Status, trends and future dynamics of biodiversity and ecosystems underpinning nature's contributions to people. In *The IPBES Regional Assessment Report on Biodiversity and Ecosystem Services for Asia and the Pacific,* 2018 (pp. 220–336). Secretariat of the Intergovernmental Science-Policy Platform on Biodiversity and Ecosystem services.

Keesing F, Ostfeld RS. Impacts of biodiversity and biodiversity loss on zoonotic diseases. *Proceedings of the National Academy of Sciences.* 2021 Apr 27;118(17):e2023540118.

Keller JM, Balazs GH, Nilsen F, Rice M, Work TM, Jensen BA. Investigating the potential role of persistent organic pollutants in Hawaiian green sea turtle fibropapillomatosis. *Environmental Science & Technology.* 2014 Jul 15;48(14):7807–16.

Kerr PJ. Myxomatosis in Australia and Europe: a model for emerging infectious diseases. *Antiviral Research.* 2012 Mar 1;93(3):387–415.

Khasim SM, Long C, Thammasiri K, Lutken H, editors. *Medicinal Plants: Biodiversity, Sustainable Utilization and Conservation.* Singapore: Springer; 2020 Apr 3.

Kilpatrick AM, Briggs CJ, Daszak P. The ecology and impact of chytridiomycosis: an emerging disease of amphibians. *Trends in Ecology & Evolution.* 2010 Feb 1;25(2):109–18.

Kinloch Jr BB. White pine blister rust in North America: past and prognosis. *Phytopathology.* 2003 Aug;93(8):1044–7.

Kline DI, Vollmer SV. White band disease (type I) of endangered Caribbean acroporid corals is caused by pathogenic bacteria. *Scientific Reports.* 2011 Jun 14;1(1):7.

Kumar N, Maurya PD, Raghuvanshi RS, Rahmani U, Verma B. Role of dispersal in different plant pathogens. Dr. Alok Ranjan Sahu. 2022;91.

Lafferty KD. Is disease increasing or decreasing, and does it impact or maintain biodiversity. *Journal of Parasitology.* 2003;89(Suppl):S101–5.

Legg JP, Jeremiah SC, Obiero HM, Maruthi MN, Ndyetabula I, Okao-Okuja G, Bouwmeester H, Bigirimana S, Tata-Hangy W, Gashaka G, Mkamilo G. Comparing the regional epidemiology of the cassava mosaic and cassava brown streak virus pandemics in Africa. *Virus Research.* 2011 Aug 1;159(2):161–70.

Libert Amico A, Ituarte-Lima C, Elmqvist T. Learning from social–ecological crisis for legal resilience building: Multi-scale dynamics in the coffee rust epidemic. *Sustainability Science.* 2020 Mar;15(2):485–501.

Lips KR. Overview of chytrid emergence and impacts on amphibians. *Philosophical Transactions of the Royal Society B: Biological Sciences.* 2016 Dec 5;371(1709):20150465.

McArthur DB. Emerging infectious diseases. *Nursing Clinics.* 2019 Jun 1;54(2):297–311.

Miner CM, Burnaford JL, Ambrose RF, Antrim L, Bohlmann H, Blanchette CA, Engle JM, Fradkin SC, Gaddam R, Harley CD, Miner BG. Large-scale impacts of sea star wasting disease (SSWD) on intertidal sea stars and implications for recovery. *PLoS One.* 2018 Mar 20;13(3):e0192870.

Mueller GM, Cunha KM, May TW, Allen JL, Westrip JR, Canteiro C, Costa-Rezende DH, Drechsler-Santos ER, Vasco-Palacios AM, Ainsworth AM, Alves-Silva G. What do the first 597 global fungal red list assessments tell us about the threat status of fungi?. *Diversity.* 2022 Sep 7;14(9):736

Neldner VJ, Laidlaw M, McDonald KR, Mathieson MT, Melzer R, McDonald WJ, Limpus CJ, Hobson R, Seaton R. *Scientific Review of the Impacts of Land Clearing on Threatened Species in Queensland.* Brisbane: Department of Science, Information Technology and Innovation; 2017.

Nguyen T, Malina R, Mokas I, Papakonstantinou A, Polyzos O, Vanhove MP. WASP: the World Archives of Species Perception. *Database.* 2023 Jan 1;2023:baad003.

Noar RD, Thomas E, Daub ME. Genetic characteristics and metabolic interactions between Pseudocercospora fijiensis and banana: Progress toward controlling black Sigatoka. *Plants.* 2022 Mar 31;11(7):948.

Ordonez N, Seidl MF, Waalwijk C, Drenth A, Kilian A, Thomma BP, Ploetz RC, Kema GH. Worse comes to worst: bananas and Panama disease—when plant and pathogen clones meet. *PLoS Pathogens.* 2015 Nov 19;11(11):e1005197.

Patil J, Sharma P, Mhatre K. Global warming induced stress and its impact on biodiversity. *Science and Technology.* 2021 May;6:21–9.

Payne JL, Bush AM, Heim NA, Knope ML, McCauley DJ. Ecological selectivity of the emerging mass extinction in the oceans. *Science.* 2016 Sep 16;353(6305):1284–6.

Peiris JM, De Jong MD, Guan Y. Avian influenza virus (H5N1): a threat to human health. *Clinical Microbiology Reviews.* 2007 Apr;20(2):243–67.

Pemberton HB, Ong K, Windham M, Olson J, Byrne DH. What is rose rosette disease?. *HortScience.* 2018 May 1;53(5):592-5.

Rawat US, Agarwal NK. Biodiversity: Concept, threats and conservation. *Environment Conservation Journal.* 2015 Dec 15;16(3):19–28.

Rigling D, Prospero S. Cryphonectria parasitica, the causal agent of chestnut blight: invasion history, population biology and disease control. *Molecular Plant Pathology.* 2018 Jan;19(1):7–20.

Ristaino JB. Worldwide migrations, host shifts, and reemergence of Phytophthora infestans, the plant destroyer. *Evolution of Virulence in Eukaryotic Microbes.* 2012 Aug 3:192–207.

Roeder P, Mariner J, Kock R. Rinderpest: the veterinary perspective on eradication. *Philosophical Transactions of the Royal Society B: Biological Sciences.* 2013 Aug 5;368(1623):20120139.

Rohr JR, Civitello DJ, Halliday FW, Hudson PJ, Lafferty KD, Wood CL, Mordecai EA. Towards common ground in the biodiversity–disease debate. *Nature Ecology & Evolution.* 2020 Jan;4(1):24–33.

Rosell JM, de la Fuente LF, Parra F, Dalton KP, Badiola Sáiz JI, Pérez de Rozas A, Badiola Díez JJ, Fernández de Luco D, Casal J, Majó N, Casas J. Myxomatosis and rabbit haemorrhagic disease: A 30-year study of the occurrence on commercial farms in Spain. *Animals*. 2019 Oct 10;9(10):780.

Saponari M, Boscia D, Altamura G, Loconsole G, Zicca S, D'attoma G, Morelli M, Palmisano F, Saponari A, Tavano D, Savino VN. Isolation and pathogenicity of Xylella fastidiosa associated to the olive quick decline syndrome in southern Italy. *Scientific Reports*. 2017 Dec 18;7(1):17723.

Sarkar S. Defining "biodiversity"; assessing biodiversity. *The Monist*. 2002 Jan 1;85(1):131–55.

Saxe H, Cannell MG, Johnsen Ø, Ryan MG, Vourlitis G. Tree and forest functioning in response to global warming. *New Phytologist*, 2001;149(3):369–399.

Scheper JA. Promoting wild bees in European agricultural landscapes. *Alterra Scientific Contribution*. 2015;47:179.

Schmeller DS, Courchamp F, Killeen G. Biodiversity loss, emerging pathogens and human health risks. *Biodiversity and Conservation*. 2020 Oct;29:3095–102.

Shahriar SA, Imtiaz AA, Hossain MB, Husna A, Eaty M, Khatun N. Rice blast disease. *Annual Research & Review in Biology*. 2020 Mar 19;35(1):50–64.

Shinwari ZK, Gilani SA, Khan AL. Biodiversity loss, emerging infectious diseases and impact on human and crops. *Pakistan Journal of Botany*. 2012 May 1;44(1):137–42.

Singh RP, Hodson DP, Huerta-Espino J, Jin Y, Bhavani S, Njau P, Herrera-Foessel S, Singh PK, Singh S, Govindan V. The emergence of Ug99 races of the stem rust fungus is a threat to world wheat production. *Annual Review of Phytopathology*. 2011 Sep 8;49:465–81.

Skerratt LF, Berger L, Speare R, Cashins S, McDonald KR, Phillott AD, Hines HB, Kenyon N. Spread of chytridiomycosis has caused the rapid global decline and extinction of frogs. *EcoHealth*. 2007;4(2):125–34.

Sonnberg S, Webby RJ, Webster RG. Natural history of highly pathogenic avian influenza H5N1. *Virus Research*. 2013 Dec 5;178(1):63–77.

Surasinghe T. Loss of biodiversity in the face of global warming. *News of Zoo and Wildlife Networks in South Asia--*. 2009;24(7):13.

Thormann I, Engels JMM. Genetic Diversity and Erosion—A Global Perspective. In: Ahuja, M., Jain, S. (eds) *Genetic Diversity and Erosion in Plants*. Sustainable Development and Biodiversity, Springer, Cham. 2015;7:263–294.

Vrancken K, Holtappels M, Schoofs H, Deckers T, Valcke R. Pathogenicity and infection strategies of the fire blight pathogen Erwinia amylovora in Rosaceae: state of the art. *Microbiology*. 2013 May;159(Pt_5):823–32.

Yap TA, Nguyen NT, Serr M, Shepack A, Vredenburg VT. Batrachochytrium salamandrivorans and the risk of a second amphibian pandemic. *EcoHealth*. 2017 Dec;14:851–64.

Young HS, Parker IM, Gilbert GS, Guerra AS, Nunn CL. Introduced species, disease ecology, and biodiversity–disease relationships. *Trends in Ecology & Evolution*. 2017;32(1):41–54.

21 Biodiversity and Ecosystem Services in India

Challenges and Opportunities

*Sheenu Sharma[1], Sabir Hussain[1], Abhishek Kumar[1], Meenu Patil[1], Pardeep Kumar[1], Isha[1], Sheenam Kamboj[1], Anand Narain Singh[1]**

[1]Soil Ecosystem and Restoration Ecology Lab, Department of Botany, Panjab University, Chandigarh, India

*Corresponding author: Anand Narain Singh, Soil Ecosystem and Restoration Ecology Lab, Department of Botany, Panjab University, Chandigarh, India

21.1 INTRODUCTION

India's natural heritage is a remarkable asset made up of invaluable ecosystem services and ultimate biodiversity. India's varied landscapes, from the sun-radiant coasts of its coastal areas to the snow-capped peaks of the Himalayas, include a wealth of flora, fauna, and ecosystems that have supported life and enhanced human cultures for millennia (Venkataraman, 2012). Every species, habitat, and entire ecosystem in the complex web are essential to the resilience and vibrancy of India's natural environment (Anil et al., 2014).

Biodiversity of India is plentiful because of the country's diverse geography and climate, which include lavish wetlands, arid deserts, tropical rainforests, and a wide range of other habitats (Chitale et al., 2014). A surprising variety of fauna can be found thriving in these settings, ranging from large, magnificent megafauna like the Indian elephant and Bengal tiger to microscopic pollinators and microbes that are the backbone of ecosystems (Ceballos et al., 2015). Millions of people across the country derive their cultural identity, subsistence, and inspiration from this biological diversity, in addition to it being a source of admiration and amazement.

However, because of the various ecosystem services that biodiversity offers, it is far more important than only its natural exquisiteness and biological significance. The benefits that nature provides for humans are known as ecosystem services, and they include provisioning, regulatory, cultural, and supporting services (Oguh et al., 2021; De Groot et al., 2002). They include provisions that improve human lives and support economic activity, such as water conservation, soil formation, pollination, climate regulation, and spiritual and recreational benefits (Dem et al., 2024). Sustainable development is based on the mutually beneficial interaction between ecosystem services and biodiversity. But because of human actions like habitat destruction, excess pollution, climate change, and the introduction of exotic species, this relationship is becoming more and more stressed (Owusu et al., 2024). All these issues threaten the natural heritage of India and the services that the country depends on.

Abundant species and ecosystems are declining as a result of habitat loss, fragmentation, and degradation caused by speedy urbanisation, industry, agricultural growth, and infrastructure development (Adla et al., 2022; Bozkurt and Basaraner, 2024). Environmental deterioration is further

DOI: 10.1201/9781032697550-21

aggravated by pollution from industrial effluents, air pollutants, plastic trash, and agricultural run-off, which impact both terrestrial and aquatic ecosystems (Bashir et al., 2020; Ajibade et al., 2021). Extreme weather, shifting climatic zones, rising sea levels, and altered precipitation patterns are just a few of the phenomena brought on by climate change that negatively influence ecosystem stability and biodiversity (Samuelson et al., 2022).

Invasive alien species also cause ecological imbalances by outcompeting native flora and fauna and upsetting native ecosystems (Yadav et al., 2024). When these risks combine, they endanger not just the integrity of ecosystems but also the services they offer, such as the regulation of disease, floods, carbon sequestration, and water quality. However, these difficulties also present opportunities for improvement. Humans and nature can coexist peacefully through conservation programmes, scientific advancements, community projects, and legislative changes (Borrini-Feyerabend and Hill, 2015; Percival et al., 2021). There are several approaches to preserve biodiversity and repair damaged ecosystems, including protected areas, wildlife conservation programmes, restoration projects, and sustainable land-use techniques (Benayas and Bullock, 2012). Traditional knowledge systems, scientific research, and technological developments all provide ideas and resources for the efficient management and conservation of natural resources.

Participatory methods, indigenous stewardship, and community involvement enable local communities to take on the role of stewards of biodiversity, fostering sustainable resource use and means of subsistence (Mulrennan et al., 2012; Wali et al., 2017). In order to promote a conservation ethic and mobilise community action for environmental sustainability, education, awareness, and advocacy are essential. Policy, legal documents, and international cooperation are important for protecting biodiversity conservation and integrating ecosystem services into decision-making processes. Sustainable development goals, such as permaculture, clean energy, water conservation, and energy efficiency, provide ways to achieve biodiversity goals while meeting people's needs and aspirations (Henfrey et al., 2023).

By recognising the details of India's biodiversity, the importance of ecosystem services, and embracing a holistic approach to conservation and development, we can open a door to a future where prosperity, people, and environment are in balance. This chapter delves into the in-depth overview of biodiversity and ecosystem services landscapes in India, highlighting the complexities, challenges, and opportunities that define this dynamic relationship. Drawing from a wide range of disciplines including ecology, economics, policy, ethics, and community engagement, we embark on a transformative journey to explore the wonders of India's natural world, address the threats it faces, and create sustainable solutions for generations to come.

21.2 BIODIVERSITY: AN OVERVIEW

Biodiversity, which includes all living forms, ecosystems, and ecological processes, is essential to human existence and economic growth. Although estimates of the overall number of species on Earth vary greatly, ranging from 5 million to over 100 million, a reasonable estimate places the number at about 13.6 million, of which only 1.78 million have been officially recognised and described (Singh, 2012). This inadequate understanding highlights how much biodiversity is yet unknown. The delicate balance between the rates of species development (speciation) and loss (extinction) that exists at any given period is represented by biodiversity (Orr and Smith, 1998; Losos and Schluter, 2000). But human activity has changed this equilibrium dramatically. There are clear latitudinal and altitudinal variations in the distribution of biodiversity on Earth, with some locations having more species richness than others. There have been significant mass extinctions on Earth throughout its history, and research suggests that we are currently living through the sixth such event. Populations of various species are diminishing in freshwater, marine, and terrestrial environments; extinction rates are predicted to be 100–1000 times faster than the background rate of natural selection. In particular, tropical forests are losing two to five species every hour due to rapid species loss (Vonlanthen et al., 2012).

In order of increasing influence, land-use changes, climate change, nitrogen deposition, biotic exchange, and rising atmospheric CO_2 levels are the key drivers of future biodiversity changes (Nicholson et al., 2009). However, determining the precise number of species and rate of extinction, as well as comprehending how the loss of a species affects the stability and function of ecosystems, continues to be difficult tasks. The two main strategies for maintaining biodiversity are ex situ conservation, which involves saving species outside of their natural habitats, and in situ conservation, which emphasises preserving ecosystems to preserve genetic variety, species, and ecological processes. Public engagement and conservation efforts can be bolstered by implementing biodiversity banking programmes and creating networks of protected places.

The creation of the Intergovernmental Science-Policy Platform for Biodiversity and Ecosystem Services (IPBES), akin to the Intergovernmental Panel on Climate Change (IPCC) for climate change, could facilitate the coordination of research, conduct assessments, and offer predictions at both regional and global scales, thereby augmenting global biodiversity conservation endeavours (Perrings et al., 2011). By emphasising the value of biodiversity and ecosystem services and directing policy decisions for their preservation, this platform would be extremely helpful.

21.2.1 Status of Biodiversity in India

India is known around the world for its historical heritage and diverse flora and fauna. India is home to over 45,000 plant species and 65,000 animal species, according to estimates. As a result of its rich flora and fauna, the country has been added to the list of mega biodiversity hotspots around the world. The Western Ghats and the Eastern Himalayas are India's mega biodiversity hotspots (MoEF, 2000; Myers et al., 2000). There are an estimated 9000 plant species in the Eastern Himalayas, with 3500 of these being endemic. There are over 5800 plant species in India's sector of the area, with approximately 2000 of them being endemic. About 3160 plant species, as well as 71 taxa, 300 animals, 977 birds, 176 reptiles, 105 amphibians, and 269 freshwater fishes, are endemic to the Himalayan hotspot (MoEF, 2000) (Table 21.1).

Climate change affects the biodiversity of the Indian Himalayan Region (IHR) from the species to ecosystem level. Habitat loss and fragmentation are the major causes of species extinction, which are enhanced in many ways by climate change. The species *Catreus wallichi* (cheer pheasants), a native of Jammu and Kashmir, is declining due to habitat loss (Wikramanayake et al., 1998). At least six endemic species of *Schizothorax* can be found in high mountain lakes and streams, while two other genera of snow trout, *Ptychobarbus* and *Gymnocypris biswasi*, are monotypic genera currently assumed to be extinct and are likewise restricted to the Himalayan hotspot (IUCN, 2004).

TABLE 21.1
Mega Biodiversity of IHR

Taxonomic Group	Total Number of Species Identified	Endemic Epecies	Percentage of Endemism
Plants	10000	3160	31.6
Fishes	269	33	12.3
Amphibians	105	42	40
Reptiles	176	48	27.3
Birds	977	15	1.5
Mammals	300	12	4

Source: (Myers et al., 2000; Gadgil, 2008).

The biggest source of current concern is the rise in greenhouse gas levels, particularly CO_2, which alters global climate and weather patterns.

21.2.2 IPBES, the Intergovernmental Platform for Science-Policy on Biodiversity and Ecosystem Services

Efforts to coordinate a global response to biodiversity loss have been impeded by the absence of a global platform for biodiversity policy support, like the IPCC for climate change. Thus, the UN General Assembly's approval and endorsement of the IPBES establishment process is a first step in this direction. Similar to the IPCC, the IPBES is an impartial, worldwide science group that coordinates and prioritises research on urgent issues (Perrings et al., 2011). The panel makes predictions and performs assessments at both the regional and global levels.

21.3 BIODIVERSITY AND ECOSYSTEM SERVICES

The basis for numerous ecosystem services that support both environmental balance and human well-being is India's biodiversity (Butler and Oluoch-Kosura, 2006; Costanza et al., 1997). Provisioning services, which cover the material resources offered by ecosystems, are among these services (MEA, 2005). India's abundant biodiversity of flora and fauna, together with its varied agricultural systems and medicinal plant resources supporting both traditional medicine and the pharmaceutical industry, greatly contribute to the country's food security (Díaz et al., 2006). The preservation of environmental stability depends on regulating services, and India's biodiversity is essential for water purification, natural hazard reduction, and climate regulation (MEA, 2005). In addition to stabilising local temperatures and removing carbon dioxide from the atmosphere, forests, wetlands, and mangroves also cleanse water, protecting ecosystems and public health. India's rich cultural legacy informs cultural services that provide spiritual, recreational, and aesthetic aspects. Rivers, mountains, and sacred groves are spiritually significant, and national parks and beautiful scenery draw visitors and support regional businesses. These cross-cultural exchanges promote a sense of self and an appreciation of the beauty of the natural world.

The fundamental mechanisms that keep ecosystems alive and make it possible to provide other services are known as supporting services (MEA, 2005). The formation of soil, nutrient cycling, and pollination are all facilitated by India's various ecosystems and are crucial for agricultural output, biodiversity preservation, and ecosystem resilience.

Costanza et al. (1997) categorised ecosystem services into 17 subtypes. Out of these 17, forests provide 13 ecosystem services. Singh (2007) quantified the annual value of individual ecosystem services of forests in Uttarakhand and also estimated the ecosystem services value of all the IHR states with the help of different methods. According to his report, the forests of Uttarakhand provide ecosystem services valued at US\$1150 $ha^{-1}yr^{-1}$ with the help of various parameters of forests given by Costanza et al. (1997) (Tables 21.2 and 21.3).

21.4 SIGNIFICANCE OF BIODIVERSITY AND ECOSYSTEM SERVICES

Understanding the connections between ecosystem services and biodiversity is essential to advancing the case for ecological restoration (Rey Benayas et al., 2009). To achieve the dual objectives of maximising the delivery of ecosystem services and assisting in the conservation of biodiversity, it might also help with the management of protected and restored areas (Bastian, 2013; Palomo et al., 2014). The impact of the complexity of biodiversity components on the ecosystem functioning that supports service provision (Bastian, 2013; Cardinale et al., 2012; Mace et al., 2012), however, is still largely unknown despite a number of meta-analyses, advances in research, and understanding of this relationship (Schröter et al., 2014). There are few studies that use empirical evidence to examine

TABLE 21.2
Annual Value of 13 Ecosystem Services of Forests of Uttarakhand

Ecosystem Services	Value (US\$ $ha^{-1}yr^{-1}$)
Climate regulation	167.6
Disturbance regulation	2.3
Water regulation	5.2
Erosion control	114.6
Soil formation	11.6
Nutrient cycling	429.6
Waste treatment	102.7
Biological control	2.3
Food production	50.7
Raw material	164.0
Genetic resource	18.5
Recreation	78.6
Cultural	2.3
Total	**1150**

Source: Singh (2007).

TABLE 21.3
Estimation of Ecosystem Services Value from Forests of IHR States

IHR States	Dense Forest Area (Km^2)	Open Forest Area (Km^2)	Total Forest Area for Ecosystem Services Estimation (Km^2)	Ecosystem Services Value (Billion US\$)	Ecosystem Services Value (Billion Rupees)
Himanchal Pradesh	6656	2892	8102	0.93	41.46
Uttarakhand	18245	5283	20887	2.40	106.89
Jammu and Kashmir	16156	13814	23063	2.65	118.02
Arunachal Pradesh	57756	11091	63302	7.28	323.95
Assam	14517	9171	19103	2.20	97.76
Manipur	5936	11448	11660	1.34	59.67
Mizoram	3786	14552	11062	1.27	56.61
Meghalaya	5925	9708	10779	1.24	55.61
Nagaland	5137	9027	9651	1.11	49.39
Tripura	2228	3517	3987	0.46	20.40
Sikkim	2363	755	2741	0.32	14.02
Total	**138706**	**91257**	**184334**	**21.20**	**943.33**

Source: Singh (2007).
Note: The estimation of forest area for ecosystem services has been calculated from dense forest area plus half of the open forest area.

the role of biodiversity in providing ecosystem services (Mertz et al., 2007), current knowledge has been poorly integrated, and the quantitative relationships between components of biodiversity and ecosystem services remain poorly understood (Carpenter et al., 2009). Widely varying biodiversity attributes and ecosystem services are also not well-represented in studies that use empirical evidence.

Because of their significant effects on environmental stability, human well-being, and sustainable development, biodiversity and ecosystem services are important. The range of life forms on Earth, including plants, animals, microbes, and ecosystems, is referred to as biodiversity. On the other hand, ecosystem services are the advantages that people receive from the activities and processes of nature. Following are some key aspects of their importance.

21.4.1 Economic Significance

The economic value of ecosystem services and biodiversity includes a broad range of concrete advantages that directly advance economic growth and well-being for people. By supplying genetic variation for fish, cattle, and crops, biodiversity promotes agricultural productivity (Thrupp, 2000). Increased crop resilience to pests, illnesses, and environmental challenges is a result of this variety, which raises yields and improves food security. Crop productivity is greatly impacted by ecosystem services including bee, butterfly, and other insect pollination. Many agricultural crops would have lower yields or perish completely in the absence of these pollinators.

Biodiversity provides natural resources that are used by several enterprises. For timber, fuelwood, and non-timber forest products including fruits, nuts, and resins, for example, the forestry sector depends on a variety of forests. Various marine and freshwater ecosystems support the fishing and aquaculture sectors by supplying fish stocks and other aquatic resources needed for survival and livelihood (Jennings et al., 2016). Biodiversity is a great source of genetic material, bioactive compounds, and medicinal plants that are utilised in biotechnology and pharmaceuticals. Natural sources are the origin of many innovative biotechnological discoveries and life-saving medications and therapies. Additionally, bioprospecting and the creation of novel medications, agricultural goods, and industrial materials depend on the genetic diversity present in ecosystems. India's wildlife, landscapes, and different ecosystems attract millions of visitors annually, which greatly boosts the country's economy and creates jobs in the tourism sector. For its attractiveness and feasibility, ecotourism, wildlife tourism, and nature-based leisure activities depend on biodiversity and natural habitats (Kuthiala, 2001).

These are facilitated by ecosystem services, which include the prevention of soil erosion, flood control, water purification, and climate regulation. In addition to reducing the chances of natural disasters, healthy ecosystems also reduce the effects of climate change and promote sustainable resource management techniques. For millions of people, including farmers, fishermen, foresters, nature guides, researchers, and conservationists, biodiversity supports a variety of livelihoods and economic endeavours. Opportunities for employment in green industries, ecotourism, wildlife management, and environmental education are created by the conservation and sustainable use of biodiversity.

21.4.2 Ecological Significance

The ecological significance of biodiversity and ecosystem services incorporates the crucial roles they play in maintaining the health, functioning, and resilience of ecosystems. Ecosystems depend on biodiversity to function properly. It is made up of different kinds of fungi, animals, plants, and microorganisms that work together to support ecosystem functions like energy flow, nutrient cycling, and water purification. As producers (plants), consumers (animals), decomposers (fungi, bacteria), and pollinators (insects, birds), different species play specific functions within ecosystems. Ecosystem production and balance are preserved by these interactions (Elmqvist et al., 2012). Nitrogen fixation, decomposition, and plant uptake are examples of nutrient cycling activities that are influenced by biodiversity (Van Groenigen et al., 2015). Many plant species contribute to soil fertility through the deposition of organic matter in the soil and their varied nutritional requirements through their root systems. Microorganisms found in soil, such as fungi and bacteria, are essential for recycling nutrients, breaking down organic debris, and preserving the structure and health of the soil.

Forests, wetlands, and oceans in particular are important ecosystems that help regulate and sequester carbon (Siikamäki et al., 2013). Through photosynthesis, they take in carbon dioxide from the atmosphere and store it in biomass, soils, and sediments. Because they stabilise temperatures, control water cycles, and lessen extreme weather events, ecosystems rich in biodiversity are more successful at sequestering carbon and lessening the effects of climate change. By controlling evapotranspiration, groundwater recharge, and water storage in soils and vegetation, biodiversity affects the water cycle (Sans-Fuentes and Meixner, 2017). Healthy ecosystems control water flow in rivers, lakes, and aquifers and preserve hydrological balance. As natural filters, riparian zones, mangroves, and wetlands remove excess nutrients, pollutants, and sediments from water. Aquatic habitats and water quality are maintained by biodiversity. Habitats rich in biodiversity offer a variety of resources and niches for different species, enhancing biodiversity resistance to invading species, environmental changes, and disturbances. The stability, resilience to disease, and ability of an ecosystem to rebound from shocks are all improved by species variety. High biodiversity ecosystems are more resilient and adaptive to the effects of climate change, habitat loss, and human disturbances, promoting ecosystem health and long-term sustainability (Oliver et al., 2015). Food production, clean water, air purification, and climate regulation are just a few of the ecosystem services that come from biodiversity and are vital to human health and way of life. Sustaining these functions requires preserving ecological integrity.

21.4.3 Cultural and Social Significance

For Indian people, biodiversity and ecosystem services are extremely valuable from a social and cultural standpoint. The rich biodiversity of India is entangled with customs, beliefs, and rituals that promote a sense of self and kinship with the natural world. Having access to parks, green spaces, and natural places enhances both physical and mental well-being and contributes to a healthy society. The traditional ecological knowledge (TEK) of indigenous and local groups is derived from centuries of interactions with biodiversity and provides important insights into conservation and sustainable resource management techniques (Dudgeon and Berkes, 2003).

21.5 CHALLENGES TO BIODIVERSITY AND ECOSYSTEM SERVICES

21.5.1 Habitat Fragmentation

The main reasons for the reductions in global biodiversity are the devastation and degradation of natural ecosystems (Pereira et al., 2010; Rands et al., 2010). The loss of habitat usually results in fragmentation, which is the division of the habitat into smaller, more isolated pieces divided by a matrix of land cover altered by humans. Long-term changes to the structure and function of the remaining pieces are brought about by area loss, increased isolation, and increased exposure to human land activities along fragment margins (Lindenmayer and Fischer, 2013). Although ecologists agree that habitat destruction is bad for preserving biodiversity, they often sharply disagree about how much fragmentation is to blame for this (Didham et al., 2012; Fahrig, 2013). A theoretical framework for comprehending the impact of fragmentation on extinction in terrestrial landscapes made up of "islands" of natural habitat dispersed across a "sea" of habitat that has been altered by humans was first proposed by early theories based on the biogeography of oceanic islands (MacArthur and Wilson, 2001). The main source of disagreement in the debate has been the persistent lack of clarity regarding how much smaller fragments and greater isolation contribute to ecosystem degradation and biodiversity decreases when compared to the ubiquitous and broad effects of habitat loss (Fahrig, 2003). Because non-manipulative research can only draw conclusions based on correlation and because observational studies have frequently only looked at one or two fragmentation-related factors (such as area, edge, and isolation), the debate surrounding the impacts of fragmentation has frequently been exacerbated (Collinge, 2009). Nevertheless, experimental investigations show that fragmentation has numerous simultaneous effects that are intricately intertwined and may act over extended periods of time in addition to these correlated data (Ibáñez, 2014). Major human activities that contribute to habitat loss and fragmentation include deforestation, urbanisation, infrastructure development, and expansion of agriculture. Fragmentation lowers habitat quality, isolates populations of species, and disrupts ecological processes. These outcomes result in biodiversity loss, extinctions of species, decreased genetic diversity, habitat degradation, increased conflicts between humans and wildlife, and disruptions to ecosystem services and functions.

21.5.2 Invasion

One of the main criticisms levelled against the idea of ecosystem services is that it frequently only takes into account the positive outcomes of ecosystems, ignoring any negative, undesirable, or economically detrimental effects (Lyytimäki, 2014; Schröter et al., 2014). Ecosystem disservices are the negative aspects of ecosystems. Ecosystem disservices are defined as "the ecosystem generated functions, processes and attributes that result in perceived or actual negative impacts on human well-being" (Shackleton et al., 2016). Ecosystem disservices can result from a variety of environmental features that are viewed as undesirable, such as biological invasions (Shackleton et al., 2016; Escobedo et al., 2011). They are caused by ecosystem activities that are dangerous to humans, such as floods and wildfires, and even while they may be considered natural processes, management can lessen or intensify them (Lyytimäki, 2014). Certain individuals may perceive an ecosystem function as ecosystem services, while others may perceive it as ecosystem disservices (Saunders and Luck, 2016). This is contingent upon various factors such as individuals' behaviours, acquired knowledge, and the broader political, economic, and social context (Rasmussen et al., 2017; Shackleton et al., 2016). The way that anthropogenic stresses are configured, as well as how ecosystem services and disservices are provided and perceived, can differ across time, space, and between individuals or social groups (Chan et al., 2012; Shackleton et al., 2016).

More recently, the applicability of ecosystem disservices has been explored in two other contexts: forests and fisheries (Shackleton et al., 2016). However, ecosystem disservices have rarely been examined in relation to more general social-ecological issues, like plant invasions (Saunders and Luck, 2016; Shackleton et al., 2016). Depending on people's preferences and the spatiotemporal context, plant invaders can be both beneficial and a nuisance for human well-being (Tassin and Kull, 2015; Simberloff et al., 2013; Kueffer and Kull, 2017). In certain situations, invasive plants improve regulating functions like soil protection and coastal sediment dynamics, or they support people's daily necessities and economic incomes (Kull et al., 2011). However, in other situations, plant invasions may have negative effects on human health. Examples include competition with another species that provides a service, wildfires in non-fire-prone locations, and health issues related to allergic chemicals or skin irritations (Fenesi et al., 2015). The extent, velocity, and scope of the invasion process might intensify the positive or negative effects of plant invasions (Eviner et al., 2012). The same species can contribute to ecosystem disservices provision in later times, such as when trees expand widely in the wild, or they can promote ecosystem services to some spatiotemporal extent, such as when they are restricted to private gardens (the "transient disservices"; Saunders and Luck, 2016). This is contingent upon the absence of human management, unavoidably (Brundu and Richardson, 2016).

21.5.3 Climate Change

The abolition of poverty and sustainable development depend on the preservation and conservation of biodiversity. India, which occupies just 2.4% of the global land area, is home to 18% of the world's population and is a nation rich in biodiversity (Venkataraman, 2012). It includes significant portions of four global biodiversity hotspots, some of the largest wild populations of large, widespread mammals, and high taxonomic values. India faces unique and difficult problems in maintaining its abundant biological resources and promoting human progress and well-being (Corvalan et al., 2005). Climate change raises the issue to a general level. Many people anticipate that climate change will negatively impact wildlife in various ways and negatively impact human nature in general. However, because it provides supportive ecosystem services, biodiversity is crucial for adaptation to climate change as well as for mitigating its effects. Water, which is essential to life on Earth, is one of the many resources that are significantly impacted by climate change. Natural resources, ecosystems, and biosphere shifts all have negative effects on human health and quality of life. In the 21st century, global warming is expected to affect India. India will experience greater seasonal variations in temperature, with winters being warmer than summers. With rising and falling nighttime temperatures, India's heatwaves have lasted longer in recent years and are expected to rise in the future. When the concentrations double, the overall temperature rise is predicted to be between 2.33°C and 4.78°C (Mulla et al., 2023).

The increasing variability of the summer monsoon brought about by these heatwaves would have a major impact on Indian agriculture. Climate models predict a continuous global rise in carbon dioxide concentration and temperature. These models, however, are not very good at predicting changes in regional weather patterns (Schlesinger and Mitchell, 1987). As long as plants can be controlled by their surroundings, locally adaptive plant diversity, cropping strategies, soil quality, and weather factors like rain, sun, and wind can all work together to maximise food production. Biological diversity must be preserved at all scales, from genes to biomes, in order to effectively increase the rapid changes needed for human populations to adapt to future climate change.

21.6 GREEN TECHNOLOGIES AS AN OPPORTUNITY FOR SUSTAINABLE DEVELOPMENT

Green technologies, also known as clean technologies or eco-friendly technologies, are creative approaches and methods that support resource economy, ecological footprint reduction including carbon emission, and environmental sustainability. Solar, wind, hydroelectric, geothermal, and biomass energy are some of the examples of renewable energy sources that are included in green technology (Rahman et al., 2022). By switching to renewable energy sources, we can lower the strain that the mining and burning of fossil fuels places on natural habitats and ecosystems, reduce greenhouse gas emissions, and improve the effects of climate change. By minimising negative environmental effects on ecosystems and biodiversity, we may reduce energy consumption, lower greenhouse gas emissions, and promote sustainable development (Luthra et al., 2015). This can be achieved through the use of green technology in buildings, transportation, industries, and appliances.

Advanced waste treatment techniques, vermicomposting, recycling of waste, and waste-to-energy conversion are examples of green technology used in waste management (Tsui and Wong, 2019). By implementing sustainable waste management strategies, we can minimise landfill waste, cut down on pollution, save resources, and shield ecosystems from the negative effects of garbage. Rainwater harvesting, water recycling, advanced and effective irrigation systems, and water-saving devices in homes and industries are examples of green technologies for water conservation and management. These inventions promote sustainable water use practices, lessen water pollution, preserve freshwater resources, and repair aquatic ecosystems. A few of the sustainable agricultural methods that are supported by green technology in agriculture are organic farming, agroforestry, sustainable agriculture, and integrated pest management (Rai et al., 2023). These methods lessen agricultural impacts on ecosystems, decrease dependency on chemical inputs, improve soil health, conserve biodiversity, and maximise crop resilience. The main areas of attention for green building technologies include indoor environmental quality, water conservation, sustainable use of materials, energy-efficient designs, and integration of renewable energy sources. Urban parks, green roofs, and permeable surfaces are examples of nature-based solutions known as "green infrastructure," which improve air quality, lower urban heat islands, increase biodiversity, and offer ecosystem services to urban environments (Jones et al., 2022). Green technologies in tourism support nature-based tourism, eco-friendly practices, eco-friendly certifications, and experiences that support local communities, cultural heritage, and biodiversity conservation while reducing environmental footprints and protecting natural habitats.

Nonetheless, there are still a number of obstacles standing in the way of the advancement and widespread use of green technology (Luthra et al., 2015; Xia et al., 2019). These obstacles include legislative frameworks (Yoshino et al., 2019; Darko et al., 2018; He et al., 2019), the market (Campiglio, 2016; Agyemang et al., 2018), public knowledge and awareness (Darko et al., 2018), and financial constraints (Ji and Zhang, 2019; Bhandari et al., 2019). Specifically, there isn't a global standard for categorising green technology (WIPO, 2019). The selection of green technology may be highly ambiguous due to the absence of a standardised classification system. Furthermore, green technology is an intricate and ever-evolving concept encompassing a broad spectrum of disciplines. Inconsistent classification standards may lead to misconceptions regarding green technologies. The different classification systems used for green technology in different places make it difficult to transfer green technology on a large scale. To improve the efficacy of green technology transfer globally, a uniform framework that encompasses a wide range of stakeholders, including technology producers, users, and investors, must be established.

21.7 POLICY AND GOVERNANCE FOR SUSTAINING BIODIVERSITY AND ECOSYSTEM SERVICES

Robust rules and regulations protecting biodiversity, proactive government activities, and coordinated international collaboration efforts are some of the ways that policy and governance are essential to conserve biodiversity. There are many biodiversity laws implemented to protect key habitats, endangered species, and ecosystems, which ultimately serve to protect protected areas, keystone species, environmental impact assessments (EIAs), and forest and wildlife conservation (Kolahi et al., 2012). These legal frameworks play a key role in preventing the degradation of habitats, the unsustainable use of resources, and the exploitation of wildlife. In order to promote sustainable biodiversity management, restore degraded ecosystems, and encourage community engagement in conservation efforts, government initiatives like National Biodiversity Strategies and Action Plans (NBSAPs), ecosystem restoration programmes, biodiversity monitoring and research endeavours, and community-based conservation initiatives are some of the major examples of proactive measures (Bawa et al., 2021).

International biodiversity agreements like the Convention on Biological Diversity (CBD) and global collaborative platforms allow knowledge exchange, capacity-building programmes, and cooperative initiatives for biodiversity conservation and sustainable development (Chandra and Idrisova, 2011). At the same time, global collaboration arises as a key component of effective biodiversity governance. Many bilateral and multilateral partnerships create synergies between governments, NGOs, international organisations, and research institutions in order to achieve common goals for biodiversity conservation. These partnerships also encourage resource mobilisation, innovation, and coordinated actions. International collaboration is strengthened by scientific networks and financial structures such as the Global Environment Facility (GEF) and Green Climate Fund (GCF), which provide global funding for biodiversity conservation programmes, technology transfer, and sustainable development efforts (Kasdan et al., 2021). These governance and policy frameworks should come in a collaboration to protect biodiversity, reduce environmental risks, and promote peaceful coexistence between human societies and natural ecosystems for a sustainable future.

21.8 FUTURE DIRECTIONS IN BIODIVERSITY CONSERVATION AND SUSTAINABLE DEVELOPMENT

21.8.1 Research and Invention

The importance of research and innovation in tackling new problems and grasping opportunities is emphasised in the path that sustainable development and biodiversity conservation are taking. Along with the use of monitoring and evaluation methods like big data analytics and remote sensing, scientists and decision-makers can determine the effectiveness of conservation efforts, monitor changes in biodiversity, and set priorities for conservation. Novel approaches to the preservation of genetic diversity, the protection of endangered species, and the improvement of ecological resilience are provided by advances in biotechnology and genetic conservation (Salgotra et al., 2019; Hoban et al., 2021). In order to build resilience plans and integrate climate science with biodiversity conservation practices, research on the effects of climate change and adaptation techniques is essential. Ecological restoration and rewilding are two examples of innovative conservation techniques that support sustainable land management and natural capital accounting, which in turn promote ecological, social, and economic sustainability (Gann et al., 2019).

21.8.2 Community Participation

Dynamic future directions include empowering local communities and encouraging their active participation in biodiversity conservation and sustainable development. Local communities, indigenous peoples, and stakeholders are empowered to participate in biodiversity conservation and natural resource management through community-based conservation projects, participatory decision-making procedures, and co-management approaches (Wasonga et al., 2010). Incorporating indigenous techniques and traditional ecological knowledge improves conservation results and fosters cultural resilience. Enhancing community skills, knowledge, and engagement in biodiversity conservation, sustainable agriculture, eco-tourism, and natural resource governance are the goals of capacity-building programmes, environmental education, and awareness campaigns (Franco and Tracey, 2019). By utilising a range of advantages, local knowledge, social justice, equal gender participation, and inclusivity in conservation programmes and benefit-sharing structures promotes social cohesiveness, lowers conflict, and improves conservation results.

21.8.3 Sustainable Development Goals (SDGs): A Paradigm for Sustainability

An inclusive framework for attaining integrated and sustainable development is provided to coordinate biodiversity conservation initiatives with the SDGs. The conservation of natural resources, restoration, and sustainable management of terrestrial and aquatic ecosystems, including forests, biodiversity, and wildlife habitats, are the key objectives of SDG 15 (Bridgewater et al., 2015). SDG 14 plays an important role in preserving marine biodiversity, protecting coastal and marine environments, sustainable fishing practices, and preventing ocean pollution (Virto, 2018). In order to mitigate the effects of climate change and adapt to them, SDG 13 promotes climate resilience behaviour and incorporates climate resilience biodiversity conservation plans (Nunes et al., 2019). SDG 6 emphasises ecosystem-based methods to water conservation, sustainable management of water resources, and protection of water quality (Vörösmarty et al., 2018). SDG 12 promotes circular economy techniques, sustainable agriculture, and responsibility for consumption and production of food systems in order to reduce environmental impact and increase resource efficiency (Bengtsson et al., 2018). A synergistic approach is achieved by coordinating biodiversity conservation activities with the SDGs, addressing interrelated concerns related to environmental, social, and economic sustainability.

21.9 CONCLUSION

The discussion in India about biodiversity and ecosystem services highlights the complex interactions that exist between human activity, natural ecosystems, and the need for preservation. This investigation has discovered a number of important points. First, India's diverse variety of plant and animal species, along with its huge biodiversity, highlights the country's ecological significance on an international level. However, a number of serious issues, including pollution, invasion, habitat loss, and climate change, pose a threat to biodiversity and ecosystem services. In order to effectively address these complications, strong biodiversity laws and regulations, government actions, and improved international collaboration are all needed. Therefore, conservation and sustainable management efforts are dynamic. The future of India's biodiversity and ecosystem services depends on cooperation, inventiveness, and community participation. In order to monitor biodiversity, create adaptation plans, and promote sustainable practices, research and innovation are essential. By involving local people and indigenous knowledge systems, inclusive conservation strategies that support cultural norms and encourage stewardship are nurtured. A comprehensive framework for integrated and sustainable development is also provided by coordinating conservation efforts with the SDGs, guaranteeing that biodiversity conservation is incorporated into larger sustainability agendas. There is still hope despite these challenges. Technological developments, when combined with political determination, can create revolutionary transformation in this direction. India can lead towards a future of flourishing biodiversity and ecosystem services, thereby promoting human well-being, ecological resilience, and sustainable development objectives. This may be achieved by seizing these opportunities, promoting cross-sector collaboration, and prioritising conservation.

REFERENCES

Adla, K., Dejan, K., Neira, D., & Dragana, Š. (2022). Degradation of ecosystems and loss of ecosystem services. In *One Health* (pp. 281–327). Academic Press.

Agyemang, M., Zhu, Q., Adzanyo, M., Antarciuc, E., & Zhao, S. (2018). Evaluating barriers to green supply chain redesign and implementation of related practices in the West Africa cashew industry. *Resources, Conservation and Recycling, 136*, 209–222.

Ajibade, F. O., Adelodun, B., Lasisi, K. H., Fadare, O. O., Ajibade, T. F., Nwogwu, N. A., ... & Wang, A. (2021). Environmental pollution and their socioeconomic impacts. In *Microbe Mediated Remediation of Environmental Contaminants* (pp. 321–354). Woodhead Publishing.

Anil, M. N. V., Kumari, K., & Wate, S. R. (2014). Loss of biodiversity and conservation strategies: an outlook of Indian scenario. *Asian Journal of Conservation Biology, 3*(2), 105–114

Bashir, I., Lone, F.A., Bhat, R.A., Mir, S.A., Dar, Z.A., Dar, S.A. (2020). Concerns and threats of contamination on aquatic ecosystems. In: Hakeem, K., Bhat, R., Qadri, H. (eds) *Bioremediation and Biotechnology*. Springer, Cham. https://doi.org/10.1007/978-3-030-35691-0_1

Bastian, O. (2013). The role of biodiversity in supporting ecosystem services in Natura 2000 sites. *Ecological Indicators, 24*, 12–22.

Bawa, K. S., Sengupta, A., Chavan, V., Chellam, R., Ganesan, R., Krishnaswamy, J., ... & Vanak, A. T. (2021). Securing biodiversity, securing our future: a national mission on biodiversity and human well-being for India. *Biological Conservation, 253*, 108867.

Benayas, J. M. R., Newton, A. C., Diaz, A., & Bullock, J. M. (2009). Enhancement of biodiversity and ecosystem services by ecological restoration: a meta-analysis. *Science, 325*(5944), 1121–1124.

Bengtsson, M., Alfredsson, E., Cohen, M., Lorek, S., & Schroeder, P. (2018). Transforming systems of consumption and production for achieving the sustainable development goals: moving beyond efficiency. *Sustainability Science, 13*, 1533–1547.

Bhandari, D., Singh, R. K., & Garg, S. K. (2019). Prioritisation and evaluation of barriers intensity for implementation of cleaner technologies: framework for sustainable production. *Resources, Conservation and Recycling, 146*, 156–167.

Borrini-Feyerabend, G., & Hill, R. (2015). Governance for the conservation of nature. *Protected Area Governance and Management, 7*, 169–206.

Bozkurt, S. G., & Basaraner, M. (2024). Spatio-temporal investigation of urbanization and its impact on habitat fragmentation in natural ecosystems of Istanbul using Shannon's entropy and landscape metrics in GIS. *Environment, Development and Sustainability, 26*, 1–29.

Bridgewater, P., Régnier, M., & García, R. C. (2015). Implementing SDG 15: Can large-scale public programs help deliver biodiversity conservation, restoration and management, while assisting human development? *Natural Resources Forum, 39*(3–4), 214–223).

Brundu, G., & Richardson, D. M. (2016). Planted forests and invasive alien trees in Europe: a code for managing existing and future plantings to mitigate the risk of negative impacts from invasions.

Butler, C. D., & Oluoch-Kosura, W. (2006). Linking future ecosystem services and future human well-being. *Ecology and Society, 11*(1).

Campiglio, E. (2016). Beyond carbon pricing: The role of banking and monetary policy in financing the transition to a low-carbon economy. *Ecological Economics, 121*, 220–230.

Cardinale, B. J., Duffy, J. E., Gonzalez, A., Hooper, D. U., Perrings, C., Venail, P., ... & Naeem, S. (2012). Biodiversity loss and its impact on humanity. *Nature*, *486*(7401), 59–67.

Carpenter, S. R., Mooney, H. A., Agard, J., Capistrano, D., DeFries, R. S., Díaz, S., ... & Whyte, A. (2009). Science for managing ecosystem services: beyond the millennium ecosystem assessment. *Proceedings of the National Academy of Sciences*, *106*(5), 1305–1312.

Ceballos, G., Ehrlich, A. H., & Ehrlich, P. R. (2015). *The Annihilation of Nature: Human Extinction of Birds and Mammals*. JHU Press.

Chan, K. M., Satterfield, T., & Goldstein, J. (2012). Rethinking ecosystem services to better address and navigate cultural values. *Ecological Economics*, *74*, 8–18.

Chandra, A., & Idrisova, A. (2011). Convention on Biological Diversity: a review of national challenges and opportunities for implementation. *Biodiversity and Conservation*, *20*, 3295–3316.

Chitale, V. S., Behera, M. D., & Roy, P. S. (2014). Future of endemic flora of biodiversity hotspots in India. *PloS One*, *9*(12), e115264.

Collinge, S. K. (2009). *Ecology of Fragmented Landscapes*. JHU Press.

Corvalan, C., Hales, S., & McMichael, A. J. (2005). *Ecosystems and Human Well-Being: Health Synthesis*. World Health Organization.

Costanza, R., d'Arge, R., De Groot, R., Farber, S., Grasso, M., Hannon, B., ... & Van Den Belt, M. (1997). The value of the world's ecosystem services and natural capital. *Nature*, *387*(6630), 253–260.

Darko, A., Chan, A. P. C., Yang, Y., Shan, M., He, B. J., & Gou, Z. (2018). Influences of barriers, drivers, and promotion strategies on green building technologies adoption in developing countries: The Ghanaian case. *Journal of Cleaner Production*, *200*, 687–703.

De Groot, R. S., Wilson, M. A., & Boumans, R. M. (2002). A typology for the classification, description and valuation of ecosystem functions, goods and services. *Ecological Economics*, *41*(3), 393–408.

Dem, P., Hayashi, K., & Fujii, M. (2024). Resources time footprint for assessment of human influence on ecosystem service from a sustainability standpoint. *Journal of Cleaner Production*, *436*, 140612.

Díaz, S., Fargione, J., Chapin III, F. S., & Tilman, D. (2006). Biodiversity loss threatens human well-being. *PLoS Biology*, *4*(8), e277.

Didham, R. K., Kapos, V., & Ewers, R. M. (2012). Rethinking the conceptual foundations of habitat fragmentation research. *Oikos*, *121*(2), 161–170.

Dudgeon, R. C., & Berkes, F. (2003). Local understandings of the land: Traditional ecological knowledge and indigenous knowledge. In *Nature Across Cultures: Views of Nature and the Environment in Non-Western Cultures* (pp. 75–96). Dordrecht: Springer Netherlands.

Elmqvist, T., Maltby, E., Barker, T., Mortimer, M., Perrings, C., Aronson, J., ... & Salles, J. M. (2012). Biodiversity, ecosystems and ecosystem services. In *The Economics of Ecosystems and Biodiversity: Ecological and Economic Foundations* (pp. 41–111). Routledge: London.

Escobedo, F. J., Kroeger, T., & Wagner, J. E. (2011). Urban forests and pollution mitigation: analyzing ecosystem services and disservices. *Environmental Pollution*, *159*(8–9), 2078–2087.

Eviner, V. T., Garbach, K., Baty, J. H., & Hoskinson, S. A. (2012). Measuring the effects of invasive plants on ecosystem services: challenges and prospects. *Invasive Plant Science and Management*, *5*(1), 125–136.

Fahrig, L. (2003). Effects of habitat fragmentation on biodiversity. *Annual review of ecology, evolution, and systematics*, *34*(1), 487–515.

Fahrig, L. (2013). Rethinking patch size and isolation effects: the habitat amount hypothesis. *Journal of Biogeography*, *40*(9), 1649–1663.

Fenesi, A., Vágási, C. I., Beldean, M., Földesi, R., Kolcsár, L. P., Shapiro, J. T., ... & Kovács-Hostyánszki, A. (2015). Solidago canadensis impacts on native plant and pollinator communities in different-aged old fields. *Basic and Applied Ecology*, *16*(4), 335–346.

Franco, I. B., & Tracey, J. (2019). Community capacity-building for sustainable development: effectively striving towards achieving local community sustainability targets. *International Journal of Sustainability in Higher Education*, *20*(4), 691–725.

Gadgil, M. (2008). *Biodiversity Profile of India*. http: //www.biodiversiy.org. ces.iisc.ernet.in/hpg/cesmg/in diabio.html.

Gann, G. D., McDonald, T., Walder, B., Aronson, J., Nelson, C. R., Jonson, J., ... & Dixon, K. (2019). International principles and standards for the practice of ecological restoration. *Restoration Ecology*, *27*(S1), S1–S46.

He, L., Zhang, L., Zhong, Z., Wang, D., & Wang, F. (2019). Green credit, renewable energy investment and green economy development: empirical analysis based on 150 listed companies of China. *Journal of Cleaner Production*, *208*, 363–372.

Henfrey, T., Feola, G., Penha-Lopes, G., Sekulova, F., & Esteves, A. M. (2023). Rethinking the sustainable development goals: learning with and from community-led initiatives. *Sustainable Development*, *31*(1), 211–222.

Hoban, S., Bruford, M. W., Funk, W. C., Galbusera, P., Griffith, M. P., Grueber, C. E., ... & Vernesi, C. (2021). Global commitments to conserving and monitoring genetic diversity are now necessary and feasible. *Bioscience*, *71*(9), 964–976.

Ibáñez, I., Katz, D. S., Peltier, D., Wolf, S. M., & Connor Barrie, B. T. (2014). Assessing the integrated effects of landscape fragmentation on plants and plant communities: the challenge of multiprocess–multiresponse dynamics. *Journal of Ecology*, *102*(4), 882–895.

IUCN (2004). *The IUCN Red list of Threatened Species*. Gland, Switzerland: The world Conservation Union.

Jennings, S., Stentiford, G. D., Leocadio, A. M., Jeffery, K. R., Metcalfe, J. D., Katsiadaki, I., ... & Verner-Jeffreys, D. W. (2016). Aquatic food security: insights into challenges and solutions from an analysis of interactions between fisheries, aquaculture, food safety, human health, fish and human welfare, economy and environment. *Fish and Fisheries*, *17*(4), 893–938.

Ji, Q., & Zhang, D. (2019). How much does financial development contribute to renewable energy growth and upgrading of energy structure in China? *Energy Policy*, *128*, 114–124.

Jones, L., Anderson, S., Læssøe, J., Banzhaf, E., Jensen, A., Bird, D. N., ... & Zandersen, M. (2022). A typology for urban Green Infrastructure to guide multifunctional planning of nature-based solutions. *Nature-Based Solutions*, *2*, 100041.

Kasdan, M., Kuhl, L., & Kurukulasuriya, P. (2021). The evolution of transformational change in multilateral funds dedicated to financing adaptation to climate change. *Climate and Development*, *13*(5), 427–442.

Kolahi, M., Sakai, T., Moriya, K., & Makhdoum, M. F. (2012). Challenges to the future development of Iran's protected areas system. *Environmental Management*, *50*, 750–765.

Kueffer, C., & Kull, C.A. (2017). Non-native species and the aesthetics of nature. In: Vilà, M., Hulme, P. (eds) *Impact of Biological Invasions on Ecosystem Services*. Invading Nature - Springer Series in Invasion Ecology, vol 12. Springer, *Cham*. https://doi.org/10.1007/978-3-319-45121-3_20

Kull, C. A., Shackleton, C. M., Cunningham, P. J., Ducatillon, C., Dufour-Dror, J. M., Esler, K. J., ... & Zylstra, M. J. (2011). Adoption, use and perception of Australian acacias around the world. *Diversity and Distributions*, *17*(5), 822–836.

Kuthiala, S. K. (2001). Tourism and hospitality industry in India. *Journal of Services Research*, *1*(1), 35.

Lindenmayer, D. B., & Fischer, J. (2013). *Habitat Fragmentation and Landscape Change: An Ecological and Conservation Synthesis*. Island Press: Washington DC.

Losos, J. B., & Schluter, D. (2000). Analysis of an evolutionary species–area relationship. *Nature*, *408*(6814), 847–850.

Luthra, S., Kumar, S., Garg, D., & Haleem, A. (2015). Barriers to renewable/sustainable energy technologies adoption: Indian perspective. *Renewable and Sustainable Energy Reviews*, *41*, 762–776.

Lyytimäki, J. (2014). Bad nature: Newspaper representations of ecosystem disservices. *Urban Forestry & Urban Greening*, *13*(3), 418–424.

MacArthur, R. H., & Wilson, E. O. (2001). *The Theory of Island Biogeography* (Vol. 1). Princeton University Press: United Kingdom.

Mace, G. M., Norris, K., & Fitter, A. H. (2012). Biodiversity and ecosystem services: a multilayered relationship. *Trends in Ecology & Evolution*, *27*(1), 19–26.

MEA (2005). *Ecosystems and Human Well-Being: Synthesis*. Millennium Ecosystem Assessment, Island Press, Washington, DC.

Mertz, O., Ravnborg, H. M., Lövei, G. L., Nielsen, I., & Konijnendijk, C. C. (2007). Ecosystem services and biodiversity in developing countries. *Biodiversity and Conservation*, *16*, 2729–2737.

MoEF (2000). *Annual Report 1999–2000*, New Delhi: Ministry of Environment and Forests, Government of India.

Mulla, S., Ahmed, R., Singh, K. K., Singh, S. K., Deshmukh, N., & Inamdar, F. K. (2023). Climate change effect on-climate parameters like temperature, rainfall and water resources sectors in India. In *Climate Change Impacts in India* (pp. 9–59). Cham: Springer International Publishing.

Mulrennan, M. E., Mark, R., & Scott, C. H. (2012). Revamping community-based conservation through participatory research. *The Canadian Geographer/Le Géographe canadien*, *56*(2), 243–259.

Myers, N., Mittermeier R.A., Mittermeier, C.G., Da Fonseca, G.A.B., Kent, J. (2000). Biodiversity hotspots for conservation priorities. *Nature*, 403, 853–858.

Nicholson, E., Keith, D. A., & Wilcove, D. S. (2009). Assessing the threat status of ecological communities. *Conservation Biology*, *23*(2), 259–274.

Nunes, A., Speranza, C. I., Matos, P., & Branquinho, C. (2019). Land restoration for achieving SDG 13: Take urgent action to combat climate change and its impacts. *Sustainable Development Goals*, 91, 93–97.

Oguh, C. E., Obiwulu, E. N. O., Umezinwa, O. J., Ameh, S. E., Ugwu, C. V., & Sheshi, I. M. (2021). Ecosystem and ecological services; need for biodiversity conservation-a critical review. *Asian Journal of Biology*, *11*(4), 1–14.

Oliver, T. H., Heard, M. S., Isaac, N. J., Roy, D. B., Procter, D., Eigenbrod, F., ... & Bullock, J. M. (2015). Biodiversity and resilience of ecosystem functions. *Trends in Ecology & Evolution*, *30*(11), 673–684.

Orr, M. R., & Smith, T. B. (1998). Ecology and speciation. *Trends in Ecology & Evolution*, *13*(12), 502–506.

Owusu, R. O., Martinez, J., & Schwarz, N. (2024). Supply and demand of ecosystem services of urban green spaces in deprived areas: perceptions from Kumasi, Ghana. *Environmental Science & Policy*, *156*, 103742.

Palomo, I., Montes, C., Martin-Lopez, B., González, J. A., Garcia-Llorente, M., Alcorlo, P., & Mora, M. R. G. (2014). Incorporating the social–ecological approach in protected areas in the Anthropocene. *BioScience*, *64*(3), 181–191.

Percival, R. V., Schroeder, C. H., Miller, A. S., & Leape, J. P. (2021). *Environmental Regulation: Law, Science, and Policy [connected EBook with study center]*. Aspen Publishing.

Pereira, H. M., Leadley, P. W., Proença, V., Alkemade, R., Scharlemann, J. P., Fernandez-Manjarrés, J. F., ... & Walpole, M. (2010). Scenarios for global biodiversity in the 21st century. *Science*, *330*(6010), 1496–1501.

Perrings C, Duraiappah A, Larigauderie A, Mooney H. The biodiversity and ecosystem services science-policy interface. (2011). *Science, 331*(6021), 1139–1140.

Rahman, A., Farrok, O., & Haque, M. M. (2022). Environmental impact of renewable energy source based electrical power plants: solar, wind, hydroelectric, biomass, geothermal, tidal, ocean, and osmotic. *Renewable and Sustainable Energy Reviews*, *161*, 112279.

Rai, A. K., Bana, S. R., Sachan, D. S., & Singh, B. (2023). Advancing sustainable agriculture: a comprehensive review for optimizing food production and environmental conservation. *International Journal of Plant & Soil Science*, *35*(16), 417–425.

Rands, M. R., Adams, W. M., Bennun, L., Butchart, S. H., Clements, A., Coomes, D., ... & Vira, B. (2010). Biodiversity conservation: challenges beyond 2010. *Science*, *329*(5997), 1298–1303.

Rasmussen, L. V., Christensen, A. E., Danielsen, F., Dawson, N., Martin, A., Mertz, O., ... & Xaydongvanh, P. (2017). From food to pest: conversion factors determine switches between ecosystem services and disservices. *Ambio*, *46*, 173–183.

Rey Benayas, J. M., & Bullock, J. M. (2012). Restoration of biodiversity and ecosystem services on agricultural land. *Ecosystems*, *15*, 883–899.

Salgotra, R. K., Sharma, M., & Pandotra, P. (2019). Biotechnological interventions for sustainable conservation of plant genetic resources in the scenario of climate change. *Natural Resources Conservation and Research*, *2*(1), 1–13.

Samuelson, O. E., Endurance, E. O., Pharm, B., Ohikhuare, O., Raphael, U. O., Michael, O. O., & Osaruyi, O. E. (2022). Climate change; our impact on the climate and its consequences; a crosssectional study. *International Journal of Scientific Advances*, *3*(3), 431–447.

Sans-Fuentes, M. A., & Meixner, T. (2017). The role of biodiversity in the hydrological cycle. *Water Bankruptcy in the Land of Plenty*, *249*

Saunders, M. E., & Luck, G. W. (2016). Limitations of the ecosystem services versus disservices dichotomy. *Conservation Biology*, *30*(6), 1363–1365.

Schlesinger, M. E., & Mitchell, J. F. (1987). Climate model simulations of the equilibrium climatic response to increased carbon dioxide. *Reviews of Geophysics*, *25*(4), 760–798.

Schröter, M., Van der Zanden, E. H., van Oudenhoven, A. P., Remme, R. P., Serna-Chavez, H. M., De Groot, R. S., & Opdam, P. (2014). Ecosystem services as a contested concept: a synthesis of critique and counter-arguments. *Conservation Letters*, *7*(6), 514–523..

Shackleton, C. M., Ruwanza, S., Sinasson Sanni, G. K., Bennett, S., De Lacy, P., Modipa, R., ... & Thondhlana, G. (2016). Unpacking Pandora's box: understanding and categorising ecosystem disservices for environmental management and human wellbeing. *Ecosystems*, *19*, 587–600.

Siikamäki, J., Sanchirico, J. N., Jardine, S., McLaughlin, D., & Morris, D. (2013). Blue carbon: coastal ecosystems, their carbon storage, and potential for reducing emissions. *Environment: Science and Policy for Sustainable Development*, *55*(6), 14–29.

Simberloff, D., Martin, J. L., Genovesi, P., Maris, V., Wardle, D. A., Aronson, J., ... & Vilà, M. (2013). Impacts of biological invasions: what's what and the way forward. *Trends in Ecology & Evolution*, *28*(1), 58–66.

Singh, J. S. (2012). Biodiversity: an overview. *Proceedings of the National Academy of Sciences, India Section B: Biological Sciences*, *82*, 239–250.

Singh, S.P. (2007). *Himalayan Forests Ecosystem Services: Incorporating in National Accounting, Central Himalayan Environment Association (CHEA).* Uttarakhand: Nainital.

Tassin, J., & Kull, C. A. (2015). Facing the broader dimensions of biological invasions. *Land Use Policy*, *42*, 165–169.

Thrupp, L. A. (2000). Linking agricultural biodiversity and food security: the valuable role of agrobiodiversity for sustainable agriculture. *International Affairs*, *76*(2), 265–281.

Tsui, T. H., & Wong, J. W. (2019). A critical review: emerging bioeconomy and waste-to-energy technologies for sustainable municipal solid waste management. *Waste Disposal & Sustainable Energy*, *1*, 151–167.

Van Groenigen, J. W., Huygens, D., Boeckx, P., Kuyper, T. W., Lubbers, I. M., Rütting, T., & Groffman, P. M. (2015). The soil N cycle: new insights and key challenges. *Soil*, *1*(1), 235–256.

Venkataraman, K. (2012). Biodiversity and its conservation. *Proceedings of the National Academy of Sciences, India Section B: Biological Sciences*, *82*, 271–282.

Venkataraman, K. (2012). Biodiversity and its conservation. *Proceedings of the National Academy of Sciences, India Section B: Biological Sciences*, *82*, 271–282.

Virto, L. R. (2018). A preliminary assessment of the indicators for Sustainable Development Goal (SDG) 14 "Conserve and sustainably use the oceans, seas and marine resources for sustainable development". *Marine Policy*, *98*, 47–57.

Vonlanthen, P., Bittner, D., Hudson, A. G., Young, K. A., Müller, R., Lundsgaard-Hansen, B., ... & Seehausen, O. (2012). Eutrophication causes speciation reversal in whitefish adaptive radiations. *Nature*, *482*(7385), 357–362.

Vörösmarty, C. J., Osuna, V. R., Cak, A. D., Bhaduri, A., Bunn, S. E., Corsi, F., ... & Uhlenbrook, S. (2018). Ecosystem-based water security and the Sustainable Development Goals (SDGs). *Ecohydrology & Hydrobiology*, *18*(4), 317–333.

Wali, A., Alvira, D., Tallman, P. S., Ravikumar, A., & Macedo, M. O. (2017). A new approach to conservation: using community empowerment for sustainable well-being. *Ecology & Society*, *22*(4), 1–13.

Wasonga, V. O., Kambewa, D., & Bekalo, I. (2010). Community-based natural resource management. In Isaac Bekalo, P. C. Sanginga, Washington Odongo Ochola (eds.) *Managing Natural Resources for Development in Africa: A Resource Book*, 165–205, Nairobi, Kenya.

Wikramanayake, E., Dinerstein, E., Allnut, T., Oucks, C., Wettengel, W. (1998). A Biodiversity Assessment and Gap Analysis of the Himalayas. World Wildlife Fund-U.S., Conservation Science Program/ UNDP Rept.

WIPO (2019). WIPO Green Strategic Plan 2019-2023: Accelerating the Transition to a Greener Global Economy. https://www.wipo.int/edocs/pubdocs/en/wipo_pub_ greenstrpl1923.pdf

Xia, D., Zhang, M., Yu, Q., & Tu, Y. (2019). Developing a framework to identify barriers of green technology adoption for enterprises. *Resources, Conservation and Recycling*, *143*, 99–110.

Yadav, S. P. S., Mehata, D. K., Pokhrel, S., Ghimire, N. P., Gyawali, P., Katel, S., & Timsina, U. (2024). A comprehensive study of banmara (invasive alien plant species): Understanding the invasive potential and ecological consequences for biodiversity conservation, and management strategies. *Journal of Agriculture and Food Research*, *15*, 101031.

Yoshino, N., Taghizadeh–Hesary, F., & Nakahigashi, M. (2019). Modelling the social funding and spill-over tax for addressing the green energy financing gap. *Economic Modelling*, *77*, 34–41.

22 Approach to Agrobiodiversity Preservation and Sustainable Agriculture in a Changing World

Bindiya Barsola[1], *Priyanka Kumari*[1], *Swati Yadav*[2], *Suman Mishra*[2], *Kamal Jaiswal*[2]

[1]School of Biological and Environmental Sciences, Shoolini University of Biotechnology and Management Sciences, Solan, Himachal Pradesh, India

[2]Department of Zoology, School of Life Sciences, Babasaheb Bhimrao Ambedkar University, Lucknow, Uttar Pradesh, India

22.1 INTRODUCTION

Agriculture serves as one of the most significant sectors of the Indian economy. It provides a living for about two-thirds of the rural population in the country. Agriculture has evolved significantly in recent years and is becoming progressively efficient (Kumar *et al.*, 2023). The idea of "biodiversity" is difficult to describe due to its comprehensive nature, as it incorporates the vast array of living forms on Earth, along with the associated natural processes. Biodiversity is an essential element of ecosystem function, playing a crucial role in both natural and human-engineered ecosystems, as well as in the provision of ecosystem services to human civilization (Sobti *et al.*, 2023). Agriculture that consistently supplies food and other resources to an expanding global population is essential to human life. Climate change, an elevated level of diminution in biodiversity, soil deterioration, pollution, higher expenses for production, a steadily declining number of farms, poverty, and a drop in the rural population are just a few of the issues that jeopardize agriculture's ability to meet human requirements both now and in generations to come (Mogni *et al.*, 2009). The growing global population, which is expected to approach 10 billion by 2050 and require a significant increase in food production, is another pressing issue (Giller *et al.*, 2021; Sekhar *et al.*, 2024). Amid these obstacles, the concept of sustainable agriculture has become more prevalent. The term "sustainable agriculture" signifies "management practices that integrate with the environment to conserve the entirety of resources, minimizing waste and impact on the environment, fostering agroecosystem resilience, evolution and consistent cultivation for the supplies and fulfillment of all people" (Velten et al., 2015). There are four main categories of sustainable agriculture: regenerative agriculture, permaculture, agroforestry, and organic farming, which include different techniques such as polycultures, biodynamic agriculture, ecological aquaculture, hydroponics, natural pest management, silvopasture, and keyline subsoiling (Ince & Karaca, 2022). The foundation of agricultural sustainability is the idea that current demands must be fulfilled without compromising the capacity of future generations to meet their own needs. Thus, the preservation of natural resources for future generations and the advancement of humankind are equally vital. The goal of agricultural sustainability is to manage soil, water, agroforestry, pests, and crops using a variety of integrated strategies. The

DOI: 10.1201/9781032697550-22

leftovers or waste from one project are used as inputs for others. Environmental impact decreases as natural processes progressively take the place of outside inputs. The sustainability of agriculture supports environmental services like biodiversity, soil preservation, watershed services, and carbon sequestration. Although agrobiodiversity is known to have many benefits, the particular assertion that it increases smallholder resilience needs closer examination (Bravo-Peña & Yoder, 2024).

Agrobiodiversity is the variety of life forms (including bacteria, plants, and animals) employed in agriculture. Agricultural biodiversity has been impacted by contemporary, intensive agriculture. There is growing evidence that diversity in general must be a key component of sustainable agricultural growth and that agricultural biodiversity as an asset offers significant benefits in a variety of contexts (Frison *et al.*, 2011). Different agroecosystem services include weed management, nitrogen usage effectiveness, environmental stress tolerance, and pest management. To interrelate crop features to agrobiodiversity services, the agroecosystem is divided into functional composition, functional identity, and functional diversity. Linking crop features to agroecosystem services may result in concrete alternatives for farmers and policymakers (Costanzo *et al.*, 2014).

Sustainable agriculture entails addressing the demands of current and future generations for agricultural products and services while also maintaining profitability, ecological wellness, and economic and social impartiality. As a result, regulatory, governance, technological, and financial frameworks that enable farmers and resource managers to engage in the innovation process are required. Sustainable agriculture aims to achieve three major goals: ecological wellness, financial viability, and equality in society and the economy. These three crucial aims are directly tied to the agricultural sector's adaptability. Furthermore, sustainable development can be divided into three fundamental elements of agricultural sustainability: economic, social, and environmental sustainability. Sustainable agriculture is built on the foundation of agrobiodiversity. Investigations in natural and microenvironment contexts have found a correlation between biodiversity and improved productivity, enhanced stability, and efficiency of ecosystems. As a result, agrobiodiversity facilitates the sustainability of agricultural operations. Agrobiodiversity conservation is crucial because cultivating plant genetic diversity must be expanded to prevent the extinction of specifically adapted ecotypes and to produce crops that have regional adaptability (Love & Spaner, 2007). As a result, the notion of "sustainable development" has arisen, serving present demands without jeopardizing future generations' needs. The United Nations has revealed 17 Sustainable Development Goals (SDGs) to accomplish synergy between human well-being and environmental resource conservation by 2030. Agricultural biodiversity contributes to several facets of sustainability and SDGs as given in Figure 22.1. A multifaceted meaning is given to agriculture by the fact that nearly all of the 17 SDGs released in 2015 to meet today's global challenges by 2030 have relevance to the agriculture sector. Agriculture must be regarded as an essential component of the SDGs since it is a vital link between humans and the environment. The world's greatest persistent challenge is undoubtedly the food crisis because of the startling rate of population expansion. Approximately 780 million individuals in developing nations suffer from malnutrition (Kumar *et al.*, 2023). Many methods have been employed in a sustainable agriculture system, including zero-tillage farming, mulching, contour agriculture, crop switching and diversification, and organic farming. Sustainable agricultural systems must find an equilibrium between environmental wellness and economic prosperity to achieve equality in both society and the economy. Genetic resources, governance policies, global social–ecological interactions, and health and nutrition are the four major integrative approaches for agrobiodiversity. Using genomic data, geospatial, and ecological appliances provides fresh perspectives on global agrobiodiversity in agricultural landscapes and food networks. This progress is driven by the growing amount of valuable information on multifaceted environmental functionality and cultural behaviors (Zimmerer & De Haan, 2017). Ecological viability, financial gain, and social justice are the three primary objectives of sustainable farming.

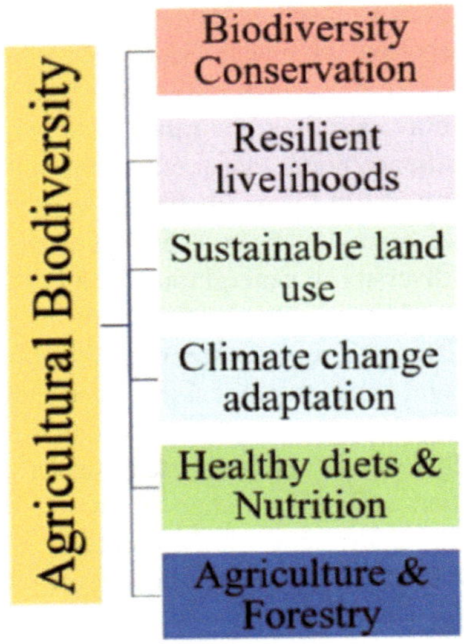

FIGURE 22.1 Contribution of agrobiodiversity to several facets of SDGs.

22.2 CHALLENGES TO SUSTAINABLE AGRICULTURE

Almost every nation in the world depends, in some way or another, on agriculture for its economic growth. The sustainability of the ecosystem is harmed by intensive farming and the employment of contemporary techniques, which assure a good output. The major threat to soil biodiversity is urbanization, which reduces the vast pool of fauna and microflora that inhabit the soil. Innovative basic and translational research is needed to provide effective and long-lasting disease and pest management solutions since plants are constantly exposed to and interact with their pests and diseases. These fields of study include the ecology and epidemiology of flora-based pathogens, plant–pest interactions in the environment, and molecular mechanisms mediating these interactions (Malini *et al.*, 2023). The following five significant issues pose a threat to the adoption of sustainable farming practices: a decline in yields in the first few years, an increase in toil, the accessibility and availability of required materials, the overall availability of food, and the economics of politics. There isn't just one specific, final aim for sustainable agriculture. Modern concerns, viewpoints, and beliefs are influencing the scientific understanding of sustainability as it relates to the environment, society, and economy. Sustainable agriculture requires the integration of economic, social, and environmental aspects, all of which are interdependent. Environmental impact has traditionally been the only focus of sustainability studies in agricultural systems. Their primary focus has been to determine areas that require improvement while highlighting ongoing shortcomings related to conveyance, waste elimination, water consumption diminution, and carbon emissions (Mogni *et al.*, 2009). Societies may only encourage more sustainable farming systems by developing policies that balance economic, environmental, and social concerns. Water resource scarcity, declining arable land, contamination of the environment, flaws in protective measures for arable land, and ineffective land tenure rights management are challenges to agriculturally sustainable development, as shown in Figure 22.2 (Parray *et al.*, 2019). For the following reasons, adopting eco-friendly practices is still challenging: enough food production, water scarcity, loss of arable land, excessive energy use, climate changes, and the affordability of sustainable methods.

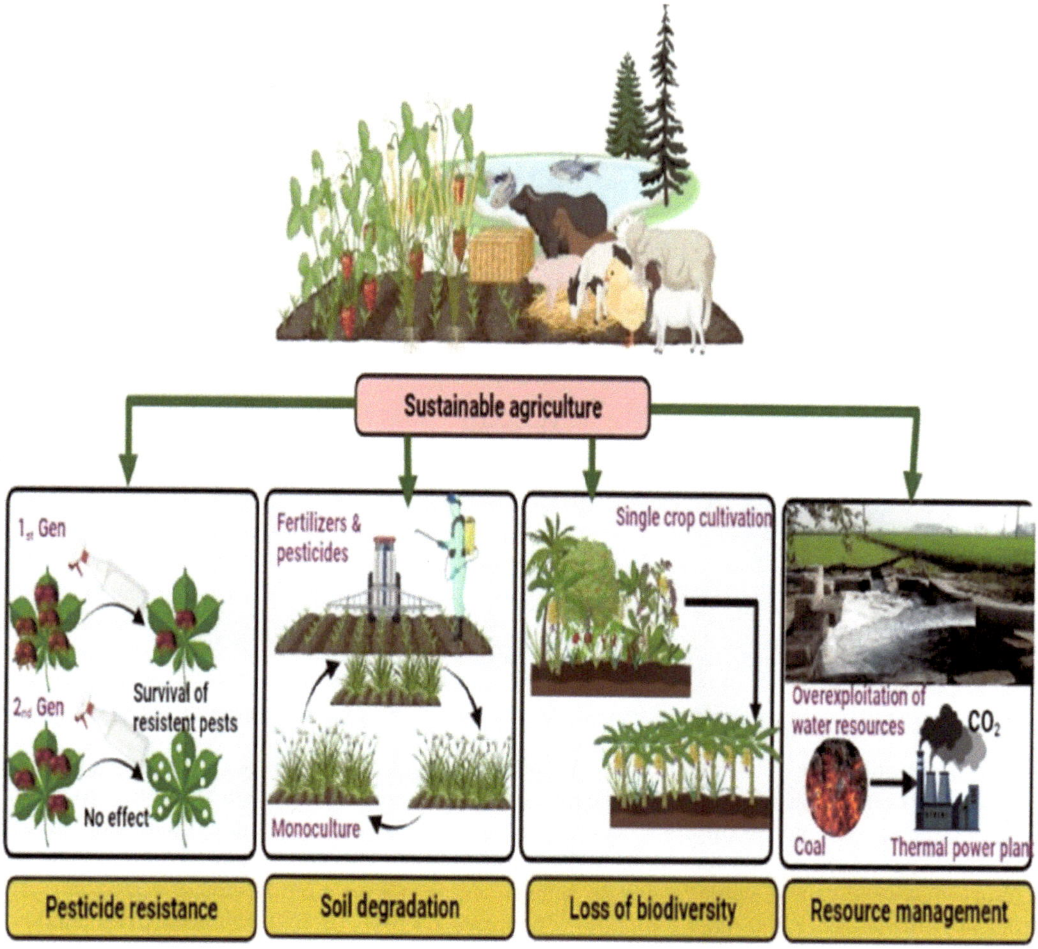

FIGURE 22.2 Challenges to sustainable agriculture.

22.3 MOLECULAR APPROACHES TO SUSTAIN AGROBIODIVERSITY

Habitat destruction, climate alterations, global warming, and intensive agricultural behaviors are the major reasons for the substantial decline in agrobiodiversity. The preservation of agrobiodiversity can be accomplished by using two strategies: in situ preservation, which is the conservation of species in their native ecosystems, and ex situ preservation, which includes preserving live organisms outside of their native settings (Belokurova, 2010). Advanced biotechnological interventions are currently being employed in the maintenance and restoration of agrobiodiversity, which encompass a variety of laboratory-based methods that can be utilized separately or in tandem (Mondal *et al.*, 2022). Molecular strategies having great potential toward agrobiodiversity restoration and its maintenance involve different markers such as chloroplast spacer sequences, molecular markers, biochemical markers, and internal transcribed spacer sequences. These strategies also involve RNAi technology, DNA microarray, and molecular farming, as given in Figure 22.3.

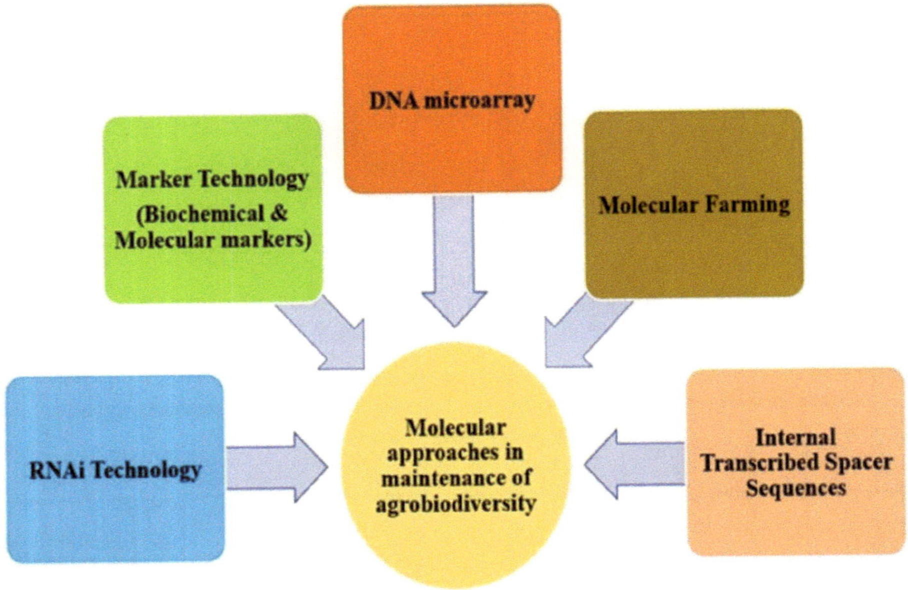

FIGURE 22.3 Different molecular approaches in the maintenance of agrobiodiversity.

22.4 BIOTECHNOLOGICAL STRATEGIES TO ENHANCE CROP SUSTAINABILITY

The use of technology in biology is known as biotechnology, and it mainly entails the transmission of genomic content carrying particular features from one crop or organism to another to increase crop yield and resilience to different ecological challenges. Numerous studies conducted on farms and small plots have shown that biotechnology intervention can enhance crop output, nutritional quality, and weed control while having no negative effects on the environment or ecosystem. Weed control is made easier and more effective by engineered crops, such as glyphosate-resistant crops (Dill *et al.*, 2005).

The following are some ways that biotechnology has helped to promote sustainable agriculture:

1. Enhanced resilience to biotic stressors, such as diseases and insect pests
2. Enhanced ability to withstand abiotic challenges such as cold, flooding, drought, and problematic soils
3. The use of biodetectors to monitor pollution and bioremediation of contaminated soils
4. Enhanced output and precision
5. Improved efficiency of nutrient absorption and utilization and enhanced nitrogen fixation
6. Enhanced technology for fermentation
7. Enhanced methods for producing energy from biomass
8. Production of high nutritional levels in staple crops lacking in nutrients

Biotechnology has played an important part in accomplishing environmental sustainability by adopting ecologically sustainable crops such as herbicide-resistant and pest-resistant plants, as well as crops that can fix nitrogen in the atmosphere, resulting in environmental cleansing. Biotechnology has drastically altered agriculture and offered numerous advantages to both cultivators and customers alike. It is the alteration of biological organisms to generate novel products, methods, and approaches to highlight various agricultural concerns.

The main barrier to sustainable agriculture is the proliferation of pests and illnesses that can drastically reduce crop yields and quality. To solve these problems, biotechnology can be utilized to produce crops that are resistant to diseases and pests. Another barrier to sustainable agriculture is a lack of nutrients in the soil, which can lower crop yields and plant health. Using biotechnology, more nutrient-rich and fruitful crops could be created. On the other hand, it is crucial to ensure that these technologies are developed prudently and that the advantages they bring about are shared fairly across groups and geographical areas (Das *et al.*, 2023). Both disease resistance and a reduction in the need for synthetic pesticides can be achieved by engineering the crop. Specific metabolic pathway-specific genes can be added to crops to biofortify them. For example, genes encoding for beta-carotene production were added to golden rice to fortify it. There are now greater prospects and optimism for creating the second or third generation of biotechnological products owing to recent advancements in OMICS techniques, particularly CRISPR genome editing. Public opinion continues to provide obstacles for biotechnology in sustainable agriculture, despite its potential. To address the questions and concerns of the public, further awareness and extension are required. A barrier to the widespread use of biotechnology in agriculture is the possible dangers of genetically modified organisms and the need to protect agrobiodiversity. In sustainable agriculture, biotechnology holds great promise for enhancing soil wellness and nutrient cycling. Certain types of microorganisms can be introduced to soil to improve soil fertility and structure, inhibit plant diseases, and encourage flora development. Instances of using natural sources to improve soil quality, plant wellness, and total yield are biofertilizers and biopesticides. Biotechnology offers numerous opportunities to improve agricultural sustainability, quality, and production, and it has enormous potential in sustainable agriculture. One of the most recognized benefits of biotechnology in farming is the production of enhanced crop varieties that are resistant to ecological stresses and hence well adapted to varied environments. Producers may grow a greater variety with fewer resources, and consumers can benefit from a more ample and cost-effective food supply. Despite the tremendous advantages of biotechnology in agriculture, there are concerns regarding the potential drawbacks of this technology. As a result, regulatory policies must keep up with the rapid advancements in biotechnology. Our environment is a significant repository of microbiological genetic variety (Robe *et al.*, 2003). Multiple interrelated elements, such as pH, moisture content, soil makeup, climatic fluctuations, and biotic activity, contribute to the intricate nature of microbial diversity. The majority of soil microbes appear to be extraordinarily well suited to their surroundings, but they cannot be cultivated under conventional laboratory conditions. According to current projections, more than 99% of the microbes prevalent in many ecosystems in nature are not easily culturable and thus unavailable for biotechnology or basic study (Schloss & Handelsman, 2003). Microbial communities are critical to the maintenance of all ecological systems, but the significance of unculturable bacteria in natural ecosystems is unknown (Tyson *et al.*, 2004). Metagenomics has the potential to explore the vast untapped microbiological variety found in nature for novel compounds with medicinal, biotechnological, and agricultural potential. This technique involves extracting DNA from an ecosystem and pooling the genomes of all organisms in the community. The pure metagenomic DNA from soil can be transcribed into a variety of plasmid vectors. High-molecular-weight DNA are often cloned into cosmid or bacterial artificial chromosome (BAC) vectors and injected into a surrogate carrier. The most popular method for creating a metagenomic library is the fosmid-based approach because of its exceptional cloning efficiency, enhanced stability in *E. coli*, and ability to create a medium-sized (40 kb) insert genomic library. The cloning efficiency is an essential component for developing an extensive clone member library because it is challenging to uncover genes for novel biological activities from a soil metagenomic library. Isolating unadulterated high-molecular-weight DNA that can be cloned into vectors is one of the most important stages in the effective building of libraries. Typically, these genomes have been divided up and cloned into culturable organisms to produce "metagenomic libraries." These databases are then analyzed using a sequence-driven method based on DNA sequences or functions that the ambient DNA offers to the host (Delmont *et al.*, 2011). Different methods for soil DNA extraction, generation of the metagenomic library, and sequence-driven, function-driven analysis of the metagenomic library are the main experimental approaches that have been involved in metagenomics. Metagenomics exhibits a very important role in the evaluation and analysis of genetic variations as well as in paleogenomics. Advances in metagenomic approaches, along with bioinformatic tools, have made it easier to tackle the problems of comparing the diversity of different environments, evaluating shifts in diversity, and pooling sequence data to figure out a specific community member's entire genome. Due to the intricacy and expense of the technologies required for metagenomics, multidisciplinary collaboration of specialists is required.

22.5 NEED FOR AGROBIODIVERSITY TO MEET GOALS OF SUSTAINABILITY

Humans rely on several other species and are a part of Earth's biodiversity. Thus, there is a tendency to reframe biological diversity in a way that views genetic variability, species diversity, and ecosystem diversity as a form of insurance against the climate crisis and other planetary changes, going beyond utilitarian arguments that support the preservation of diversity (Barbault, 1998; Mario *et al.*, 2023). Maintaining environmental quality (e.g., by reducing pesticide usage) and increasing agricultural output by 70% are prerequisites for achieving sustainable food production, even in harsher weather regions. The creation of sustainable alternatives that make use of the natural resources already in place to attain food security in the face of climate change and constrained land resources has been promoted by numerous state and international organizations (Hamim *et al.*, 2018). Microbial inoculants have thus become a useful tool in sustainable agricultural practices within the rapidly evolving subject of soil microbiology. By improving the accessibility of plant nutrients and lowering biotic stress (diseases and pests) and abiotic stress (heat, cold), microbial inoculants can increase crop production (Wang *et al.*, 2022). Drought, soil deterioration, and agricultural area invasion, owing to urban growth, are some of the severe repercussions of global warming and worldwide population growth. These concerns have a direct impact on modern agriculture and restrict agricultural productivity, as underlined by SDG15. The role of sustainable agrobiodiversity in attaining sustainable development goals is given in Table 22.1. Some advances in sustainable agricultural routines, such as vertical cultivation, hydroponics, and aquaponics, have been developed to minimize the use of land, improve irrigation, and maximize nutrient recirculation (Chalabi *et al.*, 2015).

Agricultural areas are becoming less resilient due to the increasing prevalence of modern, commercial farming and the high input–high output carbon energy paradigm. Four of these SDGs are related to the production of food and its security: reducing biodiversity (SDG 15); reduction of ecological services and agricultural ecosystem stability due to rising stress from intensifying food production and changes in the climate (SDG 13); decreasing soil quality brought on by farming practices (SDGs 2 and 6); and reliance on artificial pesticides and fertilizers for sustaining high yield (SDG 2) (Shahmohamadloo *et al.*, 2022). Food production will need to rise by 25%–75% from 2014 levels to fulfill requirements since the worldwide population is predicted to reach approximately 10 billion people by 2050 (Hunter *et al.*, 2017). Meeting the requirements of billions of people comes at the expense of the environment. To eradicate weeds and assimilate agricultural residues, several fields are planted with a similar crop year after year, exposed to regular and intense mechanical perturbation of the soil, which ultimately degrades the land. Similarly, an excessive amount of fertilizer is frequently sprayed to maximize production, but a portion of the chemical either escapes into the surrounding environment and contributes to climate alterations or these chemicals escape via runoff from the surface, resulting in hazardous algal blooms and ecological dead zones. Another issue is the widespread cultivation of crops resistant to herbicides accompanied by extensive herbicide use. Herbicide-resistant "superweeds" have evolved as a result, and herbicides have drifted onto nearby farms, posing new difficulties for certified organic farming along with other farms growing crops that are not susceptible to herbicides. The hazards to human health and the environment posed by repeated exposure to pesticides have also been brought up by intensive usage of pesticides (Shelton *et al.*, 2014; Hayes *et al.*, 2011). These problems highlight

TABLE 22.1
Roles of Sustainable Agrobiodiversity in Attaining Sustainable Development Goals

Goal Number	Sustainable Development Goal	Role of Sustainable Agrobiodiversity in Attaining SDG Targets
1	No poverty	• Increased revenue from sustainable agriculture will assist in reducing poverty in rural areas. • Establish local employment. • Aid poor and vulnerable people in developing resilient means of subsistence. • Give both men and women the opportunity to have equal access to economic and natural resources.
2	Zero hunger	• Throughout the year, sufficient, healthy, and nutritious food is produced through sustainable agriculture. • Biotechnological tools and molecular approaches help to boost crop production.
3	Good health and well-being	• The nutrition needs of pregnant and lactating women, elderly individuals, and teenage girls will be met by sustainable agriculture. • The nutritive and therapeutic properties of GM products also help to achieve this goal. • Agricultural products are used in both conventional and alternative medicine to treat cancer and other chronic diseases with potent bioactive components.
4	Quality education	• Vocational training and conventional knowledge can improve employment and entrepreneurship opportunities for men, women, and indigenous people. • Farmers can obtain the knowledge, resources, equipment, and abilities they require to succeed through agricultural extension.
5	Gender equalities	• Different agricultural opportunities can increase women's participation in economic and social decision-making. • Having equal access to assets for men and women can increase agricultural productivity.
6	Clean water and sanitation	• Water availability and sanitation for all are guaranteed by sustainable agricultural water management.
7	Affordable and clean energy	• Reducing reliance on nonrenewable energy sources can be achieved through sustainable agriculture.
8	Decent work and economic growth	• Agriculture is a key driver of employment and economic prosperity. • Local social enterprise establishment. • Boost agricultural production and GDP. • Provides both men and women living in rural areas with the opportunity to diversify their sources of income.
9	Industrial innovation and infrastructure	• Resilient infrastructure, sustainable industrialization, and innovation are all facilitated by sustainable agriculture.
10	Reduced inequalities	• An equal emphasis on economic, social, and environmental problems is provided by sustainable agriculture, which contributes to reducing inequality. • Initiatives in support of women and remote areas. • Sustainable income growth for the poor can be bolstered by the improved livelihoods made possible by agricultural farming and its contribution to GDP.
11	Sustainable communities	• Planning for human settlement sustainability and long-term sustainability are enhanced by sustainable agriculture.
12	Responsible consumption and production	• Water, soil, and other resource exploitation are minimized through sustainable agriculture.

(Continued)

TABLE 22.1 (CONTINUED)
Roles of Sustainable Agrobiodiversity in Attaining Sustainable Development Goals

Goal Number	Sustainable Development Goal	Role of Sustainable Agrobiodiversity in Attaining SDG Targets
13	Climatic actions	• Climate-smart agriculture serves as a roadmap for changing agri-food systems to adopt environmentally conscious and climate-resilient behaviors.
14	Life below water	• Sustainable agriculture controls water resources and avoids contaminating water supplies.
15	Life on land	• Preserving pristine climates and biodiversity. • Eco-citizenship will result from sustainable agriculture. • It can support reforestation, leading to poverty reduction and sustainable regional development.
16	Peace, justice and strong institutions	• Sustainable agriculture empowers people to peacefully protect their rights to their land and resources and to promote the environment without fear of retaliation or intimidation.
17	Partnerships for the goals	• Memorandums of understandings signed with associations and activist groups everywhere. • Goals for sustainable agriculture can be met by working together with governmental agencies, non-governmental organizations, and commercial banks.

the pressing need for strengthening agriculture's sustainability. Agroecological farming systems, characterized by natural farming methods, can preserve productivity while yielding long-term environmental advantages (Davis *et al.*, 2012).

Sustainable development goals can be accomplished by three main approaches, which include ecological intensification, sustainable intensification, and regenerative agriculture (Mockshell & Kamanda, 2018). The following actions can be taken to address the issues with sustainable agriculture, establishing suitable guidelines and rewards, increasing funding for innovation, offering financial support to encourage environmentally friendly farming, and novel systems of government. Rainfed areas may be the first to expand because they presently engage in low-productivity, low-resource farming and stand to benefit the most from the shift.

22.6 INTRIGUING PROSPECTS OF NANOTECHNOLOGY IN SUSTAINABLE AGRICULTURE

Retaining soil fertility and raising crop yields as well as quality are two important goals of nutrient fertilization. In the agricultural industry, nanofertilizers play a major role in increasing productivity and resilience to abiotic challenges. Minimized leaching and volatilization related to the use of conventional fertilizers are the primary economic advantages of nanofertilizers. The well-established beneficial effects on productivity and product quality have a huge potential to boost cultivators' profit margins at the same time when this technology is used (Zulfiqar *et al.*, 2019). There are many advantages to using nanofertilizers for ecologically friendly and sustainable crop development as given in Table 22.2. By minimizing nutrient loss and enhancing the greater absorption rate, nanofertilizers can be used alone or in combination with organic materials to effectively increase agricultural plant nutrition while lowering environmental pollution. A variety of studies using various nanomaterials were noted to improve the antioxidant and chlorophyll contents as well as the development of roots, plant height, and germination rate. For sustainable crop production, smart nanofertilizers distribute nutrients according to plant needs (Thavaseelan *et al.*, 2021).

TABLE 22.2
Advantages of Nanofertilizers over Conventional Fertilizers

Conventional Fertilizers	Nanofertilizers
Due to leaching, drifting, and run-off, loss rate is high	Loss of fertilizer nutrients is low
Soil imbalance and increased toxicity due to excessive nutrient release	Controlled release pattern
Low solubility	High solubility
Low bioavailability	Higher bioavailability
Lower nutrient uptake efficiency	Enhanced nutrient uptake efficiency

TABLE 22.3
Precision Farming with Agri-Nanotechnologies

Precision farming with agri-nanotechnology	Evaluating the health of the soil	Nanosensors
	Limiting the dispersal of agricultural chemicals	Nanoencapsulation for prolonged release
	Management of diseases	Nanoherbicides, nanopesticides
	Controlling the growth of the plant	Nanobiofertilizers
	Water-holding capacity	Nanoclay
	Improvement of quality	Genetic material to be released with nano assistance for improved quality

Incorporating nanotechnology into agriculture can improve agricultural productivity while providing an effective means of sensing, recognition, and bioremediation, as given in Table 22.3. The application of ceramic tools, filters, laminating techniques, and nanomediated agrochemicals has the potential to revolutionize traditional agricultural processes and improve agricultural productivity and efficiency. At ideal concentrations, nanofertilizers can easily penetrate seeds and increase the accessibility and availability of nutrients for the developing seedling, which will improve the length of the shoots and roots. When plants use nanofertilizers, their branches and leaves can more easily access nutrients, which leads to better vegetative growth, quality, and yield (Pandey, 2018).

22.7 DISCUSSION

Sustainability is a systemic issue that refers to eternal capacities. A system is a collection of several parts, components, subsystems, or products that come together to form a single whole that aids in maintaining the system's integrity and organization. The purpose of sustainability is the constancy of an activity or system. Sustainable intensification via increased technology, scientific understanding, and management might indicate the future of the agricultural sector. This entails boosting agricultural output without raising inputs and minimizing externalities (such as greenhouse gas emissions). Traditional agricultural methods, agricultural self-determination, and inclusive peer-learning networks have all been proven to contribute significantly to sustainable agriculture. Achieving agricultural sustainability requires finding an equilibrium between social responsibility, financial viability, and environmental stewardship. Environmental quality is one of the main reasons sustainability matters. To have healthy networks, we must have clean air, distinctive assets, and a non-toxic environment. For example, the quality of the air and water is legitimately linked to an array of medical issues (Patel *et al.*, 2023). Developing tailored governance and oversight mechanisms to improve cultivators' access to agrobiodiversity is a critical problem for advancing the biodiversity-based agricultural sector. Pluralistic and collaborative governance strategies are essential to enhance farmers' access to agrobiodiversity (Labeyrie *et al.*, 2021). An agroecological framework has been established for sustainable cultivation practices that include different levels such as a focus on maximizing productivity, less damaging inputs, integration of agroecological activities, reinforcement of producers and consumers' interactions, and development and integration of agrodiversity. Many farmers use non-chemical pest-control techniques, compost, and cover crops in place of chemical fertilizers or less frequent field tilling. Most often, several hazardous inputs are replaced in organic agricultural systems. Policies and incentives have been implemented to reinforce the interconnectedness of society and agriculture. Farm resource management in sustainable agriculture necessitates a comprehensive systems approach. In the age of agricultural industrialization, a component approach that concentrated on certain farming techniques, businesses, and practices would have made it acceptable. Knowledge of ecological services could be the foundation for learning. Farmers that employ sustainable farming methods will save scarce resources, use less non-renewable energy, and apply less chemical input. The combination of enhanced plant breeding, agronomic management, fertilizer technology, and farm establishment has led to a significant increase in crop productivity, which is the cause of the 21st-century agricultural revolution. However, research has shown that recent years have seen greater yield variability and yield stagnation, suggesting that climate variations, abiotic and biological stress, ecological losses, and soil degradation may have an impact on agricultural productivity. The inability to produce the same number of harvests at a single time is one of the main drawbacks of sustainable agriculture, as an additional manageable approach typically results in lower yields in a single plot to prevent plants from drawing nutrients out of the soil. For agriculture to be really sustainable, social, economic, and environmental factors must be carefully balanced. More sustainable agriculture systems could be advanced by communities simply by devising policies that balance economic, ecological, and social objectives. Due to their small dimension, vast surface area, shape, and catalytic reactivity, nanomaterials can improve the release profiles, association, and effective uptake of nutrients for agricultural fertilization. This also increases potential benefits to the environment and the economy (Thavaseelan *et al.*, 2021). The increased growth, germination, nitrogen metabolism, carbohydrate and protein synthesis, and other characteristics of crops attributable to nanofertilizers have improved crop production. In addition to being environmentally safe, both nanofertilizers and nanopesticides allow for a steady and continuous release. To meet the particular requirements of the crop, they might be tweaked. As they can withstand more stress, nanofertilizers are very beneficial in the age of climate change (Arora *et al.*, 2022).

22.8 CONCLUSION

The idea of sustainable agriculture is crucial for maintaining both our ecosystem and the world's expanding human population. Sustainable agriculture delivers an opportunity for future generations to produce and consume healthy food while protecting the environment. The social, environmental, and economic aspects of sustainability, as well as the four fundamental elements of food security—availability, utilization, access, and stability—are all impacted by sustainable food and agriculture, or SFA. Numerous novel discoveries have been made as a result of research into varied microbiomes, particularly genes that encode proteins that could be crucial to preserving the sustainability of agriculture. Achieving sustainable agriculture necessitates the correct balance between the need to produce food and preserving the natural ecosystem.

REFERENCES

Al-Chalabi M. Vertical farming: Skyscraper sustainability. *Sustainable Cities and Society*. 2015 Nov 1;18:74–7.

Arora S, Murmu G, Mukherjee K, Saha S, Maity D. A comprehensive overview of nanotechnology in sustainable agriculture. *Journal of Biotechnology*. 2022 Aug 20;355:21–41.

Barbault R. Mankind and biodiversity: lessons for sustainable development - a viewpoint. Intern. *The Journal of Environmental Studies.* 1998;55:259–270.

Belokurova VB. Methods of biotechnology in system of efforts aimed at plant biodiversity preservation. *Cytology and Genetics.* 2010 Jun;44(3):174–85.

Bravo-Peña F, Yoder L. Agrobiodiversity and smallholder resilience: a scoping review. *Journal of Environmental Management.* 2024;351:119882.

Costanzo A, Bàrberi P. Functional agrobiodiversity and agroecosystem services in sustainable wheat production: a review. *Agronomy for Sustainable Development.* 2014 Apr;34:327–48.

Das S, Ray MK, Panday D, Mishra PK. Role of biotechnology in creating sustainable agriculture. *PLOS Sustainability and Transformation.* 2023 Jul 13;2(7):e0000069.

Davis L. *Basic Methods in Molecular Biology.* Elsevier; 2012 Dec 2.

Delmont TO, Robe P, Cecillon S, Clark IM, Constancias F, Simonet P, Hirsch PR, Vogel TM. Accessing the soil metagenome for studies of microbial diversity. *Applied and Environmental Microbiology.* 2011;77(4):1315–1324

Dill GM. Glyphosate-resistant crops: history, status and future. *Pest Management Science: Formerly Pesticide Science.* 2005 Mar;61(3):219–24.

Frison EA, Cherfas J, Hodgkin T. Agricultural biodiversity is essential for a sustainable improvement in food and nutrition security. *Sustainability.* 2011 Jan;3(1):238–53.

Giller KE, Delaune T, Silva JV, Descheemaeker K, van de Ven G, Schut AG, ... van Ittersum MK. The future of farming: Who will produce our food?. *Food Security.* 2021;13(5):1073–1099.

Hamim I, Borth WB, Marquez J, Green JC, Melzer MJ, Hu JS. Transgene-mediated resistance to Papaya ringspot virus: challenges and solutions. *Phytoparasitica.* 2018 Feb;46:1–8.

Hayes SC, Villatte M, Levin M, Hildebrandt M. Open, aware, and active: Contextual approaches as an emerging trend in the behavioral and cognitive therapies. *Annual Review of Clinical Psychology.* 2011 Apr 27;7:141–68.

Hunter MC, Smith RG, Schipanski ME, Atwood LW, Mortensen DA. Agriculture in 2050: recalibrating targets for sustainable intensification. *Bioscience.* 2017 Apr 1;67(4):386–91.

Ince AG, Karaca M. Systems and Techniques for Sustainable Agriculture. 2022. https://orcid.org/0000-0003-3219-9109

Kumar S. Sustainable agriculture development in India: emerging issues, challenges and opportunities. *Asian Journal of Advances in Agricultural Research.* 2023 Oct 20;23(3):64–74.

Labeyrie V, Antona M, Baudry J, Bazile D, Bodin Ö, Caillon S, Leclerc C, Le Page C, Louafi S, Mariel J, Massol F. Networking agrobiodiversity management to foster biodiversity-based agriculture. A review. *Agronomy for Sustainable Development.* 2021 Feb;41:1–5.

Li J, Wang J, Liu H, Macdonald CA, Singh BK. Application of microbial inoculants significantly enhances crop productivity: a meta-analysis of studies from 2010 to 2020. *Journal of Sustainable Agriculture and Environment.* 2022 Sep;1(3):216–25.

Malini R, Ebanisha J, Lakshmi MA. Sustainable agriculture is a stimulator to stretch out sustainable development goals. *London Journal of Social Sciences.* 2023 Sep 17(6):96–107.

Mario XR-G, Vicente O. Agrobiodiversity: conservation, threats, challenges, and strategies for the 21st century. *AgroLife Scientific Journal*, 2023:12(1):174–185. https://doi.org/10.17930/AGL2023120

Mockshell J, Kamanda J. Beyond the agroecological and sustainable agricultural intensification debate: Is blended sustainability the way forward? *International Journal of Agricultural Sustainability*, 2018 Mar 4;16(2):127–49. https://doi.org/10.1080/14735903.2018.1448047

Mogni F, Tresoldi C, Senesi S, Palau H, Vilella F. Sustainable development in food and agribusiness: application of the theoretical model to the Argentine beekeeping sector. In *VII International PENSA Conference.* Brazil; Nov 2009 (pp. 1–16).

Mondal K, Yadav S, Raigar OP. Molecular Approaches in Agrobiodiversity Conservation. In *Agrobiodiversity and Agri-Ecosystem Management,* 2022 Jul 16 (pp. 71--84). Singapore: Springer Nature Singapore.

Pandey G. Challenges and future prospects of agri-nanotechnology for sustainable agriculture in India. *Environmental Technology & Innovation.* 2018 Aug 1;11:299–307.

Parray JA, Mir MY, Shameem N. *Sustainable Agriculture: Biotechniques in Plant Biology.* Singapore: Springer; 2019 Nov 6.

Patel S. Sustainability of sustainable agriculture. *Journal of Emerging Technologies and Innovative Research.* 2023 Aug www.jetir.org(ISSN-2349-5162).

Robe P, Nalin R, Capellano C, Vogel TM, Simonet P. Extraction of DNA from soil. *The European Journal of Soil Biology.* 2003;39(4):183–190

Schloss PD, Handelsman J. Biotechnological prospects from metagenomics. *Current Opinion in Biotechnology.* 2003;14(3):303–310

Shahmohamadloo RS, Febria CM, Fraser ED, Sibley PK. The sustainable agriculture imperative: A perspective on the need for an agrosystem approach to meet the United Nations Sustainable Development Goals by 2030. *Integrated Environmental Assessment and Management.* 2022 Sep;18(5):1199–205.

Sekhar M, Rastogi M, CM R, Saikanth DRK, Rout S, Kumar S, Patel AK. Exploring traditional agricultural techniques integrated with modern farming for a sustainable future: a review. *Journal of Scientific Research and Reports.* 2024;30(3):185–98.

Shelton JF, Geraghty EM, Tancredi DJ, Delwiche LD, Schmidt RJ, Ritz B, Hansen RL, Hertz-Picciotto I. Neurodevelopmental disorders and prenatal residential proximity to agricultural pesticides: the CHARGE study. *Environmental Health Perspectives.* 2014 Oct;122(10):1103–9.

Sobti R C, Thakur M, Kaur T, Mishra S. Biodiversity: Threats and conservation strategies. In *Biodiversity.* Boca Raton: CRC Press. 2023. https://doi.org/10.1201/9781003220398.

Thavaseelan D, Priyadarshana G. Nanofertilizer use for sustainable agriculture. The *Journal of Research Technology and Engineering.* 2021;2(1):41–59.

Tyson GW, Chapman J, Hugenholtz P, Allen EE, Ram RJ, Richardson PM, Solovyev VV, Rubin EM, Rokhsar DS, Banfield JF. Community structure and metabolism through reconstruction of microbial genomes from the environment. *Nature.* 2004;428(6978):37–43

Velten S, Leventon J, Jager N, Newig J. What is sustainable agriculture? A systematic review. *Sustainability.* 2015 Jun 18;7(6):7833–65.

Zimmerer KS, De Haan S. Agrobiodiversity and a sustainable food future. *Nature Plants.* 2017 Apr 6;3(4):1–3.

Zulfiqar F, Navarro M, Ashraf M, Akram NA, Munné-Bosch S. Nanofertilizer use for sustainable agriculture: advantages and limitations. *Plant Science.* 2019 Dec 1;289:110270.

23 Green Medicine
Unraveling the Healing Potential of Medicinal Plants and Bioactive Compounds

Ambika Choudhary[1], Aarzoo Yadav[2], Manish Thakur[3]
[1]School of Biological and Environmental Sciences, Shoolini University of Biotechnology and Management Sciences, Solan, India
[2]Fish Molecular Biology Laboratory, Department of Zoology, University of Delhi, India
[3]Department of Microbiology, School of Bio Engineering and Biosciences, Lovely Professional University, Phagwara, Punjab, India

23.1 INTRODUCTION

For a variety of reasons, more individuals are using herbal medicines throughout the world. Plants and extracts from plants are relied upon by the majority of the people in nations located in Asia and Africa for the treatment of many kinds of ailments (Thakur et al., 2024). Ancient medical knowledge has been passed down through the years and incorporated into our society since the beginning of recorded history. The herbal and natural sciences provided almost all of the early inventions in the industry of drug discovery (Boy et al., 2018). Tribal and rural populations employ ethnomedicines to cure a variety of ailments, according to a World Health Organization (WHO) report. In 2008, the global market for ethnopharmaceuticals was valued at $83 billion, with 25% of newly developed medications derived from plant species (Puri et al., 2019). For basic medical treatment, rural populations throughout the world rely on their traditional local knowledge of medicinal plants. The study of the relationship between individual pre-colonial cultures and plant habitats is known as ethnobotany. The knowledge of traditional medicinal plants has been passed down from generation to generation among rural people (Taek et al., 2019). Finding contemporary medications made from naturally occurring medicinal plant resources requires conducting ethnobotanical research (Idolo et al., 2010). The use of plant species as traditional medicines offers a good substitute for existing healthcare systems for rural communities in developing countries (Mahmood et al., 2013). The dried herb is rich in proteins, carbohydrates, fats, fibers, calcium, phosphorus, and phenolic compounds, including flavonoids and phenol carboxylic acid (Akbar, 2020).Some of the applications of the medicinal plants are mentioned in Figure 23.1. According to reports, traditional medicines provide primary healthcare for 80% of the population in developing countries. It is commonly reported that these medicinal plants are inexpensive, safe, and readily accessible in the surrounding area (Rahman et al., 2004). There are about 7500 plant species in India that are known to be used medicinally in both traditional and contemporary medical systems (Gowthami et al., 2021). India has used plants for human sustenance and medicine since the Vedic era. The first writings on the medicinal use of plants were found in the Rig Veda and Athar Veda (Wani et al., 2016). The population of India is estimated to be 75% rural. To meet their food needs during times of food scarcity,

DOI: 10.1201/9781032697550-23

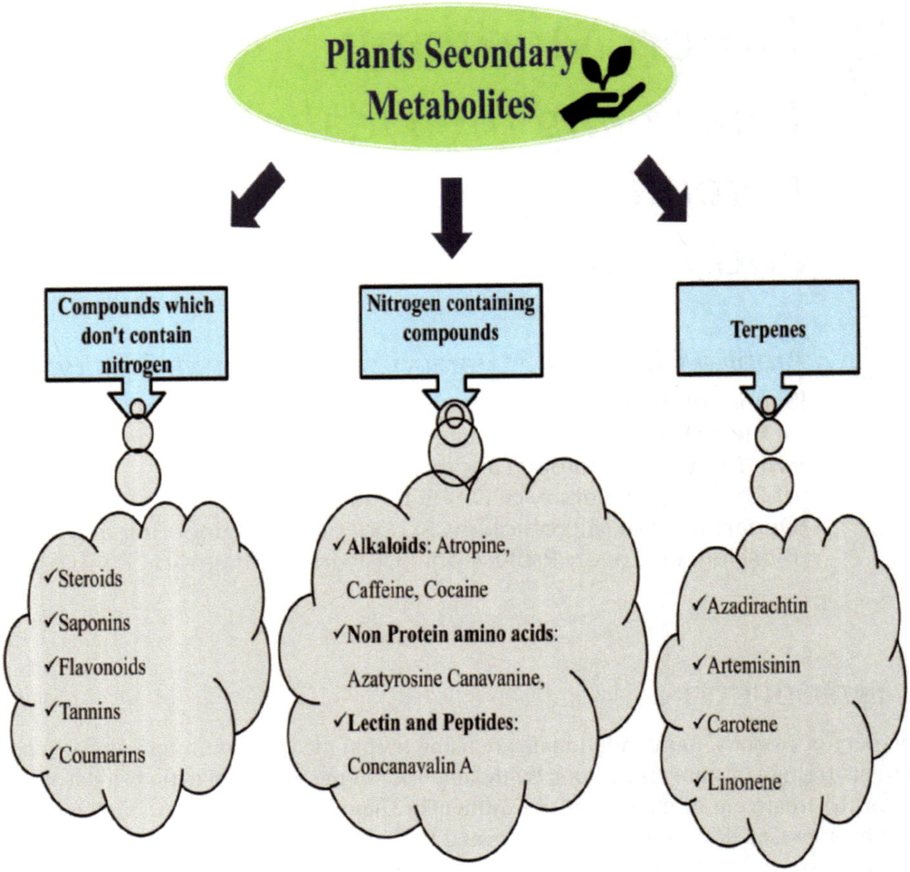

FIGURE 23.1 Applications of medicinal plants.

the majority of rural communities rely on natural resources like edible wild plants (Devendra et al., 2002).

Rural communities consume around 800 species of food plants (Kumar et al., 2021). The Cucurbitaceae family is employed in the treatment of various ailments, including epilepsy, gonorrhea, malaria, measles, smallpox, arthritis, hypertension, kidney disorders, arteriosclerosis, skin conditions, herpes, influenza, diabetes, freckle removal, and as an aphrodisiac (Avinash and Rai, 2017). Throughout history, wild plants have played a crucial role in human life, serving purposes such as food, medicine, fiber, and fodder for livestock (Bharucha and Pretty, 2010). The significance of wild edible food plants extends to keeping agricultural productivity and population growth in balance, especially within emerging nations (Vishwakarma et al., 2011). In various parts of India, an estimated 54 million tribal people inhabit different regions, relying on forests and their products to sustain their daily lives. The survival of many tribal communities hinges on traditional local medicines (Vidyarthi et al., 2013). In remote regions of India, certain medicinal plants serve as the sole healthcare resource due to the absence of medical facilities (Radha et al., 2021). Globally, the outdated uses of medicinal plants fluctuate across individuals and geographical areas (Bhat et al., 2013). The decline of old evidence regarding remedial plants among native communities in current eras suggests a potential danger of destruction. Economic development, market exposure, and infrastructure modernization in India have led to a transformation in the traditional lifestyles of indigenous groups, making their traditional knowledge more vulnerable (Agrawal et al., 1997).

Natural products or their derivatives are used to make a large number of significant medications. Approximately 39.1% of all medications approved by the Food and Drug Administration come from natural sources, and 48.6% of all cancer medications registered between the 1940s and the present day are either natural products or their derivatives. Natural products are valuable resources when looking for new drugs. The significance of natural products in the development of novel drugs is demonstrated by the existence of over 200,000 natural metabolites with a range of bioactive characteristics (Svahn, 2015; Yuca and Karakaya, 2022). There is a long history of using plants to treat a variety of diseases. Plants are one of the major sources of new, pharmacologically active compounds that hit the pharmaceutical industry in drug making. Approximately 35,000–70,000 plant species have undergone testing for potential medical applications. Historically, early drug discovery related to the original ethnopharmacological uses of plants has primarily come from these sources. Today's search for plant-based pharmaceuticals is based on the bioactivity of those plants, which helped isolate significant anticancer medications like camptothecin and paclitaxel (Veeresham, 2012).

23.2 MEDICINAL PLANTS AS SOURCE OF NATURAL PRODUCTS

Medicinal plants have been a fundamental source of natural products for centuries, offering a diverse array of bioactive compounds with potential therapeutic applications (Shoaib et al., 2023). Antioxidants, which are essential to plants' survival in the environment, are generally abundant in plants (Singh and Kumar, 2017; Srivastava and Singh, 2021). These antioxidants are frequently linked to lower health risks that precede diseases like diabetes. Natural products have been widely utilized in the cosmetic industry to avoid the side effects of conventional formulations. Herbal beauty ingredients like *Emblica officinalis* (Amla), *Acacia concinna* (Shikakai), and *Callicarpa macrophylla* (Priyangu) have been extensively used in skin and hair care (Desam and Rajab, 2021). Because plants are generally inexpensive to prepare, effective, and have few side effects when used to treat common ailments, plant-based traditional medicines are still widely used today.

As a result, plant extracts are now a priceless resource for finding potential compounds for new medication development (Boy et al., 2018). Within the villages and surrounding areas, a wide variety of plant species is used in healing practices. Many rural areas continue to be major users of traditional medicine. Many individuals commonly use a decoction of P. guajava leaves due to its antispasmodic and antimicrobial properties to treat dysentery and diarrhea (Anand et al., 2016). Because of the various records of extensive traditional ethnomedical knowledge, scientists were provided with new information (Liu et al., 2016). Piper species are potent medicinal plants commonly used in traditional medicine. They have long been employed to treat ailments such as stomach aches, rheumatoid arthritis, diarrhea, and various infections, with their effectiveness attributed to their bioactive compounds. Studies on bioactive compounds and extracts from Piper species have revealed their clinical significance in treating both malignant and non-malignant diseases (Mgbeahuruike et al., 2017).

23.3 THERAPEUTIC POTENTIAL OF MEDICINAL PLANTS ACROSS CULTURES

Given the abundant biodiversity, it is possible that discovering the latent capacities of certain plants might lead to the discovery of innovative techniques for treating a variety of diseases. A good example of this would be *Brucea mollis* Wall. The prevalence of fatal illnesses like diabetes and arthritis, along with the negative effects of synthetic medications, led to a change in focus from allopathy to complementary and alternative medicine. Plant-derived natural products are an appealing source of biologically active molecules since they are natural and more reasonably priced than chemical manufacturing (Ghosh et al., 2008). According to Boy et al. (2018), the bitter shrub known as *Brucea mollis* Wall. Ex Kurz may be found growing in mountainous regions of China, India, the Philippines, Nepal, Thailand, Malaysia, Laos, Cambodia, Myanmar, and Bhutan. Additionally, it can be found growing in thickets and bushes that are next to roadways. Patients suffering from cancer and malaria in China and India make use of it as a treatment. Yandanziolide and bruceine are examples of phytochemicals that have been found in plants. These phytochemicals have been shown to contain antimalarial and anticancer effects, which lends credence to their use. According to Bharati et al. (2012), these compounds have a number of other qualities in addition to their cytotoxic properties. These features include insecticidal, antigiardial, cytotoxic, antiviral, antileukemic, pesticidal, and antifeedant capabilities. For instance, the lignin and anthricin in *Anthriscus sylvestris* roots exhibited insecticidal properties (Kozawa et al., 1982). A broad variety of herbs, including ashwagandha, giloy, ginger, cinnamon, tulsi, black pepper, black cumin, amla, turmeric, garlic, and flax seeds, have been used for a very long time as natural medicines for a wide variety of diseases. This practice dates back to ancient times. In a broad variety of culinary and medicinal applications, these plants have been used extensively across a variety of cultural contexts. On the other hand, their culinary use is highly widespread in many nations, and they constitute a fundamental component of the kitchens of every family. As a further point of interest, Chyawanprash, Triphala, and Rooh Afza are examples of traditional formulations that are used extensively in India as daily nutritional supplements. Every Indian, regardless of their religious beliefs, the group they belong to, or their socioeconomic standing, makes use of these plants and formulas on a regular basis and incorporates them into their daily routine. It has been shown via scientific study that the herbs and formulations mentioned above contain potent immunomodulatory, antioxidant, and anti-infective effects (Ahmad et al., 2021). The leaves of *Vitex negundo* have been reported to exhibit antigenotoxic, antihistamine, CNS depressant, and anti-fertility activities (Gautam et al., 2008).

23.4 SECONDARY METABOLITES

Secondary metabolites are the organic compounds that are not essential to an organism's regular growth and development (Figure 23.2). In the absence of secondary metabolites, an organism's ability to survive is impaired long term rather than immediately killed, and this is frequently crucial for plant defense. Secondary metabolites are classified into three main groups: terpenes (such as plant volatiles, cardiac glycosides, carotenoids, and sterols), phenolics (such as phenolic acids, coumarins, lignans, stilbenes, flavonoids, tannins, and lignins), and nitrogen-containing compounds (such as alkaloids and glucosinolates). The secondary metabolites that are being developed for medical purposes are the source of the natural compounds that are being developed. According to Rahman et al. (2021) and Thakur and Kaur, (2022, 2023) these chemicals, which are classified as phenolics and consist of quinones, polyphenols, tannins, and flavonoids, are well-known for the powerful cytotoxic, antioxidative, and antibacterial activities that they possess (Saleem et al., 2016). A reevaluation of the use of antibiotics in therapy was motivated by the rise in the prevalence of multidrug-resistant microbes and the advent of infectious illnesses. This reevaluation advocated for the use of creative ways such as the utilization of secondary metabolites as resistance agents (Kumar et al., 2020). Nevertheless, the scientific community continues to face difficulties in the process of exploiting plants that are not being used to their full potential for this purpose (Bharati et al., 2012). According to Blowman et al. (2018) and Verma et al. (2015), essential oils that are loaded with a wide variety of secondary metabolites exhibit a significant amount of bioactivity throughout a wide range. These active ingredients work in a variety of ways, and they play an essential role in the existence of plants. They function as chemical defenses, affording protection against illnesses and predators (Abad et al., 2012). In addition, they operate as attractants for pollinators, which are beneficial to the economy. Plants make use of a wide variety of secondary metabolites in order to deal with the constant battle against naturally occurring difficulties such as diseases, pests, and ecological constraints (Divekar et al., 2022). Certain compounds that fall within this category include alkaloids, glucosinolates, cyanogenic glycosides, flavonoids, saponins, tannins, and terpenoids. Notably, alkaloids, steroids, tannins, terpenoids, flavonoids, saponins, cyanogenic glycosides, and glucosinolates are the secondary metabolites that are found in the kingdom of plants the most often (Mahawer et al., 2023).

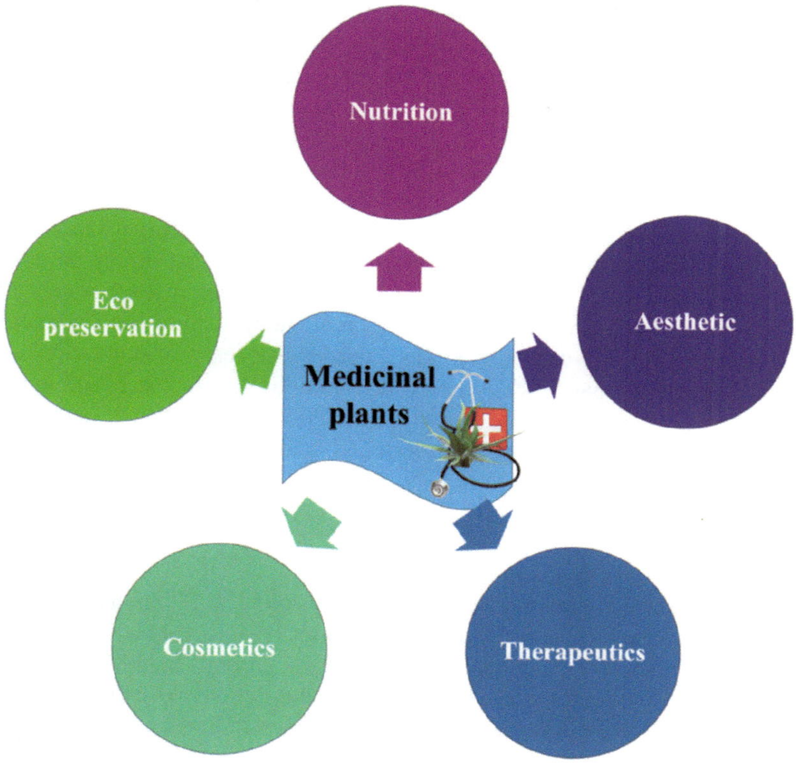

FIGURE 23.2 Plants secondary metabolites: as a natural source of drugs.

Both traditional and cutting-edge techniques can be used to extract plant bioactive metabolites (Gupta et al., 2012). Conventional methods encompass techniques such as decoction, distillation, Soxhlet extraction, steam distillation, pressurized liquid extraction, percolation, surfactant-mediated extraction, countercurrent extraction, maceration, and hydrodistillation (Gupta et al., 2012; Giacometti et al., 2018). On the other hand, modern methods involve microwave-assisted extraction (MAE), sonography-assisted extraction, warm-supported extraction, and supercritical liquid extraction (Gupta et al., 2012; Njila et al., 2017).

23.4.1 Alkaloids

Alkaloids, a class of naturally occurring compounds primarily derived from plants, assume a crucial role as integral components of the plant's defense system against both wildlife and microorganisms, as highlighted by research findings (Zaynab et al., 2018). Specifically, among alkaloids, a distinct member of the beta-carboline group has garnered attention due to its multifaceted properties. This particular alkaloid has demonstrated noteworthy capabilities, showcasing its effectiveness across various domains. First, it exhibits potent antiparasitic attributes, indicating its ability to combat parasitic infections within biological systems. This feature underscores the alkaloid's potential as a natural defense mechanism against parasitic organisms, aligning with the plant's adaptive strategies for survival (Patel et al., 2012). Second, the alkaloid displays remarkable antimicrobial properties, suggesting its efficacy in inhibiting the growth or activity of microorganisms. This antimicrobial activity positions the alkaloid as a promising candidate for potential applications in addressing microbial infections, further underscoring its significance in the plant's defense arsenal. Moreover, the alkaloid demonstrates anti-HIV characteristics, implying its potential role in combating HIV infections. This finding not only adds to the alkaloid's versatility but also highlights its relevance in the context of potential therapeutic interventions against viral infections (Patel et al., 2012). Intriguingly, the alkaloid's spectrum of activity extends beyond infectious diseases, as it has been found to possess anticancer properties (Tang et al., 2011). This discovery positions the alkaloid as a valuable subject for further exploration in the development of anticancer therapies, potentially contributing to the ongoing quest for novel and effective treatments (Yamazaki et al., 2011).

Furthermore, alkaloid exhibits anti-inflammatory activity, suggesting its potential role in modulating inflammatory responses within biological systems. This anti-inflammatory characteristic aligns with the broader trend of natural compounds with anti-inflammatory properties being investigated for their therapeutic potential in various inflammatory conditions (Yamazaki et al., 2011). Alkaloid, belonging to the beta-carboline group, emerges as a versatile and multifunctional natural compound with notable antiparasitic, antimicrobial, anti-HIV, anticancer, and anti-inflammatory properties. Its diverse range of activities underscores the complexity and adaptability of plant-derived alkaloids, opening avenues for future research and potential applications in medicine and biotechnology.

23.4.2 Steroids

Steroids, a class of organic compounds with diverse physiological functions, have become integral components in the treatment of various medical conditions, as highlighted by Janahi et al. (2018). These compounds, often associated with anti-inflammatory properties, play a crucial role in managing and alleviating a spectrum of diseases, including but not limited to rheumatoid arthritis, asthma, inflammatory bowel disease, pulmonary disease, and systemic lupus erythematosus. Rheumatoid arthritis, a chronic inflammatory disorder affecting the joints, stands out as one of the conditions effectively treated with steroids. In this context, steroids contribute significantly to reducing inflammation and alleviating symptoms associated with joint inflammation and pain.

The anti-inflammatory qualities of steroids play a pivotal role in modulating the immune response, thereby mitigating the autoimmune processes that contribute to rheumatoid arthritis (Boy et al., 2018). Asthma, a respiratory condition characterized by inflammation and constriction of the airways, is another ailment where steroids have proven efficacy. Steroids help manage asthma by exerting anti-inflammatory effects on the airways, thereby reducing inflammation and preventing bronchoconstriction, which is central to the pathology of asthma. This anti-inflammatory action aids in controlling symptoms and improving respiratory function (Boy et al., 2018). Inflammatory bowel disease, encompassing conditions like Crohn's disease and ulcerative colitis, involves chronic inflammation of the gastrointestinal tract. Steroids have demonstrated effectiveness in this context by attenuating the inflammatory response within the digestive system. By dampening the immune-mediated inflammation, steroids contribute to the control and remission of inflammatory bowel disease symptoms (Boy et al., 2018). Pulmonary diseases, including chronic obstructive pulmonary disease (COPD) and certain interstitial lung diseases, often involve inflammation and structural changes in the lungs. Steroids are employed in the treatment of these conditions to counteract inflammation, leading to improved lung function and symptom management. Their anti-inflammatory qualities are particularly beneficial in addressing the underlying inflammatory processes in the respiratory system (Boy et al., 2018). Systemic lupus erythematosus (SLE), an autoimmune disease affecting multiple organs, is another example where steroids are commonly utilized. Steroids help suppress the immune system's aberrant activity in SLE, reducing inflammation and preventing damage to various organs. This anti-inflammatory action is instrumental in controlling the disease and alleviating symptoms associated with SLE (Boy et al., 2018). Steroids play a crucial role in the treatment of diverse diseases, ranging from rheumatoid arthritis and asthma to inflammatory bowel disease, pulmonary diseases, and systemic lupus erythematosus. Their anti-inflammatory qualities are central to their therapeutic efficacy, making them valuable tools in managing conditions characterized by inflammatory processes and autoimmune responses.

23.4.3 Tannins

Tannins, a class of water-soluble polyphenolic compounds, exhibit noteworthy properties that have garnered attention in various fields of research, as elucidated by Sulaiman et al. (2013). One prominent aspect of their biological significance lies in their antioxidative properties, contributing to their potential health benefits. The antioxidative nature of tannins plays a crucial role in cellular protection against oxidative stress, as suggested by studies conducted by Sulaiman et al. (2013) and Provino (2010). These compounds are present in a wide range of foods and drinks, such as wine, coffee, tea, grapes, blueberries, pomegranates, and strawberries. They are widely dispersed throughout fruit, roots, wood, bark, and leaves (Shinjyo et al., 2020). Actually, 5–10% of the dry weight of plant leaves is made up of tannin (Hwang, 2018). Oxidative stress, characterized by an imbalance between the production of reactive oxygen species (ROS) and the body's ability to counteract their harmful effects, is implicated in various pathological processes, including carcinogenesis and mutagenesis. Tannins, with their antioxidative capabilities, are associated with potential anticarcinogenic and antimutagenic effects, indicating their role in mitigating cellular damage and reducing the risk of certain diseases.

In addition to their antioxidative properties, tannins demonstrate antimicrobial attributes, particularly through the inhibitory effects of tannic acid on foodborne and aquatic bacteria, as highlighted by Kaczmarek et al. (2020). This antimicrobial activity positions tannins as potential natural agents for controlling bacterial growth in various environments, with implications for food safety and water quality. Physiologically, tannins have been linked to several health benefits, as indicated by Walia et al. (2019). One such association is the reduction of arterial hypertension, suggesting a potential role for tannins in managing blood pressure. Furthermore, tannins have been implicated in the lowering of lipid levels in the blood, contributing to their potential cardiovascular benefits. The control of immune responses is another physiological aspect linked to tannins, highlighting their potential immunomodulatory properties. Tannins, characterized by their antioxidative nature, exhibit a range of health-related properties. Their reported anticarcinogenic and antimutagenic effects underscore their potential in preventing cellular damage and reducing the risk of certain diseases. The antimicrobial attributes of tannins, particularly their inhibitory effects on bacteria, have implications for food safety and environmental health. Additionally, the physiological associations with the reduction of arterial hypertension, lowering of lipid levels, and modulation of immune responses suggest the multifaceted health benefits of tannins, making them subjects of interest in both scientific research and potential therapeutic applications.

23.4.4 Terpenoids

Terpenoids, a diverse class of phytochemicals, contribute significantly to the food industry as natural flavoring agents, enhancing the taste and aroma of various consumables. With over 40,000 distinct compounds identified and ongoing discoveries, their abundance and diversity underscore their versatility in both biological activities and applications (Alamgir, 2018). Beyond their role in flavor enhancement, terpenoids are recognized for their antioxidant properties, crucial for neutralizing free radicals and mitigating oxidative stress-related diseases. Additionally, certain terpenoids show promise in the treatment of liver cancer (Thoppil et al., 2011), indicating potential applications in oncology therapeutics.

Studies have highlighted the pharmacological prowess of terpenoids, showcasing their analgesic, antimicrobial, antiviral, anti-inflammatory, antifungal, and antiparasitic activities (Safari et al., 2016; Sülsen et al., 2017). These diverse properties position terpenoids as valuable candidates for addressing various health conditions and infections. Notably, their antioxidant qualities make them essential in protecting cells from oxidative damage. The multifaceted nature of terpenoids extends their relevance from the food industry to potential medicinal applications, marking them as intriguing subjects for ongoing research. As natural compounds found in plants, terpenoids play a vital role in the ongoing exploration of natural products for their therapeutic potential. Their broad range of activities, coupled with their prevalence in both traditional medicine and modern pharmaceutical research, underscores their significance in drug discovery and development. Overall, terpenoids emerge as compounds of interest, not only for their aromatic contributions to the culinary world but also for their potential to address health challenges, making them subjects of exploration in the realms of both food science and medicine.

23.4.5 Flavonoids

Flavonoids, as elucidated by Jain et al. (2013), showcase a spectrum of beneficial properties within the realm of human health. Their inherent antioxidant characteristics make them instrumental in combating oxidative stress, contributing to the prevention of cellular damage caused by free radicals. This capacity holds particular significance in the context of antileukemic activities, as the antioxidative nature of flavonoids aids in mitigating factors associated with leukemia. Moreover, the role of flavonoids extends to enhancing blood circulation, a notable aspect of addressing conditions like Alzheimer's disease. Jain et al. (2013) suggest that the positive impact on blood circulation may have implications for cognitive health, potentially influencing the progression or onset of Alzheimer's disease. This aspect underscores the diverse physiological effects of flavonoids, reaching beyond their conventional antioxidant role. In the broader pharmacological landscape, flavonoids emerge as versatile compounds with free radical scavenging capabilities (Sajid et al., 2021). Emphasizes their value in scavenging free radicals, positioning flavonoids as valuable contributors to various pharmacological activities. This includes their potential in anticancer strategies, where the ability to counteract free radicals aligns with efforts to impede cancerous processes. Furthermore, the antiaging effects associated with flavonoids underscore their multifaceted contributions to human health. In essence, flavonoids represent a class of compounds that go beyond mere antioxidant agents. Their involvement in antileukemic activities, influence on blood circulation in Alzheimer's disease, and broader pharmacological implications, including anticancer and antiaging effects, highlight the extensive and interconnected roles that flavonoids play in maintaining and promoting human health (Sekikawa et al., 2021). As researchers delve deeper into the mechanisms and applications of flavonoids, their potential impact on therapeutic interventions across various health conditions continues to unfold.

23.4.6 Saponins

Saponins, as elucidated by Oleszek et al. (2020), constitute a class of bioactive compounds predominantly found in vegetables, herbs, and beans. The multifaceted nature of saponins manifests in their diverse health-promoting properties, marking them as valuable contributors to overall well-being. A significant aspect of saponins lies in their potential as anticancer agents. Research by Man et al. (2010) has highlighted their role in impeding the proliferation of cancer cells, suggesting a promising avenue for incorporating saponins into cancer prevention and treatment strategies. This anticancer property underscores the importance of exploring natural compounds, such as saponins, as potential allies in the ongoing battle against cancer. Beyond their anticancer potential, saponins exhibit immunomodulatory effects, as emphasized by Man et al. (2010). This implies that these compounds have the capacity to regulate and enhance immune system functions. By doing so, saponins contribute to bolstering the body's defense mechanisms, potentially aiding in the prevention and management of various diseases by strengthening the immune response. Additionally, saponins play a role in cardiovascular health by influencing cholesterol levels. Man et al. (2010) suggest that the consumption of saponin-rich foods may contribute to lowering cholesterol levels, thereby reducing the risk of heart disease. This dual action on both cancer prevention and cardiovascular health positions saponins as versatile bioactive compounds with wide-ranging implications for maintaining a healthy body. In essence, the presence of saponins in plant-based foods extends beyond their culinary significance. Their documented anticancer properties, immunomodulatory effects, and impact on cholesterol levels underscore their potential as functional components in promoting health and preventing diseases. As our understanding of the intricate interplay between dietary components and human health evolves, saponins emerge as noteworthy constituents with promising applications in the realms of cancer research, immunology, and cardiovascular health.

23.4.7 Glucosinolates

Numerous edible species are included in 16 families of dicotyledonous angiosperms that contain glucosinolates (Blažević et al., 2016). One of the main metabolites of the Brassicaceae family, glucosinolates are biosynthesized from amino acids and give *Brassica* plants their flavor and odor. Abiotic stressors, including temperature and heavy metal stress, have an impact on these chemicals' levels (Variyar et al.,2014). Glucosinolates are nitrogen- and sulfur-containing secondary metabolites that are widely known for helping plants maintain auxin homeostasis and develop resistance to infections. Aside from these beneficial qualities, glucosinolates also have certain negative impacts. These include the development of their distinctively bitter taste and antinutritional values, such as goiter in pigs and poultry (Baskar et al., 2012). They are well known for their fungicidal, bactericidal, nematocidal, and allelopathic qualities and are used as food flavorings to enhance appetite and the digestive process (Dhakad et al., 2018). Their cancer chemoprotective qualities have attracted a lot of attention lately (Arumugam et al., 2018). Glucosinolates inhibit the synthesis of carcinogens and cancer-causing agents from producing genetic damage that results in tumor development (Pfeifer and Fahrendorf, 2015).

23.4.8 Thioglucoside-N-Hydroxysulfates

Thioglucoside-N-hydroxysulfate is a precursor of isothiocyanates, found in 16 families of dicotyledonous angiosperms, including a significant number of comestible species (Fahey et al., 2001). Although glucosinolates and/or their breakdown products have long been recognized for their allelopathic, nematocidal, bacteriocidal, and fungicidal qualities, their cancer-chemoprotective qualities have recently sparked a great deal of interest in study (Fahey et al., 2001). However, it's crucial to note that excessive intake and improper dosage of herbal plants can pose serious health risks. While they may have the potential to cure a specific disease, the process could adversely affect other organs. Common secondary metabolites present in herbal plants also carry inherent risks. Plant poisoning may arise from incorrect dosage and improper preparation of alkaloids. Contrarily, steroids have a long list of negative side effects, such as increased appetite, high blood pressure, acne, blurred vision, cataracts, glaucoma, easy bruising, insomnia, and elevated bad cholesterol (Fardet and Fève, 2014). Patients or users may be at risk if they are unaware of the potential risks associated with herbal remedies (Karimi et al., 2015). It is important to take herbal medicine responsibly because the interactions between various phytochemicals can cause unanticipated reactions if the user is not aware of them.

23.5 CONCLUSION

To discover innovative medications derived from a wide variety of prokaryotic and eukaryotic organisms, the investigation of natural products is becoming an increasingly important field of study. There are a great number of plant species that have not yet been investigated for their prospective therapeutic uses, despite recent developments. This holds the possibility of substantial scientific discoveries. Since this is the case, there is a need for more in-depth research to be conducted on less common plant species, especially those with applications in ethnopharmacology and ethnobotany. Herbal plants, which are well-known for their medicinal capabilities, come into play when it concerns the discovery of remedies or the stabilization of various ailments. To conduct in-depth research that will reveal further advantages, uses, and possible toxicity, it is vital to have a basic understanding of their phytochemical makeup. Not only does the utilization of the world's rich herbal plants contribute to the process of industrialization, it also helps to contribute to the employment growth of the local community. Plants are being increasingly recognized by scientists as a possible source of medications that might be used to fight a variety of ailments. To comprehend the possible adverse effects of these substances on the human body and to evaluate their lethality, however, detailed investigations are required. According to experts, continued examination of herbal plants and the natural compounds they produce holds the key to breakthroughs in the prospective medication development for particular ailments. At the same time, the cultivation of a wide variety of herbal plants may simultaneously encourage local job opportunities.

REFERENCES

Abad MJ, Bedoya LM, Apaza L, Bermejo P. The Artemisia L. genus: a review of bioactive essential oils. *Molecules*. 2012 Mar 2;17(3):2542–66.

Agrawal DP. Traditional knowledge systems and western science. *Current Science*. 1997 Nov 10;73(9):731–3.

Ahmad S, Zahiruddin S, Parveen B, Basist P, Parveen A, Gaurav Parveen R, Ahmad M. Indian medicinal plants and formulations and their potential against covid-19–preclinical and clinical research. *Frontiers in Pharmacology*. 2021;11:578970.

Akbar S, Akbar S. Achillea millefolium L.(Asteraceae/Compositae). *Handbook of 200 Medicinal Plants: A Comprehensive Review of Their Traditional Medical Uses and Scientific Justifications*. 2020:57–67.

Alamgir ANM Alamgir ANM. Secondary metabolites: Secondary metabolic products consisting of C and H; C, H, and O; N, S, and P elements; and O/N heterocycles. *Therapeutic Use of Medicinal Plants and their Extracts: Volume 2: Phytochemistry and Bioactive Compounds*, 2018;165–309.

Anand V, Kumar V, Kumar S, Hedina A. Phytopharmacological overview of Psidium guajava Linn. *Pharmacognosy Journal*. 2016;8(4):314–20.

Arumugam A, Razis AF. Apoptosis as a mechanism of the cancer chemopreventive activity of glucosinolates: a review. *Asian Pacific Journal of Cancer Prevention: APJCP*. 2018;19(6):1439.

Avinash T, Rai V. An ethanobotanical investigation of cucurbitaceae from South India: a review. *Journal of Medicinal Plants Studies*. 2017;5(3):250–3.

Baskar V, Gururani MA, Yu JW, Park SW. Engineering glucosinolates in plants: current knowledge and potential uses. *Applied Biochemistry and Biotechnology*. 2012 Nov;168:1694–717.

Bharati KA, Singh HB. Brucea mollis Wall. ex Kurz.: Untapped potential medicinal plant of Northeast India. *NeBIO-International Journal of Environment & Biodiversity*. 2012;3:26–8.

Bharucha Z, Pretty J. The roles and values of wild foods in agricultural systems. *Philosophical Transactions of the Royal Society B: Biological Sciences*. 2010 Sep 27;365(1554):2913–26.

Bhat JA, Kumar M, Negi AK, Todaria NP. Informants' consensus on ethnomedicinal plants in Kedarnath Wildlife Sanctuary of Indian Himalayas. *Journal of Medicinal Plants Research*. 2013 Jan 25;7(4):148–54.

Blažević I, Montaut S, Burčul F, Rollin P. Glucosinolates: Novel sources and biological potential. *Glucosinolates*. Springer International Publishing. 2016:3–60.

Blowman K, Magalhães M, Lemos MF, Cabral C, Pires IM. Anticancer properties of essential oils and other natural products. *Evidence-Based Complementary and Alternative Medicine*. 2018 Jan 1;2018.

Boy HI, Rutilla AJ, Santos KA, Ty AM, Alicia IY, Mahboob T, Tangpoong J, Nissapatorn V. Recommended medicinal plants as source of natural products: a review. *Digital Chinese Medicine*. 2018 Jun 1;1(2):131–42.

Desam NR, Al-Rajab AJ. The importance of natural products in cosmetics. *Bioactive Natural Products for Pharmaceutical Applications*. 2021:643–85.

Devendra C, Chantalakhana C. Animals, poor people and food insecurity: opportunities for improved livelihoods through efficient natural resource management. *Outlook on Agriculture*. 2002 Sep;31(3):161–75.

Dhakad AK, Pandey VV, Beg S, Rawat JM, Singh A. Biological, medicinal and toxicological significance of Eucalyptus leaf essential oil: a review. *Journal of the Science of Food and Agriculture*. 2018 Feb;98(3):833–48.

Divekar PA, Narayana S, Divekar BA, Kumar R, Gadratagi BG, Ray A, Singh AK, Rani V, Singh V, Singh AK, Kumar A. Plant secondary metabolites as defense tools against herbivores for sustainable crop protection. *International Journal of Molecular Sciences*. 2022 Feb 28;23(5):2690.

Fahey JW, Zalcmann AT, Talalay P. The chemical diversity and distribution of glucosinolates and isothiocyanates among plants. *Phytochemistry*. 2001 Jan 1;56(1):5.

Fardet L, Fève B. Systemic glucocorticoid therapy: a review of its metabolic and cardiovascular adverse events. *Drugs*. 2014 Oct;74:1731–45.

Gautam LM, Shrestha SL, Wagle P, Tamrakar BM. Chemical constituents from Vitex negundo (Linn.) of nepalese origin. *Scientific World*. 2008;6(6):27–32.

Ghosh A, Das BK, Roy A, Mandal B, Chandra G. Antibacterial activity of some medicinal plant extracts. *Journal of Natural Medicines*. 2008 Apr;62:259–62.

Giacometti J, Kovačević DB, Putnik P, Gabrić D, Bilušić T, Krešić G, Stulić V, Barba FJ, Chemat F, Barbosa-Cánovas G, Jambrak AR. Extraction of bioactive compounds and essential oils from mediterranean herbs by conventional and green innovative techniques: a review. *Food Research International*. 2018 Nov 1;113:245–62.

Gowthami R, Sharma N, Pandey R, Agrawal A. Status and consolidated list of threatened medicinal plants of India. *Genetic Resources and Crop Evolution*. 2021 Aug;68(6):2235–63.

Gupta A, Naraniwal M, Kothari V. Modern extraction methods for preparation of bioactive plant extracts. *International Journal of Applied and Natural Sciences*. 2012 Aug;1(1):8–26.

Hwang DY. Therapeutic role of natural products containing tannin for treatment of constipation. *Constipation*. 2018 Nov 7:71–90.

Idolo M, Motti R, Mazzoleni S. Ethnobotanical and phytomedicinal knowledge in a long-history protected area, the Abruzzo, Lazio and Molise National Park (Italian Apennines). *Journal of Ethnopharmacology*. 2010 Feb 3;127(2):379–95.

Jain P, Pareek A, Ratan Y, Sharma S, Paliwal S. Free radicals and dietary antioxidants: a potential review. *International Journal of Pharmaceutical Sciences Review and Research*. 2013;18(1):34–48.

Janahi IA, Rehman A, Baloch NU. Corticosteroids and their use in respiratory disorders. *Corticosteroids*. London: InTech Open. 2018 May 23:47–57.

Kaczmarek B. Tannic acid with antiviral and antibacterial activity as a promising component of biomaterials—a minireview. *Materials*. 2020 Jul 20;13(14):3224.

Karimi A, Majlesi M, Rafieian-Kopaei M. Herbal versus synthetic drugs; beliefs and facts. *Journal of Nephropharmacology*. 2015;4(1):27.

Kozawa M, Baba K, Matsuyama Y, Kido T, Sakai M, Takemoto T. Components of the root of Anthriscus sylvestris HOFFM. II. Insecticidal activity. *Chemical and Pharmaceutical Bulletin*. 1982 Aug 25;30(8):2885–8.

Kumar M, Radha, Devi H, Prakash S, Rathore S, Thakur M, Puri S, Pundir A, Bangar SP, Changan S, Ilakiya T. Ethnomedicinal plants used in the health care system: Survey of the mid hills of solan district, Himachal Pradesh, India. *Plants*. 2021 Sep 5;10(9):1842.

Kumar SB, Arnipalli SR, Ziouzenkova O. Antibiotics in food chain: The consequences for antibiotic resistance. *Antibiotics*. 2020 Oct 13;9(10):688.

Liu B, Guo ZY, Bussmann R, Li FF, Li JQ, Hong LY, Long CL. Ethnobotanical approaches of traditional medicine studies in Southwest China: A literature review. *Journal of Ethnopharmacology*. 2016 Jun 20;186:343–50.

Mahawer S, Kumar R, Prakash O, Singh S, Singh Rawat D, Dubey SK, Oliveira M. A review of phytochemical and pharmacological properties of Alpinia malaccensis (Burm. F.) Roscoe.(Zingiberaceae). *Current Topics in Medicinal Chemistry*. 2023 Aug 1;23(20):1964–72.

Mahmood A, Mahmood A, Malik RN, Shinwari ZK. Indigenous knowledge of medicinal plants from Gujranwala district, Pakistan. *Journal of Ethnopharmacology*. 2013 Jul 9;148(2):714–23.

Man S, Gao W, Zhang Y, Huang L, Liu C. Chemical study and medical application of saponins as anti-cancer agents. Fitoterapia. 2010 Oct 1;81(7):703–14.

Mgbeahuruike EE, Yrjönen T, Vuorela H, Holm Y. Bioactive compounds from medicinal plants: focus on Piper species. *South African Journal of Botany*. 2017 Sep 1;112:54–69.

Njila NI, Mahdi E, Lembe MD, Nde Z, Nyonseu D. Review on extraction and isolation of plant secondary metabolites. *7th Int'l Conference on Agricultural, Chemical, Biological and Environmental Sciences*. 2017.

Oleszek M, Oleszek W. Saponins in food. *Handbook of Dietary Phytochemicals*. 2020:1–40.

Patel K, Gadewar M, Tripathi R, Prasad SK, Patel DK. A review on medicinal importance, pharmacological activity and bioanalytical aspects of beta-carboline alkaloid "Harmine". *Asian Pacific Journal of Tropical Biomedicine*. 2012 Aug 1;2(8):660–4.

Pfeifer BL, Fahrendorf T. Indole-3-carbinol: a glucosinolate derivative from cruciferous vegetables for prevention and complementary treatment of breast cancer. *Deutsche Zeitschrift für Onkologie*. 2015;47:20–7.

Provino R. The role of adaptogens in stress management. *Australian Journal of Medical Herbalism*. 2010 Jan;22(2):41–9.

Puri S, Chandel K, Pundir A, Thakur MS, Chauhan B, Simer K, Dhiman N, Thakur YS, Kumar S. Diversity of ethnomedicinal plants in Churdhar Wildlife Sanctuary of district Sirmour of Himachal Pradesh, India. *Journal of Applied Pharmaceutical Science*. 2019 Nov 1;9(11):48–53.

Radha Janjua S, Ali M, Thakur M, Jamwal R, Rathour S, Kumar Pubral A, Kumari N, Puri S, Pundir A, et al. Documenting traditional knowledge before they are forgotten: a study on the ethnomedicinal uses of wild plants by rural people of Jubbarhatti in District Shimla. *International Journal of Theoretical and Applied Sciences*. 2021;13:37–51.

Rahman MM, Rahaman MS, Islam MR, Hossain ME, Mannan Mithi F, Ahmed M, Saldías M, Akkol EK, Sobarzo-Sánchez E. Multifunctional therapeutic potential of phytocomplexes and natural extracts for antimicrobial properties. *Antibiotics*. 2021 Sep 6;10(9):1076.

Rahman MA, Mossa JS, Al-Said MS, Al-Yahya MA. Medicinal plant diversity in the flora of Saudi Arabia 1: a report on seven plant families. *Fitoterapia*. 2004;75:149–161.

Sajid M, Channakesavula CN, Stone SR, Kaur P. Synthetic biology towards improved flavonoid pharmacokinetics. *Biomolecules*. 2021 May 18;11(5):754.

Safari VZ, Ngugi MP, Orinda G, Njagi EM. Anti-pyretic, anti-inflammatory and analgesic activities of aqueous leaf extract of Urtica dioica L. in albino mice. *Med Aromat Plants*. 2016;5(237):2167–041.

Saleem R, Ahmad M, Naz A, Siddiqui H, Ahmad SI, Faizi S. Hypertensive and toxicological study of citric acid and other constituents from Tagetes patula roots. *Archives of Pharmacal Research*. 2004 Oct;27:1037–42.

Sekikawa T, Kizawa Y, Takeoka A, Sakiyama T, Li Y, Yamada T. The effect of consuming an anthocyanin-containing supplement derived from Bilberry (Vaccinium myrtillus) on eye function: a randomized, double-blind, placebo-controlled pParallel study. *Functional Foods in Health and Disease*. 2021 Mar 30;11(3):116–46.

Shinjyo N, Waddell G, Green J. Valerian root in treating sleep problems and associated disorders—a systematic review and meta-analysis. *Journal of Evidence-Based Integrative Medicine*. 2020 Oct 21;25:2515690X20967323.

Shoaib S, Ansari MA, Fatease AA, Safhi AY, Hani U, Jahan R, Alomary MN, Ansari MN, Ahmed N, Wahab S, Ahmad W. Plant-derived bioactive compounds in the management of neurodegenerative disorders: Challenges, future directions and molecular mechanisms involved in neuroprotection. *Pharmaceutics*. 2023 Feb 23;15(3):749

Singh V, Kumar R. Study of phytochemical analysis and antioxidant activity of Allium sativum of Bundelkhand region. *International Journal of Life-Sciences Scientific Research*. 2017 Nov;3(6):1451–8.

Srivastava AK, Singh VK. Tulsi (Ocimum sanctum): a potent adaptogen. *Clinical Research Notes*. 2021;2(2):1–5.

Sulaiman M, Tijani HI, Abubakar BM, Haruna S, Hindatu Y, Mohammed JN, Idris A. An overview of natural plant antioxidants: analysis and evaluation. *Advances in Biochemistry*. 2013;1(4):64–72.

Sülsen VP, Lizarraga E, Mamadalieva NZ, Lago JH. Potential of terpenoids and flavonoids from Asteraceae as anti-inflammatory, antitumor, and antiparasitic agents. *Evidence-Based Complementary and Alternative Medicine*. 2017 Jan 1; 2017: 1–2.

Svahn S. *Analysis of Secondary Metabolites from Aspergillus fumigatus and Penicillium nalgiovense: Antimicrobial Compounds from Filamentous Fungi Isolated from Extreme Environments* (Doctoral dissertation, Acta Universitatis Upsaliensis). 2015.

Taek MM, Banilodu L, Neonbasu G, Watu YV, EW BP, Agil M. Ethnomedicine of Tetun ethnic people in West Timor Indonesia: philosophy and practice in the treatment of malaria. *Integrative Medicine Research*. 2019 Sep 1;8(3):139–44.

Tang GH, Chen DM, Qiu BY, Sheng L, Wang YH, Hu GW, Zhao FW, Ma LJ, Wang H, Huang QQ, Xu JJ. Cytotoxic amide alkaloids from Piper boehmeriaefolium. *Journal of Natural Products*. 2011 Jan 28;74(1):45–9.

Thakur M, Guleria P, Sobti R C, Gautam A, Kaur T. Comparative analysis of the antibacterial efficacy and bioactive components of Thuja occidentalis obtained from four different geographical sites. *Molecular and Cellular Biochemistry.* 2024 Feb;479(2):283–296.

Thakur M, Kaur T. Impact of geographical location on the antibacterial activity of Cuscuta reflexa. *The Journal of Indian Botanical Society.* 2022;102(2):164–170.

Thakur M, Sobti R, Kaur T. Medicinal and biological potential of Thuja occidentalis: a comprehensive review. *Asian Pacific Journal of Tropical Medicine.* 2023;16(4):148.

Thoppil RJ, Bishayee A. Terpenoids as potential chemopreventive and therapeutic agents in liver cancer. *World Journal of Hepatology.* 2011 Sep 9;3(9):228.

Variyar PS, Banerjee A, Akkarakaran JJ, Suprasanna P. Role of glucosinolates in plant stress tolerance. In *Emerging Technologies and Management of Crop Stress Tolerance,* 2014 Jan 1 (pp. 271–291). Academic Press: US.

Veeresham C. Natural products derived from plants as a source of drugs. *Journal of Advanced Pharmaceutical Technology and Research.* 2012 Oct 1;3(4):200–1.

Verma RS, Padalia RC, Goswami P, Chauhan A. Essential oil composition of Peperomia pellucida (L.) Kunth from India. *Journal of Essential Oil Research.* 2015 Mar 4;27(2):89–95.

Vidyarthi S, Samant SS, Sharma P. Traditional and indigenous uses of medicinal plants by local residents in Himachal Pradesh, North Western Himalaya, India. *International Journal of Biodiversity Science, Ecosystem Services & Management.* 2013 Sep 1;9(3):185–200.

Vishwakarma KL, Dubey V. Nutritional analysis of indigenous wild edible herbs used ineastern chhattisgarh, India. *Emirates Journal of Food and Agriculture.* 2011 Apr 15:554–60.

Walia A, Gupta AK, Sharma V. Role of bioactive compounds in human health. *Acta Scientific Medical Sciences.* 2019 Sep;3(9):25–33.

Wani ZA, Kumar N. Akash ethnobotanical study of some threatened plants in district Baramulla, Kashmir, Jammu and Kashmir, India. *International Journal of Current Research in Biosciences and Plant Biology.* 2016;3(2):58–64.

Yamazaki Y, Kawano Y. Inhibitory effects of herbal alkaloids on the tumor necrosis factor-α and nitric oxide production in lipopolysaccharide-stimulated RAW264 macrophages. *Chemical and Pharmaceutical Bulletin.* 2011 Mar 1;59(3):388–91.

Yuca H, Karakaya S. Matricaria chamomilla L. In *Novel Drug Targets With Traditional Herbal Medicines: Scientific and Clinical Evidence,* 2022 Sep 30 (pp. 387–400). Cham: Springer International Publishing.

Zaynab M, Fatima M, Abbas S, Sharif Y, Umair M, Zafar MH, Bahadar K. Role of secondary metabolites in plant defense against pathogens. *Microbial Pathogenesis.* 2018 Nov 1;124:198–202.

24 Biodiversity and Human Health

Gagandeep Singh[1], *Sharali Sharma and Anita Rana*
[1]Department of Biosciences, University Institute of Biotechnology, Chandigarh University, Gharuan, Mohali, Punjab

24.1 INTRODUCTION

Biodiversity, which is the variety of life on Earth, is like the intricate threads that make up the fabric of our planet. It includes all the different species and the diversity within their genes. This biodiversity is not just a passive part of our world; it plays a crucial role in keeping our ecosystems healthy, which in turn affects our own well-being. Among these services, it assumes the roles of air purification, water cleansing, and climate regulation—all of which intricately and profoundly mold the landscape of human health (1). Within this labyrinthine nexus of interdependence, biodiversity's role in shaping the human experience unfolds in a multifaceted panorama. It is a silent contributor to our food and medicine resources, bestowing upon us sustenance and remedies vital to our health. However, the loss of biodiversity casts a shadow over these invaluable provisions, diminishing their availability and posing ominous threats to human health. Moreover, as we venture into an era marked by habitat destruction and environmental transformation, biodiversity preservation becomes synonymous with averting the heightened risk of disease transmission from animals to humans. The delicate equilibrium between species is disrupted, forcing closer contact between wildlife and human populations—a precarious scenario with implications for the health of our species (1). The intricacies of this relationship extend beyond the tangible. The profound link between biodiversity and mental health is well-documented, offering solace in nature and reducing the burdens of stress and anxiety (2). Additionally, in the less visible realm, diverse ecosystems serve as adept filters, meticulously removing pollutants from our air and water, enhancing the quality of the elements we depend upon for life (3). However, as we stand at the crossroads of biodiversity loss and climate change, the intertwining impacts of these two formidable forces require our undivided attention. They amplify each other's effects, creating a cascade of consequences that leave neither ecosystems nor humanity untouched (4). The imperative to explore the nexus of biodiversity and human health is undeniable. It enhances our comprehension of the far-reaching consequences of biodiversity loss on our well-being, equipping us to make informed decisions that safeguard biodiversity while promoting human health (5). Furthermore, this exploration reveals untapped reservoirs of medicines and sustenance within the rich fabric of biodiversity, offering new resources vital to our health (6). It empowers us to proactively prevent disease outbreaks by understanding the heightened risk of zoonotic disease transmission resulting from biodiversity loss (5). Additionally, it unearths the therapeutic benefits of nature exposure on mental health, paving the way for strategies that foster well-being and alleviate the burdens of stress and anxiety (7). Moreover, it underscores the indispensable role of biodiversity in maintaining ecosystem health and the services they provide, including clean air, water purification, and climate regulation, inspiring strategies to protect these services and, in turn, fortify human health (6).

This chapter embarks on an expedition into the intricate interplay between biodiversity and human health. It delves into a diverse spectrum of dimensions characterizing this relationship, scrutinizing the pivotal functions of microorganisms in nutrition, dissecting the intricate connections linking biodiversity to noncommunicable diseases, and probing the repercussions of biodiversity depletion on the contours of environmental equity. We will navigate through the challenges posed

 DOI: 10.1201/9781032697550-24

by climate change-induced impacts on biodiversity, delve into the pharmaceutical potential of biodiversity, and embrace the holistic paradigm of ecohealth. By the end, we will have traversed a rich tapestry of knowledge, shedding light on the interwoven fate of biodiversity and human health, and illuminating pathways toward a healthier, more harmonious coexistence with the natural world.

24.2 BIODIVERSITY'S CONTRIBUTION TO FOOD SECURITY AND NUTRITION

Biodiversity plays a pivotal role in sustaining food production and nutrition, serving as the bedrock upon which our agricultural systems and dietary health depend. One vital facet of this contribution is the essential service provided by animal pollinators. Native and managed pollinators, through their tireless efforts, are instrumental in fertilizing crops, thus resulting in augmented yields and enhanced quality of fruits, vegetables, and nuts (8). Their role in ensuring the availability of these essential food sources is undeniably indispensable. The practice of agroforestry is another significant avenue through which biodiversity intertwines with food security, ecosystem services, and biodiversity preservation. This method entails cultivating crops beneath the protective canopy of trees, fostering an environment that benefits birds, insects, and other wildlife that contribute to pest and disease control (9). Moreover, agroforestry bolsters soil fertility and water retention, thereby amplifying crop yields. In essence, it emerges as a harmonious convergence of agriculture and biodiversity conservation.Conservation grazing, a strategy hinged on the mindful management of rangelands, also underscores the intricate interplay between biodiversity and food production. By discerning wildlife responses to grazing, conservationists can adeptly devise strategies that strike a harmonious balance between the preservation of biodiversity and the sustenance of food production on these expansive terrains (10). Integrated spatial planning is yet another noteworthy approach to aligning biodiversity conservation with food production. This method is instrumental in identifying areas where conservation and production can coexist harmoniously. By pinpointing regions of paramount conservation importance, planners can formulate strategies that shield biodiversity while concurrently fostering food production (11). Ecological intensification and diversification are approaches that shine a light on the path toward the cohabitation of biodiversity, ecosystem services, and food production in a world undergoing transformation. By harnessing ecological processes to augment food production per unit area and reintroducing diverse vegetation into landscapes, agricultural systems become more accommodating to biodiversity. The added advantage of reducing the dependency on expensive inputs while maintaining productivity amplifies the appeal of ecological intensification methods (12). The significance of crop biodiversity is paramount in navigating the challenges posed by shifting environmental conditions. Cultivating a variety of crops diminishes reliance on a single species and enhances resilience against pests, diseases, and the vagaries of climate change. Additionally, diverse crops enrich dietary nutrition, each contributing unique profiles of essential nutrients (13).

Ecosystems, beyond their role in food production, are integral to fostering diverse and healthy diets, which are fundamental to human health and well-being. One such contribution is the modulation of the human gut microbiome. This intricate ecosystem of microorganisms holds sway over human health, with studies revealing that indigenous diets brimming with diversity can serve as biotherapeutics, shaping microbial populations to thwart stunted growth, malnutrition, bowel diseases, weakened immune responses, and infant mortality (14). Furthermore, the dietary choice of a vegan regimen has been found to influence the rich tapestry of beneficial gut bacteria (15). The promotion of a fiber-rich diet emerges as a promising biotherapeutic strategy for enhancing gut lining integrity and bolstering immune tolerance towards the microbiota. Whole grain cereals like millet, maize, brown rice, sorghum, soya, and tapioca have exhibited the capacity to fortify gut lining integrity and immune tolerance, promoting overall digestive health. Diversification of dietary fiber sources, including "nonconventional" options like konjac, bamboo fiber, or seeds fiber, adds to our understanding of how varied fibers impact human gut microbiota, potentially enhancing health outcomes (16). Seasonal floodplain wetlands, found in coastal ecosystems, represent another arena where biodiversity plays a significant role. The Yolo Bypass in California's Central Valley, USA, stands as an exemplar of how these ecosystems, when inundated, can support native fishes and serve as a refuge for endangered species, such as the Delta smelt (17). Even prepregnancy dietary choices influence human health through their impact on the vaginal microbiome. A diet rich in total carbohydrates and sugars has been linked to improved vaginal health, while a microbiota dominated by lactobacilli was inversely associated with higher prepregnancy intake of animal-sourced protein (18).

The implications of ecosystems in shaping diets resonate with global hunger and malnutrition. Ecosystem conservation assumes an indomitable significance, preserving the lifelines that bolster food production and nutrition. Pollinators, fertile soils, and thriving biodiversity stand as testaments to this imperative. Advocating for diverse diets, an amalgamation of foods from various ecosystems, becomes our clarion call against the scourge of global hunger and malnutrition. Traditional fare, rich in nourishment, aligns with local food systems to deliver sustenance to the needy. Investment in research takes center stage, unfurling the mysteries of the gut microbiome, unlocking the nutritional treasures hidden within foods, and deciphering the intricate dance of ecosystems in the realm of food and nutrition. Collaboration emerges as the linchpin of progress, uniting agriculture, health, and the environment in a harmonious symphony. By pooling their wisdom, these sectors devise holistic strategies that acknowledge the intricate interplay of ecosystems, food production, and nutrition, a powerful alliance against the specter of hunger and malnutrition (8–11).

24.3 ECOSYSTEM SERVICES AND HUMAN HEALTH

24.3.1 Psychological Benefits of Biodiversity

One of the major economic and social problems of the 21st century is poor mental health (19). Rapidly growing global urbanization poses serious problems for both biodiversity and human mental health. In recent decades, the importance of natural ecosystems to human health and well-being has attracted many researchers (20, 21) and has shown compelling evidence that the natural world fosters a variety of aspects of human health and well-being. It is evident that with the increase in biodiversity, human health also rises. Around 450 million people worldwide are affected by mental health conditions (22).

In the 1980s, different ideas were given to explain how being in contact with the environment has a positive impact on mental health. The variety of living forms in ecological systems directly affects people's quality of life. According to research, being exposed to nature and biodiversity has a variety of positive effects on mental health and well-being. Spending more time in nature has been effective in reducing stress, enhancing mood, better cognitive function, and higher levels of creativity. Ecotherapy and outdoor therapy have been proven to be effective in treating mental health issues like anxiety, depression, and attention deficit hyperactivity disorder (ADHD) (23). Green exercise, such as jogging, walking, and bicycling in parks or other natural settings, improves mental wellness (24). It increases cardiovascular health, strengthens muscles, and encourages general fitness.

The direct contact of individuals with natural surroundings and the presence of urban green areas have been linked to the potential advantages of biodiversity to physical and mental health (22). However, urbanization disrupts the direct connection of people to nature, which ultimately causes a number of diseases linked with pollution (25). Psychological benefits also increase with the species richness of urban green spaces. Hence, in order to manage urban green spaces and conserve biodiversity successfully, biological complexity must also be prioritized to improve human well-being.

24.3.2 Coral Reef and Human Health

Coral reefs form one of the most precious and diversified ecosystems on Earth. Coral reefs are made up of over 800 different species of coral. They are dynamic and lively ecosystems. Polyps, the tiny organisms that make up coral and can come in a variety of shapes, feed on the exoskeletons of their

TABLE 24.1
Drugs Derived from Natural Products/Derivatives from Coral Reef Organisms

Coral Reef Organisms	Isolated Compound	Treatment
Eucheuma/Chondrus (Edible red algae)	Iota carrageenan (Carragelose®)	Known for its antiviral properties. The nasal spray Carragelose® is useful against the common cold and traps respiratory viruses.
Conus magus (marine cone snail)	Ziconotide (Prialt®)	Severe chronic pain related with neuropathies, AIDS, or cancer block cellular channels in the brain and spinal cord.
Ecteinascidia turbinata (Caribbean sea squirt)	Trabectedin (Yondelis®)	Used as chemotherapy medication for advanced soft-tissue sarcoma and ovarian cancer.
Dolabella auricularia (sea hare)	Dolastatin 10 and brentuximab vedotin (Adcetris®)	Used to treat Hodgkin's lymphoma.
Cryptotethya crypta (Caribbean sponge)	Cytarabine (Cytostar-U®; DepoCyte®)	For treating cancers of white blood cells such as myeloid leukemia, non-Hodgkin's lymphoma, and meningeal leukemia.

predecessors and add their own to the structure of the coral as it grows. It is estimated that millions of undiscovered species of organisms are living in and near the reefs. Several drugs are presently being produced from coral reef organisms as potential treatments for cancer, rheumatoid arthritis, and human bacterial and viral infections (Table 24.1).

Around 1 billion people worldwide depend on coral reefs for their livelihood, food supply, and protection of the coast (26). Coral reefs are considered one of the most commercially significant and biologically diverse ecosystems (27), yet they have been in decline globally due to rising ocean temperatures and other pressures (28). Antibiotic-resistant bacteria have been discovered from soft corals that dwell on coral reefs. The vast biodiversity is used to make natural products and drugs. The compounds Ara-A and Ara-C were isolated from a Caribbean sponge that lives on coral reefs (29). Ara-A is used to make the antiviral medication Vira-A to treat herpes simplex infections and AZT (azidothymidine) to treat HIV, while Ara-C is used to develop the anticancer medication cytarabine. Similarly, the drug ziconotide was isolated from cone snail to treat chronic pain (30). In general, the oceans are an extremely underutilized source of products for human treatment. Coral reefs in particular provide great promise for drug development, and they should be valued as a valuable resource.

24.4 BIODIVERSITY AND PHARMACEUTICALS

24.4.1 Medicinal and Pharmaceutical Roles of Plants

Plants are considered a vital source of medicine and help in maintaining human health. About 35,000–70,000 plant species have been listed as being significant for medicinal purposes. Medicinal plants or herbs are one of the potential sources of therapeutics or curative aids. All around the world, medicinal plants are used to improve human health conditions and are known to dominate the health care system. The global use of plant-based drugs is rising. Current studies on medicinal plants have led to significant advancements in pharmacological evaluation. Plant-based drugs are used to treat cancer, diabetes, jaundice, hypertension, tuberculosis, skin conditions, and mental illness.

A variety of secondary metabolites or compounds, including tannins, terpenoids, alkaloids, and flavonoids, are found in medicinal plants that determine the therapeutic effectiveness of the plants with respect to antibacterial activity (31). The antioxidant activity of medicinal plants is found to be associated with secondary metabolites that have been proven to be beneficial in pharmaceutical research (32). Significant antibacterial, antifungal, anticancerous, antidiuretic, anti-inflammatory, and antidiabetogenic effects have also been reported in a variety of plant species (33–35). A major contribution to the treatment of various deadly diseases in humans was made by the drugs developed through the screening of biologically active substances from plant extracts. In recent years, pharmacological and phytochemical research has successfully treated a wide range of disorders for which the synthetic drug industry was unable to find effective treatments (36).

For many decades, plants have been used as medicines to treat a variety of diseases. Any disease can be mitigated, treated, and prevented through pharmaceuticals. They are crucial for drug products that improve the structure and body functioning (37). Along with medicinal plants, the pharmaceutical industry is crucial in the discovery of novel medications. Over the past few decades, medicinal plants have aided in the development of drugs to treat fatal diseases like cancer and hepatitis (38). The development of useful herbal drugs can benefit further from the combination of plant chemical knowledge with therapeutic characteristics and other biological activities (39). Recent research indicates that the majority of phytochemicals have advantageous properties, including antimicrobial, antimalarial, antidiabetic, and antiarthritic.

Plants are utilized as phytomedicine to treat numerous ailments as they contain a variety of bioactive chemicals. Vinblastine and vincristine are found in *Catharanthus roseus* and are used to treat cancer. Due to the presence of serpentine, reserpine, and ajmalicine, *Rauwolfia serpentina* is found to be hypotensive. *Papaver somniferum* has morphine and codeine, thus exhibiting sedative and analgesic properties. *Artemisia annua* is used to make the malaria-fighting drug artemisinin. *Withania somnifera* has been investigated to treat arthritis. *Allium sativum* is used to isolate allicin and is known for its cardioprotective and anti-inflammatory properties (40). Paclitaxol, an alkaloid extracted from *Taxus brevifolia,* is one of the potent medications made from higher plants and is used to treat breast and ovarian malignancies (41). Table 24.2 shows some of the important compounds isolated from a variety of plant species and their known therapeutic applications.

24.4.2 Release of Active Pharmaceutical Ingredients (APIs) into Environment

Active pharmaceutical ingredients (APIs), commonly known as pharmaceutically active substances, are complex compounds with a wide range of physical and biological properties and functions. These APIs can be released directly or indirectly into the environment. Although pharmaceutical products have likely been released into the environment for as long as they have been produced and used, environmental monitoring has made it easier to detect and evaluate pharmaceutical products in recent years. A number of ways have been reported for the release of pharmaceuticals into the environment (47, 48). During the production process, APIs are released into rivers and streams directly through accidental or uncontrolled emissions or indirectly through wastewater treatment facilities. APIs are also introduced into sewer systems by humans. In areas with good sewerage connectivity, the incompletely metabolized API molecules, after release from the human body, enter into wastewater treatment plants. However, the complex molecules, which are difficult to degrade, can also pass from treatment plants and be released into rivers and streams. These molecules can also be transported to water resources through sewage effluent, thus affecting the human body. According to multiple risk assessments, the ingestion of individual APIs through drinking water is anticipated to pose relatively little danger to the public's health. Numerous studies have reported the occurrence of human APIs, including antidepressants, hormones, antibiotics, and antifungal agents, in rivers and streams.

Human and veterinary medications are frequently assessed independently in risk assessments due to various modes of entry into the environment. Animals also excrete veterinary APIs into the soil. Direct dropping, dumping, and application of animal waste as fertilizer, and substances used in aquaculture are some of the potential routes for pharmaceuticals used in livestock (49, 50). These antibiotics finally end up in the groundwater or soil if they are not broken down and ultimately can

TABLE 24.2
Important Compounds Isolated from Plant Species

Plant Species	Family	Isolated Compound	Therapeutic Application
Cinchona officinalis	Rubiaceae	Quinine	Antimalarial
Artemisia annua	Asteraceae	Artemisinin	Antimalarial
Digitalis purpurea	Scrophulariaceae	Cardenolide digitoxin	Cough, asthma, fever, cardiac diseases, edema, and impotency
Rauwolfia serpentina	Apocynaceae	Reserpine	Antihypertensive
Physostigma venenosum	Fabaceae	Physostigmine	Memory enhancement
Taxus brevifolia, Taxus canadensis	Taxaceae	Paclitaxel	Breast, ovarian, and lung cancer
Silybum marianum	Asteraceae	Silimarin	Liver infections
Catharanthus roseus	Apocynaceae	Vinblastine	Antitumor, Aantileukemic agent
		Vincristine	Antitumor, antileukemic agent
Digitalis lanata	Plantaginaceae	Acetyldigoxin	Cardiotonic
Andrographis paniculata	Acanthaceae	Andrographolide	Treatment for baccillary dysentery
Anabasis aphylla	Amaranthaceae	Anabesine	Skeletal muscle relaxant
Betula alba	Betulaceae	Betulinic acid	Anticancerous
Ardisia japonica	Primulaceae	Bergenin	Antitussive
Berberis vulgaris	Berberidaceae	Berberine	Treatment for bacillary dysentery
Camellia sinensis	Theaceae	Caffeine	CNS stimulant
Camptotheca acuminata	Nyssaceae	Camptothecin	Anticancerous
Convallaria majalis	Asparagaceae	Convallatoxin	Cardiotonic
Colchicum autumnale	Colchicaceae	Colchicine	Antitumor, antigout
Rauvolfia serpentina	Apocynaceae	Reserpine	Antihypertensive, tranquilizer
Rhododendron molle	Ericaceae	Rhomitoxin	Antihypertensive, tranquilizer
Rorippa indica	Brassicaceae	Rorifone	Antitussive
Lonchocarpus nicou	Fabaceae	Rotenone	Piscicide, insecticide
Stephania sinica	Menispermaceae	Rotundine	Analagesic, sedative, traquilizer
Citrus species (e.g., orange, grapefruit)	Rutaceae	Rutin	Treatment for capillary fragility
Salix alba	Salicaceae	Salicin	Analgesic
Artemisia maritma	Asteraceae	Santonin	Ascaricide
Urginea maritima	Asparagaceae	Scillarin A	Cardiotonic
Datura species	Solanaceae	Scopolamine	Sedative
Cassia species	Fabaceae	Sennosides A, B	Laxative
Silybum marianum	Asteraceae	Silymarin	Antihepatotoxic
Cytisus scoparius	Fabaceae	Sparteine	Oxytocic
Taxus brevifolia	Taxaceae	Taxol	Antitumor agent
Strychnos nux-vomica	Loganiaceae	Strychnine	CNS stimulant
Podophyllum peltatum	Berberidaceae	Teniposide	Antitumor agent

Sources: Adopted from References 42–46.

also be carried into surface water by runoff. Even though there is no evidence that the majority of pharmaceuticals are harmful to human health, new drug compound policies should take this into account, as the presence of teratogens or carcinogens, even at low concentrations, is concerning for people's health, particularly in situations where there could be indirect and direct water reuse.

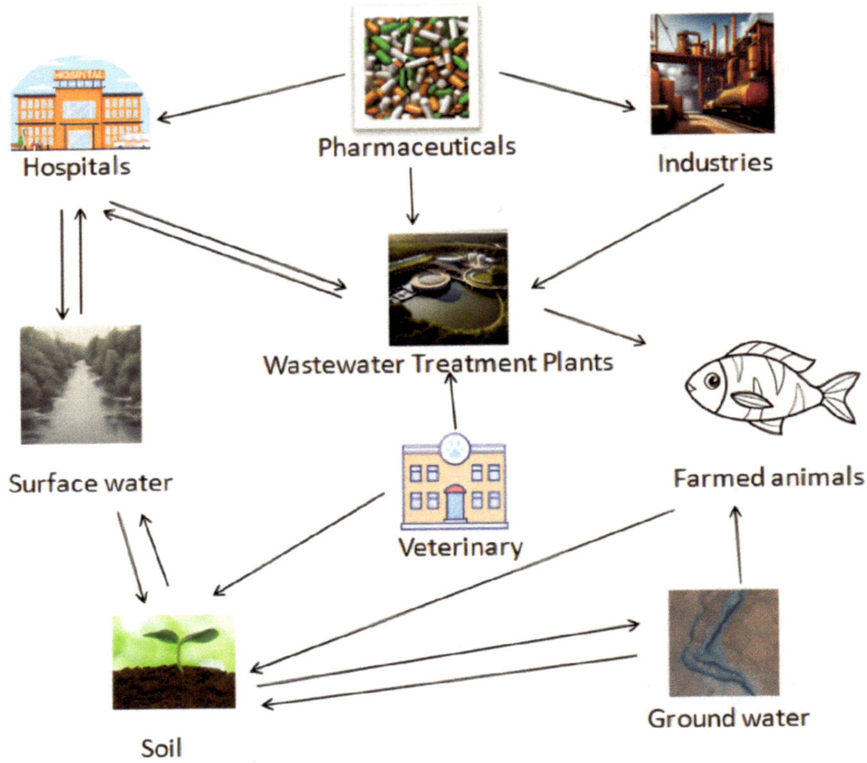

FIGURE 24.1 Representation of the pathway by which APIs are released into the environment.

24.4.3 Pharmaceuticals' Impacts on Biodiversity

Seventy percent of the freshwater supply is used by agriculture, and unsustainable farming methods can lead to problems like desertification, soil erosion, invasive species introduction, and nutrient depletion. Pharmaceuticals cause water pollution as these are discharged into the environment before being properly metabolized by patients; occasionally, up to 80% of the drug is excreted (51). Increased estrogen in the water due to the pill and hormone replacement treatment can cause fish to become more feminized, decrease fertility, have developmental consequences, and an adverse influence on the sustainability of wild fish populations (52). This disrupts the ecosystems, as well as wildlife, and potentially human health too (Figure 24.1).

24.5 STRATEGIES FOR HEALTH AND BIODIVERSITY

Recognizing the intrinsic connection between biodiversity and human health is paramount to our efforts in conserving the natural world. One effective approach to appreciating the value of biodiversity lies in quantifying its economic benefits. Numerous studies have illustrated that ecosystems offer invaluable services such as water purification, carbon sequestration, and pollination, all of which can be ascribed a monetary worth (53). To harmonize ecological preservation with economic prosperity, it is imperative to advocate for sustainable land-use practices. Practices like agroforestry, conservation agriculture, and integrated pest management not only contribute to biodiversity conservation but also bestow economic advantages upon farmers (54). It is crucial that health policies incorporate the health benefits intrinsic to biodiversity, encompassing the salutary effects of green spaces in bolstering physical activity and mental well-being (55). In parallel, public education endeavors should emphasize the significance of biodiversity and its positive impact on health, fostering greater awareness and kindling enthusiasm for conservation efforts (53). Embracing cutting-edge technology, blockchain systems can be harnessed to evaluate the value of health-focused datasets, such as health records, for legitimate research and commercial purposes (56).

A pivotal facet of safeguarding biodiversity and human health is managing ecosystems to curtail disease risks. Exercising control over the movement of animals stands out as a critical measure in mitigating the spread of infectious diseases. A notable instance can be found in the Serengeti ecosystem of Tanzania, where a vaccination campaign targeting canine distemper virus (CDV) among African lions (*Panthera leo*) can inadvertently impact the survival of a coexisting threatened species, the cheetah (*Acinonyx jubatus*) (57). Vaccination emerges as a potent tool in the arsenal against disease outbreaks within specific species, as evidenced by the exemplary CDV vaccination initiative among African lions in the Serengeti (57). Additionally, the advocacy and implementation of sustainable land-use practices, such as agroforestry, conservation agriculture, and integrated pest management, extend their protective mantle not only over biodiversity but also over the attenuation of disease transmission (58). To minimize the risk of disease transmission, it is vital to curtail human–wildlife contact, particularly in regions prone to zoonotic diseases like Ebola and COVID-19. This entails measures such as abstaining from bushmeat consumption and reducing interactions with wild animals (57).

Addressing the foundational catalysts of environmental change assumes paramount importance in safeguarding the well-being of both ecosystems and human populations. The implementation of efficacious waste management protocols occupies a central role in the mitigation of pollution, simultaneously fortifying the realms of human health and biodiversity. A case in point can be drawn from a study conducted in Antigua and Barbuda, which underscores the transformative potential of enhanced waste management, yielding substantial enhancements in both human and environmental health (59). Concurrently, the promulgation of sustainable land-use practices, encompassing agroforestry, conservation agriculture, and integrated pest management, not only buttresses the edifice of biodiversity but also mitigates the deleterious repercussions of environmental vicissitudes upon human health. Concerted endeavors aimed at curbing greenhouse gas emissions acquire an elemental status in the combat against the pernicious consequences of climate change for both human well-being and biodiversity. Emissions abatement measures, particularly within sectors such as transportation and industry, wield the capacity to effectively curtail airborne pollution, thereby preserving human health and ameliorating the toll exacted by climate change upon ecosystems. The safeguarding of natural habitats assumes a cardinal role in the preservation of biodiversity and the enhancement of human health. Exemplifying this, the preservation of wetlands concurrently serves a dual purpose: it attenuates the perils of flooding and upholds water quality, while furnishing vital habitats for a profusion of species. Raising public awareness regarding the significance of environmental conservation and its reciprocal benefits for human health and biodiversity stands as an indispensable endeavor in fostering a collective commitment to conservation (59).

24.6 FUTURE PERSPECTIVE

As we venture into the future, advancing our comprehension of the intricate interplay between biodiversity and human health is essential. To navigate this evolving landscape effectively, it is imperative that we establish common metrics and frameworks for assessing health and biodiversity. These tools will ensure consistency and comparability across diverse studies and projects, allowing for more robust conclusions and evidence-based decision-making (60). Citizen science, a potent instrument for biodiversity monitoring, has emerged as a force to reckon with. It not only aids in detecting rare events but also facilitates the assessment of long-term trends. However, harnessing the full potential of citizen science demands a deeper understanding of participation patterns. It necessitates an exploration of variations in participation across different citizen science approaches and within individual projects (60). In our quest to safeguard freshwater ecosystems, the use of benthic diatoms as bioindicators has gained prominence. To evaluate and monitor the quality of lotic environments, several metrics have come to the fore. These encompass species diversity, richness, and evenness, as well as biotic indices like the Trophic Diatom Index (TDI) and the Biological Diatom Index (BDI) (61). Turning our attention beneath the surface, soil arthropods play pivotal roles in the maintenance of soil health and multifunctionality. Community metrics and indicators rooted in soil arthropods can offer invaluable insights into the biological soil quality within diverse urban green spaces, shedding light on their health and functionality along gradients of disturbance and management intensity (62).

In our pursuit of a sustainable future, the application of precautionary policies stands as an indispensable pillar. The precautionary principle, a cornerstone of environmental policy, places the onus on those advocating for actions or policies with potential harm to demonstrate their harmlessness in the absence of scientific consensus (63). Environmental impact assessments (EIAs) provide a tangible mechanism for identifying and evaluating the potential environmental and health impacts of proposed projects or policies. These assessments can unveil risks, uncertainties, and data that guide informed decisions, factoring in the impacts on ecosystems and human health (64). Legal frameworks and regulations, such as the Environmental Management and Coordination Act of 1999 in Kenya, underscore the significance of such policies in managing and regulating the sustainable utilization of vital resources (64).

As we chart our path forward, there remain critical research gaps and uncharted territories demanding exploration. A pressing need exists for economic evaluations concerning the environmental effects on both infectious and noninfectious diseases in South Asia and Nepal (65). This research realm is instrumental in aiding policymakers in making informed decisions regarding the potential health ramifications of environmental shifts. The complex landscape of climate change adaptation strategies and their impacts on human health calls for further elucidation. Understanding the critical factors that influence the success or failure of these strategies in the face of climate variability is imperative (66). Delving into the green heart of urban landscapes, the role of biodiversity within green spaces in enhancing human health and well-being requires deeper exploration (65). Unlocking the secrets of this relationship holds the potential to reshape urban planning and design. Last, an urgent need exists for nationally representative economic evaluations concerning the environmental effects on both infectious and noninfectious diseases. This research avenue is essential in empowering policymakers with the knowledge needed to navigate the health impacts of environmental transformations.

24.7 CONCLUSION

In conclusion, this book chapter underscores the indisputable interconnectedness of biodiversity and human health. It emphasizes the invaluable role of medicinal plants, the resurgence of plant-based drugs, and the need to protect ecosystems to mitigate disease risks. Effective waste management and emissions reduction emerge as critical strategies to safeguard both human health and biodiversity. Looking ahead, standardizing metrics for assessment, harnessing citizen science, and exploring the roles of bioindicators offer promising avenues. Precautionary policies and environmental impact assessments are crucial for informed decision-making. Furthermore, this chapter underscores the urgent requirement for comprehensive economic evaluations of the environmental effects on diseases. It is only through such holistic efforts that we can forge a sustainable future where the health of our planet and its inhabitants thrive in harmonious coexistence.

REFERENCES

1. Marselle MR, Lindley SJ, Cook PA, Bonn A. Biodiversity and health in the Urban environment. *Curr Environ Health Rep.* 2021;8(2):146.
2. Mackinnon K, Ham CV, Reilly K, Hopkins J. Nature-based solutions and protected areas to improve Urban biodiversity and health. *Biodiv Health Face Clim Change* [Internet]. 2019 [cited 2023 Sep 28]; Available from: https://doi.org/10.1007/978-3-030-02318-8_16

3. Viegas S. Biodiversity and health. *Eur J Public Health* [Internet]. 2022 Oct [cited 2023 Sep 28];32(Suppl 3). Available from: https://www.ncbi.nlm.nih.gov/pmc/articles/PMC9593403/
4. Marselle M, Stadler J, Korn H, Irvine K, Bonn A. Biodiversity and health in the face of climate change. *Biodiv Health Face Clim Change* [Internet]. 2019 [cited 2023 Sep 28]; Available from: https://doi.org/10.1007/978-3-030-02318-8
5. Morand S, Lajaunie C. Outbreaks of vector-borne and zoonotic diseases are associated with changes in forest cover and oil palm expansion at global scale. *Front Veteri Sci* [Internet]. 2021 [cited 2023 Sep 28];8. Available from: https://www.ncbi.nlm.nih.gov/pmc/articles/PMC8024476/
6. Abdel-Azeem AM, Abdel-Azeem M, Darwish A, Nafady NA, Ibrahim NA. Fusarium: biodiversity, ecological significances, and industrial applications. *Recent Advan White Biotechnol Through Fungi* [Internet]. 2019 [cited 2023 Sep 28]; Available from: https://doi.org/10.1007/978-3-030-10480-1_6
7. Workshop on Bridging the Gap between Forestry and Health Sectors in Ethiopia. 2008 [cited 2023 Sep 28]; Available from: https://www.semanticscholar.org/paper/4ad69304c3150796da135e758ea07d3990cf6116
8. Potts S, Imperatriz-Fonseca V, Ngo H, Biesmeijer J, Breeze T, Dicks L, et al. Summary for policymakers of the assessment report of the Intergovernmental Science-Policy Platform on Biodiversity and Ecosystem Services (IPBES) on pollinators, pollination and food production. 2016 [cited 2023 Sep 28]; Available from: https://www.semanticscholar.org/paper/774de8e7fbda8efc9c44572f277448cab41340c0
9. Mayorga I, de Mendonça JLV, Hajian-Forooshani Z, Lugo-Pérez J, Perfecto I. Tradeoffs and synergies among ecosystem services, biodiversity conservation, and food production in coffee agroforestry. 2022 [cited 2023 Sep 28]; Available from: https://doi.org/10.3389/ffgc.2022.690164
10. Neilly H, VanDerWal J, Schwarzkopf L. Balancing biodiversity and food production: a better understanding of wildlife response to grazing will inform off-reserve conservation on rangelands☆. *Rangeland Ecol Manage* [Internet]. 2016 [cited 2023 Sep 28]; Available from: https://doi.org/10.1016/j.rama.2016.07.007
11. Fastré C, van Zeist WJ, Watson J, Visconti P. Integrated spatial planning for biodiversity conservation and food production. *One Earth* [Internet]. 2021 [cited 2023 Sep 28]; Available from: https://doi.org/10.1016/j.oneear.2021.10.014
12. Kremen C. Ecological intensification and diversification approaches to maintain biodiversity, ecosystem services and food production in a changing world. *Emerg Topi Life Sci.* 2020 Sep 9;4(2):229.
13. Falco S, Bezabih M, Yesuf M. Seeds for livelihood: Crop biodiversity and food production in Ethiopia. *Ecol Econ* [Internet]. 2010 [cited 2023 Sep 28]; Available from: https://doi.org/10.1016/J.ECOLECON.2010.03.024
14. Isibor PO, Akinduti PA, Aworunse OS, Oyewale JO, Oshamika O, Ugboko HU, et al. Significance of African Diets in Biotherapeutic Modulation of the Gut Microbiome. *Bioinform Biol Insights* [Internet]. 2021 [cited 2023 Sep 28];15. Available from: https://www.ncbi.nlm.nih.gov/pmc/articles/PMC8107938/
15. Losno EA, Sieferle K, Armando Perez-Cueto FJ, Ritz C. Vegan Diet and the Gut Microbiota Composition in Healthy Adults. Nutrients [Internet]. 2021 Jul [cited 2023 Sep 28];13(7). Available from: https://www.ncbi.nlm.nih.gov/pmc/articles/PMC8308632/
16. Calatayud M, Van den Abbeele P, Ghyselinck J, Marzorati M, Rohs E, Birkett A. Comparative effect of 22 dietary sources of fiber on gut microbiota of healthy humans in vitro. *Front Nutri* [Internet]. 2021 [cited 2023 Sep 28];8. Available from: https://www.ncbi.nlm.nih.gov/pmc/articles/PMC8282825/
17. Mahardja B, Hobbs JA, Ikemiyagi N, Benjamin A, Finger AJ. Role of freshwater floodplain-tidal slough complex in the persistence of the endangered delta smelt. *PLoS One* [Internet]. 2019 [cited 2023 Sep 28];14(1). Available from: https://www.ncbi.nlm.nih.gov/pmc/articles/PMC6314582/
18. Dall'Asta M, Laghi L, Morselli S, Re MC, Zagonari S, Patuelli G, et al. Pre-pregnancy diet and vaginal environment in caucasian pregnant women: an exploratory study. Front Mol Biosci [Internet]. 2021 [cited 2023 Sep 28];8. Available from: https://www.ncbi.nlm.nih.gov/pmc/articles/PMC8356051/
19. Marselle MR, Bowler DE, Watzema J, Eichenberg D, Kirsten T, Bonn A. Urban street tree biodiversity and antidepressant prescriptions. *Sci Rep.* 2020 Dec 31;10(1):22445.
20. Beute F, Andreucci MB, Lammel A, Davies ZG, Glanville J, Keune H, et al. Types and characteristics of urban and peri-urban green spaces having an impact on human mental health and wellbeing: a systematic review. 2020 Jun 1 [cited 2023 Oct 1]; Available from: https://kar.kent.ac.uk/89644/1/Beute%20et%20al.%20Green%20Space%20and%20Mental%20Health%202020%20Main%20Report.pdf
21. Felappi JF, Sommer JH, Falkenberg T, Terlau W, Kötter T. Green infrastructure through the lens of "One Health": a systematic review and integrative framework uncovering synergies and trade-offs between mental health and wildlife support in cities. *Sci Total Environ.* 2020 Dec 15;748:141589.

22. Dean J, van Dooren K, Weinstein P. Does biodiversity improve mental health in urban settings? *Med Hypotheses* [Internet]. 2011 Jun [cited 2023 Oct 1];76(6). Available from: https://pubmed.ncbi.nlm.nih.gov/21429673/
23. Chaudhury P, Banerjee D. "Recovering With Nature": a review of ecotherapy and implications for the COVID-19 pandemic. *Front Public Health* [Internet]. 2020 Dec 10 [cited 2023 Oct 1];8. Available from: https://pubmed.ncbi.nlm.nih.gov/33363096/
24. Zhang X, Zhang Y, Zhai J. Home Garden with eco-healing functions benefiting mental health and biodiversity during and after the COVID-19 pandemic: a scoping review. *Front Public Health* [Internet]. 2021 Nov 11 [cited 2023 Oct 1];9. Available from: https://pubmed.ncbi.nlm.nih.gov/34869154/
25. Frumkin H. Urban sprawl and public health. *Public Health Rep* [Internet]. 2002 May [cited 2023 Oct 1];117(3). Available from: https://pubmed.ncbi.nlm.nih.gov/12432132/
26. Cinner J. Coral reef livelihoods. Current Opinion in Environmental Sustainability [Internet]. 2014 [cited 2023 Oct 1]; Available from: https://doi.org/10.1016/J.COSUST.2013.11.025
27. Groot R, Brander L, Ploeg S, Costanza R, Bernard F, Braat L, et al. Global estimates of the value of ecosystems and their services in monetary units. *Ecosyst Serv* [Internet]. 2012 [cited 2023 Oct 1]; Available from: https://doi.org/10.1016/J.ECOSER.2012.07.005
28. Hughes TP, Kerry JT, Álvarez-Noriega M, Álvarez-Romero JG, Anderson KD, Baird AH, et al. Global warming and recurrent mass bleaching of corals. *Nature* [Internet]. 2017 Mar 15 [cited 2023 Oct 1];543(7645). Available from: https://pubmed.ncbi.nlm.nih.gov/28300113/
29. Issues. *Issues in Science and Technology.* 2002 [cited 2023 Oct 1]. Life-Saving Products from Coral Reefs. Available from: https://issues.org/p_bruckner-coral-reefs-importance/
30. Schmidtko A, Lötsch J, Freynhagen R, Geisslinger G. Ziconotide for treatment of severe chronic pain. *Lancet* [Internet]. 2010 [cited 2023 Oct 1]; Available from: https://www.sciencedirect.com/science/article/pii/S0140673610603546?dgcid=api_sd_search-api-endpoint
31. Trease and Evans' Pharmacognosy.
32. Prashar S, Kaur P, Sharma P, Khatun A, Shaikh N. A Study on Comparative Antioxidant Properties of Mimosa pudica, Vachellia nilotica, Leucas aspera, Phyllanthus niruri, emidesmus indicus and Adhatoda vasica. *Int J Curr Microbiol Appl Sci* [Internet]. 2020 [cited 2023 Oct 1]; Available from: https://doi.org/10.20546/IJCMAS.2020.912.100
33. Sule WF, Okonko I, Joseph TA, Ojezele M, Nwanze J, Alli JA, et al. In vitro antifungal activity of Senna alata Linn. crude leaf extract. *Res J Biol Sci* [Internet]. 2010 [cited 2023 Oct 1]; Available from: https://doi.org/10.3923/RJBSCI.2010.275.284
34. Midawa SM, Ali BD, Mshelia BZ, Johnson J. cenresinjournals. 2010 [cited 2023 Sep 28]. Cutaneous wound healing activity of the ethanolic extracts of the leaf of Senna Alata L. (FABACEAE). Available from: https://www.cenresinjournals.com/2020/03/23/cutaneous-wound-healing-activity-of-the-ethanolic-extracts-of-the-leaf-of-senna-alata-l-fabaceae/
35. Sy T, LamuF W, As R. Acute toxicity, phytochemistry and antibacterial activity of aqueous and ethanolic leaf extracts of cassia Alata Linn . 2012 [cited 2023 Oct 1]; Available from: https://www.semanticscholar.org/paper/ACUTE-TOXICITY%2C-PHYTOCHEMISTRY-AND-ANTIBACTERIAL-OF-Sy-LamuF/f4ccd116b60d42a0b79f351f642db41f44eb79c1
36. Rastogi RP, Mehrotra BN, Sinha S, Pant P, Seth R. Compendium of Indian medicinal plants. 1990 [cited 2023 Oct 1]; Available from: https://www.semanticscholar.org/paper/Compendium-of-Indian-medicinal-plants-Rastogi-Mehrotra/0942159e0856f53765635ff696468eb27a4ce27c
37. Vermeer BJ, Gilchrest BA. Cosmeceuticals. A proposal for rational definition, evaluation, and regulation. *Arch Dermatol* [Internet]. 1996 Mar [cited 2023 Oct 1];132(3). Available from: https://pubmed.ncbi.nlm.nih.gov/8607641/
38. Jain D, Chaudhary P, Kotnala A, Hossain R, Bisht K, Hossain MN. Hepatoprotective activity of medicinal plants: a mini review. *J Med Plants Stud* [Internet]. 2020 [cited 2023 Oct 1]; Available from: https://doi.org/10.22271/PLANTS.2020.V8.I5C.1212
39. Vivekraj P, Vinotha S, Vijayan A, Anand Gideon V. Preliminary phytochemical screening and GC-MS analysis of methanolic extract of Turnera subulata Smith (Passifloraceae). *J Phytopharmacol* [Internet]. 2017 [cited 2023 Sep 28]; Available from: https://doi.org/10.31254/phyto.2017.6305
40. Shakya A. Medicinal plants: Future source of new drugs. *Int J Herbal Med* [Internet]. 2016 [cited 2023 Oct 1]; Available from: https://www.semanticscholar.org/paper/Medicinal-plants%3A-Future-source-of-new-drugs-Shakya/e21da2a46b41d288238a27eb7137e72f5178b59e
41. Shaikh AM, Shrivastava B, Apte KG, Navale SD. Medicinal plants as potential source of anticancer agents: a review. *J Pharmacogn Phytochem.* 2016;5(2):291–5.

42. Plant Based Drugs and Medicines [Internet]. [cited 2023 Oct 1]. Available from: https://www.rain-tree.com/plantdrugs.htm
43. Singh S, Pathak N, Fatima E, Negi AS. Plant isoquinoline alkaloids: Advances in the chemistry and biology of berberine. *Eur J Med Chem* [Internet]. 2021 Dec 15 [cited 2023 Oct 1];226. Available from: https://pubmed.ncbi.nlm.nih.gov/34536668/
44. Lai HC, Singh NP, Sasaki T. Development of artemisinin compounds for cancer treatment. *Invest New Drugs* [Internet]. 2013 Feb [cited 2023 Oct 1];31(1). Available from: https://pubmed.ncbi.nlm.nih.gov/22935909/
45. Mamadalieva N, Mamedov N. Taxus brevifolia a High-Value Medicinal Plant. *Source Taxol.* 2020 [cited 2023 Oct 1]; Available from: https://doi.org/10.1007/978-3-030-44930-8_9
46. Veeresham C. Natural products derived from plants as a source of drugs. *J Adv Pharm Technol Res* [Internet]. 2012 Oct [cited 2023 Oct 1];3(4). Available from: https://pubmed.ncbi.nlm.nih.gov/23378939/
47. Heberer T. Occurrence, fate, and removal of pharmaceutical residues in the aquatic environment: a review of recent research data. *Toxicol Lett.* 2002 May 10;131(1–2):5–17.
48. Choi K, Kim Y, Jung J, Kim MH, Kim CS, Kim NH, et al. Occurrences and ecological risks of roxithromycin, trimethoprim, and chloramphenicol in the Han River, Korea. *Environ Toxicol Chem* [Internet]. 2008 Mar [cited 2023 Oct 1];27(3). Available from: https://pubmed.ncbi.nlm.nih.gov/17944547/
49. Ingerslev F, Toräng L, Loke ML, Halling-Sørensen B, Nyholm N. Primary biodegradation of veterinary antibiotics in aerobic and anaerobic surface water simulation systems. *Chemosphere* [Internet]. 2001 Aug [cited 2023 Oct 1];44(4). Available from: https://pubmed.ncbi.nlm.nih.gov/11482679/
50. Costello MJ, Grant A, Davies IM, Cecchini S, Papoutsoglou S. The control of chemicals used in aquaculture in Europe. *J Aquacul Res Develop* [Internet]. 2015 [cited 2023 Oct 1]; Available from: https://www.indexindex.com/journal-of-aquaculture-research-development/the-control-of-chemicals-used-in-aquaculture-in-europe-187737.html
51. Teleosis Institute. PRLog. [cited 2023 Oct 1]. Teleosis Institute Urges to Consider Environmental Consequence of Medical Treatments. Available from: https://www.prlog.org/10066588-teleosis-institute-urges-to-consider-environmental-consequence-of-medical-treatments.html
52. Kidd KA, Blanchfield PJ, Mills KH, Palace VP, Evans RE, Lazorchak JM, et al. Collapse of a fish population after exposure to a synthetic estrogen. *Proc Natl Acad Sci U S A* [Internet]. 2007 May 22 [cited 2023 Oct 1];104(21). Available from: https://pubmed.ncbi.nlm.nih.gov/17517636/
53. Shroff R, Cortés CR. The biodiversity paradigm: building resilience for human and environmental health. *Development.* 2020;63(2–4):172.
54. Chidawanyika F, Muriithi B, Niassy S, Ouya FO, Pittchar J, Kassie M, et al. Sustainable intensification of vegetable production using the cereal "push-pull technology": benefits and one health implications. *Environ Sustain* [Internet]. 2023 [cited 2023 Sep 28]; Available from: https://doi.org/10.1007/s42398-023-00260-1
55. Schmidt K, Walz A. Ecosystem-based adaptation to climate change through residential urban green structures: co-benefits to thermal comfort, biodiversity, carbon storage and social interaction. *One Ecosyst* [Internet]. 2021 [cited 2023 Sep 28]; Available from: https://pdfs.semanticscholar.org/4a53/7ec107ae6f940b7bf3c1e80f72c850954e13.pdf
56. Charles WM, Delgado BM. Health datasets as assets: Blockchain-based valuation and transaction methods. *Blockchain Healthcare Today* [Internet]. 2022 [cited 2023 Sep 28];5. Available from: https://www.ncbi.nlm.nih.gov/pmc/articles/PMC9907414/
57. Chauvenet ALM, Durant SM, Hilborn R, Pettorelli N. Unintended Consequences of Conservation Actions: Managing Disease in Complex Ecosystems. *PLoS One* [Internet]. 2011 [cited 2023 Sep 28];6(12). Available from: https://www.ncbi.nlm.nih.gov/pmc/articles/PMC3233597/
58. Mulumba JV, Nankya R, Adokorach J, Kiwuka C, Fadda C, Santis P, et al. A risk-minimizing argument for traditional crop varietal diversity use to reduce pest and disease damage in agricultural ecosystems of Uganda. *Agric Ecosyst Environ* [Internet]. 2012 [cited 2023 Sep 28]; Available from: https://doi.org/10.1016/J.AGEE.2012.02.012
59. Spencer R. Improvement of human and environmental health through waste management in Antigua and Barbuda. *Foster Trans Change Sustain Context Socio-Ecol Prod Landscapes Seascapes (SEPLS)* [Internet]. 2021 [cited 2023 Sep 28]; Available from: https://doi.org/10.1007/978-981-33-6761-6_12
60. Pocock M, August TA, Fox R, Logie M. The information content of biodiversity datasets is affected by patterns of citizen science participation both between and within projects. *Biodivers Inf Sci Stand* [Internet]. 2019 [cited 2023 Sep 28]; Available from: https://pdfs.semanticscholar.org/6148/24224e433f753ac50a58102da4911fa04cc7.pdf

61. Costa APT, Schneck F. Diatoms as indicators in running waters: trends of studies on biological assessment and monitoring. *Environ Monit Assess* [Internet]. 2022 Aug 19 [cited 2023 Sep 28];194(10). Available from: https://pubmed.ncbi.nlm.nih.gov/35986195/
62. Tóth Z, Dombos M, Hornung E. Urban soil quality deteriorates even with low heavy metal levels: an arthropod-based multi-indices approach. *Ecol Appl* [Internet]. 2023 Jun [cited 2023 Sep 28];33(4). Available from: https://pubmed.ncbi.nlm.nih.gov/36938710/
63. Wali M, Evrendilek F, Fennessy M. The environment: science, issues, and solutions. 2009 [cited 2023 Sep 28]; Available from: https://doi.org/10.1201/9781420007336
64. Mwangi-Gachau E. Legal requirements for geothermal development in Kenya. 2016 [cited 2023 Sep 28]; Available from: https://www.semanticscholar.org/paper/ef5c9f497776d95b71eb343ce8329f408e8460ed
65. Paudel U. A review of effects of environmental change on human health. *Am J Environ Sci* [Internet]. 2018 [cited 2023 Sep 28]; Available from: https://doi.org/10.3844/AJESSP.2018.95.109
66. Nhemachena C, Chakwizira J. Meta analysis of research on climate change adaptation and health in southern Africa: Identifying research gaps. 2010 [cited 2023 Sep 28]; Available from: https://www.semanticscholar.org/paper/96ecc94a728acac0dfb8ac917053451c070bc844

25 The Green Revolution Reimagined

Agricultural Biodiversity and Sustainable Food Security

Rashi Miglani[1]*, Gaurav Rawat*[1]*, Swati Joshi*[1]*, Vishal Singh*[1]*, Nagma Parveen*[1]*, Uzma Siddiqui*[1]*, Netrapal Sharma*[1]*, Ankit Kumar*[2]*, Seeta Dewali*[1]*, Satpal Singh Bisht*[1]

[1]Department of Zoology, DSB Campus, Kumaun University, Nainital, Uttarakhand-India

[2]Department of Pharmaceutical Sciences, Sir J.C Bose Technical Campus, Bhimtal, Kumaun University, Nainital, Uttarakhand

25.1 INTRODUCTION

The phrase "green revolution" refers to the significant rise in wheat and rice yields in developing nations that are developing brought about by better cultivars together with greater usage of fertilizers and other chemical inputs. This phenomenon has had a huge influence on the availability of food and earnings in many developing nations (John and Babu, 2021; Milani et al., 2022). Since the second half of the 19th century, food systems have been affected by the phenomenon of globalization, which promoted the involvement of several intermediaries between producers and consumers. Scientists from various nations endeavored to convert conventional staple crops, mainly wheat, into semi-dwarf high-yielding varieties (HYVs) by means of hybridization (Tauger, 2023). Simultaneously, it offers protection against shocks to regional output, including, but not limited to, conflicts, diseases, and environmental hazards hastened by climate change. Further benefits of this market globalization include increased international trade in goods and services, which gives local food producers access to a larger market and allows them to spend more capital on their operations. Therefore, by permitting regional specialization in the production of the most appropriate foods at the local level, it raises the overall efficiency of food production (Soria-Lopez et al., 2023).

Sato, 2013 identifies three distinct paradigm shifts in the domain of food security that have transpired since the 1970s. From a macroscopic perspective, including the global and national levels, down to a microscopic one focusing on the family and individual, from the standpoint of prioritizing food security and considering its implications on people's livelihoods to the transition from empirical indicators to subjective perception.

Based on the green revolution, the practice of agriculture has resulted in the depletion of soil nutrients (Tilman, 1998), the exhaustion of water resources (Davis et al., 2018), the decline in agrobiodiversity (Perfecto et al., 2019), and the escalation of greenhouse gas emissions (Sapkota et al., 2019). The consequences on public health, as well as the ongoing financial commitments required for purchasing seeds and chemical inputs, have contributed to the emergence of debt cycles among several agricultural households, hence exacerbating the environmental ramifications (Wakchaure, 2020).

DOI: 10.1201/9781032697550-25

In order to mitigate greenhouse gas emissions, ensure sustainable production levels, and preserve ecosystem services, it is imperative to implement transformative reforms within India's agricultural sector. India must enhance the management of soils, water, and crop residues, minimize food waste, enhance biodiversity in and around agricultural areas, and intensify carbon sequestration efforts (Bawa and Seidler, 2023).

25.2 SUSTAINABLE FOOD SYSTEMS WITH AGRICULTURAL BIODIVERSITY

The current state of global food production stands as the primary catalyst for environmental degradation and the decline of biodiversity (Willett et al., 2019). The increasing worldwide need for food, coupled with the constraints of limited arable land, necessitates the expansion of agricultural frontiers and output. This phenomenon frequently occurs irrespective of the surrounding ecological context, resulting in the depletion of biodiversity as well as the degradation of land and water resources (Biodiversity International, 2017). The phenomenon of climate change is contributing to the rapid depletion of biodiversity. Elevated temperatures have been found to disturb the process of pollination and the natural regulation of pests, exerting an impact on the overall quality of food (Food and Agriculture Organization of the UN, 2017). Similarly, the imperative to sustainably nourish an extra 2 billion individuals by the year 2050 is compelling us to enhance productivity in some essential crops, thus diminishing both the range of food sources and genetic variation. The reduction of biodiversity within food systems results in a diminished range of choices available to farmers in addressing the risks associated with crop failure, deteriorating soil fertility, and unpredictable weather patterns (Biodiversity International, 2017). Consequently, this leads to decreased output, heightened food insecurity, and an increased prevalence of malnutrition (FAO, IFAD, UNICEF, WFP, WHO, 2018).

Food and agricultural biodiversity play a crucial role in fostering sustainable food systems. The enhancement of agrobiodiversity has been found to have positive effects on both production and the nutritional quality of crops. Additionally, it has been observed to contribute to improvements in soil and water quality while also reducing the reliance on synthetic fertilizers. Using this approach enhances the resilience of farmers' livelihoods by mitigating the adverse effects of climate change and minimizing yield losses caused by insect damage. Expanding the range of cultivated plant species also yields environmental advantages, such as augmenting the population of pollinators and beneficial soil organisms, while concurrently mitigating the likelihood of insect outbreaks. To effectively utilize and preserve agrobiodiversity in a sustainable manner, it is imperative for governments to implement specific policies and plans that encompass several sectors and are informed by empirical research. The significance of food producers, ranging from smallholder farmers to multinational enterprises, in the preservation of genetic resources and the adoption of sustainable agricultural practices is steadily growing. It is important for consumers to enhance their awareness of the ecological consequences of their dietary decisions and their responsibility in safeguarding the ecosystem (Remans et al., 2019).

25.3 AGRICULTURAL BIODIVERSITY: ITS ROLE AND IMPORTANCE

The roles of agrobiodiversity have been aided by a wide variety of disciplinary disciplines. Crop and animal breeding may select for genes to boost agricultural output, as demonstrated by the agricultural sciences. A diversified soil community may control tighter nutrient cycling, lower pest and disease levels, and improve soil structure. Crop variety, both in area and time, can also increase fertilizer usage efficiency (Lachat et al., 2018). According to ecological studies, biodiversity can make grassland ecosystems more stable (Burlingame and Dernini, 2010). This has consequences for managing pastures for the production of animals. Agricultural biodiversity arises from the interplay of genetic resources, environmental factors, and management strategies and practices, such as multiple cropping and mixed cropping, which are embraced by the agricultural community

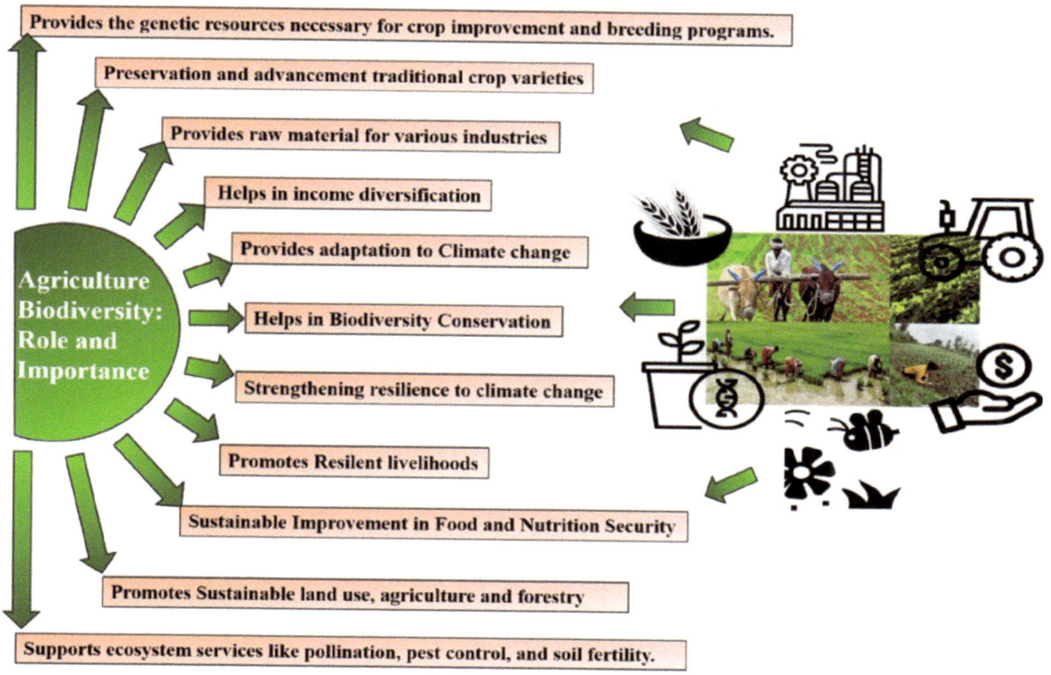

FIGURE 25.1 Agricultural biodiversity promotes several Sustainable Development Goals (SDGs) and aspects of sustainability.

(Figure 25.1). These elements are widely recognized as crucial for ensuring nutrition security. The concept encompasses the diverse range and fluctuation of plant, animal, and microbe species that are essential for maintaining crucial processes within agro-ecosystems, which are responsible for the production of food (Singh, 2022). The presence of a diverse range of genetic resources has the potential to enhance agricultural output, preserve and enhance soil fertility, and mitigate the negative impacts of pests and diseases. Additionally, this technology has the potential to provide farmers an increased gene pool from which they may choose adaptive populations that are better suited to changing climatic circumstances. In a similar vein, the diversification of dietary patterns among individuals according to various agricultural systems can contribute to nutritional security by ensuring the availability of vital nutrients in sufficient amounts.

Additionally, it expands the dietary composition to encompass crucial nutrients, micronutrients, vitamins, fiber, and minerals that are vital for maintaining optimal health. Nevertheless, the presence of agricultural biodiversity and the various food production systems is now being negatively impacted as a result of the loss or reduction in numbers of harvested and unharvested species, variations, animals, and fish breeds (Agricultural Biodiversity, 1999; Singh et al., 2022). Therefore, there is a pressing requirement to control and implement agro-ecosystems that can effectively preserve biodiversity to the appropriate degree, while simultaneously enhancing sustainable food and nutrition security (Srivastava, 2022).

25.4 ECOSYSTEM SERVICES OF SUSTAINABLE AGRICULTURE

Agricultural ecosystem services encompass various functions that contribute to the well-being of ecosystems. These services can be categorized into support functions, such as nutrient cycling and soil formation; provision functions, including the production of food, fuel, fiber, and freshwater;

regulation functions, such as flood, disease, and climate regulation; and cultural functions, which encompass aesthetic, spiritual, and recreational aspects. Implementing sustainable agricultural practices can play a crucial role in enhancing these services (Power, 2010).

25.4.1 Pollination

Pollination is essential for food and livelihoods. About 90% of blooming plants develop seeds by biotic pollination, primarily insects. Insect pollination is crucial to terrestrial ecosystems and provides economic, aesthetic, and cultural advantages to civilization (Abrol, 2015). Insect pollination boosts seed production, productivity, food quality, and genetic diversity (Fijen et al., 2018). Habitat fragmentation, land-use patterns, pesticides, invasive species, predation and competition, and climate change can reduce insect pollinator numbers. Sustainable agriculture reduces pollinator danger by diversifying the agricultural landscape and using natural processes for food production.

25.4.2 Soil Structure and Fertility

The fertility of soil may be influenced by several variables, including physical, chemical, and biological elements, which eventually exert an influence on the development of plants. Plants do not directly uptake nutrients such as nitrogen, phosphorus, sulfur, and carbon in their original forms. Instead, they rely on microbial processes and nutrient cycling to transform these substances into their standardized forms (Javed et al., 2022). The proficient administration of soil nutrients is imperative due to its fundamental role in sustaining human existence. Fertilization serves to augment the soil by introducing supplementary nutrients, hence enhancing soil quality, crop output, and economic gains. Soil fauna is integral to the formation of soil structure through their activities, which include the construction of channels and pores, the aggregation of small soil particles, and the fragmentation and dispersion of organic matter within the soil matrix. The soil microbiomes play a crucial role in agroecosystems by influencing essential processes such as soil fertility, crop yield, and stress tolerance. The microbiome exhibits a complex interdependence with soil structure, encompassing factors such as aggregation and pore connectivity. This structural framework plays a pivotal role in governing the movement of water, oxygen, and nutrients throughout the system (Hartmann and Six, 2023). The maintenance of soil fertility is of paramount importance in ensuring the long-term sustainability of agricultural productivity. The implementation of effective soil management practices is crucial for achieving high levels of productivity that are economically viable and for maintaining soil fertility (Fonte et al., 2010).

25.5 ROLE OF GENETIC DIVERSITY IN CROP IMPROVEMENT

Increased diversity in the field ensures the harvest and minimizes the risk of crop failure due to both biotic and abiotic factors (Joshi et al., 2020; Begna, 2021). Crop diversity contributes to yield stability and insect and pest resistance/tolerance due to a broad genetic base derived from diverse germplasm sources (Furman et al., 2021; He et al., 2021). Success in crop enhancement is predominantly dependent on the extensive range of genetic diversity at the degree level (Mohammadi and Prasanna, 2003; Begna, 2021). The production of improved cultivars in crop improvement greatly relies on the significant role of genetic variation. When distinct genetic materials are crossed, it is anticipated that there will be enhanced performance and the production of desired hybrids, in contrast to the crossover of comparable genetic materials (Govindaraj et al., 2015; Bhandari et al., 2017; Begna, 2021; Salgotra and Chauhan, 2023). Due to climatic variations and the requirements of different products to meet the preferences of people, the majority of farmers in Nepal grow a large number of crop species and their varieties, as well as landraces. Growing varieties of the same crops is easy and also provides multiple benefits. This is very important for sustainable agricultural businesses maintaining a functional agroecosystem. Many farmers are very familiar with the advantages of increased crop diversity in the fields. Farmers keep seeds from such mixed populations for the next season, which favors natural selection. Alongside this, they are resilient to climate and other environmental factors, including weeds, and perform relatively better than other newly developed narrow genetic base varieties (Joshi et al., 2020).

25.6 CHALLENGES IN GENETIC CROP IMPROVEMENT AND THE GREEN REVOLUTION

25.6.1 Genetic Erosion and Loss of Traditional Crop Varieties

The productivity, resilience, and adaptive ability of agriculture are fundamentally supported by crop variety. The loss of biodiversity in crops, commonly referred to as crop genetic erosion, is a matter of significant concern. The issue of declining crop diversity has been a subject of concern for more than a century; however, there is still a lack of comprehensive understanding regarding the extent, direction, causes, and importance of these losses (Khoury et al., 2022). The loss of this diversity, commonly referred to as crop genetic erosion, is consequently problematic (Renard and Tilman, 2019; Egli et al., 2020). The phenomenon of genetic erosion and the subsequent decline in traditional crop types provide substantial challenges to the preservation of agricultural biodiversity and the maintenance of global food security. The aforementioned phenomena pertain to the reduction in genetic variation observed within a given species, with a specific emphasis on crops. This reduction occurs as a consequence of the substitution of a wide range of locally adapted varieties with a restricted selection of economically useful cultivars that exhibit high yield potential (Brown, 2008).

India had a significant reduction in its indigenous rice varieties, amounting to almost 1 lakh, as a consequence of the green revolution (Nelson et al., 2019). Since the advent of the green revolution, there has been a decline in the cultivation of native strains of rice, millets, lentils, and other crops. Consequently, there was a notable augmentation in the yield of hybrid crops, characterized by their accelerated growth rate (Taylor, 2019). There has been a significant surge in the cultivation of wheat, soybeans, and rice. Furthermore, there has been a significant decline in the production of sorghum, as well as other millets, barley, and groundnuts. The rise in the cultivation of certain crops can be attributed to the widespread availability of HYVs of seeds, along with an expansion in the overall cultivation area dedicated to these crops (Singh, 2019).

25.6.2 Pests and Pesticide

The use of pesticides has become an essential component of contemporary society to satisfy the needs of a burgeoning population projected to reach 10 billion individuals by the year 2050 (Miglani and Bisht, 2019). India became one of Asia's major pesticide manufacturers as pesticide use increased (Narayanan et al., 2016). Although this has led to economic growth (Gollin et al., 2018), many pesticides are superfluous in both industrialized and developing nations. Overuse of pesticides and anthropogenic climate change raise crop losses and pesticide resistance, compromising 21st-century food security and safety (Palumbi, 2001; Carvalho, 2006; Deutsch et al., 2018). For instance, chemicals in freshwater surpass specified limits, causing significant problems (Choudhary et al., 2018). Despite using less pesticide than many other nations, India has substantial pesticide residue. Large amounts of water contamination and soil degradation result. Pest attacks, caused by pest imbalance, are another big concern. Due to pesticide use, predator and prey pests are out of balance, resulting in an abundance of one insect that attacks specific crops. The production of certain crops becomes imbalanced. These crops require stronger or novel insecticides to combat pests. This has also disrupted the food chain (Narayanan et al., 2016).

FIGURE 25.2 Application of Insecticides without safety gears (Miglani et al., 2021)

25.6.3 Impact on Farmers' Health

To feed a growing population, agricultural practices strive to boost crop output sustainably. Farmers' illogical usage and improper mixing of pesticides to boost productivity harm health and the environment. In many developing and developed nations, pesticide application techniques have not improved. India has a higher pesticide danger than other industrialized nations.

Damalas and Eleftherohorinos (2016) report that improper pesticide handling and spraying raise health risks. Operator expertise, health, protective equipment (such as gloves, boots, and glasses), and improper storage and disposal of leftover or left-out chemicals are crucial in pesticide application (Figure 25.2).

25.7 GREEN REVOLUTION TO GENOMIC REVOLUTION: IMPROVING AGRICULTURAL RESILIENCE TO ENVIRONMENTAL INSTABILITY

Crop enhancement initiatives have been initiated since the advent of agriculture, mostly through the utilization of conventional breeding techniques. Farmers and plant breeders persist in employing these tactics for crop enhancement owing to their extensive use in altering the genetic compositions of crops. However, traditional breeding is associated with notable drawbacks in terms of the required level of work and time. The yield of crops appears to be reaching a plateau due to environmental factors and the limited availability of agricultural land. Hence, the ongoing endeavor to enhance agricultural development is imperative. In recent times, there have been notable advancements in technology that have brought about a significant transformation in breeding practices, with a growing emphasis on molecular approaches. Recently, marker-assisted selection (Rana et al., 2019; Tian et al., 2019; Xu et al., 2020; Sadia et al., 2021), QTL mapping (Bouchez et al., 2002; Malik and Malik, 2015; Tian et al., 2019; Xu et al., 2020), omics-assisted breeding (Barrera-Redondo et al., 2020), genome-wide association studies, and genome editing have emerged as notable methodologies in the field. Numerous nations have gradually eased their regulations pertaining to genome editing. This study focuses on the evolutionary and revolutionary strategies employed in crop development to develop climate-resilient crops, while considering the impact of climate change (Gupta, 2004). Moreover, it will enhance the understanding of the progression of plant breeding thus far. The allocation of resources toward the use of modern sequencing technology and bioinformatics has the potential to enhance our comprehension of genetic differences and their consequential effects, hence facilitating significant advancements in the domains of agricultural enhancement and biodiversity preservation (Chawla et al., 2023).

25.8 INTERNATIONAL AND NATIONAL POLICIES: AGRICULTURAL BIODIVERSITY AND SUSTAINABLE FOOD SECURITY

Policies on both the international and national levels play an essential part in the process of fostering agricultural biodiversity and achieving long-term food security. These policies address a variety of issues, including the preservation of biological diversity, the use of environmentally responsible farming practices, the adjustment to the effects of climate change, and the provision of equal access to resources (Lee, 2005; Ogundari, 2014; Ramphul et al., 2016; Pawlak and Kolodziejczak, 2020). Programs and policies aim to enhance agricultural efficiency by enhancing extension services and implementing comprehensive training programs for farmers on the use of contemporary technology, among other strategies (Ogundari, 2014; Prus, 2017; Prus, 2018; Prus, 2019).

25.8.1 International Policies

(a) **Convention on Biological Diversity (CBD)**: Initiatives to conserve biodiversity are centered on the Convention on Biological Diversity (CBD). Through locally developed solutions and measures, it provides an opportunity to address global concerns at the national level (Chandra and Idrisova, 2011). The CBD is a globally recognized legal framework that aims to promote the preservation of biological diversity, the sustainable utilization of its various components, and equitable distribution of benefits derived from the exploitation of genetic resources. This convention has been formally adopted by 196 nations worldwide (UN, 2023a).

The CBD acknowledges that biological variety encompasses not only flora, fauna, and microorganisms with their associated ecosystems but also encompasses the human population and its fundamental requirements for sustenance, pharmaceutical resources, unpolluted air and water, adequate housing, and a pristine and salubrious environment conducive to habitation (CBD, 2023). The primary aim of this initiative is to promote behaviors that will ultimately contribute to the attainment of a sustainable future.

(b) **International Treaty on Plant Genetic Resources for Food and Agriculture (ITPGRFA)**: The preservation and responsible utilization of plant genetic resources for the purpose of nourishment and agriculture play a crucial role in guaranteeing adequate food production to maintain the expanding global population in the forthcoming years. The establishment of the Commission on Genetic Resources for Food and Agriculture took place in 1983, alongside the adoption of the voluntary International Undertaking on Plant Genetic Resources. A significant milestone occurred in 1996 when the Global Plan of Action was embraced during the Leipzig International Technical Conference on Plant Genetic Resources. The culmination of these efforts occurred in 2001 when the legally binding International Treaty on Plant Genetic Resources for Food and Agriculture was adopted, marking a significant milestone. The treaty came into effect on June 29, 2004 (FAO, 2023) and aims at:

- acknowledging farmers' important contributions to the diversity of foods that feed the globe
- developing a worldwide system to enable access to plant genetic resources for farmers, plant breeders, and scientists
- ensuring that recipients enjoy the advantages derived from the use of these genetic resources

(c) **Sustainable Development Goals (SDGs)**: Agricultural biodiversity and food security are closely intertwined with a number of Sustainable Development Goals (SDGs), notably Goal 2 (Zero Hunger) and Goal 15 (Life on Land). The aforementioned objectives underscore the significance of sustainable agriculture, the preservation of biodiversity, and the eradication of hunger (UN, 2023b).

25.8.2 National Policies

(a) **National Biodiversity Strategies and Action Plans (NBSAPs)**: Numerous nations have formulated National Biodiversity Strategies and Action Plans (NBSAPs) to delineate their approaches and initiatives in the preservation of biodiversity, encompassing agricultural biodiversity as well. The plans incorporate strategies aimed at safeguarding indigenous crop types, advancing sustainable agricultural methods, and improving the economic well-being of farmers (MoNRE, 2016).

(b) **National seed policies**: Seed policies at the national level encompass the oversight and control of seed production, distribution, and trade. The implementation of policies that promote the use of traditional and regionally adapted crop types has the potential to augment agricultural biodiversity. Furthermore, it is of utmost importance to advocate for the preservation of farmers' entitlements to save, employ, interchange, and vend seeds (De Jonge et al., 2021). For example, standard operational procedures were created in 2021 by the Plant Genetic Resources Centre (PGRC) of the National Agricultural Research Organization (NARO) to provide guidance for the creation of community seed banks across Uganda in accordance with national policy (Vernooy et al., 2022).

(c) **Agricultural extension and research programs**: The agricultural extension services and research programs that educate farmers about sustainable agricultural practices, such as crop diversification, agroecology, and organic farming approaches, are frequently supported by the government (Gamage et al., 2023). There are many government projects in progress, such as in India the KisanMitr and mKisan Portal. The KisanMitr initiative can be accessed at https://kisanmitr.gov.in, while the mKisan Portal can be found at http://www.mkisan.gov.in/aboutmkisan.aspx. Civil society organizations, such as Digital Green and e-Kisaan, have shown promising capabilities in reaching a large number of farmers and offering impactful information and services throughout the entire crop cycle, including product marketing. Digital Green's website can be accessed at https://www.digitalgreen.org/, while e-Kisaan's website is available at http://www.ekisaan.com/. Additionally, the company Ag Revolution provides further insights into the transformative nature of these organizations' initiatives (https://agrevolution.in/company) (Bawa and Seidler, 2023).

The formulation and implementation of policies and governance mechanisms pertaining to agricultural biodiversity are of utmost importance in influencing the long-term sustainability and resilience of our food systems. These policies involve a variety of strategies intended to preserve and enhance the genetic diversity of crops and livestock that are crucial for ensuring food security. They encompass seed restrictions, intellectual property rights, land-use planning, and the availability of genetic resources. The implementation of effective governance mechanisms facilitates the inclusion of small-scale farmers, indigenous groups, and local stakeholders in decision-making processes, hence promoting agricultural systems characterized by equity and inclusivity. Through the cultivation of a conducive policy framework and the establishment of robust governance structures, it is possible to protect agricultural biodiversity, bolster the resilience of ecosystems, and effectively tackle the obstacles presented by climate change, pests, and diseases. This concerted effort will eventually ensure the long-term sustainability of our global food production (Leventon et al., 2019; Valujeva et al., 2023; Schaal et al., 2023; Achieng et al., 2023).

25.9 CHALLENGES AND FUTURE PERSPECTIVE

Innovative farming practices that boost productivity while protecting various crops and their wild equivalents are required to strike a fair balance between the pursuit of high crop yields and the preservation of agricultural biodiversity. Safeguarding genetic resources and fostering sustainable food systems requires strong regulations and governance structures. Still, it's not only difficult to come up with these regulations, it's also difficult to make sure they're consistently enforced and standardized across borders. Since agriculture faces a serious risk from climate change, it is crucial to produce climate-resilient crops using modern breeding and biotechnology. At the same time, we need to educate shoppers so they can make more informed decisions about where their food comes from. Although scientific inquiry and technical progress are crucial, it remains a problem to make them available to and affordable for small-scale farmers in underdeveloped areas. Promoting sustainable practices within large-scale agriculture is another challenge, as is making the move economically viable for farmers. Effective management of agricultural biodiversity requires extensive data and monitoring systems. Lastly, resolving the complexity of these challenges needs cross-sector collaboration, bridging the gap between agriculture, environment, health, trade, and other important sectors to provide a comprehensive and sustainable approach to food security in a changing world.

REFERENCES

Abrol, D. P. (2015). Pollination Biology, Vol. Cham, Switzerland: Springer Science and Business Media.

Achieng, A.O., Arhonditsis, G.B., Mandrak, N., Febria, C., Opaa, B. et al. (2023). Monitoring biodiversity loss in rapidly changing Afrotropical ecosystems: an emerging imperative for governance and research Phil. *Trans. R. Soc.* B37820220271 20220271. http://doi.org/10.1098/rstb.2022.0271

Agricultural Biodiversity (1999). FAO multifunctional character of agriculture and land: Conference background paper no.1 Maastrichet.

Barrera-Redondo J., Pinero D., Eguiarte L. E. (2020). Genomic, transcriptomic and epigenomic tools to study the domestication of plants and animals: A field guide for beginners. *Front. Genet.* 11, 742. https://doi.org/10.3389/fgene.2020.00742

Bawa, K. S., Seidler, R. (2023). Sustainable pathways toward reimagining India's agricultural systems. *Commun. Earth Environ.* 4, 262. https://doi.org/10.1038/s43247-023-00902-6

Begna T. (2021). Role and economic importance of crop genetic diversity in food security. J. Agric. Sci. Food Technol. 7(1), 164–169.

Bhandari, H.R., Bhanu, A.N., Srivastava, K., Singh, M.N., Shreya, H.A. (2017). Assessment of genetic diversity in crop plants—an overview. *Adv. Plants Agric. Res.* 7, 279–286.

Biodiversity International (2017) *Mainstreaming Agrobiodiversity in Sustainable Food Systems: Scientific Foundations for an Agrobiodiversity Index.* Biodiversity International, Rome, Italy. [ISBN 978-92-9255-070-7]

Bouchez A., Hospital F., Causse M., Gallais A., Charcosset A. (2002). Marker-assisted introgression of favorable alleles at quantitative trait loci between maize elite lines. *Genetics.* 162(4), 1945–1959. https://doi.org/10.1093/genetics/162.4.1945

Brown, A.H.D. (2008). *Indicators of Genetic Diversity, Genetic Erosion and Genetic Vulnerability for Plant Genetic Resources for Food and Agriculture.* pp. 1–26. chrome-extension://efaidnbmnnnibpcajpcglclefindmkaj/https://www.fao.org/fileadmin/templates/agphome/documents/PGR/SoW2/PGRFA_Indicators_Thematic_Study.pdf

Burlingame, B., Dernini, S. (2010).Sustainable diets and biodiversity directions and solutions for policy, research and action. *Biodiversity and Sustainable Diets United Against Hunger.* FAO Headquarters, Rome. pp. 1–307.

Carvalho, F. P. (2006). Agriculture, pesticides, food security and food safety. *Environ. Sci. Policy.* 9, 685–692.

CBD (2023). The Convention on Biological Diversity. The Convention on Biological Diversity (cbd.int).

Chandra, A., Idrisova, A. (2011). Convention on biological diversity: a review of national challenges and opportunities for implementation. *Biodiver. Conser.* 20, 3295–3316.

Chawla, R., Poonia, A., Samantara, K., Mohapatra, S.R., Naik, S.B., Ashwath, M.N., Djalovic, I.G., Prasad, P. V. V. (2023). Green revolution to genome revolution: driving better resilient crops against environmental instability. *Front Genet.* 14, 1204585. https://doi.org/10.3389/fgene.2023.1204585.

Choudhary, S., Yamini, N. R., Yadav, S. K., Kamboj, M., and Sharma, A. (2018). A review: pesticide residue: cause of many animal health problems. *J. Entomol. Zool. Stud.* 6, 330–333. https://www.researchgate.net/profile/Sanjay_Choudhary10/publication/325314814_A_review_Pesticide_residue_cause_of_many_animal_health_problems/links/5b063a710f7e9b1ed7e8314f/A-review-Pesticide-residue-cause-of-many-animal-health-problems.pdf

Damalas, C.A., Eleftherohorinos. I.G. (2016). Pesticide exposure, safety issues, and risk assessment indicators. *Int. J. Environ. Res. Public Health.* 8(5), 1402–1419.

Davis, K. F. et al. (2018). Alternative cereals can improve water use and nutrient supply in India. *Sci. Adv.* 4, eaao1108.

De Jonge, B., López Noriega, I., Otieno, G., Cadima, X., Terrazas, F., Hpommalath, S., van Oudenhoven, F., Shrestha, S., Pudasaini, N., Singh Shrestha, D., et al. (2021). Advances in the registration of farmers' varieties: Four cases from the Global South. *Agronomy.* 11, 2282

Deutsch, C. A. et al. (2018). Increase in crop losses to insect pests in a warming climate. *Science.* 361, 916–919.

Egli, L., Schröter, M., Scherber, C., Tscharntke, T., Seppelt, R. (2020). Crop asynchrony stabilizes food production. *Nature.* 588, E7–E12.

FAO, IFAD, UNICEF, WFP, WHO (2018) The State of Food Security and Nutrition in the World 2018. Building climate resilience for food security and nutrition. FAO, Rome. [ISBN978-92-5-130571-3]

FAO (2023). Food and Agricultural Organization of the United States. https://www.fao.org/plant-treaty/overview/en/

Fijen TP, Scheper JA, Boom TM, Janssen N, Raemakers I, Kleijn D. Insect pollination is at least as important for marketable crop yield as plant quality in a seed crop. *Ecology Letters.* 2018 Nov;21(11):1704–1713.

Food and Agriculture Organization of the UN (2017) The future of food and agriculture –Trends and challenges. FAO, Rome. [ISBN 978-92-5-109551-5]

Fonte, S. J., Barrios, E. & Six, J. (2010). Earthworms, soil fertility and aggregate-associated soil organic matter dynamics in the Quesungual agroforestry system. *Geoderma.* 155, 320–328.

Furman, B., Noorani, A., Mba, C., Furman, B., Noorani, A. and Mba, C. (2021). On-farm crop diversity for advancing food security and nutrition. In *Landraces—Traditional Variety and Natural Breed,* Elkelish, A., Ed.; Intech Open: London, UK.

Gamage, A., Gangahagedara, R., Gamage, J., Jayasinghe, N., Kodikara, N. et al. (2023). Role of organic farming for achieving sustainability in agriculture. *Farming Syst.* 1(1), 100005. https://doi.org/10.1016/j.farsys.2023.100005

Gollin, D., Hansen, C. W., Wingender, A. (2018). *Two Blades of Grass: The Impact of the Green Revolution.* Cambridge, CA: National Bureau of Economic Research. https://doi.org/10.3386/w24744

Govindaraj, M., Vetriventhan, M., Srinivasan, M. (2015). Importance of genetic diversity assessment in crop plants and its recent advances: aAn overview of its analytical perspectives. *Genet. Res. Int.* 2015, 431487

Gupta A. K. (2004). Origin of agriculture and domestication of plants and animals linked to early Holocene climate amelioration. *Curr. Sci.* 87, 54–59. https://www.jstor.org/stable/24107979 .

Hartmann, M., Six, J. (2023). Soil structure and microbiome functions in agroecosystems. *Nat. Rev. Earth Environ.* 4, 4–18. https://doi.org/10.1038/s43017-022-00366-w

He, P., Wang, C., Zhang, N., Liu, B., Yang, Y., Zhu, Y., Li, X., Yu, X., Han, G. and Wang, Y.-Y. (2021). Multi-genotype varieties reduce rice diseases through enhanced genetic diversity and show stability and adaptability in the field. *Phytopathol. Res.* 3, 28.

Javed, A., Ali, E., Afzal, K.B., Osman, A. and Riaz, D.S. (2022). Soil Fertility: Factors Affecting Soil Fertility, and Biodiversity Responsible for Soil Fertility. *Int. J. Plant Animal Environ. Sci.* 12: 021-033.

John, D.A., Babu, G.R. (2021). Lessons from the aftermaths of green revolution on food system and health. *Front. Sustain. Food Syst.* 5, 644559. https://doi.org/10.3389/fsufs.2021.644559

Joshi, B.K., Vista, S.P., Gurung, S.B., Ghimire, K.H., Gurung, R., Pant, S., Gautam, S., Paneru, P.B. (2020). Cultivar mixture for minimizing risk in farming and conserving agrobiodiversity. In *Traditional Crop Biodiversity for Mountain Food and Nutrition Security in Nepal,* Gauchan, D., Joshi, B.K., Bhandari, B., Manandhar, H.K., Jarvis, D., Eds., NAGRC, LI-BIRD and the Alliance of Biodiversity International and CIAT: Kathmandu, Nepal, pp. 14–25.

Khoury, C. K., Brush, S., Costich, D.E., Curry, H.A., De Haan, S., Engels, J.M., ... Thormann, I. (2022). Crop genetic erosion: understanding and responding to loss of crop diversity. *New Phytologist,* 233(1), 84–118.

Lachat, C., Raneri, J.E., Smith, K.W., Kolsteren, P., van Damme, P., Verzelen, K., Penafiel, D., Vanhove, W., Kennedy, G., Hunter, D., et al. (2018). Dietary species richness as a measure of food biodiversity and nutritional quality of diets. *Proc. Natl. Acad. Sci.* 115, 127–132.

Lee, D.R. (2005). Agricultural sustainability and technology adoption: issues and policies for developing countries. *Am. J. Agric. Econ.* 87, 1325–1334.

Leventon, J., Schaal, T., Velten, S., Loos, J., Fischer, J., & Newig, J. (2019). Landscape-scale biodiversity governance: scenarios for reshaping spaces of governance. *Environ. Policy Governance.* 29(3), 170–184.

Malik S., Malik T. A. (2015). Genetic mapping of potential QTLs associated with drought tolerance in wheat. *J. Anim. Plant Sci.* 25(4), 1032–1040.

Miglani, R., Bisht, S.S. (2019). World of earthworms with pesticides and insecticides. *Interdiscip. Toxicol.* 12(2), 71–82. https://doi.org/10.2478/intox-2019-0008.

Miglani, R., Parveen, N., Kumar, A., Sharma, N., Bisht. S. S. (2021). Insecticide Toxicity: Consumption, Assessment Methodologies and Safety Issues in India. Intern. *J. Zool. Invest.* 7 (2), 621–628. https://doi.org/10.33745/ijzi.2021.v07i02.046

Milani, P., Torres-Aguilar, P., Hamaker, B., Manary, M., Abushamma, S. et al. (2022). The whole grain manifesto: From green revolution to grain evolution. *Global Food Security*, 34, 100649. https://doi.org/10.1016/j.gfs.2022.100649

Mohammadi, S.A., Prasanna, B.M. (2003). Analysis of genetic diversity in crop plants-salient statistical tools and considerations. *Crop Sci.* 43, 12351248. https://bit.ly/3drY0xH

MoNRE (2016). Ministry of Natural Resources and the Environment. National Biodiversity Strategy and Action Plan for Lao PDR 2016-2025. Department of Forest Resource Management. Vientiane Lao PDR.

Narayanan, J., Sanjeevi, V., Rohini, U., Trueman, P., and Viswanathan, V. (2016). Postprandial glycaemic response of foxtail millet dosa in comparison to a rice dosa in patients with type 2 diabetes. *Indian J. Med. Res.* 144, 712. https://doi.org/10.4103/ijmr.IJMR_551_15

Nelson AR, Ravichandran K, Antony U. The impact of the Green Revolution on indigenous crops of India. *Journal of Ethnic Foods.* 2019 Dec;6(1):1–0.

Ogundari, K. (2014). The paradigm of agricultural efficiency and its implication on food security in Africa: what does meta-analysis reveal? *World Dev.* 64, 690–702.

Palumbi, S. R. (2001). Humans as the world's greatest evolutionary force. *Science,* 293, 1786–1790.

Pawlak, K., Kołodziejczak, M. (2020). The role of aAgriculture in ensuring food security in developing countries: considerations in the context of the problem of sustainable food production. *Sustainability.* 12, 5488. https://doi.org/10.3390/su12135488

Perfecto, I. et al. (2019). *Nature's Matrix: Linking Agriculture, Biodiversity Conservation and Food Sovereignty,* (2nd edn.) NYC: Routledge.

Power AG. Ecosystem services and agriculture: tradeoffs and synergies. *Philosophical Transactions of the Royal Society B: Biological Sciences.* 2010 Sep 27;365(1554):2959–2971.

Prus, P. (2017). Sustainable farming production and its impact on the natural environment-case study based on a selected group of farmers. In *Proceedings of the 8th International Scientific Conference Rural Development,* 2017: Bioeconomy Challenges, Aleksandras Stulginskis University, Kaunas, Lithuania, 23–24 November 2017; Raupeliene, A., Ed.; VDU Research Management System: Kaunas, Lithuania, pp. 1280–1285.

Prus, P. (2018). Farmers' Opinions about the Prospects of Family Farming Development in Poland. In *Proceedings of the 2018 International Conference "Economic Science for Rural Development"* No. 47, Jelgava, Latvia, pp. 267–274.

Prus, P. (2019). The role of higher education in promoting sustainable agriculture. *J. East Eur. Manag. Studies Corp. Soc. Responsib. Bus. Ethics Cent. Eastern Eur.* 2019, 99–119.

Ramphul, N., Nowbutsing, K.B., Chittoo, H.B. (2016). An analysis of government policies in ensuring food security in small island economies: a case study of mauritius. *IOSR J. Humanit. Soc. Sci.* 21, 43–59.

Rana, M. M., Takamatsu, T., Baslam, M., Kaneko, K., Itoh, K., Harada, N., et al. (2019). Salt tolerance improvement in rice through efficient SNP marker-assisted selection coupled with speed-breeding. *Int. J. Mol. Sci.* 20(10), 2585. https://doi.org/10.3390/ijms20102585

Renard, D., Tilman, D. (2019). National food production stabilized by crop diversity. *Nature.* 571, 257–260.

Remans, R., Villani, C., Jones, S., Smith, A., Laporte, M.A. et al. (2019). Measuring agricultural biodiversity for sustainable food systems. *Biodivers. Inf. Sci. Stand.* 3, e46785. https://doi.org/10.3897/biss.3.46785.

Sadia B., Awan F. S., Saleem F., Altaf J., Umar A. B., Nadeem M., et al. (2021). Exploring plant genetic variations with morphometric and molecular markers. In *Genetic Variation* (Intech Open).

Salgotra, R.K., Chauhan, B.S. (2023). Genetic diversity, conservation, and utilization of plant genetic resources. *Genes.* 14(1), 174. https://doi.org/10.3390/genes14010174

Sapkota, T. B. et al. (2019). Cost-effective opportunities for climate change mitigation in Indian agriculture. *Sci. Total Environ.* 655, 1342–1354.

Sato K. Genetically modified food in France: Symbolic transformation and the policy paradigm shift. *Theory and Society.* 2013 Sep;42:477–507.

Schaal, T., Mitchell, M., Scheele, B.C. et al. (2023). Using the three horizons approach to explore pathways towards positive futures for agricultural landscapes with rich biodiversity. *Sustain Sci.* 18, 1271–1289. https://doi.org/10.1007/s11625-022-01275-z

Singh, I. (2019). Changes of agriculture production in India after Green Revolution. *J. Gujarat Res. Soc.* 21, 2290–2294. http://gujaratresearchsociety.in/index.php/JGRS/article/view/2967

Singh, K. (2022). Agrobiodiversity, status, and conservation strategies. In *Agro-Biodiversity and Agri-Ecosystem Management.* Singapore: Springer Nature Singapore. pp. 27–52.

Singh, S., Thakur, A., Tomar, R.S., Tiwari, S., Sharma, R.A. (2022). Agro-biodiversity, status, and conservation strategies: an Indian perspective. In *Towards Sustainable Natural Resources*, Rani, M., Chaudhary, B.S., Jamal, S., Kumar, P., Eds, . Springer, Cham. https://doi.org/10.1007/978-3-031-06443-2_5

Soria-Lopez, A., Garcia-Perez, P., Carpena, M., Garcia-Oliveira, P., Otero, P. et al. (2023). Challenges for future food systems: from the Green Revolution to food supply chains with a special focus on sustainability. *Food Front.* 4(1): 9–20. https://doi.org/10.1002/fft2.173

Srivastava, G. P. (2022). Role of agricultural biodiversity in sustainable food and nutrition security. *Acta Sci. Agri.* 6(12), 28–31.

Tauger, M. B. (2023). Nazareno Strampelli and the first Green Revolution. *Journal of International Development*, 35(8), 2485–2512.

Taylor, M. (2019). Hybrid realities: making a new Green Revolution for rice in south India. *J. Peasant Stud.* 47, 483–502. https://doi.org/10.1080/03066150.2019.1568246

Tian D., Guo X., Zhang Z., Wang M., Wang F. (2019). Improving blast resistance of the rice restorer line, Hui 316, by introducing Pi9 or Pi2 with marker-assisted selection. *Biotechnol. Biotechnol. Equip.* 33 (1), 1195–1203. https://doi.org/10.1080/13102818.2019.1649095

Tilman, D. (1998). The greening of the green revolution. *Nature,* 396, 211–212.

UN (2023a). United Nations. Convention on Biological Diversity, key international instrument for sustainable development. https://www.un.org/en/observances/biological-diversity-day/convention

UN (2023b). United Nation, Sustainable Development Goals. https://www.un.org/sustainabledevelopment/hunger/

Valujeva, K., Freed, E.K., Nipers, A., Jauhiainen, J., Schulte, R.P. (2023). Pathways for governance opportunities: Social network analysis to create targeted and effective policies for agricultural and environmental development. *J. Environ. Manage.* 325, 116563.

Vernooy, R., Rana, J., Otieno, G., Mbozi, H., Shrestha, P. (2022). Farmer-led seed production: community seed banks enter the national seed market. *Seeds.* 1, 164–180. https://doi.org/10.3390/seeds1030015

Wakchaure, S. D. Analytical study of agricultural crisis and farmers suicides in India. *Int. Res. Jour. Hum. Res. Soc. Sci.* 7, 97–104.

Willett, W, Rockström, J, Loken, B, Springmann, M, Lang, T. et al. (2019) Food in the Anthropocene: the EAT–Lancet Commission on healthy diets from sustainable food systems. The Lancet, 393(10170), 447–492. https://doi.org/10.1016/s0140-6736(18)31788-4

Xu, Z., Hua, J., Wang, F., Cheng, Z., Meng, Q., Chen, Y., et al. (2020). Marker-assisted selection of qMrdd8 to improve maize resistance to rough dwarf disease. *Breed. Sci.* 70(2), 183–192. https://doi.org/10.1270/jsbbs.19110

26 Recent Advances in In Situ and Ex Situ Germplasm Conservation Approaches

An Approach for Conservation of Medicinal Plants

Tripti Singh[1], *Ashwani Mathur*[1]

[1]Department of Biotechnology, Jaypee Institute of Information Technology, Noida, Uttar Pradesh, India

26.1 INTRODUCTION

According to the World Health Organization, about 88% of people worldwide rely on the use of traditional medicines for some aspect of their primary health care needs. Medicinal plants have been explored for ages in the treatment of various diseases such as heart disease, respiratory issues, jaundice, facial paralysis, joint disorders, and asthma (https://www.dabur.com/amp/in/en-us/about/science-of-ayurveda/herbal-medicinal-plants). These plants have the capacity to affect the entire health of an individual by boosting their immune system. The nutritive nature of these plants makes them non-toxic and harmless, thus fit for human consumption. Apart from these benefits, medicinal plants deal with the overall well-being of an individual, creating harmony between body, soul, and mind, and thus they play a crucial role in socio-economic development. With the increasing population and decreasing land, the cultivation of medicinal plants has become challenging in the present scenario.

Limited cultivable land often creates competition between cash crops, staple crops, and medicinal plants for their decreasing cultivable land. This encompasses negative pressure on cultivation, prevention of overexploitation, and loss of biodiversity from their natural habitat. The impact is evident as per the International Union for the Conservation of Nature (IUCN) list, indicating the majority of the medicinal plants are in the category of vulnerable, critically endangered, endangered, or extinct, as shown in Table 26.1.

26.1.1 Need for Conservation

Maintenance of the genetic diversity of the species with decreasing cultivable land is a key point for the need for conservation of medicinal plants. Genetic erosion, which refers to the loss of individual genes or in combination over time in a given region, and the persistent reduction of common alleles have raised the alarm for conservation (Guarino, 2003). Not only this, but when there is a reduction in geographical diversity, it affects the overall gene pool, which further becomes vulnerable and later extinct (Upadhyaya, 2009). Various factors such as overexploitation of species, changing land use, overgrazing, population pressure, environmental degradation, and climate change all are key causes of genetic erosion.

DOI: 10.1201/9781032697550-26

TABLE 26.1
IUCN List of Some Medicinal Plants

Botanical Name	Common Name	IUCN Red List Category	Native	Threats	Uses	References
Myristica malabarica	Malabar nutmeg, rampatri, Bombay mace	Vulnerable	India (Tamil Nadu, Kerala, Maharashtra, Karnataka)	There is evidence of low seed viability because of unfavorable soil conditions and pollination difficulties.	Anti-ulcer, sedatives hypnotics, antimicrobial, nematicidal, and anti-inflammatory.	Chelladurai a et al. (2017) https://www.iucnredlist.org/species/31219/9615327#taxonomy
Gardenia gummifera	Gummy gardenia	Least concern	India	The gum is tapped and traded.	Hemorrhoids, bone fracture, nervous disorders, diarrhea, wounds, skin diseases, and stomach ulcers.	Prasad et al. (2007) https://www.iucnredlist.org/species/34622/9879266
Ilex Khasiana	Holly	Critically endangered	India (Meghalaya)	Regeneration is reported to be poor and the habitat has declined in extent and quality.	Analgesic, anti diarrheal, antipyretic, astringent, hematostatic, and hepatoprotective properties.	Hao et al. (2013) https://www.iucnredlist.org/species/31239/9618655
Saraca asoca	Ashoka	Vulnerable	India, Sri Lanka	This small tree has become threatened in some parts of its range mainly through the loss of its habitat.	Treatment of uterine, genital, and other reproductive disorders in women; fever; pain; and inflammation.	Gupta et al. (2014) https://www.iucnredlist.org/species/34623/9879360
Prioria pinnate	Malabar mahogany	Endangered	India (Tamil Nadu, Kerala, Karnataka)	The population has declined considerably because of overexploitation and habitat degradation. Regeneration appears to be very poor.	Gonorrhea, catarrhal conditions of genito-urinary and respiratory tracts, and in curing sores in elephants.	Shahid et al. (2017) https://www.iucnredlist.org/species/33647/9800386

26.1.2 Methods of Conservation

Various strategies have been proposed for scientific intervention focused on exploring new and innovative techniques for the conservation of biological diversity. The methods that have been used can be segregated into in situ and ex situ. They are primarily focused on the preservation of plant material or their reproductive products, such as buds or seeds.

26.2 EX SITU CONSERVATION

Ex situ conservation literally means "off-site conservation." In a simpler form, this type of conservation is also known as outside conservation as it is carried out away from the natural location.

26.2.1 Conventional Methods

26.2.1.1 Seed Storage

Seed storage is cost-effective as it involves regular viability testing of the material along with timely recovery, depending on the nature of the plant and its reproductive biology (International Seed Testing Association, 2018). Seed banks play an important role in preserving and restoring biodiversity of wild species of plants (Roberts, 1973). Management of these seed banks is very much important for the production of viable seeds at the time of utilization. Seeds can be divided into two classes depending upon their storage capacity, namely, orthodox and recalcitrant seeds (Harrington, 1960). Orthodox seeds, having much greater tolerance, will survive drying or freezing during conservation, while recalcitrant seeds, when stored for longer durations, lose their viability due to less tolerance to drying and freezing.

26.2.2 Advanced Methods

26.2.2.1 In Vitro Conservation

In vitro conservation of plants has been carried out since 1970 and is generally used for vegetatively propagating species and plants often producing recalcitrant seeds (Engelmann, 2011). In vitro conservation of plants relies on the fundamentals of plant tissue culture, which is propagation of tissue/cells using explant from a donor plant under a controlled environment (including synthetic media) and aseptic conditions (Ding, 2008). Figure 26.1 represents the in vitro cultivation of one of the medicinally important plants, *Humulus lupulus*. Meristem culture (El-Dougdoug and El-Shamy, 2011) or cryotherapy (Bettoni et al., 2016) can be used to protect such conserved plant species from viral infections.

26.2.2.1.1 Various Methods Used for In Vitro Conservation of Germplasm

Different in vitro conservation strategies have been used for germplasm (Figure 26.2). Among the strategies used, cryopreservation and in vitro active gene bank (IVAG) have proven to be the most successful.

26.2.2.1.1.1 In Vitro Active Gene Bank (IVAG) In the case of in vitro gene banks, plants can be conserved from several months to a few years under normal growth conditions, including standard culture room, low temperature, and light storage (Agrawal et al., 2022). However, this technique cannot be used for long-term conservation, as extensive collections using this approach are quite impossible.

26.2.2.1.1.2 Cryopreservation Cryopreservation involves the preservation of plant tissues/cells in a frozen state under low temperatures. This strategy uses solid carbon dioxide (–79°C), deep freezers (–80°C), vapor nitrogen (–150°C), or liquid nitrogen (–196°C) for preservation of tissues. Plant

FIGURE 26.1 In vitro culture of medicinally important plant, *Humulus lupulus.*

tissues, including calli, protoplasts, meristems, eggs, endosperms, plants themselves, and ovules can be cryopreserved (Kaviani, 2011). Various advanced cryopreservation methods are also used, focusing on the reduction of cell damage caused by the formation of ice crystals on the surface. One

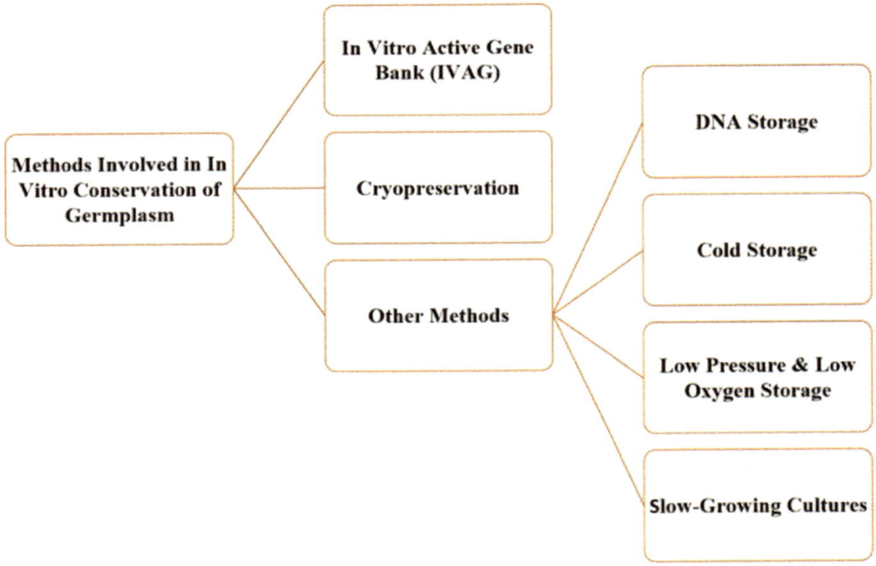

FIGURE 26.2 Schematic eepresentation of methods involved in in vitro conservation of germplasm.

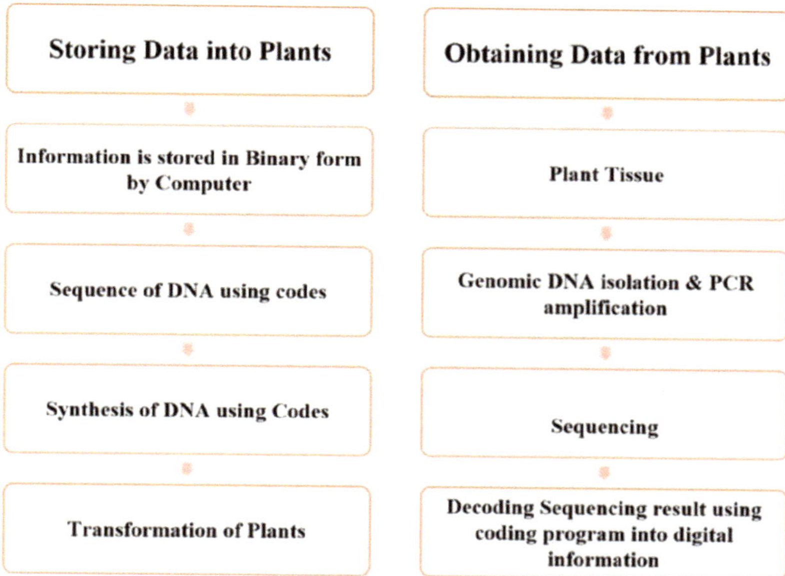

FIGURE 26.3 The processes involved in storing and obtaining data from plants.

approach involves encasing specimens using alginate gel followed by dehydration, while the other approach uses cryoprotective products, thus vitrifying cellular water.

26.2.2.1.1.3 Other Methods

26.2.2.1.1.3.1 DNA Storage One of the latest advanced technologies suggested for ex situ conservation is DNA storage in a facility (Dulloo et al., 2010). Figure 26.3 represents the processes involved when data of the plants need to be stored and how data can be obtained from plants (Fister et al., 2017).

However, this approach is not commonly used as it demands the usage of isolated and cloned DNA, later producing transgenic plants. Although this approach complements germplasm conservation, it is also useful in explaining the origin of species and diversity in the population.

26.2.2.1.1.3.2 Low-Pressure and Low-Oxygen Storage These approaches are useful for both long- and short-term storage of the plants. As the name suggests, low-pressure storage (LPS) involves reduction in the atmospheric pressure surrounding the plant material, which causes a partial decrease in the pressure exerted by gases around the germplasm, thus decreasing the growth of plants using the in vitro method. Apart from this, LPS prevents spore formation by decreasing the activity of pathogenic organisms. In the case of low-oxygen storage, the concentration of oxygen is decreased, but atmospheric pressure is maintained via the addition of some inert gases like nitrogen. If the partial pressure of oxygen is below a certain level (50 mm Hg), it leads to a reduction in plant tissue growth, as the photosynthetic activity is decreased.

26.2.2.1.1.3.3 Slow Growing Cultures Slow growing cultures is an alternative approach to cryopreservation, maintaining the viability and is a cost-effective, straightforward technique in which contamination via gene alterations is reduced (Chauhan et al., 2019). Subculture cycles may take place up to two years, shortening time, effort, and equipment required for their maintenance. Slower growth reduces the rate of cell division, thus increasing the spontaneous mutation in the growing culture. Besides this, collections preserved using this approach are often susceptible to infections

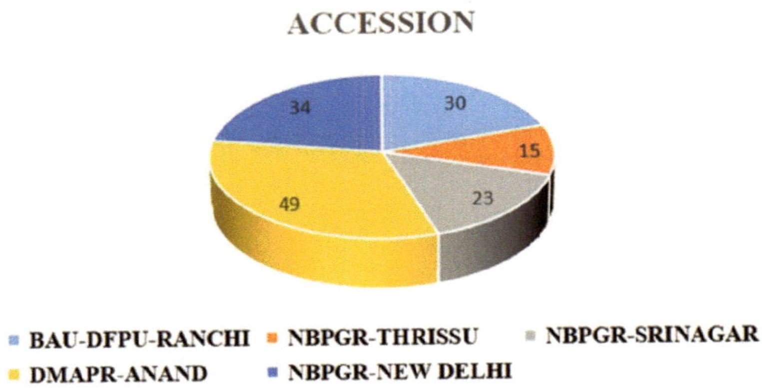

FIGURE 26.4 Pie chart representation of field gene bank data of medicinal and aromatic plants found in India (ICAR-NBPGR GeneBank Database).

and genetic instability. Factors such as light, temperature, nutritional constraints, osmotic concentration, growth regulation, concentration of oxygen, and type of propagation vessel play a key role in culture development. Stress is also another factor affecting the genotype of the population, sometimes somaclonal variation (Silva et al., 2016) over the desired culture, thus contributing to genetic instability.

26.2.2.1.1.3.4 Cold Storage Cold storage involves storing the germplasm at moderate and non-freezing temperatures (1–9°C). It is a slow growth preservation process and falls under the category of short-term storage. The advantage of this method is that it accelerates growth of the plant in cold conditions compared to cryopreservation, which causes cryogenic damage (West et al., 2006). The technique is inexpensive and produces germplasm exhibiting higher survival rates.

26.2.2.1.1.4 Field Gene Banks, Botanical Gardens, and Arboreta In this type of conservation approach, germplasm is grown in nurseries depending on the species type and requires maintenance. One of the advantages of field gene banks is that they provide easy access to plant genetic resources for either utilization or characterization. Also, they are helpful in conserving plants propagated using vegetative means, as genetic variation is quite evident in the field in comparison to in vitro (https://cropgenebank.sgrp.cgiar.org/index.php/procedures-mainmenu242/conservation-mainmenu-198/field-bank-mainmenu-203). The field gene bank data of medicinal plants in India is represented in Figure 26.4.

Other approaches to ex situ management involving whole plant collection are botanical gardens and arboreta. Botanical gardens and arboreta play a vital role in providing a scientific basis for sustainable use, thus conserving the plant biodiversity.

26.3 IN SITU CONSERVATION

In situ conservation deals with the preservation of germplasm within their naturally occurring habitat. National parks, gene sanctuaries, and natural reserves all fall under this category of conservation. This conservation approach is unique as it has the advantage of maintaining species continuity along with the evolution of new recombination types. However, the lack of protection in the absence of managed surveillance, environmental contaminants degrading the germplasm, and high cost are some of the factors that make this approach disadvantageous (Meir et al., 1989; Hoyt, 1988; Plucknett et al., 2014). Also, conserved substances are not readily usable, and the longevity period of conserved germplasm is unknown.

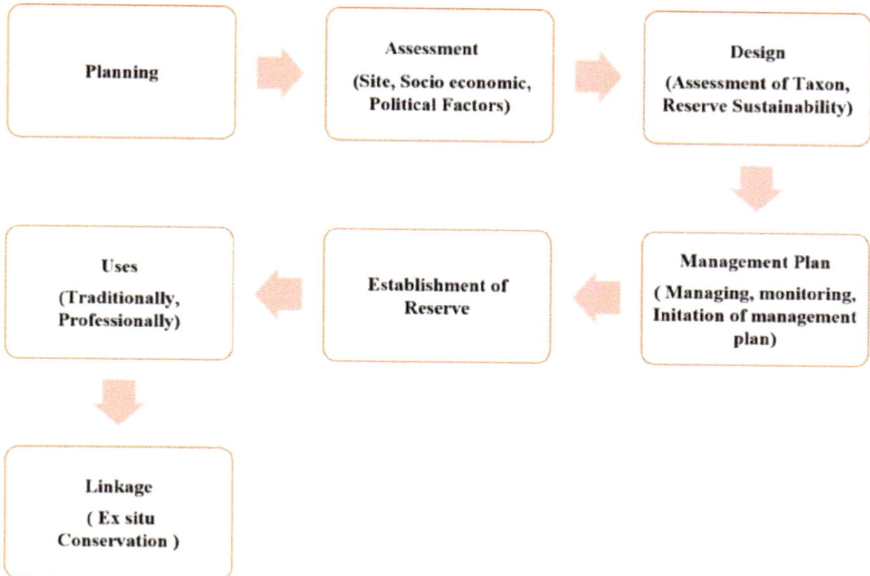

FIGURE 26.5 Basic design for establishment for natural reserve conservation.

26.3.1 Genetic Reserves

Genetic reserves' main objective is the maintenance of the individual and its population-level variation along its natural habitats (Hawkes, 1998). The establishment of genetic reserves is not so easy and requires planning and execution, as depicted in Figure 26.5. To obtain knowledge about target taxa, information such as genetic diversity, breeding mechanisms, geographical distribution, composition of the population, and requirements of habitat are essential.

26.3.2 Home Garden and On-Farm Conservation

With the help of traditional farming systems, medicinal plants are maintained by farmers using this approach, thus forming a conservation technique. In the case of home gardens, it is the farmer or the individual who is working who is solely responsible for the conservation of germplasm either deliberately or accidentally. However, the conservationist can only keep an eye on the actual work, which is not done by them but by farmers (Maxted et al.,2002). However, the approach may not be beneficial in the long run as farmers may switch to modern cultivars and, as a result, we may lose such an important resource for our future (Swaminathan, 2000).

26.4 CONCLUSION

Germplasm of a species provides knowledge about the genetic makeup of that species and thus is valuable and of keen importance and hence should be preserved and conserved. A large population of the world is dependent on the usage of medicinal plants, and thus conservation of these plants with limited land is of great concern. Two of the most common approaches known for germplasm conservation are ex situ and in situ. These approaches are useful as they help in the conservation of plants that are either vulnerable or headed toward extinction. Recent advances in conservation approaches such as DNA banks and genetic reserves have made the work of conservationists easier and convenient. In vitro culture technology is effective in increasing germplasm numbers and is quite useful in species producing recalcitrant seeds. However, in the present scenario, ex situ and in situ conservation are considered complementary conservation strategies, as neither of the approaches is sufficient individually to conserve existing and evolving diversity of species (Dulloo et al., 2005; Gepts, 2006; Hunter and Heywood, 2011; Maxted et al., 1997). The choice of the type of conservation strategy, either specific to in situ or ex situ conservation actions, depends on factors such as the biology of the species, its performance under storage, and the intended use of the germplasm that is being conserved.

REFERENCES

Agrawal, A.; Gowthami, R.; Chander, S.; Srivastava, V. Sustainability of in vitro genebanks and cryogenebanks. *Ind. J. Plant Gene. Res.* **2022**, *35,* 180–184. https:// doi.org/10.5958/0976-1926.2022.00065.1.

Bettoni, J.C.; Costa, M.D.; Gardin, J.P.P.; Kretzschmar, A.A.; Pathirana, R. Cryotherapy: A new technique to obtain grapevine plants free of viruses. *Rev. Bras. Frutic.* **2016**, *38*, 833.

Chauhan, R.; Singh, V.; Quareshi, A. In vitro conservation through slow-growth storage. *Synth. Seeds Germplasm Regen. Preserv. Prospect.* **2019**, *1*, 482.

Chelladurai, P.K.; Ramalingam, R. Myristica malabarica: a comprehensive review. *J. Pharm. Phytochem.* **2017,** *6*(2), 255–258

Ding, F.; Jin, S.; Hong, N.; Zhong, Y.; Cao, Q.; Yi, G.; Wang, G. Vitrifification-cryopreservation, an efficient method for eliminating *Candidatus Liberobacter asiaticus*, the citrus Huanglongbing pathogen, from in vitro adult shoot tips. *Plant Cell Rep.* **2008**, *27*, 241–250.

El-Dougdoug, K.A.; El-Shamy, M.M. Plant tissue culture: current status and opportunities. *Afr. J. Microbiol.* **2011**.

Engelmann, F. Use of biotechnologies for the conservation of plant biodiversity. *In Vitro Cell. Dev. Biol. Plant* **2011**, *47*, 5–16.

Dulloo, M.E.; Hunter, D.; Borelli, T. Ex situ and in situ conservation of agricultural biodiversity: major advances and research needs. *Not. Bot. Horti Agrobot. Cluj Napoca* **2010**, *38*, 123–135.

Dulloo, M. E.; Rao, V. R.; Engelmann, F.; Engels, J.M.M. Complementary conservation of coconuts. In Batugal P., Rao V. R., & Olive J. (Eds.), *Coconut Genetic Resources* (pp. 75–90). Serdang, Malaysia: IPGRI-APO.

Fister, K.; Fister jr, I.; Murovec, J. The potential of plants and seeds in DNA-based information storage, *Understanding Information: From the Big Bang to Big Data.* **2017**, 69–81. https:// doi.org/10.1007/978-3-319-59090-5_4.

Gepts, P. Plant genetic resources conservation and utilization: the accomplishments and future of a societal insurance policy. *Crop Sci.* **2006**, *46*, 2278–2292.

Guarino, L. Approaches to measuring genetic erosion. PGR Documentation and Information in Europe—Towards a sustainable and user-oriented information infrastructure. In *Proceedings of the EPGRIS Final Conference Combined with a Meeting of the ECP/GR Information and Documentation Network*, Prague, Czech Republic, 11–13 September 2003.

Gupta, M.; Sasmal, S.; Mukherjee, A. Therapeutic effects of acetone extract of saraca asoca seeds on rats with adjuvant-induced arthritis via attenuating inflammatory responses. *ISRN Rheumatology*, **2014**, *2014*, 959687. https://doi.org/10.1155/2014/959687.

Hao, D.; Gu, X.; Xiao, P.; Liang, Z.; Xu, L.; Peng, Y. Research progress in the phytochemistry and biology of Ilex pharmaceutical resources. *Acta Pharm. Sinica B.* **2013,** *3*(1), 8–19.

Harrington, J.F. Thumb rules of drying seed. *Crops Soils* **1960**, *13*, 16–17.

Hawkes JG. *Plant Genetic Conservation: The in situ Approach.* 1998. Springer.

Hoyt, E. *Conserving the Wild Relatives of Crops*; International Board for Plant Genetic Resources: Rome, Italy, 1988.

https://cropgenebank.sgrp.cgiar.org/index.php/procedures-mainmenu-242/conservation-mainmenu-198/field-bank-mainmenu-203.

Hunter, D.; Heywood, V.H. (Eds.) *Crop Wild Relatives. A Manual of In-Situ Conservation.* Washington, DC: Earthscan.

International Seed Testing Association. *International Rules for Seed Testing 2018*; ISTA: Bassersdorf, Switzerland, 2018.

Kaviani, B. Conservation of plant genetic resources by cryopreservation. *Aust. J. Crop Sci.* **2011**, *5*, 778–800.

Maxted, N.; Ford-Lloyd, B.V.; Hawkes, G. (Eds.) *Plant Genetic Conservation. In Situ Approach.* London, UK: Chapman & Hall.

Maxted, N.; Guarino, L.; Myer, L.; Chiwona, E.A. Towards a methodology for on-farm conservation of plant genetic resources. *Genet. Resour. Crop Evol.* **2002**, *49*, 31–46.

Meir, N.; Anikster, Y.; Waldman, M.; Ashri, A. Population dynamics research for in situ conservation: Wild wheat in Israel. *Plant Genet. Resour. Newsl.* **1989**, *75/76*, 9–11.

Plucknett, D.L.; Smith, N.J.H. *Gene Banks and The World's Food*; Princeton University Press: Princeton, NJ, USA, 2014; ISBN 1400858119.

Prasad, M.N.V.; Padmalatha, K.; Jayaram, K.; Raju, N.L.; da Silva, J.A.T. Medicinal plants from Deccan ecoregion, India: traditional knowledge, ethnopharmacology, cultivation, utilization, conservation and biotechnology-opportunities and impediments, *Med. Aromatic. Plant Sci. Biotechnol.* **2007,** *1,* 155–208.

Roberts, E.H. Predicting the storage life of seeds. *Proceedings* **1973**, *1*, 499–514.

Shahid, A. P.; Sasidharan, N.; Salini, S.; Padikkala, J.; Meera, N.; Raghavamenon, A.C.; Babu, T.D. *Kingiodendron pinnatum*, a pharmacologically effective alternative for *Saraca asoca* in an Ayurvedic preparation, *Asokarishta. J. Trad. Complement. Med.*, **2017**, *8*(1), 244–250. https://doi.org/10.1016/j.jtcme.2017.06.005.

Silva, R.L.; Ferreira, C.F.; Ledo, C.A.S.; Souza, E.H.; Silva, P.H.; Costa, M.A.P.C.; Souza, F.V.D. Viability and genetic stability of pineapple germplasm after 10 years of in vitro conservation. *Plant Cell. Tissue Organ Cult.* **2016**, *127*, 123–133.

Swaminathan, M.S. The past, present and future contributions of farmers to the conservation and development of genetic diversity. In *Managing Plant Genetic Diversity, Proceedings of an International Conference,* Kuala Lumpur, Malaysia, 12–16 June 2000.

Upadhyaya, H.D., Gowda, C.L.L. *Managing and Enhancing the Use of Germplasm—Strategies and Methodologies*; Technical Manual No. 10; ICRISAT: Patancheru, India, 2009.

West, T.P.; Ravindra, M.B.; Preece, J.E. Encapsulation, cold storage, and growth of Hibiscus moscheutos nodal segments. *Plant Cell. Tissue Organ Cult.* **2006**, *87*, 223–231.

Glossary/Index

	Page No.
Abiotic stress: Non-living factors like drought or heat affecting plant health.	298
Acclimatisation: Adapting to new environmental conditions.	5
Active pharmaceutical ingredients: Biologically active components of drugs.	322
Afforestation: Planting trees in barren or deforested areas to restore ecosystems.	144, 145
Agricultural biodiversity: The variety of life forms and ecosystems related to farming	331, 332, 333, 335, 337, 338, 339
Agrobiodiversity: The variety of plants, animals, and microorganisms in agriculture.	292–302
Agroclimatic zones: Regions with specific climatic conditions affecting agriculture.	147
Agroforestry: Combining trees with crops or livestock to enhance sustainability.	284, 319, 324, 325
Agronomic management: Farming practices enhancing crop productivity.	302
Algal blooms: Rapid algae growth in water, often harmful to ecosystems.	298
Alien species: Non-native species introduced to an area.	3
Alkaloids: Nitrogen-containing compounds with therapeutic properties.	61, 62, 63, 308–310, 313
Allelic: Related to different versions of a gene.	21
Allelopathic: Plants affecting others via biochemical interactions.	313
Alpine forests: High elevation woodlands.	147
Alpine meadows: High-altitude ecosystems with diverse vegetation.	147
Alteration: Change or modification in structure or function.	96
Alzheimer's disease: Neurodegenerative disorder causing memory loss.	312
Ammonia/Sulphur oxidation: Microbial processes converting ammonia or sulphur compounds in nutrient cycles.	109, 112
Anaerobic digestion: Microbial process converting organic waste into biogas.	125
Analgesics: Drugs or agents used to relieve pain.	59
Annealing: Binding of primers to DNA during replication.	44
Anoxic/Anaerobic: Environments or processes occurring without oxygen.	111
Anthropogenic: Human-caused effects on the environment.	1, 9
Antibacterial: Agents preventing bacterial growth.	321, 322
Anticancerous: Agents that combat or prevent cancer.	322, 323
Anticarcinogenic effects: Prevents or reduces cancer risk.	311
Antidiabetic: Substances or treatments that help manage diabetes.	59

	Page No.
Conservation genetics: Using genetics to preserve biodiversity.	1, 8, 28
Conservation: Preserving natural resources and biodiversity.	68, 69, 71, 72, 97, 98, 99, 155, 261, 262, 265
Conservationists: Individuals dedicated to preserving biodiversity.	38, 281
Consumers: Organisms that eat plants or other animals for energy.	281
Contour agriculture: Farming along natural land contours to reduce soil erosion.	293
Convention on biological diversity: An international agreement to conserve biodiversity.	337
Cosmic rays: High-energy radiation from space.	21
Crohn's disease: Chronic inflammatory bowel condition.	310
Crop breeding: Developing better crop varieties through genetics.	38
Crop diversity: The range of different crops cultivated to promote resilience.	334, 335
Crop genetics: Study of genetic traits in agricultural crops.	267
Crop switching: Changing crop types to adapt to environmental conditions.	293
Crossbreeding: Combining genetic traits from different parent plants.	77
Cryogenic: Pertaining to the use of extremely low temperatures.	348
Cryopreservation: Preservation of cells or tissues at extremely low temperatures.	68, 69, 72, 345, 347, 348
Cryopreservatives: Substances used to freeze and preserve cells or tissues.	6
Cultivation: Growing crops or plants.	21
Deciduous forests: Forests with broad-leaved trees that shed all their leaves during one season.	147
Decomposers: Organisms breaking down dead material, recycling nutrients.	281
Defoliation: Shedding or removal of plant leaves.	268
Degradome: The collection of degraded RNA molecules in a system.	63
Deletion: Loss of DNA sequences from a genome.	47
Delta smelt: Endangered fish species found in Californian waters.	320
Denaturation: Separation of DNA strands during replication or analysis.	50
Desertification: Degradation of land into arid conditions.	324
Diarrhoea: Frequent, loose bowel movements.	307
Diatoms: Microscopic algae with silica deposited cell walls.	269
Divergence: Genetic differences developing between populations or species.	47, 52
Diverse ecosystems: Complex ecological communities with various species.	147
Diversification: Increasing variety to improve resilience.	319
Diversity: The variety of species, genes, and ecosystems in an area.	107, 108, 110, 114
DNA barcode: Short DNA sequence for species identification.	15
DNA barcoding gap: Clear genetic difference between species for identification.	46

	Page No.
Epidemiology: Study of disease distribution and control in populations.	294
Epidermis: The outermost layer of skin or plant tissue.	271
Epigenomics: Study of changes in gene activity without altering DNA sequence.	12, 27
Epiphytes: Plants growing on other plants for support, not parasitic.	76
Erosions: Loss of surface layers due to damage or wear.	107, 271
Ethnobotany: Study of plant uses in cultures and traditions.	189, 193, 305, 314
Ethnopharmacology: Study of traditional medicine and its applications.	314
Etiolation therapies: Techniques using light deprivation to stimulate specific plant growth.	69
Eutrophication: Excessive nutrient enrichment in water bodies, causing algae blooms and oxygen depletion.	111
Evolution: Changes in species over generations through genetic variation.	30
***Ex situ* preservation**: Conserving genetic material outside its natural habitat.	295
Exoskeletons: Hard external coverings supporting some organisms.	320
Exotic species: Non-native organisms affecting native ecosystems.	251, 252, 255
Extension: Adding nucleotides to form a new DNA strand.	44
Extinct: A species that no longer exists.	343
Facial paralysis: Loss of facial muscle movement, often caused by nerve damage.	343
Fermentation: Microbial process producing energy and byproducts like alcohol.	296
Fertilizer technology: Advances in creating efficient and sustainable fertilizers.	302
Fertilizer: Substances boosting plant growth and productivity.	183, 322
Flavonoids: Plant compounds with antioxidant benefits.	305, 308, 312
Flora and fauna: Plant and animal life in different ecosystems.	147
Flora-based pathogens: Plant-associated microbes causing diseases.	294
Flow cytometry: Technique to measure cells and their components.	5
Fluorescence: Emission of light by a substance after absorbing light.	18
Fluorescent labelling: Tagging molecules with fluorescence for detection.	28
Food web: Network of interconnected feeding relationships in an ecosystem.	108
Forensic identification: Applying genetics to solve criminal cases.	28
Forest ecosystems: Biodiverse habitats dominated by trees and vegetation.	140
Fragmentation: Division of habitats into smaller, disconnected pieces.	96, 247, 248, 249, 254, 263
Functional diversity: Variety of species roles in an ecosystem.	293
Gasification/Pyrolysis: Converting biomass into energy-rich gases or liquids using heat.	125, 129, 133
Gene isoforms: Different versions of the same gene.	15, 20
Gene pool: The collection of genetic material in an interbreeding population.	343

	Page No.
Glucosinolates: Sulphur-containing compounds in plants with health benefits.	308, 313
Glycosylation: The enzymatic addition of sugars to proteins or lipids.	63
Grafting: Joining plant parts to grow as one.	69
Grassland ecosystems: Biomes dominated by grasses, home to large herbivores and predators.	332
Green climate fund: Fund supporting climate change mitigation in developing nations.	285
Green medicine: Using natural substances for healthcare.	305
Green revolution: Technological advancements in agriculture to increase crop yields.	335, 336
Green technologies: Innovations minimizing environmental impact and resource use.	284, 285
Greenhouse gases: Atmospheric gases causing global warming (e.g., CO2, CH4).	99
Greenhouses: Controlled environments for plant growth.	20
Gut microbiome: Microorganisms living in the digestive tract.	319, 320
Habitat fragmentation: The breaking up of natural habitats due to human activity.	334
Habitat rehabilitation: Repairing damaged ecosystems.	101
Habitat restoration: Rehabilitating ecosystems to their natural state.	95
Habitat: Natural home of a species.	12
Healing potential: Ability of natural or synthetic agents to restore health.	305
Herbarium: Collection of preserved plant specimens.	52
Herbicide-resistant plants: Crops genetically modified to withstand herbicides.	70, 296
Heterogeneity: Variation within ecosystems or populations.	269
Heterologous: Genes or proteins expressed in a foreign host.	61, 94
Heterotrophic: Organisms that obtain energy by consuming others.	263
Heterozygosity: Presence of different alleles at a gene location.	8
Histone Modifications: Changes in histone proteins affecting gene expression.	30
Horticultural: Relating to the cultivation of gardens or plants.	85, 147
Hybrid: Offspring from two different species or varieties.	7
Hybridization: Breeding between different species or varieties.	30, 61, 62, 77, 331
Hydrodistillation: Extraction using water or steam.	309
Hydroelectric: Power generated by harnessing energy from flowing water.	284
Hydroponics: Growing plants in nutrient-rich water without soil.	292, 298
Hyperglycemia: High blood sugar levels, often associated with diabetes.	64
Hyperplasia: Excessive growth of cells or tissues.	271
Hypersaline: Highly saline environments where specialized organisms thrive.	107
Hypertension: Elevated blood pressure leading to health risks.	306, 311
Immunology: Study of the immune system.	313

	Page No.
Mineralization: Conversion of organic matter into inorganic nutrients.	112
Mini-barcode: Short DNA sequences for species identification.	52
Minimal amplification: Efficient DNA replication with fewer starting materials.	43
Mitochondrial DNA: DNA from mitochondria, often used in ancestry studies.	36
Molecular farming: Using plants to produce pharmaceuticals or industrial compounds.	295
Molecular taxonomy: Classifying organisms based on molecular data.	30
Monocots: Flowering plants with one seed leaf.	42
Monocultures: Large-scale cultivation of a single crop species.	267
Monophyletic: Group of organisms sharing a common ancestor.	49
Mulching: Covering soil with organic materials to retain moisture and prevent erosion.	293
Multigenome sequencing: Sequencing DNA from multiple species or genomes.	18
Multilocus: Involving multiple gene locations for analysis.	51, 53
Mutations: Changes in DNA sequence that may alter function or structure.	19, 20, 62
Nanobiofertilizers: Nano-sized biofertilizers for improved crop growth.	301
Nanofertilizers: Nano-sized nutrients improving crop uptake and reducing waste.	300–302
Nanoherbicides: Herbicides designed using nanotechnology for precision application.	301
Nanopesticides: Nano-enhanced pesticides targeting pests efficiently.	301, 302
Nanopore: Technology for DNA sequencing using tiny pores.	14, 16, 17
Nanotechnology: Replicating natural processes at the nanoscale for technological advancements.	210
Nanotherapy: Therapeutics using nanoparticles to deliver drugs.	64
National parks: Protected areas established to conserve wildlife and natural habitats.	348
National seed policies: Policies regulating the production, distribution, and use of seeds.	338
Natural habitats: Environments where species naturally live and thrive.	261, 263, 264
Natural pest management: Controlling pests using ecological methods like predators.	292
Natural reserves: Protected areas set aside for conservation purposes.	348
Natural selection: The process by which organisms better suited to their environment survive.	272, 335
Necrotic: Dead tissue resulting from injury or disease.	266, 268–271
Nematocidal: Substances killing nematodes.	313
Next-generation sequencing: Advanced DNA sequencing technology.	12, 13, 15, 16, 19
Niche: Role or function of a species in an ecosystem.	1, 96
Nitrogen deposition: Accumulation of nitrogen compounds in ecosystems from the atmosphere.	278

	Page No.
Viral metagenomics: Study of genetic material from viruses in environmental samples.	32
Virions: Complete virus particles.	32
Viruses: Infectious agents that replicate inside host cells.	13
Volatilization: Evaporation of nutrients like nitrogen into the atmosphere.	300
Vulnerable: Species at risk of becoming endangered without conservation efforts.	343, 344, 349
Warm-supported extraction: Heat-aided method for compound isolation.	309
Water cleansing: Removal of impurities to purify water.	318
Water purification: The process of removing contaminants from water.	279, 281
Water retention: Soil's ability to hold water for plant growth.	319
Wetlands: Ecosystems saturated with water, supporting unique flora and fauna.	276, 279, 281, 282
White-nose syndrome: Fungal disease affecting bat populations.	270, 271
Wildlife conservation programmes: Initiatives to protect and restore animal populations.	277
Wildlife habitats: Natural environments where animals live and breed.	286
Wilting: Loss of rigidity in plants due to water deficiency or stress.	266, 267, 269
Zero-tillage farming: Growing crops without ploughing to preserve soil structure.	293
Zoonotic disease: Diseases transmitted between animals and humans.	318, 325
Zoonotic: Diseases transmitted from animals to humans.	263